MW01030760

FORMULAS FOR DYNAMIC ANALYSIS

MECHANICAL ENGINEERING

A Series of Textbooks and Reference Books

Founding Editor

L. L. Faulkner

Columbus Division, Battelle Memorial Institute
and Department of Mechanical Engineering
The Ohio State University
Columbus, Ohio

1. *Spring Designer's Handbook*, Harold Carlson
2. *Computer-Aided Graphics and Design*, Daniel L. Ryan
3. *Lubrication Fundamentals*, J. George Wills
4. *Solar Engineering for Domestic Buildings*, William A. Himmelman
5. *Applied Engineering Mechanics: Statics and Dynamics*, G. Boothroyd and C. Poli
6. *Centrifugal Pump Clinic*, Igor J. Karassik
7. *Computer-Aided Kinetics for Machine Design*, Daniel L. Ryan
8. *Plastics Products Design Handbook, Part A: Materials and Components; Part B: Processes and Design for Processes*, edited by Edward Miller
9. *Turbomachinery: Basic Theory and Applications*, Earl Logan, Jr.
10. *Vibrations of Shells and Plates*, Werner Soedel
11. *Flat and Corrugated Diaphragm Design Handbook*, Mario Di Giovanni
12. *Practical Stress Analysis in Engineering Design*, Alexander Blake
13. *An Introduction to the Design and Behavior of Bolted Joints*, John H. Bickford
14. *Optimal Engineering Design: Principles and Applications*, James N. Siddall
15. *Spring Manufacturing Handbook*, Harold Carlson
16. *Industrial Noise Control: Fundamentals and Applications*, edited by Lewis H. Bell
17. *Gears and Their Vibration: A Basic Approach to Understanding Gear Noise*, J. Derek Smith
18. *Chains for Power Transmission and Material Handling: Design and Applications Handbook*, American Chain Association
19. *Corrosion and Corrosion Protection Handbook*, edited by Philip A. Schweitzer
20. *Gear Drive Systems: Design and Application*, Peter Lynwander
21. *Controlling In-Plant Airborne Contaminants: Systems Design and Calculations*, John D. Constance
22. *CAD/CAM Systems Planning and Implementation*, Charles S. Knox
23. *Probabilistic Engineering Design: Principles and Applications*, James N. Siddall
24. *Traction Drives: Selection and Application*, Frederick W. Heilich III and Eugene E. Shube

25. *Finite Element Methods: An Introduction*, Ronald L. Huston and Chris E. Passerello
26. *Mechanical Fastening of Plastics: An Engineering Handbook*, Brayton Lincoln, Kenneth J. Gomes, and James F. Braden
27. *Lubrication in Practice: Second Edition*, edited by W. S. Robertson
28. *Principles of Automated Drafting*, Daniel L. Ryan
29. *Practical Seal Design*, edited by Leonard J. Martini
30. *Engineering Documentation for CAD/CAM Applications*, Charles S. Knox
31. *Design Dimensioning with Computer Graphics Applications*, Jerome C. Lange
32. *Mechanism Analysis: Simplified Graphical and Analytical Techniques*, Lyndon O. Barton
33. *CAD/CAM Systems: Justification, Implementation, Productivity Measurement*, Edward J. Preston, George W. Crawford, and Mark E. Coticchia
34. *Steam Plant Calculations Manual*, V. Ganapathy
35. *Design Assurance for Engineers and Managers*, John A. Burgess
36. *Heat Transfer Fluids and Systems for Process and Energy Applications*, Jasbir Singh
37. *Potential Flows: Computer Graphic Solutions*, Robert H. Kirchhoff
38. *Computer-Aided Graphics and Design: Second Edition*, Daniel L. Ryan
39. *Electronically Controlled Proportional Valves: Selection and Application*, Michael J. Tonyan, edited by Tobi Goldoftas
40. *Pressure Gauge Handbook*, AMETEK, U.S. Gauge Division, edited by Philip W. Harland
41. *Fabric Filtration for Combustion Sources: Fundamentals and Basic Technology*, R. P. Donovan
42. *Design of Mechanical Joints*, Alexander Blake
43. *CAD/CAM Dictionary*, Edward J. Preston, George W. Crawford, and Mark E. Coticchia
44. *Machinery Adhesives for Locking, Retaining, and Sealing*, Girard S. Haviland
45. *Couplings and Joints: Design, Selection, and Application*, Jon R. Mancuso
46. *Shaft Alignment Handbook*, John Piotrowski
47. *BASIC Programs for Steam Plant Engineers: Boilers, Combustion, Fluid Flow, and Heat Transfer*, V. Ganapathy
48. *Solving Mechanical Design Problems with Computer Graphics*, Jerome C. Lange
49. *Plastics Gearing: Selection and Application*, Clifford E. Adams
50. *Clutches and Brakes: Design and Selection*, William C. Orthwein
51. *Transducers in Mechanical and Electronic Design*, Harry L. Trietley
52. *Metallurgical Applications of Shock-Wave and High-Strain-Rate Phenomena*, edited by Lawrence E. Murr, Karl P. Staudhammer, and Marc A. Meyers
53. *Magnesium Products Design*, Robert S. Busk
54. *How to Integrate CAD/CAM Systems: Management and Technology*, William D. Engelke
55. *Cam Design and Manufacture: Second Edition*; with cam design software for the IBM PC and compatibles, disk included, Preben W. Jensen

56. *Solid-State AC Motor Controls: Selection and Application*, Sylvester Campbell
57. *Fundamentals of Robotics*, David D. Ardayfio
58. *Belt Selection and Application for Engineers*, edited by Wallace D. Erickson
59. *Developing Three-Dimensional CAD Software with the IBM PC*, C. Stan Wei
60. *Organizing Data for CIM Applications*, Charles S. Knox, with contributions by Thomas C. Boos, Ross S. Culverhouse, and Paul F. Muchnicki
61. *Computer-Aided Simulation in Railway Dynamics*, by Rao V. Dukkipati and Joseph R. Amyot
62. *Fiber-Reinforced Composites: Materials, Manufacturing, and Design*, P. K. Mallick
63. *Photoelectric Sensors and Controls: Selection and Application*, Scott M. Juds
64. *Finite Element Analysis with Personal Computers*, Edward R. Champion, Jr., and J. Michael Ensminger
65. *Ultrasonics: Fundamentals, Technology, Applications: Second Edition, Revised and Expanded*, Dale Ensminger
66. *Applied Finite Element Modeling: Practical Problem Solving for Engineers*, Jeffrey M. Steele
67. *Measurement and Instrumentation in Engineering: Principles and Basic Laboratory Experiments*, Francis S. Tse and Ivan E. Morse
68. *Centrifugal Pump Clinic: Second Edition, Revised and Expanded*, Igor J. Karassik
69. *Practical Stress Analysis in Engineering Design: Second Edition, Revised and Expanded*, Alexander Blake
70. *An Introduction to the Design and Behavior of Bolted Joints: Second Edition, Revised and Expanded*, John H. Bickford
71. *High Vacuum Technology: A Practical Guide*, Marsbed H. Hablanian
72. *Pressure Sensors: Selection and Application*, Duane Tandeske
73. *Zinc Handbook: Properties, Processing, and Use in Design*, Frank Porter
74. *Thermal Fatigue of Metals*, Andrzej Weronski and Tadeusz Hejwowski
75. *Classical and Modern Mechanisms for Engineers and Inventors*, Preben W. Jensen
76. *Handbook of Electronic Package Design*, edited by Michael Pecht
77. *Shock-Wave and High-Strain-Rate Phenomena in Materials*, edited by Marc A. Meyers, Lawrence E. Murr, and Karl P. Staudhammer
78. *Industrial Refrigeration: Principles, Design and Applications*, P. C. Koelet
79. *Applied Combustion*, Eugene L. Keating
80. *Engine Oils and Automotive Lubrication*, edited by Wilfried J. Bartz
81. *Mechanism Analysis: Simplified and Graphical Techniques, Second Edition, Revised and Expanded*, Lyndon O. Barton
82. *Fundamental Fluid Mechanics for the Practicing Engineer*, James W. Murdock
83. *Fiber-Reinforced Composites: Materials, Manufacturing, and Design, Second Edition, Revised and Expanded*, P. K. Mallick
84. *Numerical Methods for Engineering Applications*, Edward R. Champion, Jr.

85. *Turbomachinery: Basic Theory and Applications, Second Edition, Revised and Expanded*, Earl Logan, Jr.
86. *Vibrations of Shells and Plates: Second Edition, Revised and Expanded*, Werner Soedel
87. *Steam Plant Calculations Manual: Second Edition, Revised and Expanded*, V. Ganapathy
88. *Industrial Noise Control: Fundamentals and Applications, Second Edition, Revised and Expanded*, Lewis H. Bell and Douglas H. Bell
89. *Finite Elements: Their Design and Performance*, Richard H. MacNeal
90. *Mechanical Properties of Polymers and Composites: Second Edition, Revised and Expanded*, Lawrence E. Nielsen and Robert F. Landel
91. *Mechanical Wear Prediction and Prevention*, Raymond G. Bayer
92. *Mechanical Power Transmission Components*, edited by David W. South and Jon R. Mancuso
93. *Handbook of Turbomachinery*, edited by Earl Logan, Jr.
94. *Engineering Documentation Control Practices and Procedures*, Ray E. Monahan
95. *Refractory Linings Thermomechanical Design and Applications*, Charles A. Schacht
96. *Geometric Dimensioning and Tolerancing: Applications and Techniques for Use in Design, Manufacturing, and Inspection*, James D. Meadows
97. *An Introduction to the Design and Behavior of Bolted Joints: Third Edition, Revised and Expanded*, John H. Bickford
98. *Shaft Alignment Handbook: Second Edition, Revised and Expanded*, John Piotrowski
99. *Computer-Aided Design of Polymer-Matrix Composite Structures*, edited by Suong Van Hoa
100. *Friction Science and Technology*, Peter J. Blau
101. *Introduction to Plastics and Composites: Mechanical Properties and Engineering Applications*, Edward Miller
102. *Practical Fracture Mechanics in Design*, Alexander Blake
103. *Pump Characteristics and Applications*, Michael W. Volk
104. *Optical Principles and Technology for Engineers*, James E. Stewart
105. *Optimizing the Shape of Mechanical Elements and Structures*, A. A. Seireg and Jorge Rodriguez
106. *Kinematics and Dynamics of Machinery*, Vladimír Stejskal and Michael Valášek
107. *Shaft Seals for Dynamic Applications*, Les Horve
108. *Reliability-Based Mechanical Design*, edited by Thomas A. Cruse
109. *Mechanical Fastening, Joining, and Assembly*, James A. Speck
110. *Turbomachinery Fluid Dynamics and Heat Transfer*, edited by Chunill Hah
111. *High-Vacuum Technology: A Practical Guide, Second Edition, Revised and Expanded*, Marsbed H. Hablanian
112. *Geometric Dimensioning and Tolerancing: Workbook and Answerbook*, James D. Meadows
113. *Handbook of Materials Selection for Engineering Applications*, edited by G. T. Murray

114. *Handbook of Thermoplastic Piping System Design*, Thomas Sixsmith and Reinhard Hanselka
115. *Practical Guide to Finite Elements: A Solid Mechanics Approach*, Steven M. Lepi
116. *Applied Computational Fluid Dynamics*, edited by Vijay K. Garg
117. *Fluid Sealing Technology*, Heinz K. Muller and Bernard S. Nau
118. *Friction and Lubrication in Mechanical Design*, A. A. Seireg
119. *Influence Functions and Matrices*, Yuri A. Melnikov
120. *Mechanical Analysis of Electronic Packaging Systems*, Stephen A. McKeown
121. *Couplings and Joints: Design, Selection, and Application, Second Edition, Revised and Expanded,* Jon R. Mancuso
122. *Thermodynamics: Processes and Applications*, Earl Logan, Jr.
123. *Gear Noise and Vibration*, J. Derek Smith
124. *Practical Fluid Mechanics for Engineering Applictions,* John J. Bloomer
125. *Handbook of Hydraulic Fluid Technology*, edited by George E. Totten
126. *Heat Exchanger Design Handbook*, T. Kuppan
127. *Designing for Product Sound Quality*, Richard H. Lyon
128. *Probability Applications in Mechanical Design*, Franklin E. Fisher and Joy R. Fisher
129. *Nickel Alloys,* edited by Ulrich Heubner
130. *Rotating Machinery Vibration: Problem Analysis and Troubleshooting,* Maurice L. Adams, Jr.
131. *Formulas for Dynamic Analysis,* Ronald L. Huston and C. Q. Liu
132. *Handbook of Machinery Dynamics*, Lynn L. Faulkner and Earl Logan, Jr.
133. *Rapid Prototyping Technology: Selection and Application,* Kenneth G. Cooper

Additional Volumes in Preparation

Reciprocating Machinery Dynamics, Abdulla S. Rangwala

Reliability Verification, Testing, and Analysis of Engineering Design, Gary S. Wasserman

Maintenance Excellence: Optimizing Equipment Life Cycle Decisions, edited by John D. Campbell and Andrew K. S. Jardine

Mechanical Engineering Software

Spring Design with an IBM PC, Al Dietrich

Mechanical Design Failure Analysis: With Failure Analysis System Software for the IBM PC, David G. Ullman

FORMULAS FOR DYNAMIC ANALYSIS

RONALD L. HUSTON
University of Cincinnati
Cincinnati, Ohio

C. Q. Liu
DaimlerChrysler Corporation
Auburn Hills, Michigan

MARCEL DEKKER, INC. NEW YORK • BASEL

Library of Congress Cataloging-in-Publication Data

Huston, Ronald L.,
Formulas for dynamic analysis / Ronald L. Huston, C.Q. Liu.
p. cm. -- (Mechanical engineering ; 131)
Includes bibliographical references and index.
ISBN 0-8247-9564-4 (alk. paper)
1. Dynamics. 2. Engineering mathematics--Formulae. I. Liu, C. Q.. II. Title. III. Mechanical engineering (Marcel Dekker, Inc.); 131.
TA352 .H87 2000
620.1'04'0151--dc21
00-047595

This book is printed on acid-free paper.

Headquarters
Marcel Dekker, Inc.
270 Madison Avenue, New York, NY 10016
tel: 212-696-9000; fax: 212-685-4540

Eastern Hemisphere Distribution
Marcel Dekker AG
Hutgasse 4, Postfach 812, CH-4001 Basel, Switzerland
tel: 41-61-261-8482; fax: 41-61-261-8896

World Wide Web
http://www.dekker.com

The publisher offers discounts on this book when ordered in bulk quantities. For more information, write to Special Sales/Professional Marketing at the headquarters address above.

Current printing (last digit):
10 9 8 7 6 5 4 3 2 1

PRINTED IN THE UNITED STATES OF AMERICA

Preface

This is a reference book summarizing the principal equations of dynamics and their underlying theoretical bases. The book is intended to be a ready source of information on dynamics for students, practitioners, and researchers in all branches of engineering and science. The book can also serve as a self-study text or as a reference text for course adoption.

The book summarizes topics ranging from elementary particle dynamics to multibody system dynamics. The book uses classical vector/ matrix notation, and it is intended to be accessible to anyone having had a first course in dynamics.

Although the objective of the book is to summarize dynamics equations, the development of these equations from first principles is provided as well. In addition, numerous examples are presented to illustrate the use of the equations.

The book itself is divided into 14 chapters, with the first two providing introductory remarks and a review of vector methods. The third and fourth chapters summarize particle kinematics and kinetics. The fifth chapter then treats particle dynamics.

The kinematic equations for rigid bodies are developed in considerable detail in Chapters 6 and 7, with particular attention given to angular velocity expressions, transformation matrices, and Euler parameters. Chapter 8 then summarizes the classical concepts of inertia, including inertia vectors ("second movement vectors"), moments and products of inertia, inertia dyadics, transformation matrices, and principal moments of inertia. The next two chapters then develop rigid body kinetics and dynamics.

Although all the major dynamics principles, equations, and "laws" are presented, there is a focus on Kane's equations, which have received increasingly broad application in the past 25 years — particularly for large, complex systems.

The remainder of the chapters are devoted to application and illustrations, with the eleventh chapter presenting some classical problems in dynamics. The final two chapters summarize results in multibody kinematics, kinetics, and dynamics.

The authors appreciate the patience and encouragement of the editors and the assistance of Charlotte Better in preparing the manuscript. Also, the assistance of Xiaobo Liu and Madhusudhan Raghusnathan is acknowledged.

Ronald L. Huston
C.Q. Liu

Contents

Chapter 1

INTRODUCTION

1.1 Preparatory Remarks

This book is intended to be a reference summarizing the principal equations of dynamics and the underlying theoretical bases of the equations. In these first two chapters we summarize and review some definitions, terminology, notation and methodology useful in the sequel. Readers already familiar with this material may want to simply skim through it and go on to Chapter 3 or to other topics of more immediate interest in the later chapters.

1.2 Postulates

Dynamics analyses are based upon a number of intuitive concepts and ideas which are not generally rigorously defined but instead are simply described with common language. These concepts and ideas are [1.1][*]:

- **Time** Time is a measure of change. It is sometimes described as a measure of the passing or the succession of events. In dynamics time is a non-negative, increasing quantity or variable. In dynamics analyses time is usually measured in seconds.

- **Distance** Intuitively, distance is a measure of separation of two points in space. Distance is sometimes defined as a "norm" or "metric." In dynamics distance is then a Euclidean norm. For example, if points P_1 and P_2 have Cartesian coordinates (x_1, y_1, z_1) and (x_2, y_2, z_3) then the distance d between P_1 and P_2 is defined as $[(x_1 - x_2)^2 + (y_1 - y_2)^2 + (z_1 - z_2)^2]^{1/2}$. Distance is usually measured in meters, centimeters, feet, and inches.

---

[*]Numbers in brackets refer to references listed at the end of the chapters.

- **Particle** A particle is a "small" body — so small that its dimensions are unimportant in dynamics analyses. Thus particles can be identified with points.

- **Body** A body may be regarded as a collection or set of particles. Intuitively a body is made up of a large number of particles.

- **Rigid Body** A rigid body is a body whose particles remain at fixed distances from one another.

- **Mass** For a particle, mass is a "strength" assigned to the particle measuring the amount of matter associated with the particle. In dynamics, the mass of a particle is a measure of the resistance to change in movement of a particle and also of the gravitational attraction of the particle to other particles or bodies and especially the gravitational attraction to the earth (the "weight" of the particle).

  The mass of a body is simply the sum of the masses of its particles.

  In dynamics, mass is usually measured in kilograms or slugs.

- **Force** A force is often described as a "push" or a "pull." Force is characterized by how hard the push or pull is and by the direction of its application. Forces are thus conveniently represented by vectors. Graphically, these vectors in turn are represented by line segments whose lengths are proportional to the magnitudes of the forces and whose senses are determined by an "arrow head." The extended line segment is called the "line of action" of the force. The line of action passes through the "point of application" of the particle or body on which the force is applied.

  The force units are usually Newtons or pounds.

1.3 Newton's Laws and Other Fundamental Principles

Analyses in dynamics are based upon Newton's laws and a few other fundamental principles which like the postulates of Section 1.2 are based upon intuition, experience, and experimental evidence. These laws and principles are [1.2]:

- **Newton's First Law** If a particle has no force acting on it, the particle will either be at rest or be moving with a constant speed in a straight line. ("Speed" is the magnitude of the velocity of the particle — defined in Chapter 3.)

- **Newton's Second Law** If a particle has a force acting on it the particle will accelerate (see Chapter 3) at a rate proportional to the force. Analytically, this law may be written as:

$$\mathbf{F} = m\mathbf{a} \tag{1.3.1}$$

  where **F** is the force and the particle m is its mass and **a** is the acceleration of the particle [see Equation (3.7.2)].

- **Newton's Third Law** If a particle P_1 exerts a force **F** on a particle P_2, either by contact or by gravity at a distance, then P_2 exerts an equal but oppositely directed force $-\mathbf{F}$ on P_1. This is sometimes called the "law of action and reaction."

- **Newton's Law of Gravitation** Two particles having masses m_1 and m_2 are attracted to each other by equal and opposite forces **F** and $-\mathbf{F}$. The magnitude F of these forces is:

$$F = Gm_1m_2/d^2 \tag{1.3.2}$$

  where d is the distance between the particles and G is the universal gravity constant with value [1.1]

$$G = 6.67 \times 10^{-11} \text{m}^3/\text{kg} \bullet \text{s}^2 \\ = 3.34 \times 10^{-8} \text{ft}^3/\text{slug} \bullet \text{s}^2 \tag{1.3.3}$$

- **Inertial Reference Frame** An inertial reference frame is a frame (or space) where Newton's laws are valid. Inertial reference frames may not exist in the strict sense of this definition. However, for most problems of practical importance, the earth approximates an inertial reference frame. An inertial reference frame is sometimes called a "Newtonian reference frame."

- **Addition and Superposition of Forces** Forces are represented by vectors and as such they obey the "parallelogram law of addition." If two or more forces are exerted on a particle or a rigid body these forces may be replaced by an "equivalent force systems" (see Chapter 2) without affecting the dynamics of the particle or body.

References [1.6] to [1.22] provide additional information on the origin, development, and application of these principles.

1.4 Gravity and Weight

Just as the earth approximates an inertial reference frame, the earth may also be approximately represented as a sphere with a spherically symmetric mass distribution. Then for gravitational analyses the earth may be considered to be a particle E located at the center of the approximating sphere with the entire earth mass concentrated in E [1.3]. Objects near the surface of the earth are then attracted toward the central particle E with a force magnitude given by Equation (1.3.2).

Specifically, consider a particle P with mass m near the surface of the earth. Then the distance d from P to the earth center E is approximately equal to the earth radius R. If the earth mass is M, Equation (1.3.2) shows the gravitational force magnitude F to be

$$F = GMm/R^2 \tag{1.4.1}$$

The earth mass M is approximately 5.976×10^{24} kg or 4.096×10^{23} slug; and the earth radius R is approximately 6.371×10^6 m or 3960 miles (2.09×10^7 ft) [1.1]. Using these values, Equation (1.4.1) may be written as:

$$F = mg \tag{1.4.2}$$

where g is defined as

$$g = GM/R^2 \tag{1.4.5}$$

By substituting the above stated values for G, M, and R, g is found to be approximately:

$$g = 9.81 \text{ m/s}^2 = 32.2 \text{ ft/sec}^2 \tag{1.4.4}$$

The gravitational force on a particle near the earth's surface is directed toward the earth's center (that is, toward the center of a sphere approximating the earth). This force is also called the "weight" w of the particle. Then we have the simple relation between mass and weight

$$w = mg \quad \text{or} \quad m = w/g \tag{1.4.5}$$

By comparing Equations (1.3.1) and (1.4.2) we see that if a particle P is held near the surface of the earth and then released from rest it will accelerate toward the center of the earth at the rate:

$$a = g \tag{1.4.6}$$

Thus, g is generally called "the acceleration due to gravity."

1.5 Systems of Units/Conversion Factors

Dynamics analyses generally employ the International System of units (the SI or metric system) or the English system (used primarily in the United States and New Zealand). The principal quantities being measured are: time, distance, mass, and force. Table 1.5.1 lists the most commonly used units for these quantities:

Table 1.5.1 Commonly Used Quantities and Units in Dynamics Analyses.

Quantity	SI Unit (symbol)	English Unit (symbol)
Time	second (s)	second (sec)
Distance	meter (m)	foot (ft)
Mass	kilogram (kg)	slug (slug)
Force	Newton (N)	pound (lb)

The conversion between these systems of units is readily obtained by multiplying by conversion factors as listed in most textbooks on dynamics (see, for example, References [1.1, 1.2, 1.4, 1.5]). Table 1.5.2 summarizes the frequently used conversion factors.

Table 1.5.2 Conversion Factors for Commonly Used Units in Dynamics Analyses

I. Time

To convert from:	to:	multiply by:
seconds	minutes	1.667×10^{-2}
seconds	hours	2.777×10^{-4}
minutes	seconds	60
minutes	hours	1.667×10^{-2}

To convert from:	to:	multiply by:
hours	seconds	3600
hours	minutes	60

II. Distance

To convert from:	to:	multiply by:
centimeters	meters	10^{-2}
centimeters	inches	0.3937
centimeters	feet	3.281×10^{-2}
meters	centimeters	100
meters	kilometers	10^{-3}
meters	inches	39.37
meters	feet	3.281
kilometers	meters	1000
kilometers	feet	3.281×10^{3}
kilometers	miles	0.6214
inches	centimeters	2.54
inches	meters	2.54×10^{-2}
inches	feet	8.333×10^{-2}
feet	centimeters	30.48
feet	meters	0.3048
feet	kilometers	3.048×10^{-4}
feet	inches	12
feet	miles	1.894×10^{-4}
miles	meters	1.609×10^{3}
miles	kilometers	1.609
miles	feet	5280

III. Mass

To convert from:	to:	multiply by:
kilograms	pound mass	2.2046
kilograms	slugs	6.8466×10^{2}

To convert from:	to:	multiply by:
pound mass	kilograms	0.4536
pound mass	slug	3.1056×10^{-2}
slugs	kilograms	14.6056
slugs	pound mass	32.2

IV. Force

To convert from:	to:	multiply by:
Newtons	pounds	0.2248
pounds	Newtons	4.448

V. Velocity

To convert from:	to:	multiply by:
meters/second (m/s)	kilometers/hour (km/hr)	3.6
meters/second (m/s)	miles/hour (mph)	2.237
kilometers/hour (km/hr)	meters/second (m/s)	0.2778
kilometers/hour (km/hr)	miles/hour (mph)	0.6214
kilometers/hour (km/hr)	feet/second (ft/sec)	0.9113
feet/second (ft/sec)	kilometers/hour (km/hr)	1.097
feet/second (ft/sec)	miles/hour (mph)	0.6818
miles/hour (mph)	kilometers/hr (km/hr)	1.609
miles/hour (mph)	feet/second (ft/sec)	1.4667
miles/hour (mph)	meters/second (m/s)	0.447

1.6 Dimensions

As noted in the foregoing section the principal variables in dynamics are those involving time, distance, mass, and force, with the conventional units being those listed in Table 1.5.2. These four principal variables may also be regarded as providing "dimensions" to dynamic analyses. In any expression or equation in dynamics analysis the dimension must be consistent (that is, the same or homogeneous) in each term of the expression or equation.

Let the dimensions of time, distance (or length), mass, and force be designated as: [T], [L], [M], and [F] respectively. Then, for example, the kinematic quantities position, velocity, and acceleration would have the dimensions [L], [L/T], and $[L/T^2]$ respectively.

These four dimensions are not independent but instead are related by the analytical expression of Newton's second law of Equation (1.3.1). That is

$$[F] = [M][L/T^2] \tag{1.6.1}$$

Equation (1.6.1) may be used to provide relations between some of the units of Table 1.5.2. Specifically, the force units Newtons (SI) and pounds (English) are related to the mass, length, and time units as:

$$1 \text{ Newton} = 1 \text{ kilogram meter/second}^2 \quad \text{or} \quad 1\text{ N} = 1 \text{ kg m/s}^2 \tag{1.6.2}$$

and

$$1 \text{ pound} = 1 \text{ slug foot/second}^2 \quad \text{or} \quad 1 \text{ lb} = 1 \text{ slug ft/sec}^2 \tag{1.6.3}$$

Finally, observe that if the dimensions of the terms of an equation are not consistent there is an error in the equation. However, even if the dimensions are consistent the equation or expression is not necessarily correct. That is, dimensional homogeneity is a *necessary* but *not sufficient* condition for correctness of an equation.

References

1.1 J. L. Meriam and L. G. Kraige, *Engineering Mechanics*, Third Edition, Wiley, New York, NY, 1992.

1.2 F. P. Beer and E. R. Johnson, Jr., *Vector Mechanics for Engineers*, McGraw Hill, NY, 1984.

1.3 T. R. Kane, *Analytical Elements of Mechanics*, Vol. 1, Academic Press, New York, NY, 1959.

1.4 T. Baumeister (Editor-in-Chief), *Marks' Standard Handbook for Mechanical Engineers*, Eight Edition, McGraw Hill, NY, 1978.

1.5 M. Kutz (Editor), *Mechanical Engineers' Handbook*, Wiley, NY, 1986.

1.6 E. T. Whittaker, *A Treatise on the Analytical Dynamics of Particles and Rigid Bodies*, Cambridge, London, 1937.

1.7 L. Brand, *Vectorial Mechanics*, Wiley, New York, NY, 1947.

1.8 G. Hamel, *Theoretische Mechanik*, Spring, 1949.

1.9 R. L. Halfman, *Dynamics*, Addison-Wesley, Reading, MA, 1959.

1.10 G. W. Housner and D. E. Hudson, *Applied Mechanics — Dynamics*, D. van Nostrand, Princeton, NJ, 1959.

1.11 H. Yeh and J. I. Adams, *Principles of Mechanics of Solids and Fluids*, Volume 1, McGraw Hill, New York, NY, 1960.

1.12 T. R. Kane, *Analytical Elements of Mechanics*, Vol. 2, Academic Press, New York, NY, 1961.

1.13 D. T. Greenwood, *Principles of Dynamics*, Prentice Hall, Englewood Cliffs, NJ, 1965.

1.14 T. R. Kane, *Dynamics*, Holt, Rinehart and Winston, New York, NY, 1968.

1.15 L. Meirovitch, *Methods of Analytical Dynamics*, McGraw Hill, New York, NY, 1970.

1.16 J. J. Tuma, *Dynamics*, Quantum Publishers, Inc., New York, NY, 1974.

1.17 L. A. Pars, *A Treatise on Analytical Dynamics*, Ox Bow Press, Wood Bridge, CT, 1979.

1.18 H. Goldstein, *Classical Mechanics*, Addison-Wesley, Reading, MA, 1980.

1.19 B. J. Torby, *Advanced Dynamics for Engineers*, Holt, Rinehart and Winston, New York, NY, 1984.

1.20 T. R. Kane and D. A. Levinson, *Dynamics: Theory and Applications*, McGraw Hill, New York, NY, 1985.

1.21 J. B. Marion and S. T. Thornton, *Classical Dynamics of Particles and Systems*, Harcourt, Brace and Jovanovich, San Diego, CA, 1988.

1.22 E. J. Haug, *Intermediate Dynamics*, Prentice Hall, Englewood Cliffs, NJ, 1992.

Chapter 2

VECTOR ANALYSIS AND PRELIMINARY CONSIDERATIONS

2.1 Introduction

In this chapter we briefly review and present some operational formulas, primarily from elementary vector analysis, which form the basis for the dynamics formulations of the subsequent chapters. We develop and illustrate many of the formulas through a series of examples. The references at the end of this chapter provide a more comprehensive review.

2.2 Fundamental Concepts

- **Vectors** — Mathematically, a vector is an element of a vector space [2.1, 2.2, 2.3]. Since dynamics is fundamentally a geometric subject it is helpful to think of vectors as being directed line segments. Symbolically, vectors are written in bold face type (for example, **v**).

- **Vector Characteristics** — The characteristics of a vector are its magnitude (length) and direction (orientation and sense). The units of a vector are the same units of those of its magnitude. The magnitude of a vector, say **v**, is written as $|\mathbf{v}|$.

- **Equality of Vectors** — Two vectors **a** and **b** are said to be "equal" ($\mathbf{a} = \mathbf{b}$) if they have the same characteristics (magnitude and direction).

- **Scalar** — A scalar is simply a variable or parameter. Scalars may be either real or complex numbers, although in dynamics they are generally real numbers. They may be positive or negative. Frequently appearing scalars in dynamics are time, distance, mass, and force, velocity and acceleration magnitudes.

- **Multiplication of Scalars and Vectors** — Let D be a scalar and let **V** be a vector. Then the product s**V** is a vector with the same orientation and sense as **V** if s is positive, and with the same orientation but opposite sense of **V** if s is negative. The magnitude of s**V** is $|s|\ |\mathbf{V}|$. (See Section 2.5.)

- **Negative Vector** — The negative of vector **V**, written as -**V**, is a vector with the same magnitude and orientation as **V**, but with opposite sense to that of **V**. Also,

$$-\mathbf{V} = (-1)\mathbf{V} \tag{2.2.1}$$

- **Zero Vector** — A zero vector, written as either **0** or 0 is a vector with magnitude zero. The direction of a zero vector is undefined.

- **Unit Vectors** — A unit vector is a vector with magnitude: 1. A unit vector has no units.

- **Separation of Characteristics** — Any non-zero vector **V** may be expressed as a product of a scalars and a unit vector **n**. That is

$$\mathbf{V} = s\mathbf{n} \tag{2.2.2}$$

If the unit vector **n** has the same direction (orientation and sense) as **V**, then the scalar s is the magnitude of **V**. That is

$$s = |\mathbf{V}| \tag{2.2.3}$$

Hence, we also have

$$\mathbf{n} = \frac{\mathbf{V}}{|\mathbf{V}|} \tag{2.2.4}$$

The right side of Equation (2.2.2) represents a separation of the characteristics of **V**, where s is the magnitude of **V** and **n** represents the direction of **V**.

Equation (2.2.4) shows that we can always construct a unit vector with the same direction as any given non-zero vector **V**.

- **Addition of Vectors** — In a geometric representation, vectors obey the "parallelogram law of addition" (see Section 2.3). The sum of the vectors is called their "resultant." The vectors making up the sum are called "components." If these vector components are expressed as products of scalar and unit vectors, the scalars are also called "components" or "scalar components."

- **Fixed, Bound, or Sliding Vectors** — Vectors acting through a specific point are called fixed, bound, or sliding vectors. The coincident line of a fixed vector is called its "line of action." Forces are usually fixed vectors.

- **Free Vectors** — Free vectors are not associated with any particular point. Hence, free vectors may be placed in any convenient location so long as their characteristics are not changed. Unit vectors are examples of free vectors.

- **Angle Between Two Vectors** — When two free vectors are positioned graphically such that their starting ends, or "tails", are connected (see Figure 2.2.1), the included angle is called "the angle between the vectors."

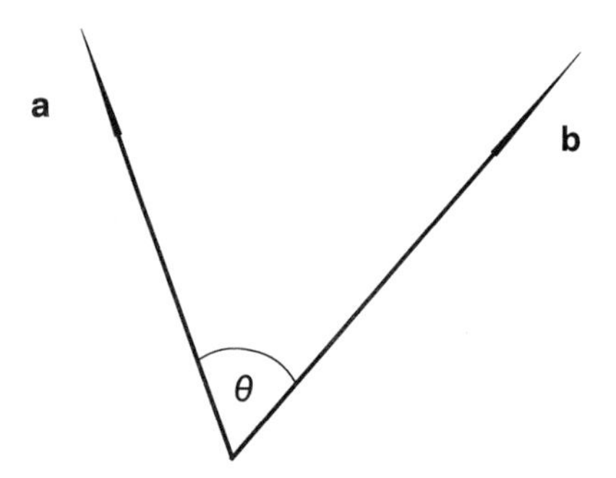

Fig. 2.2.1 Angle Between Vectors **a** and **b**

2.3 Addition of Vectors — Geometric Method

The sum of two vectors **a** and **b** may be obtained by using the parallelogram law (see Fig. 2.3.1) or the triangle law (see Fig. 2.3.2). The sum of three or more vectors may be obtained by using polygon law (see Fig. 2.3.2).

For the parallelogram law, let the vectors be connected "tail" to "tail" as in Figure 2.3.1. Then the sum **a** + **b** is the diagonal of the parallelogram developed by the vectors and passing through the connection point as shown.

For the triangle law, let the vectors be connected "head to tail" as in Figure 2.3.2. The sum **a** + **b** is then the third side of the developed triangle as shown.

By inspection it is seen that with these laws vector addition is commutative. That is:

$$\mathbf{a} + \mathbf{b} = \mathbf{b} + \mathbf{a} \tag{2.3.1}$$

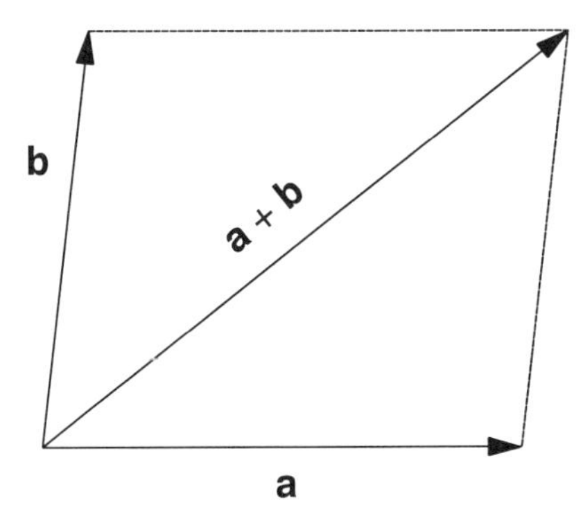

Fig. 2.3.1 Addition of Two Vectors (Parallelogram Law)

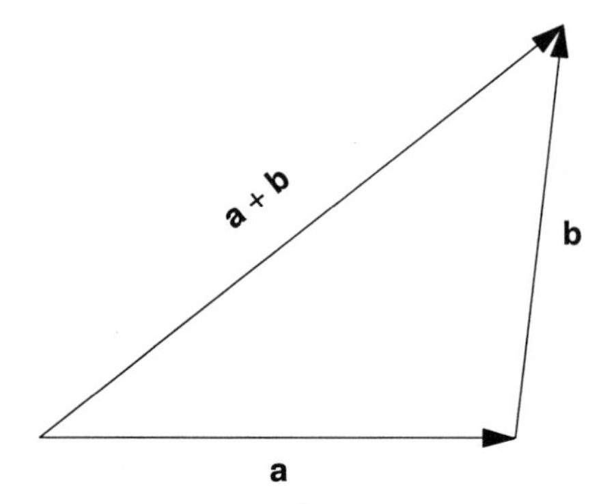

Fig. 2.3.2 Addition of Two vectors (Triangle law)

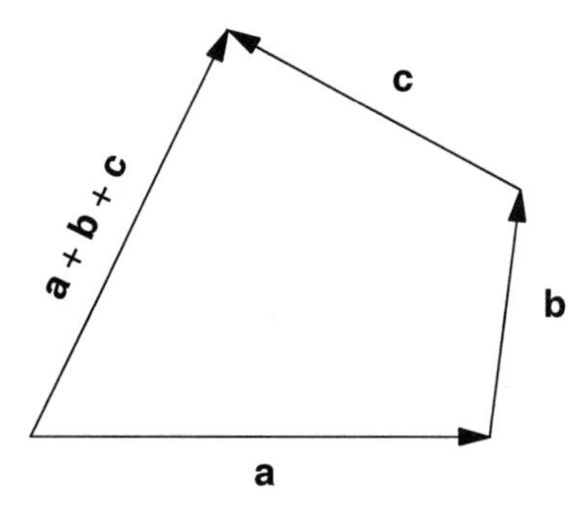

Fig. 2.3.3 Addition of Three Vectors (Polygon law)

By extending the triangle law, three or more vectors may be added as in Figure 2.3.3 (polygon law). By inspection it is seen that vector addition obeys the associative law:

$$\mathbf{a} + (\mathbf{b} + \mathbf{c}) = (\mathbf{a} + \mathbf{b}) + \mathbf{c} \qquad (2.3.2)$$

Finally, from the parallelogram law or the triangle law we see that:

$$\mathbf{a} + 0 = 0 + \mathbf{a} \tag{2.3.3}$$

$$\mathbf{a} + (-\mathbf{a}) = 0 \tag{2.3.4}$$

2.4 Difference of Vectors

The sum of **a** and -**b** is the difference of **a** and **b**, and is written as **a** - **b** (see Fig. 2.4.1). That is,

$$\mathbf{a} - \mathbf{b} = \mathbf{a} + (-\mathbf{b}) \tag{2.4.1}$$

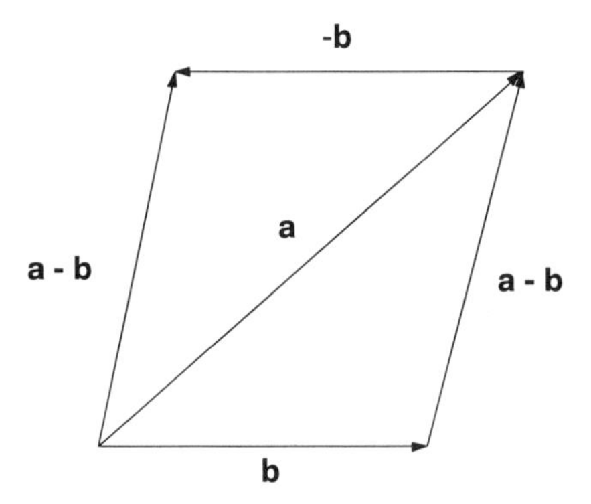

Fig. 2.4.1 Difference of **a** and **b**

2.5 Multiplication of Vectors by Scalars

Let **a** be a vector and let s be a scalar. Then s**a** is a vector (see Fig. 2.5.1) parallel to **a**, with the same sense as **a** if s is positive; and in the opposite sense of a if s is negative.

- **Operational Formulae**:

$$s\mathbf{a} = \mathbf{a}s \tag{2.5.1}$$

$$s(\mathbf{a} + \mathbf{b}) = s\mathbf{a} + s\mathbf{b} \tag{2.5.2}$$

$$(s + t)\mathbf{a} = s\mathbf{a} + t\mathbf{a} \tag{2.5.3}$$

$$0\mathbf{a} = 0 \tag{2.5.4}$$

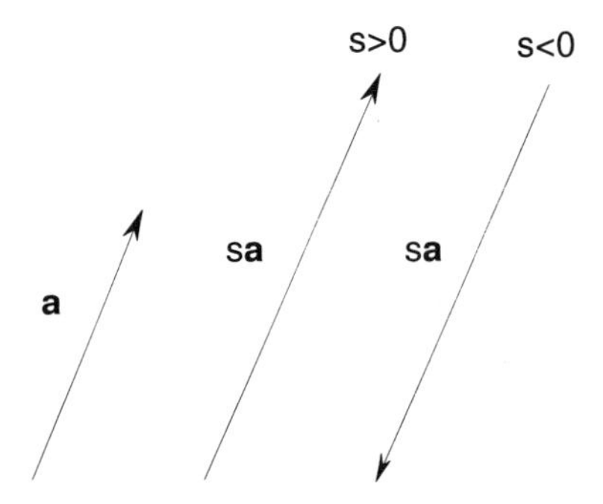

Fig. 2.5.1 Multiplication of Vector by a Scalar

2.6 Addition of Vectors — Analytical Method

The geometric and intuitive concepts of Sections 2.3, 2.4, and 2.5 form the basis for developing an analytical method of vector addition. Indeed, the conversion of the geometric methods into an analytical method is the basis for the utility of vector methods in mechanics (see [2.4] to [2.8]).

Consider again the addition of vectors as in Figure 2.3.1 in Section 2.3. Let the sum $\mathbf{a} + \mathbf{b}$ be designated by $\mathbf{c}$. That is, let

$$\mathbf{c} = \mathbf{a} + \mathbf{b} \tag{2.6.1}$$

As noted earlier, **c** is called the "resultant" of **a** and **b**, and **a** and **b** are said to be "components" of **c**.

The utility of Equation (2.6.1) is not in determining the resultant of vectors **a** and **b**, but instead in determining components of a given vector **c**.

Suppose, for example, that we are given a vector **V** parallel to an X-Y coordinate plane. It is always possible to find components $\mathbf{V}_x$ and $\mathbf{V}_y$ of **V** which are perpendicular to each other and parallel to the X and Y axes, respectively. To see this, let **V**, as depicted in Figure 2.6.1 (a), have an inclination θ relative to the X-axis as shown. Then the desired perpendicular components may be constructed as shown in Figure 2.6.1 (b) where

$$|\mathbf{V}_x| = |\mathbf{V}|\cos\theta \quad \text{and} \quad |\mathbf{V}_y| = |\mathbf{V}|\sin\theta \tag{2.6.2}$$

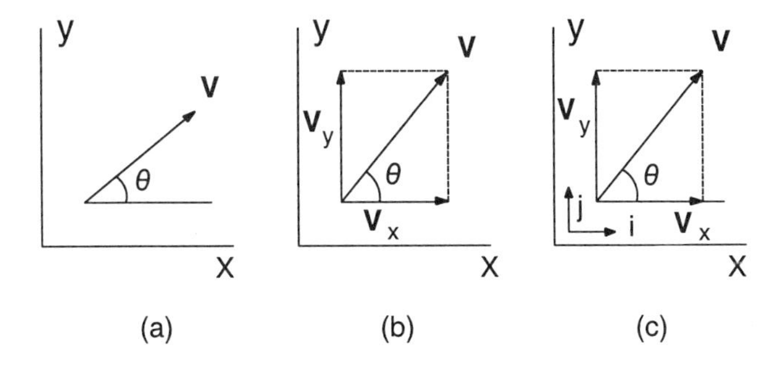

Fig. 2.6.1 Expression of a Given Vector **V** in Terms of Perpendicular Components $\mathbf{V}_x$ and $\mathbf{V}_y$

Finally, if **i** and **j** are unit vectors parallel to the X and Y axes as in Figure 2.6.1 (c) we have

$$\mathbf{V}_x = V_x \mathbf{i} \quad \text{and} \quad \mathbf{V}_y = V_y \mathbf{j} \tag{2.6.3}$$

and then

$$\mathbf{V} = V_x \mathbf{i} + V_y \mathbf{j} \tag{2.6.4}$$

where V_x and V_y are $|\mathbf{V}_x|$ and $|\mathbf{V}_y|$ if $\mathbf{V}_x$ and $\mathbf{V}_y$ have the same sense as the positive X and Y direction, respectively. (V_x is negative $|\mathbf{V}_x|$ if $\mathbf{V}_x$ is directed along the negative X-axis and $\mathbf{V}_y$ is negative $|\mathbf{V}_y|$ if $\mathbf{V}_y$ is directed along the negative Y-axis.)

In a similar manner, a vector **W** may be expressed as

$$\mathbf{W} = W_x \mathbf{i} + W_y \mathbf{j} \tag{2.6.5}$$

The sum **V** + **W** is then seen to be

$$\begin{aligned} \mathbf{V} + \mathbf{W} &= V_x \mathbf{i} + V_y \mathbf{j} + W_x \mathbf{i} + W_y \mathbf{j} \\ &= (V_x + W_x)\mathbf{i} + (V_y + W_y)\mathbf{j} \end{aligned} \tag{2.6.6}$$

In three dimensions, if we have

$$\mathbf{V} = V_x \mathbf{i} + V_y \mathbf{j} + V_z \mathbf{k} \quad \text{and} \quad \mathbf{W} = W_x \mathbf{i} + W_y \mathbf{j} + W_z \mathbf{k} \tag{2.6.7}$$

then

$$\mathbf{V} + \mathbf{W} = (V_x + W_x)\mathbf{i} + (V_y + W_y)\mathbf{j} + (V_z + W_z)\mathbf{k} \tag{2.6.8}$$

Also we have

$$|\mathbf{V}| = (V_x^2 + V_y^2 + V_z^2)^{1/2}, \quad |\mathbf{W}| = (W_x^2 + W_y^2 + W_z^2)^{1/2} \tag{2.6.9}$$

and

$$|\mathbf{V} + \mathbf{W}| = [(V_x + W_x)^2 + (V_y + W_y)^2 + (V_z + W_z)^2]^{1/2} \tag{2.6.10}$$

2.7 Vector Representations

Consider a position vector **p** locating a point P in an XYZ Cartesian reference frame as in Figure 2.7.1. Let (x,y,z) be the Cartesian coordinates of P. Then **p** may be expressed as:

$$\mathbf{p} = x\mathbf{i} + y\mathbf{j} + z\mathbf{k} \tag{2.7.1}$$

where **i**, **j**, and **k** are unit vectors parallel to the X, Y, and Z axes as shown.

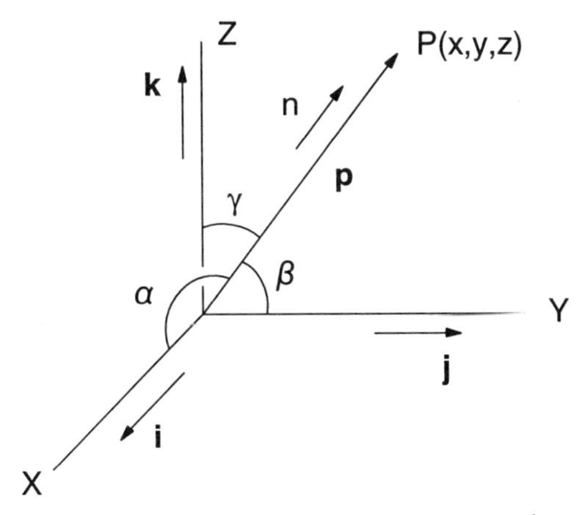

Fig. 2.7.1 Position Vector Representation

It is often convenient to identify P, **p** and (x,y,z) with one another.

Next, let α, β, and γ be the angles (called "direction cosines") that **p** makes with the X, Y, and Z axes. (See Fig. 2.7.1.) Let **n** be a unit vector parallel to **p** as shown. Then **n** may be expressed as:

$$\mathbf{n} = \frac{\mathbf{p}}{|\mathbf{p}|} = \cos\alpha\,\mathbf{i} + \cos\beta\,\mathbf{j} + \cos\gamma\,\mathbf{k} \tag{2.7.2}$$

These expressions may be generalized and adapted to any vector **V**. The results are summarized in Table 2.7.1.

Table 2.7.1 Summary of Vector Expressions

Quantity	Expressions
Vector **V** in Component Form	$\mathbf{V} = V_x\mathbf{i} + V_y\mathbf{j} + V_z\mathbf{k}$
Vector **V** in Coordinate Form	$\mathbf{V}$: (V_x, V_y, V_z)
Unit Vector	**i**: (1, 0, 0) **j**: (0, 1, 0) **k**: (0, 0, 1)
Magnitude of **V**	$\lvert\mathbf{V}\rvert = V = (V_x^2 + V_y^2 + V_z^2)^{1/2}$
Direction of Cosines of **V**	$\frac{\cos\alpha}{V_x} = \frac{\cos\beta}{V_y} = \frac{\cos\gamma}{V_z} = \frac{1}{V}$
Unit Vector **n** Parallel to **V**	$\mathbf{n} = \frac{V_x}{V}\mathbf{i} + \frac{V_y}{V}\mathbf{j} + \frac{V_z}{V}\mathbf{k}$ $= \cos\alpha\,\mathbf{i} + \cos\beta\,\mathbf{j} + \cos\gamma\,\mathbf{k}$

2.8 Examples: Addition of Force Vectors

One of the most common uses of vector addition is in obtaining the resultant of a set of force vectors. To illustrate this consider the box shown in Figure 2.8.1.

Let the box be subjected to a set of eight forces as shown where $\mathbf{F}_1$ is directed along the diagonal OA and $\mathbf{F}_2$ is directed along the diagonal AB. Let the magnitudes of these forces be:

$$|\mathbf{F}_1| = 39\text{ N} \quad |\mathbf{F}_2| = 15\text{ N} \quad |\mathbf{F}_3| = 10\text{ N}$$
$$|\mathbf{F}_4| = 10\text{ N} \quad |\mathbf{F}_5| = 7\text{ N} \quad |\mathbf{F}_6| = 8\text{ N} \tag{2.8.1}$$
$$|\mathbf{F}_7| = 8\text{ N} \quad |\mathbf{F}_8| = 5\text{ N}$$

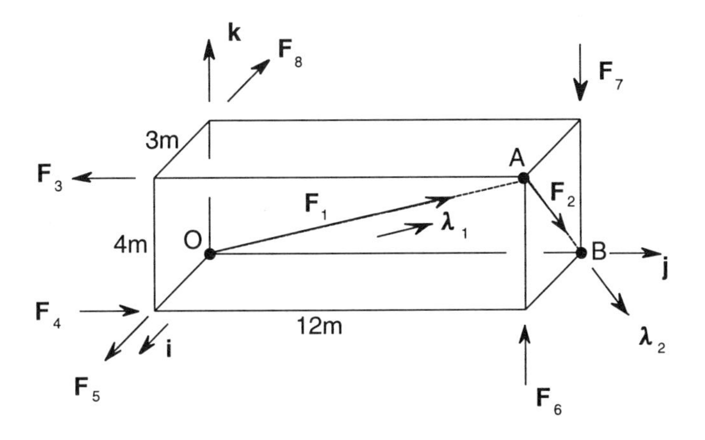

Fig. 2.8.1 Box Subjected to a Force System

Finally, let the dimensions of the box be 12m long, 4m high, and 3m deep, and let **i**, **j**, and **k** be mutually perpendicular unit vectors parallel to the edges of the box as also shown in Figure 2.8.1. The objective is to find the resultant (or sum) of the eight force vectors.

<u>*Solution*</u>: We may readily obtain the resultant by expressing each of the forces $\mathbf{F}_i$ $(i = 1,\ldots,8)$ in terms of the unit vectors **i**, **j**, and **k**. In this regard, observe that, except for $\mathbf{F}_1$ and $\mathbf{F}_2$, the force vectors are parallel to an edge of the box and thus from Equation (2.6.3) may be immediately expressed in terms of **i**, **j**, and **k** as:

$$\mathbf{F}_3 = -10\mathbf{j}\text{N} \qquad \mathbf{F}_4 = 10\mathbf{j}\text{N} \qquad \mathbf{F}_5 = 7\mathbf{i}\text{N}$$
$$\mathbf{F}_6 = 8\mathbf{k}\text{N} \qquad \mathbf{F}_7 = -8\mathbf{k}\text{N} \qquad \mathbf{F}_8 = -5\mathbf{i}\text{N} \tag{2.8.2}$$

Regarding $\mathbf{F}_1$ and $\mathbf{F}_2$ let $\boldsymbol{\lambda}_1$ and $\boldsymbol{\lambda}_2$ be unit vectors parallel to $\mathbf{F}_1$ and $\mathbf{F}_2$ as shown in Figure 2.8.1. Then from Equation (2.6.3) we have

$$\mathbf{F}_1 = 39\boldsymbol{\lambda}_1\text{N} \quad \text{and} \quad \mathbf{F}_2 = 15\boldsymbol{\lambda}_2\text{N} \tag{2.8.3}$$

But from Equations (2.7.1) and (2.7.2) we may express $\boldsymbol{\lambda}_1$ and $\boldsymbol{\lambda}_2$ in terms of **i**, **j**, and **k** as:

$$\boldsymbol{\lambda}_1 = \mathbf{OA}/|\mathbf{OA}| \quad \text{and} \quad \boldsymbol{\lambda}_2 = \mathbf{AB}/|\mathbf{AB}| \tag{2.8.4}$$

From Figure 2.8.1 we have

$$\mathbf{OA} = 3\mathbf{i} + 12\mathbf{j} + 4\mathbf{k}\ \text{m} \quad \text{and} \quad \mathbf{AB} = -3\mathbf{i} - 4\mathbf{k}\ \text{m} \tag{2.8.5}$$

Then

$$|\mathbf{OA}| = [(3)^2 + (12)^2 + (4)^2]^{1/2} = 13\text{m} \quad \text{and} \quad |\mathbf{AB}| = [(-3)^2 + (-4)^2]^{1/2} = 5\text{m} \tag{2.8.6}$$

and then from Equations (2.8.4) $\boldsymbol{\lambda}_1$ and $\boldsymbol{\lambda}_2$ become:

$$\boldsymbol{\lambda}_1 = (3/13)\mathbf{i} + (12/13)\mathbf{j} + (4/13)\mathbf{k} \quad \text{and} \quad \boldsymbol{\lambda}_1 = -(3/5)\mathbf{i} - (4/5)\mathbf{k} \tag{2.8.7}$$

Hence, from Equations (2.8.3) $\mathbf{F}_1$ and $\mathbf{F}_2$ have the forms:

$$\mathbf{F}_1 = 9\mathbf{i} + 36\mathbf{j} + 12\mathbf{k}\ \text{N} \quad \text{and} \quad \mathbf{F}_2 = -9\mathbf{i} - 12\mathbf{k} \tag{2.8.8}$$

The resultant **R** of the force system is [see Equation (2.6.1)]:

$$\mathbf{R} = \mathbf{F}_1 + \mathbf{F}_2 + \ldots + \mathbf{F}_8 = \sum_{i=1}^{8} \mathbf{F}_i \tag{2.8.9}$$

The resultant may quickly be determined by listing the forces in component form as follows:

$$
\begin{aligned}
\mathbf{F}_1 &= \ \ 9\mathbf{i} + 36\mathbf{j} + 12\mathbf{k} \\
\mathbf{F}_2 &= -9\mathbf{i} + \ \ 0\mathbf{j} - 12\mathbf{k} \\
\mathbf{F}_3 &= \ \ 0\mathbf{i} - 10\mathbf{j} + \ \ 0\mathbf{k} \\
\mathbf{F}_4 &= \ \ 0\mathbf{i} + 10\mathbf{j} + \ \ 0\mathbf{k} \\
\mathbf{F}_5 &= \ \ 7\mathbf{i} + \ \ 0\mathbf{j} + \ \ 0\mathbf{k} \\
\mathbf{F}_6 &= \ \ 0\mathbf{i} + \ \ 0\mathbf{j} + \ \ 8\mathbf{k} \\
\mathbf{F}_7 &= \ \ 0\mathbf{i} + \ \ 0\mathbf{j} - \ \ 8\mathbf{k} \\
\mathbf{F}_8 &= -5\mathbf{i} + \ \ 0\mathbf{j} + \ \ 0\mathbf{k}
\end{aligned}
\tag{2.8.10}
$$

Then from Equation (2.6.8) by adding the columns of components we immediately obtain the resultant to be

$$\mathbf{R} = 2\mathbf{i} + 36\mathbf{j}\ \text{N} \tag{2.8.11}$$

2.9 Vector Multiplication — Scalar Product

Vectors may be multiplied in several ways. One of these products, called "scalar" or "dot" product, produces a scalar measuring a projection of one vector upon the other. The scalar product is defined as:

- **Definition**:

$$\mathbf{a}\cdot\mathbf{b} \triangleq |\mathbf{a}|\,|\mathbf{b}|\cos\theta \tag{2.9.1}$$

where θ is the angle between **a** and **b**.

This definition leads to the following operational expressions:

- **Operational Formulae**:

$$\mathbf{a} \cdot \mathbf{b} = \mathbf{b} \cdot \mathbf{a} \qquad \text{(Commutativity)}$$
$$\mathbf{a} \cdot (\mathbf{b} + \mathbf{c}) = \mathbf{a} \cdot \mathbf{b} + \mathbf{a} \cdot \mathbf{c} \qquad \text{(Distributivity)} \tag{2.9.2}$$
$$\mathbf{a} \cdot \mathbf{a} = \mathbf{a}^2 = |\mathbf{a}|^2$$

For unit vectors, we have

$$\mathbf{i} \cdot \mathbf{i} = \mathbf{j} \cdot \mathbf{j} = \mathbf{k} \cdot \mathbf{k} = 1$$
$$\mathbf{i} \cdot \mathbf{j} = \mathbf{j} \cdot \mathbf{k} = \mathbf{k} \cdot \mathbf{i} = 0 \tag{2.9.3}$$

In three dimensions, any vector may be represented as:

$$\mathbf{a} = (\mathbf{a} \cdot \mathbf{i})\mathbf{i} + (\mathbf{a} \cdot \mathbf{j})\mathbf{j} + (\mathbf{a} \cdot \mathbf{k})\mathbf{k} \tag{2.9.4}$$

The scalar product may be represented in terms of vector components as follows:

If $\mathbf{a} = a_x\mathbf{i} + a_y\mathbf{j} + a_z\mathbf{k}$, $\mathbf{b} = b_z\mathbf{i} + b_y\mathbf{j} + b_z\mathbf{k}$, then
$$\mathbf{a} \cdot \mathbf{b} = a_x b_z + a_y b_y + a_z b_z \tag{2.9.5}$$
If $\mathbf{a} \cdot \mathbf{b} = 0$, then if $\mathbf{a} \neq 0$ and $\mathbf{b} \neq 0$, then **a** is perpendicular to **b**.

Conversely, if **a** is perpendicular to **b**, then $\mathbf{a} \cdot \mathbf{b} = 0$.

2.10 Vector Multiplication - Vector Product

A second way of multiplying vectors, called the "vector" or "cross" product, produces a vector. The vector product is defined as:

- **Definition** (see Fig. 2.10.1):

$$\mathbf{a} \times \mathbf{b} \triangleq |\mathbf{a}|\,|\mathbf{b}| \sin\theta\, \mathbf{n} \tag{2.10.1}$$

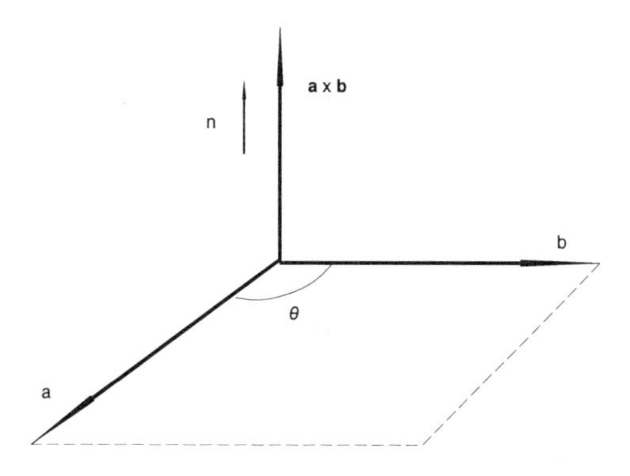

Fig. 2.10.1 Vector Product of Two Vectors

where θ is the angle between **a** and **b**, and where **n** is normal to the plane of **a** and **b**. The sense of **n** is defined by the "right-hand-rule," that is, the direction of advancement of a right hand thread screw, whose axis is normal to the plane of **a** and **b**, and is turned in the direction of **a** rotated toward **b** so as to diminish the angle between **a** and **b**.

This definition leads to the following operational expressions:

- **Operational Formulae**:

$$\mathbf{a} \times \mathbf{b} = -(\mathbf{b} \times \mathbf{a}) \quad \text{(anti-commutative)}$$
$$\mathbf{a} \times (\mathbf{b} + \mathbf{c}) = \mathbf{a} \times \mathbf{b} + \mathbf{a} \times \mathbf{c} \quad \text{(distributivity)} \tag{2.10.2}$$
$$\mathbf{a} \times \mathbf{a} = 0$$

For unit vectors we have

$$\mathbf{i} \times \mathbf{i} = \mathbf{j} \times \mathbf{j} = \mathbf{k} \times \mathbf{k} = 0$$
$$\mathbf{i} \times \mathbf{j} = \mathbf{k}, \quad \mathbf{j} \times \mathbf{k} = \mathbf{i}, \quad \mathbf{k} \times \mathbf{i} = \mathbf{j} \tag{2.10.3}$$

If $\mathbf{a} = a_x\mathbf{i} + a_y\mathbf{j} + a_z\mathbf{k}$, and $\mathbf{b}_x\mathbf{i} + b_y\mathbf{j} + b_z\mathbf{k}$, then

$$\mathbf{a} \times \mathbf{b} = (a_y b_z - a_z b_y)\mathbf{i} + (a_z b_x - a_x b_z)\mathbf{j} + (a_x b_y - a_y b_x)\mathbf{k} \tag{2.10.4}$$

$$= \begin{bmatrix} \mathbf{i} & \mathbf{j} & \mathbf{k} \\ a_x & a_y & a_z \\ b_x & b_y & b_z \end{bmatrix}$$

If $\mathbf{a} \times \mathbf{b} = 0$ and if $\mathbf{a} \neq 0$, and $\mathbf{b} \neq 0$, then $\mathbf{a}$ is parallel to $\mathbf{b}$.

2.11 Examples: Moments of Force Systems

Example 2.11.1 Moment of a Force About a Point

Probably the most common application of the vector product is in the computation of the moment of a force about a point: In Figure 2.11.1 let O be a point and let **F** be a force.

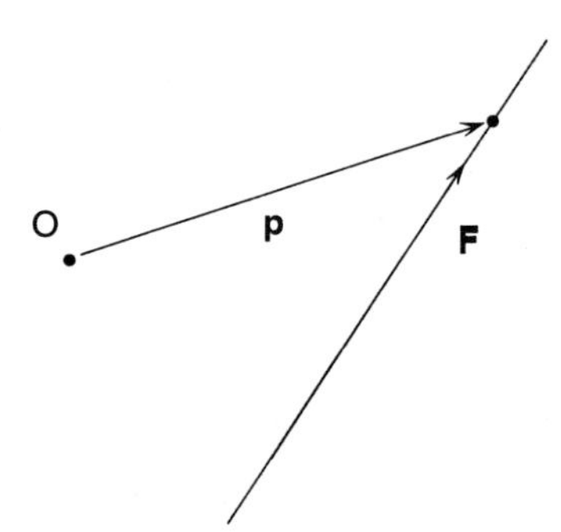

Fig. 2.11.1 Force **F**, Point O, and Position Vector **p**

Let **p** be a position vector from O to any point on the line of action of **F**. Then the moment of **F** about O is defined as:

$$\mathbf{M}_O = \mathbf{p} \times \mathbf{F} \tag{2.11.1}$$

Example 2.11.2 Moment of a System of Forces About a Point

As an extension of the definition of Example of 2.11.1, the moment of a system of forces about a point is defined as the sum of the moments of the individual forces of the system about the point. Consider the force system S depicted in Figure 2.11.2. Let O be an object point. Then the moment of S about O is defined as

$$\mathbf{M}_O = \mathbf{p}_1 \times \mathbf{F}_1 + \mathbf{p}_2 \times \mathbf{F}_2 + \cdots + \mathbf{p}_N \times \mathbf{F}_N \tag{2.11.2}$$

or

$$\mathbf{M}_O = \sum_{i=1}^{N} \mathbf{p}_i \times \mathbf{F}_i \tag{2.11.3}$$

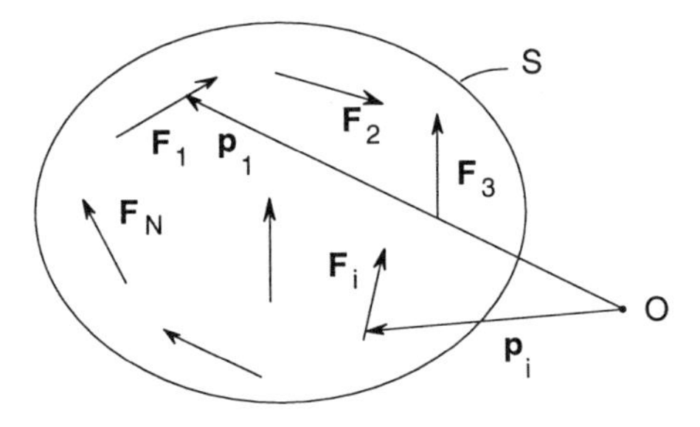

Fig. 2.11.2 A System of Forces and an Object Point O

where N is the number of forces and as before $\mathbf{p}_i$ locates a point on the line of action of a typical force $\mathbf{F}_i$ ($i = 1,\ldots,N$) of S relative to O.

Example 2.11.3 Relation Between the Moments of a System of Forces About Two Distinct Points

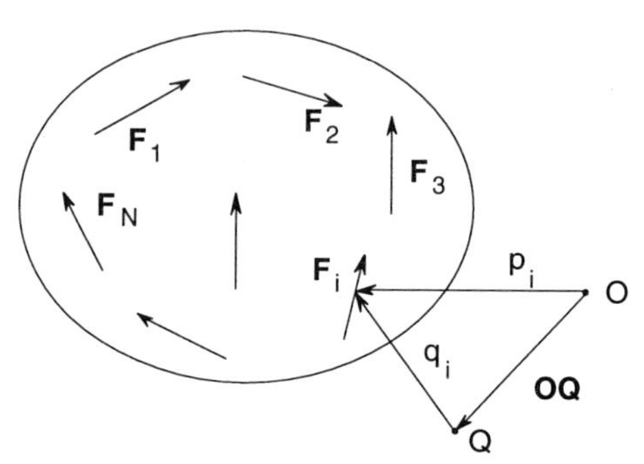

Fig. 2.11.3 A System of Forces and Object Points O and Q

Consider again a system of forces S of N forces as in Figure 2.11.3. Let O and Q be object reference points and let $\mathbf{p}_i$ and $\mathbf{q}_i$ locate a point on the line of action of typical force $\mathbf{F}_i$ of S. Then from Fig. 2.11.3 and the principles of vector addition of Section 2.3 we have

$$\mathbf{p}_i = \mathbf{q}_i + \mathbf{OQ} \tag{2.11.4}$$

where **OQ** locates Q relative to O. Then from Equation (2.11.3) we have

$$\begin{aligned} \mathbf{M}_O &= \sum_{i=1}^{N} \mathbf{p}_i \times \mathbf{F}_i = \sum_{i=1}^{N} (\mathbf{q}_i + \mathbf{OQ}) \times \mathbf{F}_i \\ &= \sum_{i=1}^{N} \mathbf{q}_i \times \mathbf{F}_i + \sum_{i=1}^{N} \mathbf{OQ} \times \mathbf{F}_i \\ &= \mathbf{M}_Q + \mathbf{OQ} \times \sum_{i=1}^{N} \mathbf{F}_i \end{aligned} \tag{2.11.5}$$

Hence, from Equation (2.8.9) we have

$$\mathbf{M}_O = \mathbf{M}_Q + \mathbf{OQ} \times \mathbf{R} \tag{2.11.6}$$

where **R** is the resultant of S.

Example 2.11.4 Verification of Equation (2.11.6)

Consider again the force system acting on the box of the example of Section 2.8 and shown again in Fig. 2.11.4.

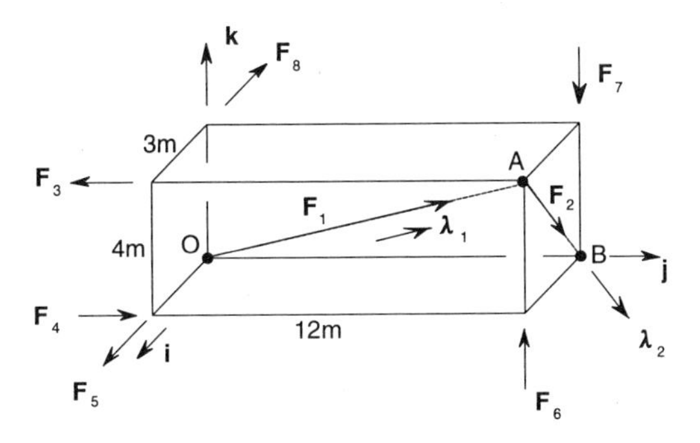

Fig. 2.11.4 Box Subjected to a Force System

As before, let the magnitudes of the forces be:

$$\begin{array}{lll} |\mathbf{F}_1| = 39\text{N} & |\mathbf{F}_2| = 15\text{N} & |\mathbf{F}_3| = 10\text{N} \\ |\mathbf{F}_4| = 10\text{N} & |\mathbf{F}_5| = 7\text{N} & |\mathbf{F}_6| = 8\text{N} \\ |\mathbf{F}_7| = 8\text{N} & |\mathbf{F}_8| = 5\text{N} & \end{array} \tag{2.11.7}$$

Then from Equations (2.8.2) and (2.8.8) we can express the forces as:

$$\begin{array}{l} \mathbf{F}_1 = 9\mathbf{i} + 36\mathbf{j} + 12\mathbf{k}\ \text{N}\ , \quad \mathbf{F}_2 = -9\mathbf{i} - 12\mathbf{k}\ \text{N} \\ \mathbf{F}_3 = -10\mathbf{j}\ \text{N}\ , \quad \mathbf{F}_4 = 10\mathbf{j}\ \text{N}\ , \quad \mathbf{F}_5 = 7\mathbf{i}\ \text{N} \\ \mathbf{F}_6 = 8\mathbf{k}\ \text{N}\ , \quad \mathbf{F}_7 = -8\mathbf{k}\ \text{N}\ , \quad \mathbf{F}_8 = -5\mathbf{i}\ \text{N} \end{array} \tag{2.11.8}$$

The objective is to find the moments of the force system about two points, say O and A, and then to verify Equation (2.11.6).

Solution: By use of Equations (2.11.1) and (2.11.2) and by inspection of Figure 2.11.4 we have

$$\begin{aligned}\mathbf{M}_O &= 0 \times \mathbf{F}_1 + 12\mathbf{j} \times \mathbf{F}_2 + (3\mathbf{i} + 4\mathbf{k}) \times \mathbf{F}_3 \\ &\quad + 3\mathbf{i} \times \mathbf{F}_4 + 0 \times \mathbf{F}_6 + (3\mathbf{i} + 12\mathbf{j}) \times \mathbf{F}_6 \\ &\quad + 12\mathbf{j} \times \mathbf{F}_7 + 4\mathbf{k} \times \mathbf{F}_8\end{aligned} \tag{2.11.9}$$

or

$$\mathbf{M}_O = -104\mathbf{i} - 44\mathbf{j} + 108\mathbf{k} \text{ Nm} \tag{2.11.10}$$

Similarly, we have

$$\begin{aligned}\mathbf{M}_A &= 0 \times \mathbf{F}_1 + 0 \times \mathbf{F}_2 + 0 \times \mathbf{F}_3 - 4\mathbf{k} \times \mathbf{F}_4 \\ &\quad + (-12\mathbf{j} - 4\mathbf{k}) \times \mathbf{F}_5 + 0 \times \mathbf{F}_6 \\ &\quad + (-3\mathbf{i}) \times \mathbf{F}_7 + (-12\mathbf{j}) \times \mathbf{F}_8\end{aligned} \tag{2.11.11}$$

or

$$\mathbf{M}_A = 40\mathbf{i} - 52\mathbf{j} + 24\mathbf{k} \text{ Nm} \tag{2.11.12}$$

From Equation (2.8.11) the resultant **R** is

$$\mathbf{R} = 2\mathbf{i} + 36\mathbf{j} \text{ N} \tag{2.11.13}$$

Then by substituting from Equations (2.11.12) and (2.11.13) into the right side of Equation (2.11.6) we have

$$\begin{aligned}\mathbf{M}_A + \mathbf{OA} \times \mathbf{R} &= 40\mathbf{i} - 52\mathbf{j} + 24\mathbf{k} \\ &\quad + (3\mathbf{i} + 12\mathbf{j} + 4\mathbf{k}) \times (2\mathbf{i} + 36\mathbf{j}) \\ &= -104\mathbf{i} - 44\mathbf{j} + 108\mathbf{k} \text{ Nm}\end{aligned} \tag{2.11.14}$$

This result is seen to be identical with that of Equation (2.11.10), thus verifying Equation (2.11.6).

Example 2.11.5 Special Force System 1: Zero System

A force system is said to be a "zero system" if: 1) it has a zero resultant and 2) it has a zero moment about some point.

From Equation (2.11.6) we see that if the resultant **R** is zero then the system has a zero moment about all points. That is,

$$\mathbf{M}_O = \mathbf{M}_Q = \mathbf{M}_A = \mathbf{M}_B = \cdots = 0 \tag{2.11.15}$$

Therefore, if the system has a zero moment about some point, it has a zero moment about all points. This is a basis for elementary statics analysis.

Example 2.11.6 Special Force Systems 2: Couples

A force system is said to be a "couple" if: 1) it has a zero resultant and 2) it has a non-zero moment about some point.

As in Example 2.11.5 we see from Equation (2.11.6) that if the resultant **R** is zero, the system has the same moment about all points. This moment is commonly called the "torque" of the couple.

Figure 2.11.5 depicts a couple.

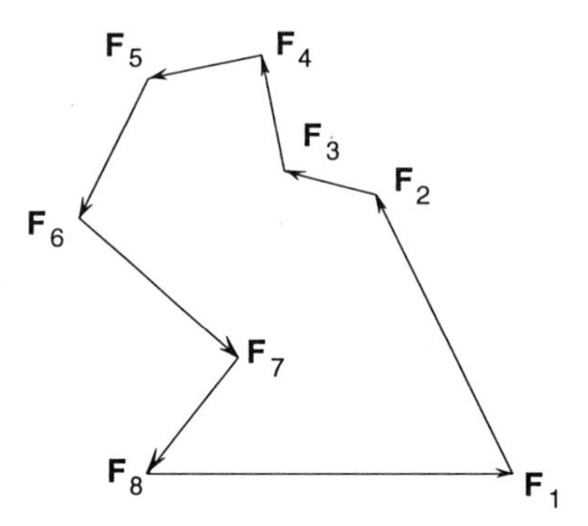

Fig. 2.11.5 A Couple — A Force System with Zero Resultant

If a couple has only two forces it is called a "simple couple." Fig. 2.11.6 depicts a simple couple. Since the resultant of a simple couple is zero, the forces must have equal magnitudes but be oppositely directed as in Fig. 2.11.6.

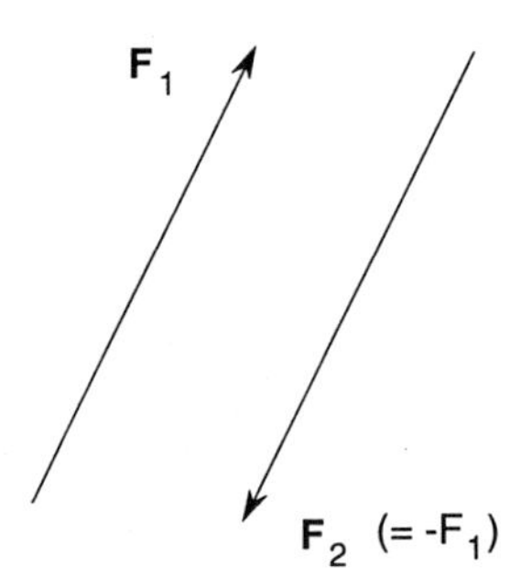

Fig. 2.11.6 A Simple Couple

Example 2.11.7 <u>Special Force Systems 3: Equivalent Force Systems</u>

Two force systems are said to be "equivalent" if: 1) they have equal resultants and 2) they have equal moments about some point.

From Equation (2.11.6) we can show that if two forces systems, say S_1 and S_2, have equal resultants (say $\mathbf{R}_1 = \mathbf{R}_2$) and equal moments about some point Q (say $\mathbf{M}_Q^{S_1} = \mathbf{M}_Q^{S_2}$), then they have equal moments about all points. To see this let Equation (2.11.6) be written in the forms:

$$\mathbf{M}_O^{S_1} = \mathbf{M}_Q^{S_1} + \mathbf{OQ} \times \mathbf{R}_1 \tag{2.11.16}$$

and

$$\mathbf{M}_O^{S_2} = \mathbf{M}_Q^{S_2} + \mathbf{OQ} \times \mathbf{R}_2 \tag{2.11.17}$$

Hence, with $\mathbf{R}_1 = \mathbf{R}_2$ and $\mathbf{M}_Q^{S_1} = \mathbf{M}_Q^{S_2}$, subtraction of the equations leads to

$$\mathbf{M}_O^{S_1} = \mathbf{M}_O^{S_2} \tag{2.11.18}$$

Since O is an arbitrary point, the premise of equal moments about all points is established.

Example 2.11.8 Special Force Systems 4: Reduction

If two force systems, say S_1 and S_2, are equivalent and if one of the force systems, say S_2 has fewer forces than the other system S_1, then S_2 is said to be a "reduction" of S_1.

Given any force system, say S_1, there exists an equivalent force system S_2, consisting of a single force **F** passing through an arbitrary point O together with a couple having a torque **T**. To see this, let **F** be equal to the resultant $\mathbf{R}_1$ of S_1 and let **T** be equal to the moment of S_1 about O. Fig. 2.11.7 then depicts S_1 and S_2. These forces are seen to be equivalent since they have equal resultants ($\mathbf{R}_2 = \mathbf{F} = \mathbf{R}_1$) and equal moments about O ($\mathbf{M}_O^{S_2} = \mathbf{T} = \mathbf{M}_O^{S_1}$).

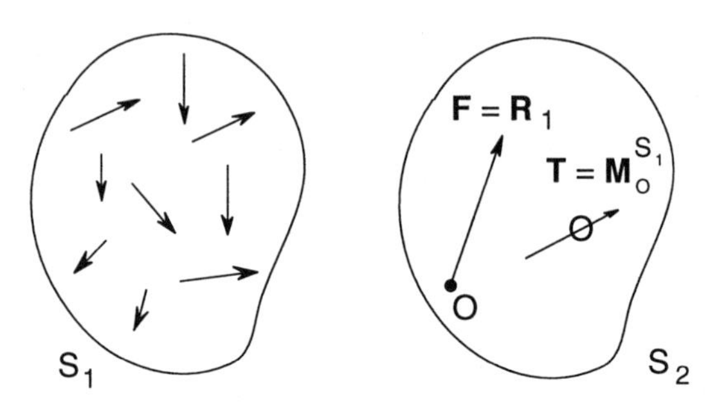

Fig. 2.11.7 Equivalent Force Systems

2.12 Included Angle Between Two Vectors

The scalar and vector products may be used to obtain the angle between vectors. From their definitions we have

$$\cos\theta = \frac{\mathbf{a} \bullet \mathbf{b}}{|\mathbf{a}|\;|\mathbf{b}|} \tag{2.12.1}$$

and

$$|\sin\theta| = \frac{|\mathbf{a} \times \mathbf{b}|}{|\mathbf{a}|\;|\mathbf{b}|} \tag{2.12.2}$$

2.13 Example: Projection of a Vector Along a Line

Given a vector **V** and a line L, the scalar product may be used to obtain the component of **V** parallel to L. Let **n** be a unit vector parallel to L. Then

$$\mathbf{V} \cdot \mathbf{n} = |\mathbf{V}|\;|\mathbf{n}|\cos\theta = |\mathbf{V}|\cos\theta \tag{2.13.1}$$

where θ is the angle between **V** and L. Let **V** be resolved into components $\mathbf{V}_{\parallel}$ and $\mathbf{V}_{\perp}$

parallel and perpendicular to L as in Figure 2.13.1. Then from the triangle geometry we see that

$$|\mathbf{V}_{\parallel}| = |\mathbf{V}| \cos\theta \tag{2.13.2}$$

Therefore, $\mathbf{V}_{\parallel}$ may be expressed as:

$$\mathbf{V}_{\parallel} = |\mathbf{V}_{\parallel}|\mathbf{n} = |\mathbf{V}| \cos\theta\,\mathbf{n} = (\mathbf{V}\cdot\mathbf{n})\mathbf{n} = \mathbf{V}\cdot\mathbf{n}\mathbf{n} \tag{2.13.3}$$

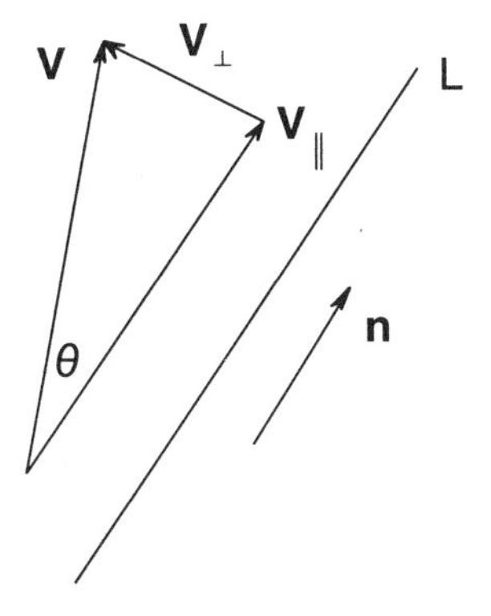

Fig. 2.13.1 A Vector **V**, a Line L, and Components of **V** Parallel and Perpendicular to L

2.14 Multiple Products of Vectors

Three or more vectors may be multiplied together using the scalar and vector products. From their definition and from component representations we have

$$(\mathbf{abc}) \triangleq (\mathbf{a}\times\mathbf{b})\cdot\mathbf{c} = (\mathbf{b}\times\mathbf{c})\cdot\mathbf{a} = (\mathbf{c}\times\mathbf{a})\cdot\mathbf{b} = \begin{vmatrix} a_x & a_y & a_z \\ b_x & b_y & b_z \\ c_x & c_y & c_z \end{vmatrix} \tag{2.14.1}$$

$$(\mathbf{a} \times \mathbf{b}) \times \mathbf{c} = (\mathbf{a} \cdot \mathbf{c})\mathbf{b} - (\mathbf{b} \cdot \mathbf{c})\mathbf{a} \tag{2.14.2}$$

$$\mathbf{a} \times (\mathbf{b} \times \mathbf{c}) = (\mathbf{a} \cdot \mathbf{c})\mathbf{b} - (\mathbf{a} \cdot \mathbf{b})\mathbf{c} \tag{2.14.3}$$

$$\mathbf{a} \times (\mathbf{b} \times \mathbf{c}) + \mathbf{b} \times (\mathbf{c} \times \mathbf{a}) + \mathbf{c} \times (\mathbf{a} \times \mathbf{b}) = 0 \tag{2.14.4}$$

$$(\mathbf{a} \times \mathbf{b}) \times (\mathbf{c} \times \mathbf{d}) = (\mathbf{abd})\mathbf{c} - (\mathbf{abc})\mathbf{d} = (\mathbf{cda})\mathbf{b} - (\mathbf{cdb})\mathbf{a} \tag{2.14.5}$$

$$(\mathbf{a} \times \mathbf{b} \cdot \mathbf{c})(\mathbf{d} \times \mathbf{e} \cdot \mathbf{f}) = \begin{vmatrix} \mathbf{a}\cdot\mathbf{d} & \mathbf{a}\cdot\mathbf{e} & \mathbf{a}\cdot\mathbf{f} \\ \mathbf{b}\cdot\mathbf{d} & \mathbf{b}\cdot\mathbf{e} & \mathbf{b}\cdot\mathbf{f} \\ \mathbf{c}\cdot\mathbf{d} & \mathbf{c}\cdot\mathbf{e} & \mathbf{c}\cdot\mathbf{f} \end{vmatrix} \tag{2.14.6}$$

$$(\mathbf{a} \times \mathbf{b}) \cdot (\mathbf{c} \times \mathbf{d}) = (\mathbf{a} \cdot \mathbf{c})(\mathbf{b} \cdot \mathbf{d}) - (\mathbf{a} \cdot \mathbf{d})(\mathbf{b} \cdot \mathbf{c}) = \begin{vmatrix} \mathbf{a}\cdot\mathbf{c} & \mathbf{a}\cdot\mathbf{d} \\ \mathbf{b}\cdot\mathbf{c} & \mathbf{b}\cdot\mathbf{d} \end{vmatrix} \tag{2.14.7}$$

$$(\mathbf{a} \times \mathbf{b} \quad \mathbf{b} \times \mathbf{c} \quad \mathbf{c} \times \mathbf{a}) = (\mathbf{abc})^2 \tag{2.14.8}$$

$$(\mathbf{a} \times \mathbf{b} \quad \mathbf{c} \times \mathbf{d} \quad \mathbf{e} \times \mathbf{f}) = (\mathbf{abd})(\mathbf{cef}) - (\mathbf{abc})(\mathbf{def}) \tag{2.14.9}$$

$$\mathbf{a} \times [\mathbf{b} \times (\mathbf{c} \times \mathbf{d})] = (\mathbf{b} \cdot \mathbf{d})(\mathbf{a} \times \mathbf{c}) - (\mathbf{b} \cdot \mathbf{c})(\mathbf{a} \times \mathbf{d}) \tag{2.14.10}$$

2.15 Examples: Multiple Products of Vectors

Example 2.15.1 Projection of a Vector Perpendicular to a Line

The triple vector products of Equations (2.14.2) and (2.14.3) may be used to find the component $\mathbf{V}_\perp$ of a vector perpendicular to a line (see Figure 2.13.1). Specifically,

$$\mathbf{V}_\perp = (\mathbf{n} \times \mathbf{V}) \times \mathbf{n} = \mathbf{n} \times (\mathbf{V} \times \mathbf{n}) \tag{2.15.1}$$

To see this we simply expand either triple product as in Equations (2.14.2) and (2.14.3):

$$\begin{aligned} \mathbf{n} \times (\mathbf{V} \times \mathbf{n}) &= (\mathbf{n} \cdot \mathbf{n})\mathbf{V} - (\mathbf{n} \cdot \mathbf{V})\mathbf{n} \\ &= \mathbf{V} - \mathbf{V}_\parallel = \mathbf{V}_\perp \end{aligned} \tag{2.15.2}$$

Example 2.15.2 Special Force Systems 5: Equivalent Force System with a Minimum Couple Torque

In Example 2.11.8 the equivalent force system having the single force **F** and couple torque **T** was such that **F** could pass through an arbitrary point O. The choice of O however, affected the torque **T** of the couple (since $\mathbf{T} = M_O^{S_1}$). This raises the question as to whether there is a point, say O*, for which the magnitude of **T** is a minimum, and if so, what is the minimum magnitude? To answer these questions, consider first, from Equation (2.11.6) that the projection of $\mathbf{M}_O$ along the direction of the resultant **R** is the same as the projection of $\mathbf{M}_Q$ along **R**. Specifically, let **n** be a unit vector parallel to **R**, defined as

$$\mathbf{n} = \mathbf{R}/|\mathbf{R}| \tag{2.15.3}$$

Then from Equation (2.11.6) we have

$$\mathbf{M}_O \bullet \mathbf{n} = \mathbf{M}_Q \bullet \mathbf{n} + \mathbf{OQ} \times \mathbf{R} \bullet \mathbf{n} \tag{2.15.4}$$

Since Q is an arbitrary point we conclude that if there is a point O* for which the magnitude of the equivalent torque is a minimum, the magnitude of that minimum torque is: $|\mathbf{n} \bullet \mathbf{M}_O|$.

Next, suppose that the moment of a force system about some point is parallel to **R** (and **n**). Then from Equation (2.15.4) and the reasoning of the foregoing paragraph, we see that the magnitude of this moment is equal to the minimum torque magnitude. Let this moment be **M***. Then we have

$$\mathbf{M}_{O^*} = \mathbf{M}^* = (\mathbf{M}_O \bullet \mathbf{n})\mathbf{n} \tag{2.15.5}$$

The questions remaining, however, are: Does O* exist? and if so, what is its location? We can answer both questions by finding O*. Specifically, let **P*** locate O* relative to O. That is,

$$\mathbf{P}^* = \mathbf{OO}^* \tag{2.15.6}$$

It is readily seen that $\mathbf{P}^*$ is

$$\mathbf{P}^* = \mathbf{R} \times \mathbf{M}_O / \mathbf{R}^2 \tag{2.15.7}$$

That is, by again using Equation (2.11.6) and the triple vector product of Equation (2.14.2), we have

$$\begin{aligned}
\mathbf{M}_O &= \mathbf{M}_{O^*} + \mathbf{OO}^* \times \mathbf{R} \\
&= \mathbf{M}_{O^*} + (\mathbf{R} \times \mathbf{M}_O) \times \mathbf{R}/\mathbf{R}^2 \\
&= \mathbf{M}_{O^*} + (\mathbf{n} \times \mathbf{M}_O) \times \mathbf{n} \\
&= \mathbf{M}_{O^*} + (\mathbf{n} \bullet \mathbf{n})\mathbf{M}_O - (\mathbf{n} \bullet \mathbf{M}_O)\mathbf{n} \\
&= \mathbf{M}_{O^*} + \mathbf{M}_O - \mathbf{M}^*
\end{aligned}$$

or

$$\mathbf{M}_{O^*} = \mathrm{M}^* \tag{2.15.8}$$

This confirms both the existence of O* and its location by Equation (2.15.7).

Example 2.15.3 Special Force Systems 6: Wrench

If a force system consists of a single force $\mathbf{F}$ together with a couple whose torque $\mathbf{T}$ is parallel to $\mathbf{F}$ that force system is called a "wrench."

For any given force system S there exists a wrench which is equivalent to S. This is seen as follows: Let $\hat{S}$ be a force system equivalent to S consisting of a single force $\hat{\mathbf{F}}$ together with a couple having torque $\hat{\mathbf{T}}$, as with the reduction of Example 2.11.8. Let $\hat{\mathbf{F}}$ pass through the point O* where the minimum couple torque occurs as in Example 2.15.2 [Equation (2.15.7) locates O*]. Then $\hat{\mathbf{T}}$ will be parallel to $\hat{\mathbf{F}}$ and thus $\hat{S}$ is a wrench.

2.16 Vector Functions and Their Derivatives

- **Definition.** If either the magnitude of a vector **V** or the direction of **V** in a reference frame R depends upon a parameter t (time), then **V** is called a "vector function of t in R." The time derivative of **V** in R is then:

$$\frac{{}^{R}d\mathbf{V}}{dt} \triangleq \lim_{\Delta t \to 0} \frac{\mathbf{V}(t+\Delta t)-\mathbf{V}(t)}{\Delta t} \tag{2.16.1}$$

where the limiting process must be conducted in the reference frame R.

By using the procedures developed in elementary calculus related to limits and derivatives, we obtain the following expressions and

- **Operational Formulas:**

$$\frac{{}^{R}d}{dt}(\mathbf{A}+\mathbf{B}) = \frac{{}^{R}d\mathbf{A}}{dt} + \frac{{}^{R}d\mathbf{B}}{dt} \tag{2.16.2}$$

$$\frac{{}^{R}d}{dt}(\mathbf{A}\cdot\mathbf{B}) = \frac{{}^{R}d\mathbf{A}}{dt}\cdot\mathbf{B} + \mathbf{A}\cdot\frac{{}^{R}d\mathbf{B}}{dt} \tag{2.16.3}$$

$$\frac{{}^{R}d}{dt}(\mathbf{A}\times\mathbf{B}) = \frac{{}^{R}d\mathbf{A}}{dt}\times\mathbf{B} + \mathbf{A}\times\frac{{}^{R}d\mathbf{B}}{dt} \tag{2.16.4}$$

$$\frac{{}^{R}d}{dt}(\mathbf{ABC}) = \left(\frac{{}^{R}d\mathbf{A}}{dt}\mathbf{BC}\right) + \left(\mathbf{A}\frac{{}^{R}d\mathbf{B}}{dt}\mathbf{C}\right) + \left(\mathbf{AB}\frac{{}^{R}d\mathbf{C}}{dt}\right) \tag{2.16.5}$$

$$\frac{{}^{R}d}{dt}(\phi\mathbf{A}) = \frac{d\phi}{dt}\mathbf{A} + \phi\frac{{}^{R}d\mathbf{A}}{dt} \quad (\phi \text{ is a scalar function}) \tag{2.16.6}$$

If the length (magnitude) of **A** in R remains unaltered, then $\frac{{}^{R}d\mathbf{A}}{dt}$ is either perpendicular to **A** or is zero.

If the direction of **A** in R remains unaltered, then $\frac{{}^{R}d\mathbf{A}}{dt}$ is parallel to **A**.

We can obtain additional operational formulas by using the concept of angular velocity as follows:

If a vector **A** is **fixed** in a moving reference from $\hat{R}$, and the angular velocity of $\hat{R}$ relative to reference frame R is ${}^{R}\omega^{\hat{R}}$, then

$$\frac{{}^{R}d\mathbf{A}}{dt} = {}^{R}\omega^{\hat{R}} \times \mathbf{A} \tag{2.16.7}$$

The derivatives of a vector **A** in two distinct reference frames R and $\hat{R}$ are related by

$$\frac{{}^{R}d\mathbf{A}}{dt} = \frac{{}^{\hat{R}}d\mathbf{A}}{dt} + {}^{R}\omega^{\hat{R}} \times \mathbf{A} \tag{2.16.8}$$

2.17 Examples: Vector Differentiation

Example 2.17.1 Velocity of a Particle Moving on a Curve

Probably the most widespread application of vector differentiation in mechanics is with the computation of velocity vectors. Velocity is defined as the "time rate of change of position" or simply as the time derivative of the position vector. To illustrate this geometrically consider a particle P moving on a curve C shown in Figure 2.17.1. Let **p** be a position vector locating P relative to a fixed point O. Then in the figure the position of P at two times: t and $t + \Delta t$ is shown. Let $\Delta\mathbf{p}$ be the difference in the respective position vectors. That is,

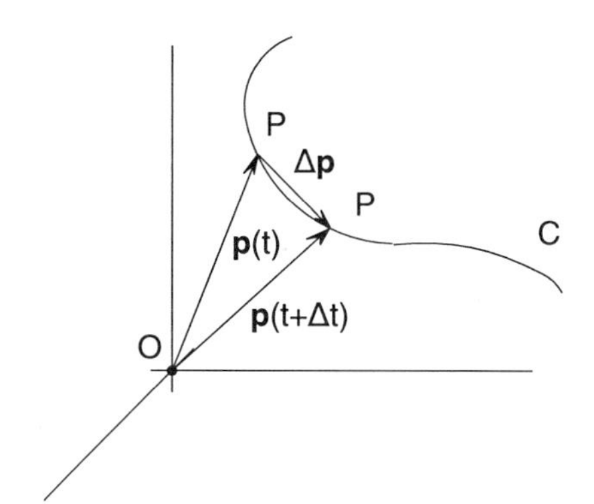

Fig. 2.17.1 A Particle P Moving on a Curve

$$\Delta \mathbf{p} = \mathbf{p}(t + \Delta t) - \mathbf{p}(t) \tag{2.17.1}$$

Then from Equation (2.16.2) we can express the velocity $\mathbf{V}$ of P as

$$\mathbf{V} \triangleq \frac{d\mathbf{p}}{dt} = \lim_{\Delta t \to 0} \frac{\mathbf{p}(t + \Delta t) - \mathbf{p}(t)}{\Delta t} = \lim_{\Delta t \to 0} \frac{\Delta \mathbf{p}}{\Delta t} \tag{2.17.2}$$

Observe in Figure 2.17.1 that the vector $\Delta\mathbf{p}$ is a chord vector along the curve C. Then in the limiting process in Equation (2.17.2) as Δt gets smaller, $\Delta\mathbf{p}$ becomes increasingly close to C until in the limit it is colinear with or tangent to C. This in turn means that the velocity vector of P is tangent to the path or curve upon which P moves.

Example 2.17.2 Differentiation of Vector Characteristics

Referring to Equations (2.2.3) and (2.2.4) let a vector $\mathbf{v}$ be expressed as:

$$\mathbf{v} = |\mathbf{v}|\mathbf{n} = s\mathbf{n} \tag{2.17.3}$$

where s is the scalar magnitude. Then from Equation (2.16.8), the derivative of $\mathbf{V}$ is:

$$\frac{d\mathbf{V}}{dt} = \frac{d}{dt}(s\mathbf{n}) = \frac{ds}{dt}\mathbf{n} + s\frac{d\mathbf{n}}{dt} \tag{2.17.4}$$

Here we see that the derivative of a vector is made up of two parts: one due to the change in magnitude of the vector and the other due to its change in direction. In Example 2.17.6 we will see that these two parts are perpendicular to each other.

Example 2.17.3 Differentiation of Constant Vectors

Let **c** be a constant vector having a constant magnitude and a fixed direction. Then the characteristics of **c** do not change and thus we have

$$d\mathbf{c}/dt = 0 \tag{2.17.5}$$

[See also Equations (2.16.1) (2.17.2).]

Example 2.17.4 Differentiation of Vectors Expressed in Terms of Fixed Unit Vectors

Let $\mathbf{n}_1$, $\mathbf{n}_2$, and $\mathbf{n}_3$ be mutually perpendicular unit vectors, having fixed directions. Let **V** be any vector and let **V** be expressed as

$$\mathbf{V} = v_1\mathbf{n}_1 + v_2\mathbf{n}_2 + v_3\mathbf{n}_3 \tag{2.17.6}$$

If **V** is a function of t then its dependence upon t will occur through the scalar components v_i. That is,

$$\mathbf{V} = \mathbf{V}(t) = v_1(t)\mathbf{n}_1 + v_2(t)\mathbf{n}_2 + v_3(t)\mathbf{n}_3 \tag{2.17.7}$$

Then from Equation (2.16.2) we have:

$$\begin{aligned} d\mathbf{V}/dt &= (dv_1/dt)\mathbf{n}_1 + (dv_2/dt)\mathbf{n}_2 + (dv_3/dt)\mathbf{n}_3 \\ &= \dot{v}_1\mathbf{n}_1 + \dot{v}_2\mathbf{n}_2 + \dot{v}_3\mathbf{n}_3 \end{aligned} \tag{2.17.8}$$

Comment: Equation (2.17.8) is an algorithm for vector differentiation. It shows that if a vector is expressed in terms of constant unit vectors, the derivative is obtained by simply differentiating the components.

Example 2.17.5 Differentiation of Variable Direction Unit Vectors

If a unit vector **n** does not have a fixed direction, then its derivative is perpendicular to itself. That is, if $d\mathbf{n}/dt$ is not zero, then $d\mathbf{n}/dt$ is perpendicular to **n**. Thus,

$$d\mathbf{n}/dt \cdot \mathbf{n} = 0 \tag{2.17.9}$$

We can see this by noting that since **n** is a unit vector, we have

$$\mathbf{n} \cdot \mathbf{n} = 1 \tag{2.17.10}$$

then by differentiating

$$\begin{aligned} d(\mathbf{n}\cdot\mathbf{n})/dt = 0 &= (d\mathbf{n}/dt)\cdot\mathbf{n} + \mathbf{n}\cdot d\mathbf{n}/dt \\ &= 2\mathbf{n}\cdot d\mathbf{n}/dt \end{aligned} \tag{2.17.11}$$

By inspection, Equations (2.17.9) and (2.17.11) are seen to be equivalent.

Example 2.17.6 Differentiation of a Unit Vector Which Remains Perpendicular to a Fixed Line [2.10]

As a specialization of the foregoing example, consider a unit vector with a variable direction such that it remains perpendicular to a fixed line, or equivalently, parallel to a fixed plane. Specifically, let Z be a fixed line and let L be a line perpendicular to Z as in Figure 2.17.2. Let X and Y also be fixed lines perpendicular to Z and to each other so that X, Y, and Z form a mutually perpendicular set of lines. Let L lie in the X-Y plane and also pass through the intersection of X, Y, and Z. Then let the orientation of L be defined by the angle $\theta(t)$ as shown. Finally, let $\mathbf{n}$ be a unit vector parallel to L and let $\mathbf{n}_x$, $\mathbf{n}_y$, and $\mathbf{n}_z$ be unit vectors parallel to X, Y, and Z. Then the derivative of $\mathbf{n}$ may be expressed as:

$$d\mathbf{n}/dt = \mathbf{n}_3 \times \mathbf{n} \; d\theta/dt \tag{2.17.12}$$

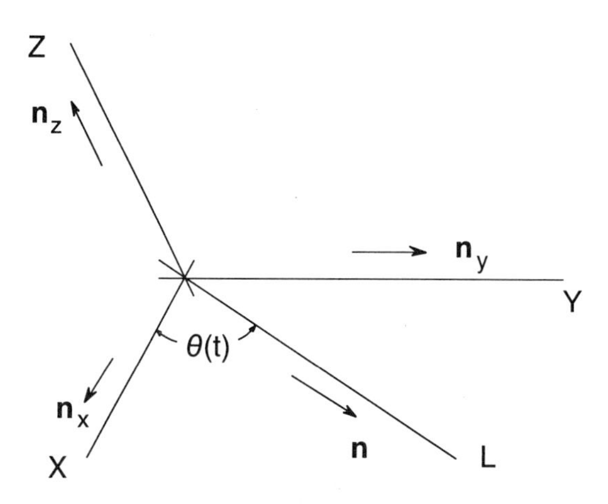

Fig. 2.17.2 A Rotating Line L with Parallel Unit Vector **n**

The validity of Equation (2.17.12) may be seen by expressing $\mathbf{n}$ in terms of the fixed unit vectors $\mathbf{n}_x$, $\mathbf{n}_y$, and $\mathbf{n}_z$ and then differentiating: Specifically, if we look at the X-Y plane as in Figure 2.17.3, we have

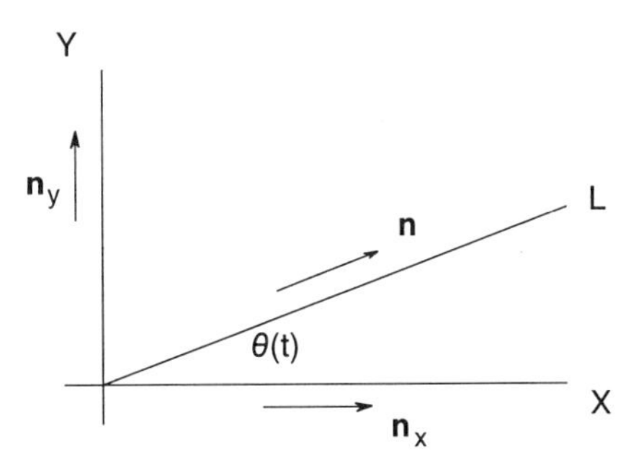

Fig. 2.17.3 Variable Direction Unit Vector and Line

$$\mathbf{n} = \cos\theta\,\mathbf{n}_x + \sin\theta\,\mathbf{n}_y \tag{2.17.13}$$

Then by differentiating with $\mathbf{n}_x$ and $\mathbf{n}_y$ being constant we have

$$d\mathbf{n}/dt = -\dot{\theta}\sin\theta\,\mathbf{n}_x + \dot{\theta}\cos\theta\,\mathbf{n}_y \tag{2.17.14}$$

But also from Equation (2.17.13) we have

$$\mathbf{n}_3 \times \mathbf{n} = -\sin\theta\,\mathbf{n}_x + \cos\theta\,\mathbf{n}_y \tag{2.17.15}$$

Then by comparison of Equations (2.17.14) and (2.17.15), (2.17.12) is established.

Comment: Observe that Equation (2.17.12) is consistent with Equation (2.17.9). Equation (2.17.12) is an algorithm for differentiation. The derivative is determined by a multiplication. Equation (2.17.12) is useful both for computation and for derivation of kinematic expressions. Finally, by using the chain rule we have

$$\mathbf{dn}/dt = (\mathbf{dn}/d\theta)(d\theta/dt) \tag{2.17.16}$$

and then by comparison with Equation (2.17.12) we have

$$\mathbf{dn}/d\theta = \mathbf{n}_z \times \mathbf{n} \tag{2.17.17}$$

2.18 Kronecker's Delta and Permutation Symbols

The mutually perpendicular, dextral unit vectors $\mathbf{n}_j$ (j=1,2,3) may be used to define the useful Kronecker's delta δ_{ij} function and the permutation symbol e_{ijk} as:

$$\mathbf{n}_i \bullet \mathbf{n}_j = \delta_{ij} \tag{2.18.1}$$

and

$$\mathbf{n}_i \times \mathbf{n}_j \bullet \mathbf{n}_k = e_{ijk} \tag{2.18.2}$$

Thus

$$\delta_{ij} \triangleq \begin{cases} 1 & \text{if } i = j \\ 0 & \text{if } i \neq j \end{cases} \tag{2.18.3}$$

and

$$e_{ijk} \triangleq \begin{cases} 1 & \text{if } i, j, k \text{ is a cyclic permutation of } 1, 2, 3; \\ -1 & \text{if } i, j, k \text{ is a cyclic permutation of } 1, 3, 2; \\ 0 & \text{otherwise} \end{cases} \tag{2.18.4}$$

or alternatively

$$e_{ijk} \triangleq \frac{1}{2}(i - j)(j - k)(k - i) \tag{2.18.5}$$

From Equation (2.14.6) we have

$$e_{ijk} e_{rst} = (e_i \times e_j \cdot e_k)(e_r \times e_s \cdot e_t) = \begin{vmatrix} \delta_{ir} & \delta_{is} & \delta_{it} \\ \delta_{jr} & \delta_{js} & \delta_{jt} \\ \delta_{kr} & \delta_{ks} & \delta_{kt} \end{vmatrix} \tag{2.18.6}$$

Hence, by expanding the determinant we have

$$\begin{aligned} e_{ijk} e_{rst} = & \ \delta_{ir}\delta_{js}\delta_{kt} - \delta_{ir}\delta_{ks}\delta_{jt} + \delta_{is}\delta_{kr}\delta_{jt} \\ & - \delta_{is}\delta_{jr}\delta_{kt} + \delta_{it}\delta_{jr}\delta_{ks} - \delta_{it}\delta_{kr}\delta_{js} \end{aligned} \tag{2.18.7}$$

Then by letting r = i, we obtain the relation:

$$e_{ijk} e_{ist} = \delta_{js}\delta_{kt} - \delta_{jt}\delta_{ks} \tag{2.18.8}$$

(Regarding notation, repeated indices [such as i in Equation (2.18.8)] indicates a sum over the range of the index.)

Next, by also letting s = j, we have

$$e_{ijk} e_{ijt} = 2\delta_{kt} \tag{2.18.9}$$

Finally, by also letting t = k, we have

$$e_{ijk} e_{ijk} = 6 = 3! \tag{2.18.10}$$

Since any row or column of a 3×3 array A may be thought of a vector, we express the determinant a of A as

$$a = (a_{1i}\mathbf{e}_i) \times (a_{2j}\mathbf{e}_j) \bullet (a_{3k}\mathbf{e}_k) = e_{ijk} a_{1i} a_{2j} a_{3k} \tag{2.18.11}$$

Hence, we also have the expressions

$$e_{rst}\,a = e_{ijk}\,a_{ir}\,a_{js}\,a_{kt} \tag{2.18.12}$$

$$e_{rst}\,e_{rst}\,a = e_{ijk}\,e_{rst}\,a_{ir}\,a_{js}\,a_{kt} \tag{2.18.13}$$

or

$$a = \frac{1}{3!}\,e_{ijk}\,e_{rst}\,a_{ir}\,a_{js}\,a_{kt} \tag{2.18.14}$$

Finally, let the elements A_{ir} of the "cofactor array" of A be defined as:

$$A_{ir} = \frac{1}{2!}\,e_{ijk}\,e_{rst}\,a_{js}\,a_{kt} \tag{2.18.15}$$

Then from Equations (2.18.12) and (2.18.14) we have

$$a_{ip}\,A_{ir} = \frac{1}{2!}\,e_{ijk}\,e_{rst}\,a_{ip}\,a_{js}\,a_{kt} = \delta_{pr}\,a \tag{2.18.16}$$

or

$$a = \frac{1}{3}\,a_{ir}\,A_{ir} \quad \text{and} \quad a_{ip}^{-1} = \frac{1}{a}\,A_{ip} \tag{2.18.17}$$

where a_{ip}^{-1} are the elements of A^{-1}, the inverse of A.

2.19 Dyads, Dyadics and Second Order Tensors

- **Definition**: If **a** and **b** are vectors, then their "indeterminate product" or "dyadic product" defined as

$$\underline{\underline{d}} \triangleq \mathbf{ab} \tag{2.19.1}$$

is called a "second order dyad."

- **Operational Rules:**

$$s\mathbf{ab} = (s\mathbf{a})\mathbf{b} = \mathbf{a}(s\mathbf{b}) = \mathbf{ab}s \tag{2.19.2}$$

where s is a scalar.

$$(\mathbf{a} + \mathbf{b})\mathbf{c} = \mathbf{ac} + \mathbf{bc} \tag{2.19.3}$$

$$\mathbf{a}(\mathbf{b} + \mathbf{c}) = \mathbf{ab} + \mathbf{ac} \tag{2.19.4}$$

$$\mathbf{v} \cdot \underline{\underline{\mathbf{d}}} = (\mathbf{v} \cdot \mathbf{a})\mathbf{b} \quad \text{and} \quad \underline{\underline{\mathbf{d}}} \cdot \mathbf{v} = \mathbf{a}(\mathbf{b} \cdot \mathbf{v}) \tag{2.19.5}$$

$$\mathbf{v} \times \underline{\underline{\mathbf{d}}} = (\mathbf{v} \times \mathbf{a})\mathbf{b} \quad \text{and} \quad \underline{\underline{\mathbf{d}}} \times \mathbf{v} = \mathbf{a}(\mathbf{b} \times \mathbf{v}) \tag{2.19.6}$$

If $\underline{\underline{\mathbf{d}}} = \mathbf{ab}$, and $\underline{\underline{\mathbf{e}}} = \mathbf{uv}$, then

$$\underline{\underline{\mathbf{d}}} \cdot \underline{\underline{\mathbf{e}}} = \mathbf{ab} \cdot \mathbf{uv} = \mathbf{a}(\mathbf{b} \cdot \mathbf{u})\mathbf{v} \tag{2.19.7}$$

$$\underline{\underline{\mathbf{d}}} \times \mathbf{e} = \mathbf{ab} \times \mathbf{uv} = \mathbf{a}(\mathbf{b} \times \mathbf{u})\mathbf{v} \tag{2.19.8}$$

- **Definition**: If $\underline{\underline{\mathbf{d}}}$ is the dyad **ab**, then the dyad **ba** is called the "transpose of $\underline{\underline{\mathbf{d}}}$," and is denoted by $\underline{\underline{\mathbf{d}}}^T$. If $\underline{\underline{\mathbf{d}}}$ and $\underline{\underline{\mathbf{d}}}^T$ are equal, then $\underline{\underline{\mathbf{d}}}$ is said to be symmetric.

- **Definition**: Let $\mathbf{n}_1$, $\mathbf{n}_2$ and $\mathbf{n}_3$ be dextral, mutually perpendicular unit vectors. Then $\mathbf{n}_1\mathbf{n}_1$, $\mathbf{n}_1\mathbf{n}_2$,..., are called "elementary dyads." They may be listed in an array as

$$[\mathbf{n}_i\mathbf{n}_j] = \begin{bmatrix} \mathbf{n}_1\mathbf{n}_1 & \mathbf{n}_1\mathbf{n}_2 & \mathbf{n}_1\mathbf{n}_3 \\ \mathbf{n}_2\mathbf{n}_1 & \mathbf{n}_2\mathbf{n}_2 & \mathbf{n}_2\mathbf{n}_3 \\ \mathbf{n}_3\mathbf{n}_1 & \mathbf{n}_3\mathbf{n}_2 & \mathbf{n}_3\mathbf{n}_3 \end{bmatrix} \tag{2.19.9}$$

• **Definition**: A sum of scalar multiples of the elementary dyads of the form

$$\underline{\underline{d}} = \mathbf{n}_i d_{ij} \mathbf{n}_j \tag{2.19.10}$$

is called a "dyadic." A dyadic is the same as a "second order tensor." The nine scalar d_{ij} are called "dyadic or tensor components." They can be listed in an array as

$$[d_{ij}] = \begin{bmatrix} d_{11} & d_{12} & d_{13} \\ d_{21} & d_{22} & d_{23} \\ d_{31} & d_{32} & d_{33} \end{bmatrix} \tag{2.19.11}$$

Equation (2.19.10) may thus be written in the matrix form:

$$\underline{\underline{d}} = \begin{bmatrix} \mathbf{n}_1 & \mathbf{n}_2 & \mathbf{n}_3 \end{bmatrix} \begin{bmatrix} d_{11} & d_{12} & d_{13} \\ d_{21} & d_{22} & d_{23} \\ d_{31} & d_{32} & d_{33} \end{bmatrix} \begin{bmatrix} \mathbf{n}_1 \\ \mathbf{n}_2 \\ \mathbf{n}_3 \end{bmatrix} \tag{2.19.12}$$

• **Definition**: The dyadic

$$\underline{\underline{U}} = \mathbf{n}_1\mathbf{n}_1 + \mathbf{n}_2\mathbf{n}_2 + \mathbf{n}_3\mathbf{n}_3 \quad \text{or} \quad \underline{\underline{U}} = \mathbf{n}_i\mathbf{n}_i \tag{2.19.13}$$

is called "the unit dyadic" or "the identity dyadic."

If $\mathbf{v}$ is any vector, then the unit dyadic, as an operator, leaves the vector unchanged:

$$\underline{\underline{U}} \bullet \mathbf{v} = \mathbf{v} \bullet \underline{\underline{U}} = \mathbf{v} \tag{2.19.14}$$

2.20 Direction Cosines and Transformation Matrices

- **Definition**: Consider two reference frames R and $\hat{R}$. Let $\mathbf{n}_i$ and $\hat{\mathbf{n}}_i$ (i = 1,2,3) be dextral perpendicular unit vectors fixed in R and $\hat{R}$, respectively. Then the nine quantities defined as

$$S_{ij} \triangleq \mathbf{n}_i \cdot \hat{\mathbf{n}}_j \tag{2.20.1}$$

are called "direction cosines" between R and $\hat{R}$. The square matrix

$$S \triangleq \begin{bmatrix} S_{11} & S_{12} & S_{13} \\ S_{21} & S_{22} & S_{23} \\ S_{31} & S_{32} & S_{33} \end{bmatrix} \tag{2.20.2}$$

is called the "direction cosine matrix" or "transformation matrix" between R and $\hat{R}$.

The transformation matrix can be used to describe the relative orientation of the two reference frames. If the reference frames are themselves fixed in rigid bodies B_j and B_k, then S can also be used to define the relative orientation of B_j and B_k. In this latter context, it is advantageous to replace the symbol S with the more elaborate symbol SJK.

- **Transformation Rules**

Transformation Between Unit Vectors:

Let the $\mathbf{n}_i$ be expressed in terms of the $\hat{\mathbf{n}}_j$ as

$$\mathbf{n}_i = (\hat{\mathbf{n}}_i \cdot \hat{\mathbf{n}}_j)\hat{\mathbf{n}}_j \tag{2.20.3}$$

Then from Equation (2.20.1) we have

$$\mathbf{n}_i = S_{ij}\hat{\mathbf{n}}_j \tag{2.20.4}$$

In matrix form, Equation (2.20.4) becomes

$$\begin{bmatrix} \mathbf{n}_1 \\ \mathbf{n}_2 \\ \mathbf{n}_3 \end{bmatrix} = \begin{bmatrix} S_{11} & S_{12} & S_{13} \\ S_{21} & S_{22} & S_{23} \\ S_{31} & S_{32} & S_{33} \end{bmatrix} \begin{bmatrix} \hat{\mathbf{n}}_1 \\ \hat{\mathbf{n}}_2 \\ \hat{\mathbf{n}}_3 \end{bmatrix} \tag{2.20.5}$$

or simply

$$\{\mathbf{n}\} = S\{\hat{\mathbf{n}}\} \tag{2.20.6}$$

Similarly, let $\hat{\mathbf{n}}_j$ be expressed in terms of the $\mathbf{n}_i$ as

$$\hat{\mathbf{n}}_j = (\mathbf{n}_i \cdot \hat{\mathbf{n}}_j)\mathbf{n}_i \tag{2.20.7}$$

Then from Equation (2.20.1) we have

$$\hat{\mathbf{n}}_j = S_{ij}\mathbf{n}_i \tag{2.20.8}$$

or simply

$$\{\hat{\mathbf{n}}\} = S^T\{\mathbf{n}\} \tag{2.20.9}$$

- **Transformation Between Vector Components:**

Let **V** be any vector, and let V_i and $\hat{V}_i$ $(i = 1,2,3)$ be components of **V** relative to $\mathbf{n}_i$ and $\hat{\mathbf{n}}_i$. That is

$$V_i \triangleq \mathbf{V} \cdot \mathbf{n}_i \quad \text{and} \quad \hat{V}_i \triangleq \mathbf{V} \cdot \hat{\mathbf{n}}_i \tag{2.20.10}$$

Then from Equation (2.20.4) we have

$$\mathrm{V}_i = \mathbf{V} \cdot \mathbf{n}_i = \mathbf{V} \cdot (\mathrm{S}_{ij}\hat{\mathbf{n}}_j) = \mathrm{S}_{ij}\hat{\mathrm{V}}_j \tag{2.20.11}$$

In matrix form this becomes

$$\begin{bmatrix} \mathrm{V}_1 \\ \mathrm{V}_2 \\ \mathrm{V}_3 \end{bmatrix} = \begin{bmatrix} \mathrm{S}_{11} & \mathrm{S}_{12} & \mathrm{S}_{13} \\ \mathrm{S}_{21} & \mathrm{S}_{22} & \mathrm{S}_{23} \\ \mathrm{S}_{31} & \mathrm{S}_{32} & \mathrm{S}_{33} \end{bmatrix} \begin{bmatrix} \hat{\mathrm{V}}_1 \\ \hat{\mathrm{V}}_2 \\ \hat{\mathrm{V}}_3 \end{bmatrix} \quad \text{or} \quad \{\mathbf{V}\} = \mathrm{S}\{\hat{\mathbf{V}}\} \tag{2.20.12}$$

Similarly, we have

$$\hat{\mathrm{V}}_j = \mathrm{S}_{ij}\mathrm{V}_i \tag{2.20.13}$$

In matrix form this becomes

$$\begin{bmatrix} \hat{\mathrm{V}}_1 \\ \hat{\mathrm{V}}_2 \\ \hat{\mathrm{V}}_3 \end{bmatrix} = \begin{bmatrix} \mathrm{S}_{11} & \mathrm{S}_{21} & \mathrm{S}_{31} \\ \mathrm{S}_{12} & \mathrm{S}_{22} & \mathrm{S}_{32} \\ \mathrm{S}_{13} & \mathrm{S}_{23} & \mathrm{S}_{33} \end{bmatrix} \begin{bmatrix} \mathrm{V}_1 \\ \mathrm{V}_2 \\ \mathrm{V}_3 \end{bmatrix} \quad \text{or} \quad \{\hat{\mathbf{V}}\} = \mathrm{S}^{\mathrm{T}}\{\mathbf{V}\} \tag{2.20.14}$$

- **Transformation of Dyadics**

Let $\underline{\underline{\mathbf{D}}}$ be a dyadic, and let D_{ij} and $\hat{\mathrm{D}}_{ij}$ $(i,j = 1,2,3)$ be components of $\underline{\underline{\mathbf{D}}}$ relative to unit vectors $\mathbf{n}_i$ and $\hat{\mathbf{n}}_j$. That is

$$\mathrm{D}_{ij} \triangleq \mathbf{n}_i \cdot \underline{\underline{\mathbf{D}}} \cdot \mathbf{n}_j \quad \text{and} \quad \ddot{\mathrm{D}}_{ij} \triangleq \hat{\mathbf{n}}_i \cdot \underline{\underline{\mathbf{D}}} \cdot \hat{\mathbf{n}}_j \tag{2.20.15}$$

Then we have

$$\begin{aligned} \mathrm{D}_{ij} &\triangleq \mathbf{n}_i \cdot \underline{\underline{\mathbf{D}}} \cdot \mathbf{n}_j = (\mathrm{S}_{ik}\hat{\mathbf{n}}_k) \cdot \underline{\underline{\mathbf{D}}} \cdot (\mathrm{S}_{jl}\hat{\mathbf{n}}_\ell) \\ &= \mathrm{S}_{ik}(\hat{\mathbf{n}}_k \cdot \underline{\underline{\mathbf{D}}} \cdot \hat{\mathbf{n}}_\ell)\mathrm{S}_{j\ell} = \mathrm{S}_{ik}\hat{\mathrm{D}}_{k\ell}\mathrm{S}_{j\ell} \end{aligned} \tag{2.20.16}$$

In matrix form, this becomes

$$D = S\hat{D}S^T \tag{2.20.17}$$

Similarly, for the $\hat{D}_{ij}$ we have

$$\begin{aligned}\hat{D}_{ij} &\triangleq \hat{\mathbf{n}}_i \bullet \underline{\underline{\mathbf{D}}} \bullet \hat{\mathbf{n}}_j = (S_{ki}\mathbf{n}_k) \bullet \underline{\underline{\mathbf{D}}} \bullet (S_{\ell j}\mathbf{n}_\ell) \\ &= S_{ki}(\mathbf{n}_k \bullet \underline{\underline{\mathbf{D}}}\mathbf{n}_\ell) S_{\ell j} = S_{ki} D_{k\ell} S_{\ell j}\end{aligned} \tag{2.20.18}$$

In matrix form this may be written as

$$\hat{D} = S^T D S \tag{2.20.19}$$

- **Properties of Transformation Matrices**

1. Transformation matrices are orthogonal, that is, the inverse is equal to the transpose:

$$S^T = S^{-1} \tag{2.20.20}$$

To see this observe that

$$\mathbf{n} = S\hat{\mathbf{n}} = SS^T\mathbf{n} \tag{2.20.21}$$

and

$$\hat{\mathbf{n}} = S^T\mathbf{n} = S^T S\hat{\mathbf{n}} \tag{2.20.22}$$

These imply that

$$SS^T = S^T S = I \tag{2.20.23}$$

where I is the unit matrix.

2. The determinant of a transformation matrix is equal to 1.

From Equation (2.20.23) we obtain

$$(\det S)^2 = 1 \tag{2.20.24}$$

Hence

$$\det S = \pm 1 \tag{2.20.25}$$

Observe that when the $\hat{\mathbf{n}}_i$ are parallel to $\mathbf{n}_i$, S is an identity transformation and detS is 1. When $\hat{\mathbf{n}}_i$ are inclined relative to the $\mathbf{n}_i$, they may be brought to their general orientation by a continuous rotation from $\mathbf{n}_i$. Hence, to maintain continuity detS is 1.

3. Transformation matrices obey a transitive law or chain rule as follows: Suppose we have three bodies with reference frames J, K and L. Let SJK be a transformation matrix between K and L. Then the transformation matrix SJL between J and L may be found by

$$SJL = SJK\ SKL \tag{2.20.26}$$

4. For any transformation matrix there exists at least one real eigenvalue with value 1: Consider the eigenvalue problem (see Section 2.23) for transformation matrix S:

$$S\{u\} = \mu\{u\} \tag{2.20.27}$$

where μ is a scalar and $\{u\}$ is the array of components of vector $\mathbf{u}$. The characteristic polynomial is then

$$\det(S - \mu I) = 0 \tag{2.20.28}$$

Expanding this equation, and simplifying, we obtain

$$(\mu - 1)[\mu^2 - (\mathrm{tr}\,S - 1)\mu + 1] = 0 \tag{2.20.29}$$

where trS represents the trace (sum of diagonal elements) of S. Defining θ by the expression

$$\cos\theta \triangleq \frac{\mathrm{tr}\,S - 1}{2} \tag{2.20.30}$$

then the eigenvalues may be written as

$$\mu = 1, \quad \mu = e^{i\theta}, \quad \mu = e^{-i\theta} \tag{2.20.31}$$

From this it follows that for any transformation matrix there exists at least one real eigenvalue with value 1. Substitution of this value into Equation (2.20.27) yields

$$S\{u\} = \{u\} \tag{2.20.32}$$

This equation states that for any transformation matrix there exists at least one real eigenvector $\mathbf{u}$ whose coordinate components are the same in both $\mathbf{n}_i$ and $\hat{\mathbf{n}}_i$.

2.21 Rotation Dyadics

- **Definition**: Consider the simple rotation of a body B through an angle θ about a line L which is fixed in a reference frame R (see Figure 2.21.1). Let $\boldsymbol{\lambda}$ be a unit vector parallel to L. Let $\mathbf{n}_i$ and $\hat{\mathbf{n}}_i$ $(i = 1,2,3)$ be mutually perpendicular unit vector sets fixed in R and B, respectively. Let the $\mathbf{n}_i$ be aligned with the $\hat{\mathbf{n}}_i$ $(i = 1,2,3)$ prior to the motion of B in R. Next, let $\mathbf{p}$ be a vector fixed in B, let $\mathbf{p}^*$ represent $\mathbf{p}$ after rotation. Then $\mathbf{p}^*$ and $\mathbf{p}$ are related by equation: [2.9]

$$\mathbf{p}^* = \underline{\underline{\mathbf{R}}} \cdot \mathbf{p} \tag{2.21.1}$$

where

$$\underline{\underline{\mathbf{R}}} \triangleq (1 - \cos\theta)\boldsymbol{\lambda\lambda} + \cos\theta\,\underline{\underline{\mathbf{U}}} + \sin\theta\,\boldsymbol{\lambda} \times \underline{\underline{\mathbf{U}}} \tag{2.21.2}$$

is called "rotation dyadic."

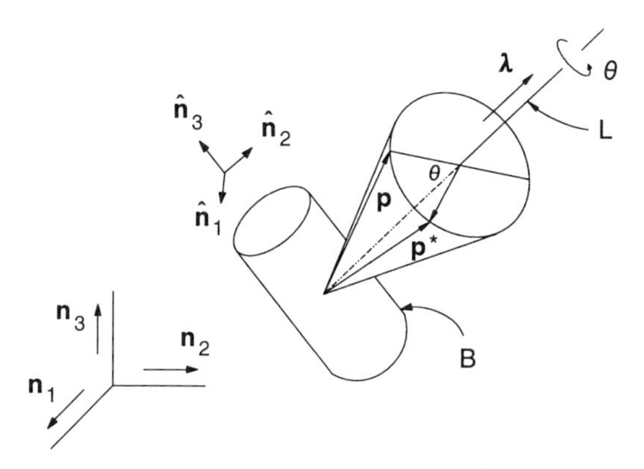

Fig. 2.21.1 A Body B Performing Simple Rotation about a Line L

- **Properties of $\underline{\underline{\mathbf{R}}}$:**

1. The elements of the matrix $\underline{\underline{\mathbf{R}}}$ are identical to the elements of the transformation matrix S between $\mathbf{n}_i$ and $\hat{\mathbf{n}}_i$. To see this, recall that, prior to rotation, $\hat{\mathbf{n}}_j$ and $\mathbf{n}_j$ (j = 1,2,3) are coincident. Thus we can let

$$\mathbf{p} = \mathbf{n}_j \quad \text{and} \quad \mathbf{p}^* = \hat{\mathbf{n}}_j \tag{2.21.3}$$

Then Equation (2.21.1) has the form:

$$\hat{\mathbf{n}}_j = \underline{\underline{\mathbf{R}}} \cdot \mathbf{n}_j \tag{2.21.4}$$

Hence S_{ij} becomes

$$S_{ij} \triangleq \mathbf{n}_i \bullet \hat{\mathbf{n}}_j = \mathbf{n}_i \bullet \underline{\underline{\mathbf{R}}} \bullet \mathbf{n}_j \tag{2.21.5}$$

From this equation it follows that

$$S_{ij} = R_{ij} \tag{2.21.6}$$

This equation thus shows that the direction cosine matrix (transformation matrix) between $\mathbf{n}_i$ and $\mathbf{n}_j$ is identical to the tensor component matrix of $\underline{\underline{\mathbf{R}}}$.

2. The rotation of dyadic $\underline{\underline{\mathbf{R}}}$ is orthogonal. Since, from Equation (2.21.6), we have

$$R_{ij}R_{jk} = S_{ij}S_{jk} = \delta_{ik} \tag{2.21.7}$$

or

$$\underline{\underline{\mathbf{R}}} \bullet \underline{\underline{\mathbf{R}}}^T = \underline{\underline{\mathbf{R}}}^T \bullet \underline{\underline{\mathbf{R}}} = \underline{\underline{\mathbf{U}}} \tag{2.21.8}$$

Hence the rotation dyadic changes the orientation of a vector while holding its magnitude constant. That is

$$|\mathbf{p}^*|^2 = (\underline{\underline{\mathbf{R}}} \bullet \mathbf{p}) \bullet (\underline{\underline{\mathbf{R}}} \bullet \mathbf{p}) = \mathbf{p} \bullet \underline{\underline{\mathbf{R}}}^T \bullet \underline{\underline{\mathbf{R}}} \bullet \mathbf{p} = \mathbf{p} \bullet \mathbf{p} = |\mathbf{p}|^2 \tag{2.21.9}$$

- **Components of $\underline{\underline{\mathbf{R}}}$:**

  In terms of the unit vectors $\mathbf{n}_i$ of Fig. 2.21.1, $\underline{\underline{\mathbf{R}}}$ may be written as:

$$\underline{\underline{\mathbf{R}}} = R_{ij}\mathbf{n}_i\mathbf{n}_j \tag{2.21.10}$$

where the components R_{ij} are components of $\underline{\underline{\mathbf{R}}}$, which may be found by using

Equation (2.21.2) as:

$$R_{ij} = \mathbf{n}_i \bullet \underline{\underline{\mathbf{R}}} \bullet \mathbf{n}_j = (1 - \cos\theta)\lambda_i\lambda_j + \cos\theta\,\delta_{ij} - \sin\theta\,\epsilon_{ijk}\lambda_k \tag{2.21.11}$$

where the λ_i ($i = 1,2,3$) are the $\mathbf{n}_i$ components of $\boldsymbol{\lambda}$, where δ_{ij} is Kronecker's delta function, and e_{ijk} is the permutation symbol [see Equations (2.18.3) and (2.18.4)].

Therefore the matrix $R \triangleq [R_{ij}]$ may be written as:

$$R = \begin{bmatrix} \lambda_1^2(1 - c\theta) + c\theta & \lambda_1\lambda_2(1 - c\theta) - \lambda_3 s\theta & \lambda_1\lambda_3(1 - c\theta) + \lambda_2 s\theta \\ \lambda_1\lambda_2(1 - c\theta) + \lambda_3 s\theta & \lambda_2^2(1 - c\theta) + c\theta & \lambda_2\lambda_3(1 - c\theta) - \lambda_1 s\theta \\ \lambda_1\lambda_3(1 - c\theta) - \lambda_2 s\theta & \lambda_2\lambda_3(1 - c\theta) + \lambda_1 s\theta & \lambda_3^2(1 - c\theta) + c\theta \end{bmatrix} \tag{2.21.12}$$

where $s\theta$ and $c\theta$ represent $\sin\theta$ and $\cos\theta$, respectively.

From this expression it is readily seen that

$$R\{\lambda\} = \{\lambda\} \tag{2.21.13}$$

This equation shows that $\boldsymbol{\lambda}$ is the eigenvector associated with the eigenvalue: $\mu = 1$, of matrix R.

Equation (2.21.12) shows that if we know the rotation angle θ and the direction of the rotation axis L, we can construct the component matrix R of the rotation dyadic $\underline{\underline{\mathbf{R}}}$. Conversely, if R is known, then θ and the direction of the rotation axis L (or of λ) may be found as follows:

$$\cos\theta = \frac{\operatorname{tr} R - 1}{2} \tag{2.21.14}$$

By subtracting nondiagonal elements, we find that if $\sin\theta \neq 0$:

$$\begin{aligned} \lambda_1 &= \frac{(R_{32} - R_{23})}{2\sin\theta} \\ \lambda_2 &= \frac{(R_{13} - R_{31})}{2\sin\theta} \\ \lambda_3 &= \frac{(R_{21} - R_{12})}{2\sin\theta} \end{aligned} \tag{2.21.15}$$

If $\sin\theta = 0$, then

$$\begin{aligned} \lambda_1 &= \pm\left[\frac{1 + R_{11}}{2}\right]^{\frac{1}{2}} \\ \lambda_2 &= \pm\left[\frac{1 + R_{22}}{2}\right]^{\frac{1}{2}} \\ \lambda_3 &= \pm\left[\frac{1 + R_{33}}{2}\right]^{\frac{1}{2}} \end{aligned} \tag{2.21.16}$$

where the signs are chosen to be consistent with the companion expressions:

$$\lambda_1\lambda_2 = \frac{R_{12}}{2} \qquad \lambda_2\lambda_3 = \frac{R_{23}}{2} \qquad \lambda_3\lambda_1 = \frac{R_{31}}{2} \tag{2.21.17}$$

- **Special Cases**:

(i) **Rotation about an axis parallel to $\mathbf{n}_1$ through an angle α.**

In this case $\lambda_1 = 1$, $\lambda_2 = \lambda_3 = 0$, $\theta = \alpha$. Substituting these relations into Equation (2.21.11) and denoting R by $R_1(\alpha)$, we obtain

$$R_1(\alpha) = \begin{bmatrix} 1 & 0 & 0 \\ 0 & \cos\alpha & -\sin\alpha \\ 0 & \sin\alpha & \cos\alpha \end{bmatrix} \tag{2.21.18}$$

(ii) **Rotation about an axis parallel to $\mathbf{n}_2$ through an angle β.**

In this case $\lambda_1 = \lambda_3 = 0$, $\lambda_2 = 1$, and $\theta = \beta$. Substituting these into Equation (2.21.11) and denoting R by $R_2(\beta)$, we have

$$R_2(\beta) = \begin{bmatrix} \cos\beta & 0 & \sin\beta \\ 0 & 1 & 0 \\ -\sin\beta & 0 & \cos\beta \end{bmatrix} \tag{2.21.19}$$

(iii) **Rotation about an axis parallel to $\mathbf{n}_3$ through an angle γ.**

In this case $\lambda_1 = \lambda_2 = 0$, $\lambda_3 = 1$ and $\theta = \gamma$. Similarly from Equation (2.21.12) and by denoting R by $R_3(\gamma)$ we have

$$R_3(\gamma) = \begin{bmatrix} \cos\gamma & -\sin\gamma & 0 \\ \sin\gamma & \cos\gamma & 0 \\ 0 & 0 & 1 \end{bmatrix} \tag{2.21.20}$$

2.22 Derivatives of Transformation Matrices

The elements of transformation matrix S_{ij} are given by

$$S_{ij} = \mathbf{n}_i \bullet \hat{\mathbf{n}}_j \tag{2.22.1}$$

where $\mathbf{n}_i$ (i = 1,2,3) are unit vectors fixed in R; and where $\hat{\mathbf{n}}_j$ (j = 1,2,3) are unit vectors fixed in $\hat{R}$. Thus the $\mathbf{n}_i$ are constant in R, the derivative of S_{ij} in R may be written as

$$\frac{{}^R dS_{ij}}{dt} = \mathbf{n}_i \bullet \frac{d\hat{\mathbf{n}}_j}{dt} \tag{2.22.2}$$

However, since the $\hat{\mathbf{n}}_j$ are fixed in $\hat{R}$, their derivatives may be written as [see Equation 2.16.7]:

$$\frac{{}^R d\hat{\mathbf{n}}_j}{dt} = \boldsymbol{\omega} \times \hat{\mathbf{n}}_j \tag{2.22.3}$$

where $\boldsymbol{\omega}$ is the angular velocity of $\hat{R}$ in R. Hence, ${}^R dS_{ij}/dt$ becomes

$$\frac{{}^R dS_{ij}}{dt} = \mathbf{n}_i \bullet (\boldsymbol{\omega} \times \hat{\mathbf{n}}_j) \tag{2.22.4}$$

Let $\boldsymbol{\omega}$ be expressed in the form:

$$\boldsymbol{\omega} = \omega_k \mathbf{n}_k \tag{2.22.5}$$

Substitution into Equation (2.22.4) leads to

$$\frac{{}^R dS_{ij}}{dt} = \mathbf{n}_i \bullet \omega_k \mathbf{n}_k \bullet \hat{\mathbf{n}}_j = \omega_k \mathbf{n}_i \times \mathbf{n}_k \bullet \hat{\mathbf{n}}_j = -e_{imk}\omega_k S_{mj} \tag{2.22.6}$$

Introduce the matrix W whose elements W_{im} are

$$W_{im} = -e_{imk}\omega_k \tag{2.22.7}$$

then $\frac{{}^R dS_{ij}}{dt}$ may be written as

$$\frac{{}^R dS_{ij}}{dt} = W_{im} S_{mj} \tag{2.22.8}$$

or in matrix form as

$$\frac{{}^R dS}{dt} = WS \tag{2.22.9}$$

where W is called the "dual matrix" of the angular velocity vector $\boldsymbol{\omega}$, and

$$W = \begin{bmatrix} 0 & -\omega_3 & \omega_2 \\ \omega_3 & 0 & -\omega_1 \\ -\omega_2 & \omega_1 & 0 \end{bmatrix} \tag{2.22.10}$$

2.23 Eigenvalues and Eigenvectors

- **Eigenvalue Problem**:

The terms "eigenvalue" and "eigenvector" are often associated with a dyadic $\underline{\underline{I}}$ and a vector $\mathbf{n}_x$ such that

$$\underline{\underline{I}} \cdot \mathbf{n}_x = \lambda \mathbf{n}_x \tag{2.23.1}$$

where λ is a scalar called an "eigenvalue" and the vector $\mathbf{n}_x$ is called an "eigenvector." Let $\mathbf{n}_x$ and $\underline{\underline{I}}$ be expressed as

$$\mathbf{n}_x = x_i \mathbf{n}_i \quad \text{and} \quad \underline{\underline{I}} = I_{ij} \mathbf{n}_i \mathbf{n}_j \tag{2.23.2}$$

Then Equation (2.22.10) may be written as

$$(I_{ij} x_j - \lambda x_i) \mathbf{n}_i = 0 \quad \text{or} \quad (I_{ij} - \lambda \delta_{ij}) x_j \mathbf{n}_i = 0 \tag{2.23.3}$$

This is a vector equation; and, if a vector is zero, its scalar components must be zero. Hence we have the eigensystem of equations:

$$(I_{ij} - \lambda \delta_{ij}) x_j = 0 \quad (i = 1,2,3) \tag{2.23.4}$$

or

$$\begin{bmatrix} I_{11} - \lambda & I_{12} & I_{13} \\ I_{21} & I_{22} - \lambda & I_{23} \\ I_{31} & I_{32} & I_{33} - \lambda \end{bmatrix} \begin{bmatrix} x_1 \\ x_2 \\ x_3 \end{bmatrix} = 0 \tag{2.23.5}$$

Equations (2.23.4) are homogeneous linear algebraic equations, hence a non-trivial solution exists only if

$$\begin{vmatrix} I_{11} - \lambda & I_{12} & I_{13} \\ I_{21} & I_{22} - \lambda & I_{23} \\ I_{31} & I_{32} & I_{33} - \lambda \end{vmatrix} = 0 \tag{2.23.6}$$

or (in more compact form):

$$\det[I - \lambda E] = 0 \tag{2.23.7}$$

where E is the third order identity matrix. Equation (2.23.7) is called the "Hamilton-Cayley" equation [2.6]. By expanding we obtain

$$\lambda^3 - I_I\lambda^2 + I_{II}\lambda - I_{III} = 0 \tag{2.23.8}$$

where I_I, I_{II}, and I_{III} are defined as:

$$I_I = I_{11} + I_{22} + I_{33} \triangleq \mathrm{tr}\, I \tag{2.23.9}$$

$$I_{II} = I_{11}I_{22} - I_{12}I_{21} + I_{22}I_{33} - I_{23}I_{32} + I_{11}I_{33} - I_{13}I_{31} \tag{2.23.10}$$

$$I_{III} = I_{11}I_{22}I_{33} - I_{11}I_{32}I_{23} + I_{12}I_{31}I_{23} - I_{12}I_{21}I_{33}$$
$$+ I_{21}I_{32}I_{13} - I_{31}I_{13}I_{22} \tag{2.23.11}$$

where I_I, I_{II} and I_{III} are called the "invariants." They may be shown to be independent of the choice of unit vector sets.

Since Equation (2.23.8) is cubic in λ, we can rewrite it as

$$(\lambda - \lambda_1)(\lambda - \lambda_2)(\lambda - \lambda_3) = 0 \tag{2.23.12}$$

where λ_1, λ_2, and λ_3 are the roots (or eigenvalues) of the equation. By expanding Equation (2.23.12) and by comparing the coefficients with those of Equation (2.23.8) we see that the invariants may be expressed in terms of the eigenvalues as

$$I_I = \lambda_1 + \lambda_2 + \lambda_3 \tag{2.23.13}$$

$$I_{II} = \lambda_1\lambda_2 + \lambda_2\lambda_3 + \lambda_3\lambda_1 \tag{2.23.14}$$

$$I_{III} = \lambda_1\lambda_2\lambda_3 \tag{2.23.15}$$

Once we have obtained the eigenvalues λ from Equation (2.23.8), we can return to Equation (2.23.4) to determine the components x_i ($i = 1,2,3$) of the eigenvector $\mathbf{n}_x$. However, as we have noted, a nontrivial solution exists only when the determinant in Equation (2.23.6) is zero. This in turn means that Equations (2.23.5) are not independent. Hence, at most, two of the three equations are useful for determining the x_i ($i = 1,2,3$). To uniquely determine the x_i we need a fourth equation. This fourth equation can be found by recalling the $\mathbf{n}_x$ is a unit vector. That is

$$x_1^2 + x_2^2 + x_3^2 = 1 \tag{2.23.16}$$

This equation together with any two Equations (2.23.4) form three independent equations for the three x_i.

References

2.1 L. J. Paige and J. D. Swift, *Elements of Linear Algebra*, Ginn and Company, Boston, MA, 1961.

2.2 B. Noble, *Applied Linear Algebra*, Prentice Hall, Englewood Cliffs, NJ, 1969.

2.3 G. Strang, *Linear Algebra and Its Applications*, Second Edition, Academic Press, New York, NY, 1980.

2.4 H. P. Hsu, *Vector Analysis*, Simon and Schuster, New York, NY, 1969.

2.5 M. R. Spiegel, *Theory and Problems of Vector Analysis and an Introduction to Tensor Analysis*, Shaum, New York, NY 1959.

2.6 L. Brand, *Vector and Tensor Analysis*, John Wiley & Sons, London, 1964.

2.7 R. C. Wrede, *Introduction to Vector and Tensor Analysis*, John Wiley & Sons, New York, 1963.

2.8 H. Jeffreys, *Cartesian Tensors*, Cambridge, London, 1957.

2.9 R. L. Huston, *Multibody Dynamics*, Butterworth-Heinemann, Stoneham, MA, 1990.

2.10 T. R. Kane, *Analytical Elements of Mechanics*, Vol. 2 *Dynamics*, Academic Press, New York, 1961.

Chapter 3

KINEMATICS OF PARTICLES

3.1 Fundamental Concepts

A "particle" is considered to be a small physical entity which may be represented (or "modeled") by a point with an associated mass. "Small" is, of course, a relative term. Many objects such as projectiles, automobiles, space vehicles, and even heavenly bodies are commonly regarded as being particles even though they are hardly small or minute. Hence, for such representations to produce useful results, the object, be it a ball, a car, an airplane, or even a planet, must be small relative to its surroundings. In any event the modeling of an object as a particle, and then as a point with an associated mass, becomes increasingly accurate the smaller the object is.

The "kinematics of a particle" is the study of a motion of a particle without regard to the forces causing the motion. In this context, the mass of the particle is irrelevant. Therefore, a kinematical analysis of particles may be reduced to a mathematical study of the movement of points.

The principal kinematic quantities of interest for particles are: position, velocity, and acceleration. "Position" is simply the location of the particle (or point) in a reference frame, or relative to other particles (or points). The position of a particle is usually represented by a vector — called a "position vector." "Velocity" is the time rate of change of position and "acceleration" is the time rate of change of velocity. Velocity and acceleration are thus also vector quantities.

3.2 Position Vectors — Cartesian Representations

If the position of a particle or point is located in a Cartesian reference frame R by its coordinates, say (x,y,z), then these coordinates also serve as the components of a position vector locating P relative to the origin O or R. That is,

$$\mathbf{p} = x\mathbf{i} + y\mathbf{j} + z\mathbf{k} \tag{3.2.1}$$

where **i**, **j**, and **k** are unit vectors parallel to the X, Y, Z axes of R as in Figure 3.2.1.

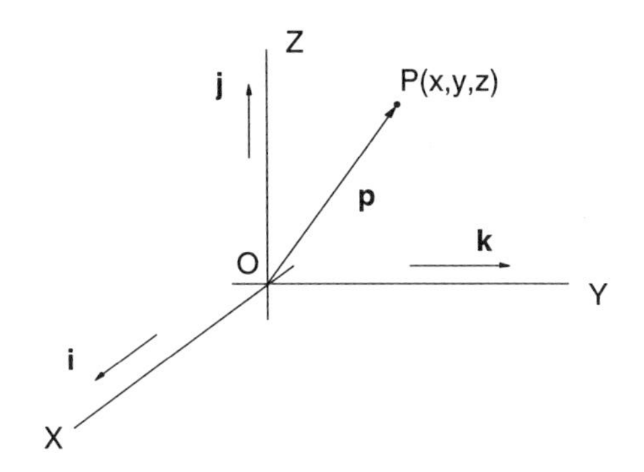

Fig. 3.2.1 Position Vector **p** Locating Point P in a Cartesian Reference Frame R

If x, y, and z are time dependent, we can think of x, y, and z as functions of time t and write them in the form:

$$x = x(t) \quad y = y(t) \quad z = z(t) \tag{3.2.2}$$

Then from Equation (3.2.1) we see that the position vector **p** will also be a function of time. That is,

$$\mathbf{p} = \mathbf{p}(t) = x(t)\mathbf{i} + y(t)\mathbf{j} + z(t)\mathbf{k} \tag{3.2.3}$$

where the unit vectors **i**, **j**, and **k** remain independent of t and are thus constants.

As t changes in Equations (3.2.2) and (3.2.3), the coordinates, position, and the position vector of P change. The locus of points of R which are coincident with P during a continuous change of t form a curve C, called the "path of motion of P in R." See Figure 3.2.2.

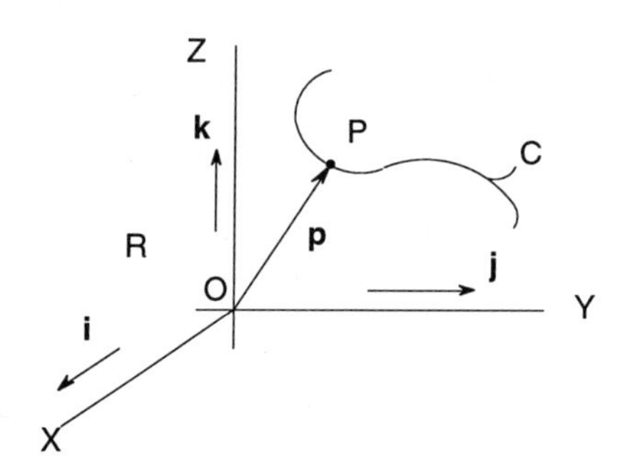

Fig. 3.2.2 Path of Motion C of a Point P in a Cartesian Reference Frame R

Alternatively, we may think of Equations (3.2.2) as parametric equations defining C, with time t being the varying parameter.

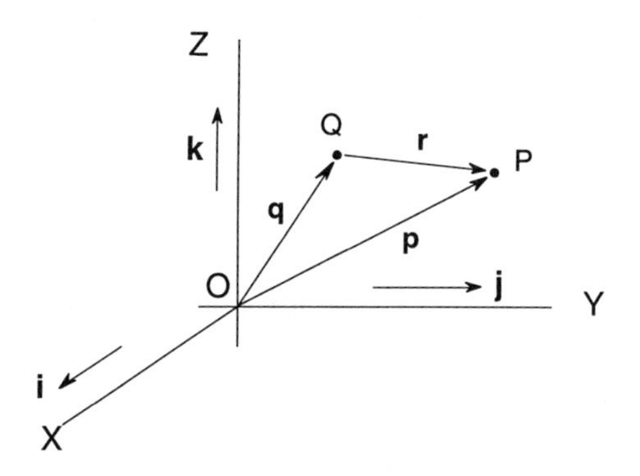

Fig. 3.2.3 Relative Position Vectors

If a particle (or point) P is located relative to another particle Q, the position vectors of P and Q and their relative position vectors may be represented as in Figure 3.2.3, where **p** locates P relative to the origin O, **q** locates Q relative to O, and **r** locates P relative to Q. We then immediately obtain the expressions:

$$\mathbf{p} = \mathbf{q} + \mathbf{r} \tag{3.2.4}$$

$$\mathbf{p} = x_p\mathbf{i} + y_p\mathbf{j} + z_p\mathbf{k} \tag{3.2.5}$$

$$\mathbf{q} = x_Q\mathbf{i} + y_Q\mathbf{j} + z_Q\mathbf{k} \tag{3.2.6}$$

$$\mathbf{r} = \mathbf{q} - \mathbf{p} = (x_Q - x_P)\mathbf{i} + (y_Q - y_P)\mathbf{j} + (z_Q - z_P)\mathbf{k} \tag{3.2.7}$$

where (x_P, y_P, z_P) and (x_Q, y_Q, z_Q) are the X,Y,Z coordinates of P and Q and are in general functions of time.

3.3 Position Vectors — Polar, Cylindrical and Spherical Representations

Dynamic systems frequently contain rotating bodies. Particles of these bodies then have circular motion. For such systems, and others as well, it is often convenient to locate these particles using polar, cylindrical, or spherical coordinates. For a particle or point P moving in a plane, its polar position may be represented as in Figure 3.3.1, where (r, θ) are the polar coordinates of P. We may then express the position vector **p** locating P relative to the origin O as:

$$\mathbf{p} = r\mathbf{n}_r \tag{3.3.1}$$

where $\mathbf{n}_r$ is a "radial" unit vector parallel to vector **OP** or **P**. Observe that unlike **i** and **j**, $\mathbf{n}_r$ does not have a fixed orientation, although its magnitude remains constant. Observe that also unlike position vectors in Cartesian coordinates [as in Equation (3.2.1)], the polar coordinates are not components of the position vector (that is, θ does not appear).

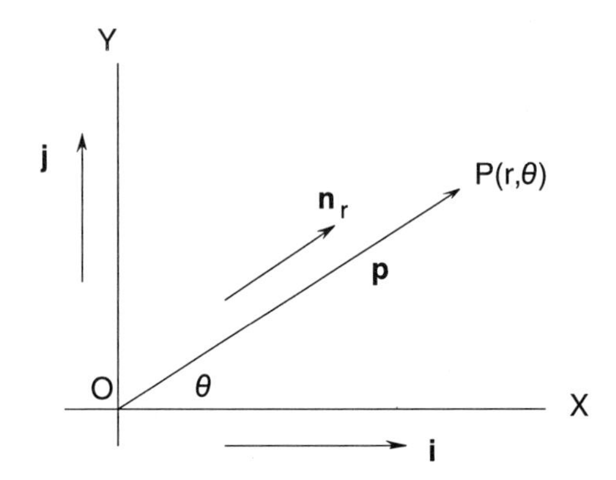

Fig. 3.3.1 Position Vector with Polar Coordinates

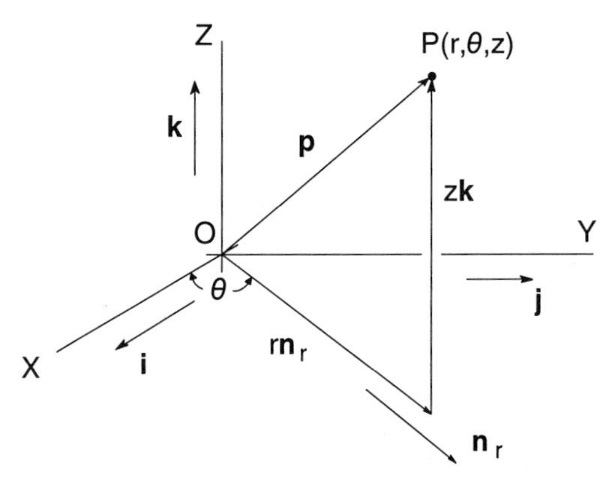

Fig. 3.3.2 Position Vector in Cylindrical Coordinates

If a particle is not restricted to planar motion, but is still generally moving in a circular manner it may be convenient to use cylindrical coordinates. In this case a position vector **p** locating P may be represented as in Figure 3.3.2. Then **p** may be expressed as:

$$\mathbf{p} = r\mathbf{n}_r + z\mathbf{k} \tag{3.3.2}$$

where z is now an axial or cylindrical coordinate.

Finally, if a particle P has general motion, or motion on or in the vicinity of a spherical surface, it may be convenient to use spherical coordinates. In this case a position vector **p** locating P may be represented as in Figure 3.3.3. Then **p** may be expressed as:

$$\mathbf{p} = \rho \mathbf{e}_\rho \tag{3.3.3}$$

where ρ is the distance from O to P, and $\mathbf{e}_\rho$ is a unit vector along OP.

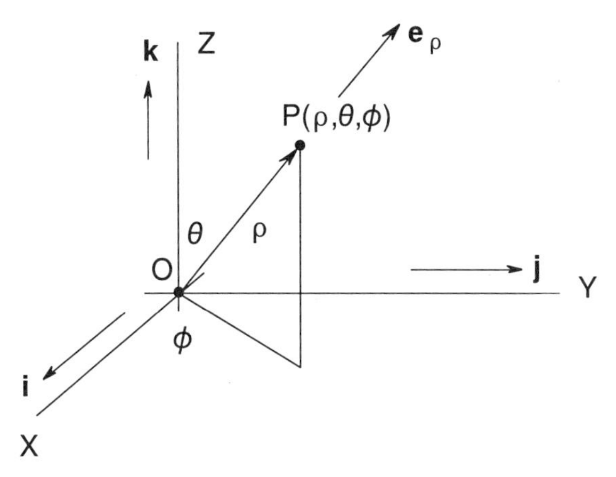

Fig. 3.3.3 Position Vector in Spherical Coordinates

Observe also in Equations (3.3.2) and (3.3.3) that the unit vectors $\mathbf{n}_r$ and $\mathbf{e}_\rho$ have variable directions and that unlike Cartesian coordinates all of the cylindrical and spherical coordinates do not appear in the position vector expressions.

3.4 Position Vector Summary Data

Table 3.4.1 provides a summary listing of the position vector data of the foregoing sections.

Table 3.4.1 Summary of Position Vector Data

Coordinate System	Position Vector	Equation	Figure
Rectangular Cartesian	$\mathbf{p} = x\mathbf{i} + y\mathbf{j} + z\mathbf{k}$	Eq. (3.2.1)	Fig. (3.2.1)
Plane Polar	$\mathbf{p} = r\mathbf{n}_r$	Eq. (3.3.1)	Fig. (3.3.1)
Cylindrical	$\mathbf{p} = r\mathbf{n}_r + z\mathbf{k}$	Eq. (3.3.2)	Fig. (3.3.2)
Spherical	$\mathbf{p} = \rho\mathbf{e}_r$	Eq. (3.3.3)	Fig. (3.3.3)

3.5 Unit Vector Derivatives

The velocity and acceleration of particles may be obtained by differentiating the position vectors of the foregoing section. A glance at the form of these vectors (see Table 3.4.1) shows that their derivatives in turn will involve derivatives of unit vectors. These unit vector derivatives may be obtained using expressions such as those of Examples 2.17.6 and 2.17.7. Table 3.5.1 provides a listing of these derivatives as well as derivatives of other related unit vectors.

Table 3.5.1 Unit Vector Derivatives for Various Coordinate Systems

Coordinate Systems and Figure	Unit Vectors	Unit Vector Derivatives
Cartesian Coordinate System	Cartesian Unit Vectors $\mathbf{i}, \mathbf{j}, \mathbf{k}$	Cartesian Unit Vector Derivatives $\frac{d\mathbf{i}}{dt} = 0$ $\frac{d\mathbf{j}}{dt} = 0$ $\frac{d\mathbf{k}}{dt} = 0$

Continued

Table 3.5.1 continued

Coordinate Systems and Figure	Unit Vectors	Unit Vector Derivatives
Plane Polar Coordinate System 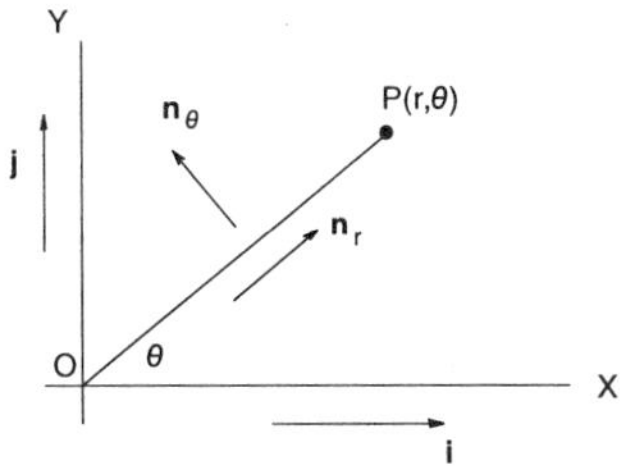 Y, j, n_θ, P(r,θ), n_r, O, θ, X, i	Plane Polar Unit Vectors $\mathbf{n}_r$, $\mathbf{n}_\theta$	Plane Polar Unit Vector Derivatives[1] $\frac{d\mathbf{n}_r}{dt} = \dot{\theta}\,\mathbf{n}_\theta$ $\frac{d\mathbf{n}_\theta}{dt} = -\dot{\theta}\,\mathbf{n}_r$
Plane Radial and Transverse[2] Coordinate System 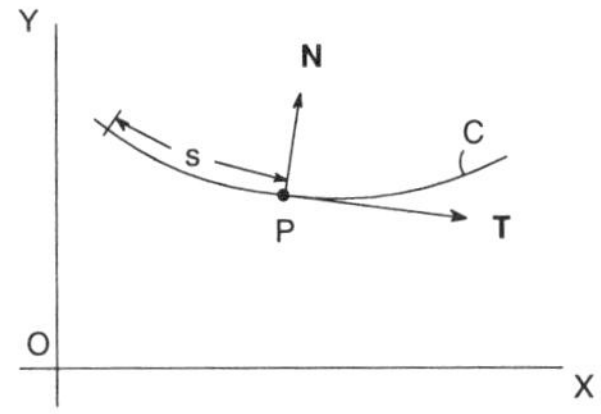 Y, N, C, s, P, T, O, X	Plane Radial and Transverse[2] Unit Vectors $\mathbf{T}$, $\mathbf{N}$	Plane Radial and Transverse[2] Unit Vector Derivatives $\frac{d\mathbf{T}}{dt} = (\dot{s}/\rho)\mathbf{n}$ $\frac{d\mathbf{N}}{dt} = -(\dot{s}/\rho)\mathbf{T}$
Spatial Radial and Transverse[2] Coordinate System 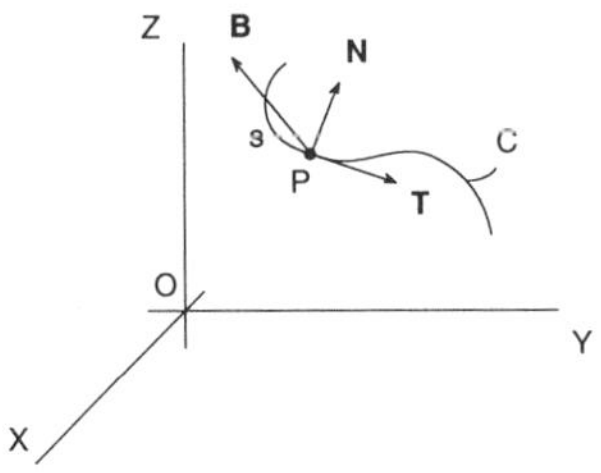 Z, B, N, s, C, P, T, O, Y, X	Spatial Radial and Transverse[2] Unit Vectors $\mathbf{T}$, $\mathbf{N}$, $\mathbf{B}$	Spatial Radial and Transverse[2] Unit Vector Derivatives $\frac{d\mathbf{T}}{dt} = \left(\frac{\dot{s}}{\rho}\right)\mathbf{N}$ $\frac{d\mathbf{N}}{dt} = -\left(\frac{\dot{s}}{\rho}\right)\mathbf{T} + \tau\mathbf{B}$ $\frac{d\mathbf{B}}{dt} = -\tau\mathbf{N}$

Continued

Table 3.5.1 Continued

Coordinate Systems and Figure	Unit Vector	Unit Vector Derivatives
Cylindrical Coordinate System 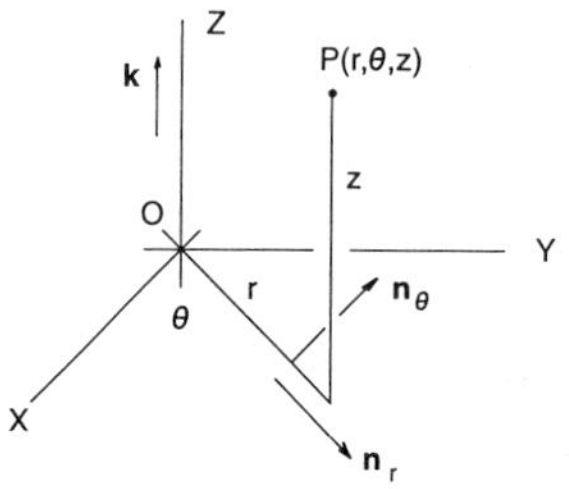	Cylindrical Unit Vectors $\mathbf{n}_r\ ,\ \mathbf{n}_\theta\ ,\ \mathbf{k}$	Cylindrical Unit Vector Derivatives $\frac{d\mathbf{n}_r}{dt} = \dot{\theta}\,\mathbf{n}_\theta$ $\frac{d\mathbf{n}_\theta}{dt} = -\dot{\theta}\,\mathbf{n}_r$ $\frac{d\mathbf{k}}{dt} = 0$
Spherical Coordinate System 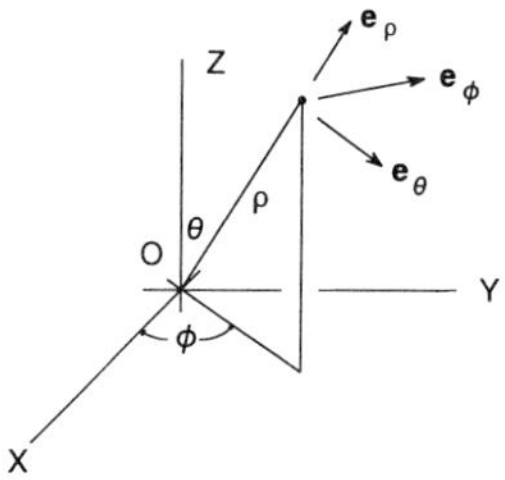	Spherical Unit Vectors $\mathbf{e}_\rho\ ,\ \mathbf{e}_\theta\ ,\ \mathbf{e}_\phi$	Spherical Unit Vector Derivatives $\frac{d\mathbf{e}_\rho}{dt} = \dot{\theta}\mathbf{e}_\theta + \dot{\phi}\sin\theta\,\mathbf{e}_\phi$ $\frac{d\mathbf{e}_\theta}{dt} = -\dot{\theta}\mathbf{e}_\rho + \dot{\phi}\cos\theta\,\mathbf{e}_\phi$ $\frac{d\mathbf{e}_\phi}{dt} = -\dot{\phi}\cos\theta\,\mathbf{e}_\theta - \dot{\phi}\sin\theta\,\mathbf{e}_\rho$

Notes (Table 3.5.1)

1. The overdot (as in $\dot{\theta}$) represents time differentiation (that is, $\dot{\theta} = d\theta/dt$).
2. The differentiation formulas for radial and transverse coordinates are obtained using principles of differential geometry [References 3.1 to 3.3]. In the expressions for the derivatives, s is length measured along the curve C ("arc length"), ρ is the radius of curvature of C at P [see Equation (3.5.9)], and τ is the torsion of C at P [see Equation (3.5.10)].

As noted, the expressions for the unit vector derivatives with radial and transverse coordinates are obtained using the principles of differential geometry. Interestingly, all of the terms in the differentiation formulas may be obtained from the position vector **p** locating a point P on a curve C. Specifically, let C be defined by the parametric equations:

$$x = x(t) \qquad y = y(t) \qquad z = z(t) \tag{3.5.1}$$

where t is any convenient parameter (including time). The position vector **p** may then be expressed in terms of fixed Cartesian unit vectors as

$$\mathbf{p} = \mathbf{p}(t) = x(t)\mathbf{i} + y(t)\mathbf{j} + z(t)\mathbf{k} \tag{3.5.2}$$

Then the arc length s, the unit vectors themselves (**T**, **N**, **B**), the radius of curvature ρ, and the torsion τ may be expressed as [3.1, 3.2]

$$s = [(dx/dt)^2 + (dy/dt)^2 + (dz/dt)^2]^{1/2} \tag{3.5.3}$$

$$\mathbf{T} = \frac{d\mathbf{p}}{dt} \Bigg/ \left|\frac{d\mathbf{p}}{dt}\right| = \frac{d\mathbf{p}}{ds} \quad \text{(unit tangent)} \tag{3.5.4}$$

$$\begin{aligned} \mathbf{N} &= \left[\left(\frac{d\mathbf{p}}{dt} \times \frac{d^2\mathbf{p}}{dt^2}\right) \times \frac{d\mathbf{p}}{dt}\right] \Bigg/ \left|\left(\frac{d\mathbf{p}}{dt} \times \frac{d^2\mathbf{p}}{dt^2}\right) \times \frac{d\mathbf{p}}{dt}\right| \\ &= \rho \frac{d^2\mathbf{p}}{ds^2} \qquad \text{(unit normal)} \end{aligned} \tag{3.5.5}$$

$$\mathbf{B} = \left(\frac{d\mathbf{p}}{dt} \times \frac{d^2\mathbf{p}}{dt^2}\right) \Bigg/ \left|\frac{d\mathbf{p}}{dt} \times \frac{d^2\mathbf{p}}{dt^2}\right| = \left(\frac{d\mathbf{p}}{ds} \times \frac{d^2\mathbf{p}}{ds} \times \frac{d^2\mathbf{p}}{ds^2}\right) \Bigg/ \left|\frac{d^2\mathbf{p}}{ds^2}\right| \tag{3.5.6}$$

(unit binormal)

$$\mathbf{T} = \mathbf{N} \times \mathbf{B} \tag{3.5.7}$$

$$\rho = \left|\frac{d\mathbf{p}}{dt}\right|^3 \Big/ \left|\frac{d\mathbf{p}}{dt} \times \frac{d^2\mathbf{p}}{dt^2}\right| = 1 \Big/ \left|\frac{d^2\mathbf{p}}{ds^2}\right| \tag{3.5.8}$$

$$\begin{aligned} \tau &= \rho^2\left(\frac{d\mathbf{p}}{dt} \times \frac{d^2\mathbf{p}}{dt^2} \cdot \frac{d^3\mathbf{p}}{dt^3}\right) \Big/ \left|\frac{d\mathbf{p}}{dt}\right|^6 \\ &= \rho^2 \frac{d\mathbf{p}}{ds} \times \frac{d^2\mathbf{p}}{ds^2} \cdot \frac{d^3\mathbf{p}}{ds^3} \end{aligned} \tag{3.5.9}$$

3.6 Velocity

The velocity of a particle is defined as its time rate of change of position. Hence, for a particle P (represented by a point P) moving on a curve C, as in Figure 3.6.1, the velocity ${}^R\mathbf{V}^P$ of P in a reference frame R is

$${}^R\mathbf{V}^P = \mathbf{V} = d\mathbf{p}/dt \tag{3.6.1}$$

where **p** locates P relative to a fixed point O in R.

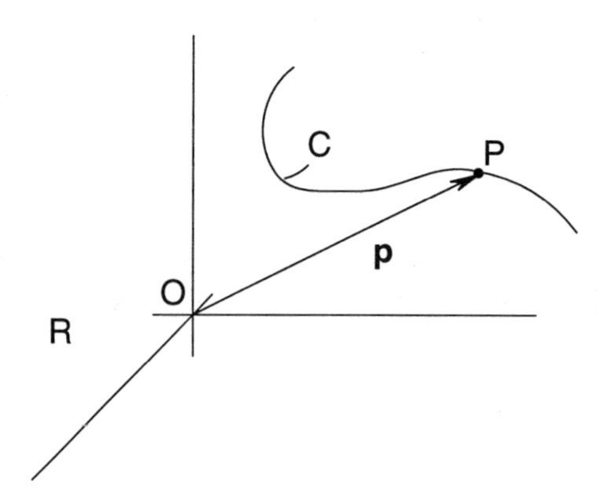

Fig. 3.6.1 Position Vector **p** Locating Particle P in Reference Frame R

From the definition of vector differentiation as in Section 2.16 and from Example 2.17.1 we see that **V** is tangent to C at P. This means that **V** may be expressed as

$$\mathbf{V} = v\mathbf{T} \tag{3.6.2}$$

where **T** is a unit vector tangent to C at P, as in Figure 3.6.2 and v is the magnitude of **V**.

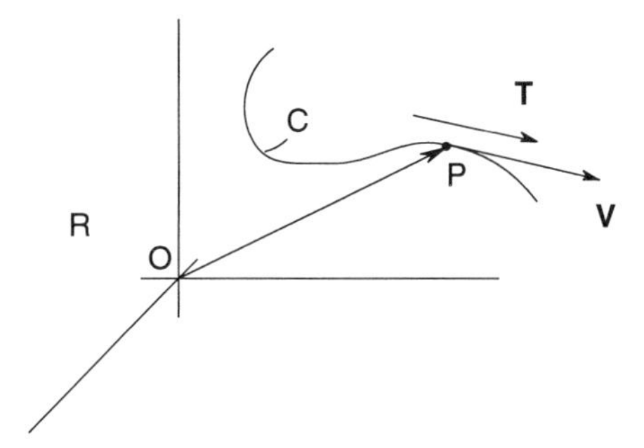

Fig. 3.6.2 Velocity and Tangent Vectors

By using Equation (3.6.1) and the listings of Tables 3.4.1 and 3.5.1 for the position vectors and the unit vector derivatives in the various coordinate systems, we can immediately obtain the corresponding velocity expressions. The results are listed in Table 3.6.1.

Table 3.6.1 Velocity Vectors in Various Coordinate Systems

Coordinate System	Velocity
Rectangular Cartesian	$\mathbf{V} = \dot{x}\mathbf{i} + \dot{y}\mathbf{j} + \dot{z}\mathbf{k}$
Plane Polar	$\mathbf{V} = \dot{r}\mathbf{n}_r + r\dot{\theta}\mathbf{n}_\theta$
Plane Radial and Transverse	$\mathbf{V} = v\mathbf{T} = \dot{s}\mathbf{T}$
Spatial Radial and Transverse	$\mathbf{V} = v\mathbf{T} = \dot{s}\mathbf{T}$
Cylindrical	$\mathbf{V} = \dot{r}\mathbf{n}_r + r\dot{\theta}\mathbf{n}_\theta + \dot{z}\mathbf{k}$
Spherical	$\mathbf{V} = \dot{\rho}\mathbf{e}_\rho + \rho\dot{\theta}\mathbf{e}_\theta + \rho\dot{\phi}\sin\theta\mathbf{e}_\phi$

3.7 Acceleration

The acceleration of a particle is defined as the time rate of change of its velocity. Hence, for a particle P moving in a reference frame R its acceleration ${}^{R}\mathbf{a}^{P}$ is given by

$$ {}^{R}\mathbf{a}^{P} = \mathbf{a} = d\,{}^{R}\mathbf{V}^{P}/dt = d\mathbf{V}/dt \tag{3.7.1} $$

Observe that unlike velocity the acceleration of a particle is not generally parallel to the curve of motion.

By using Equation (3.7.1) and the listings of Tables 3.5.1 and 3.6.1 we can immediately obtain expressions for particle acceleration in various coordinate systems. The results are listed in Table 3.7.1.

Table 3.7.1 Acceleration Vectors in Various Coordinate Systems

Coordinate System	Acceleration
Rectangular Cartesian	$\mathbf{a} = \ddot{x}\mathbf{i} + \ddot{y}\mathbf{j} + \ddot{z}\mathbf{k}$
Plane Polar	$\mathbf{a} = (\ddot{r} - r\dot{\theta}^2)\mathbf{n}_r + (r\ddot{\theta} + 2\dot{r}\dot{\theta})\mathbf{n}_\theta$
Plane Radial and Transverse	$\mathbf{a} = \dot{v}\mathbf{T} + (v^2/\rho)\mathbf{N} = \ddot{s}\mathbf{T} + (\dot{s}^2/\rho)\mathbf{N}$
Spatial Radial and Transverse	$\mathbf{a} = \dot{v}\mathbf{T} + (v^2/\rho)\mathbf{N} = \ddot{s}\mathbf{T} + (\dot{s}^2/\rho)\mathbf{N}$
Cylindrical	$\mathbf{a} = (\ddot{r} - r\dot{\theta}^2)\mathbf{n}_r + (r\ddot{\theta} + 2\dot{r}\dot{\theta})\mathbf{n}_\theta + \ddot{z}\mathbf{k}$
Spherical	$\mathbf{a} = (\ddot{\rho} - \rho\dot{\theta}^2 - \rho\dot{\phi}^2\sin^2\theta)\mathbf{e}_\rho + (\rho\ddot{\theta} + 2\dot{\rho}\dot{\theta} - \rho\dot{\phi}^2\sin\theta\cos\theta)\mathbf{e}_\theta + (\rho\ddot{\phi}\sin\theta + 2\dot{r}\dot{\phi}\sin\theta + 2r\dot{\phi}\dot{\theta}\cos\theta)\mathbf{e}_\phi$

3.8 Summary Data

The results of the foregoing sections are summarized in Table 3.8.1.

Table 3.8.1 Summary of Particle Kinematic Data in Various Coordinate Systems

Coordinate Systems, Coordinates, and Unit Vectors	Figure	Kinematics Expressions
<u>Rectangular</u> (x,y,z) $(\mathbf{i}, \mathbf{j}, \mathbf{k})$	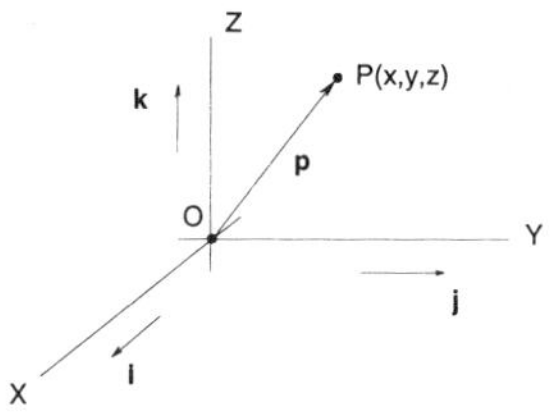	Position $\mathbf{p} = x\mathbf{i} + y\mathbf{j} + z\mathbf{k}$ Velocity $\mathbf{V} = \dot{x}\mathbf{i} + \dot{y}\mathbf{j} + \dot{z}\mathbf{k}$ Acceleration $\mathbf{a} = \ddot{x}\mathbf{i} + \ddot{y}\mathbf{j} + \ddot{z}\mathbf{k}$
<u>Plane Polar</u> (r, θ) $(\mathbf{n}_r, \mathbf{n}_\theta)$	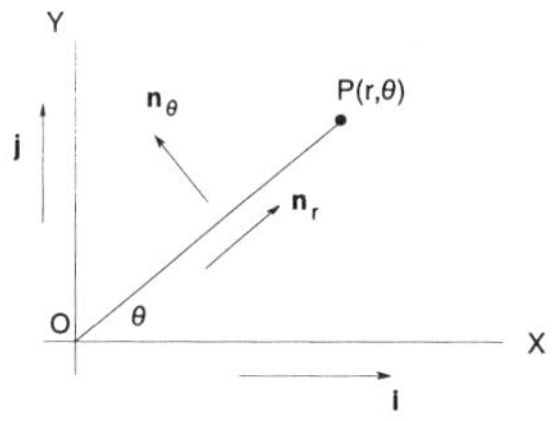	Position $\mathbf{p} = r\mathbf{n}_r$ Velocity $\mathbf{V} = \dot{r}\mathbf{n}_r + r\dot{\theta}\mathbf{n}_\theta$ Acceleration $\mathbf{a} = (\ddot{r} - r\dot{\theta}^2)\mathbf{n}_r$ $+ (r\ddot{\theta} + 2\dot{r}\dot{\theta})\mathbf{n}_\theta$
<u>Plane Radial and Transverse</u> (ρ, s) $(\mathbf{T}, \mathbf{N})$	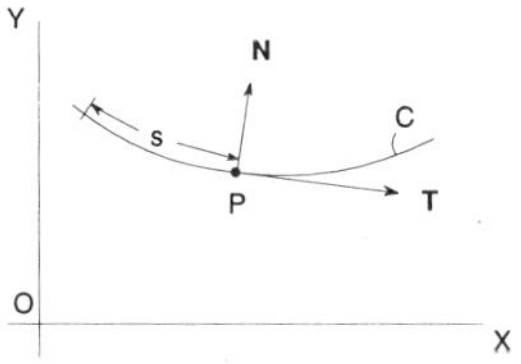	Velocity $\mathbf{V} = \dot{s}\mathbf{T}$ Acceleration $\mathbf{a} = \ddot{s}\mathbf{T} + (\dot{s}^2/\rho)\mathbf{N}$

Continued

Table 3.8.1 Continued

Coordinate Systems, Coordinates, and Unit Vectors	Figure	Kinematics Expressions
<u>Spatial Radial and Transverse</u> (ρ, τ, s) $(\mathbf{T}, \mathbf{N}, \mathbf{B})$	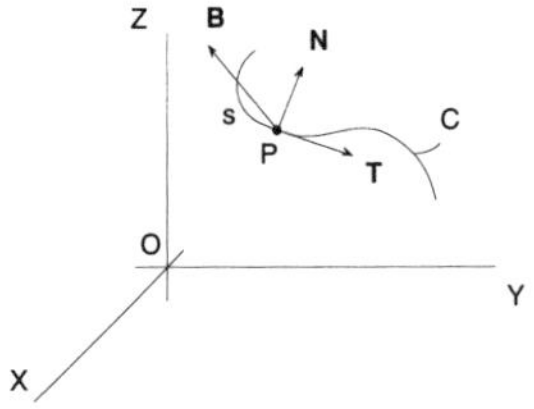	Velocity $\mathbf{V} = \dot{s}\mathbf{T}$ Acceleration $\mathbf{a} = \ddot{s}\mathbf{T} + (\dot{s}^2/\rho)\mathbf{N}$
<u>Cylindrical</u> (r, θ, z) $(\mathbf{n}_r, \mathbf{n}_\theta, \mathbf{k})$	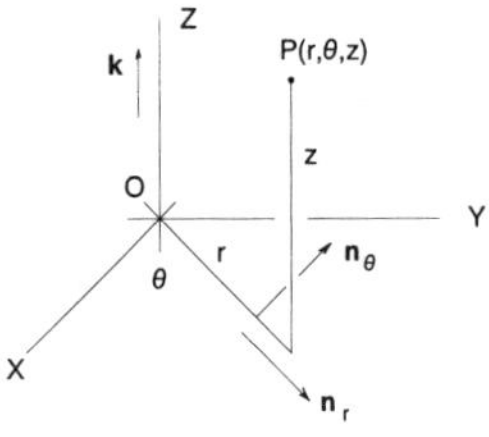	Position $\mathbf{p} = r\mathbf{n}_r + z\mathbf{k}$ Velocity $\mathbf{V} = \dot{r}\mathbf{n}_r + r\dot{\theta}\,\mathbf{n}_\theta + \dot{z}\mathbf{k}$ Acceleration $\mathbf{a} = (\ddot{r} - r\dot{\theta}^2)\,\mathbf{n}_r$ $+ (r\ddot{\theta} + 2\dot{r}\dot{\theta})\,\mathbf{n}_\theta + \ddot{z}\mathbf{k}$
<u>Spherical</u> (ρ, θ, ϕ) $(\mathbf{n}_\rho, \mathbf{n}_\theta, \mathbf{n}_\phi)$	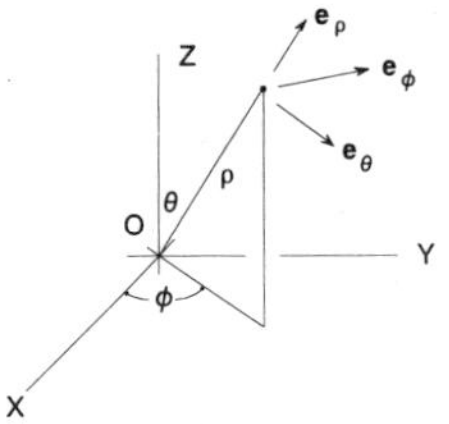	Position $\mathbf{p} = \rho\mathbf{n}_\rho$ Velocity $\mathbf{V} = \dot{\rho}\mathbf{e}_\rho + \rho\dot{\theta}\mathbf{e}_\theta + \rho\dot{\phi}\sin\theta\,\mathbf{e}_\phi$ Acceleration $\mathbf{a} = (\ddot{\rho} - \rho\dot{\theta}^2 - \rho\dot{\phi}^2\sin^2\theta)\mathbf{e}_\rho$ $+ (\rho\ddot{\theta} + 2\dot{\rho}\dot{\theta} - \rho\dot{\phi}^2\sin\theta\cos\theta)\mathbf{e}_\theta$ $+ (\rho\ddot{\phi}\sin\theta + 2\dot{r}\dot{\phi}\sin\theta$ $+ 2r\dot{\phi}\dot{\theta}\cos\theta)\mathbf{e}_\phi$

3.9 Angular Velocity

In this section we will briefly look at some aspects of angular velocity.

In Example 2.17.7 for the derivative of a unit vector **n** rotating parallel to a plane, it is enlightening to think of **n** as being parallel to an axis of a rotating reference frame. Specifically, let X,Y,Z be axes of a fixed Cartesian reference frame R with origin O as in Figure 3.9.1 Let $\mathbf{n}_x$, $\mathbf{n}_y$, and $\mathbf{n}_z$ be unit vectors parallel to X, Y, and Z as shown.

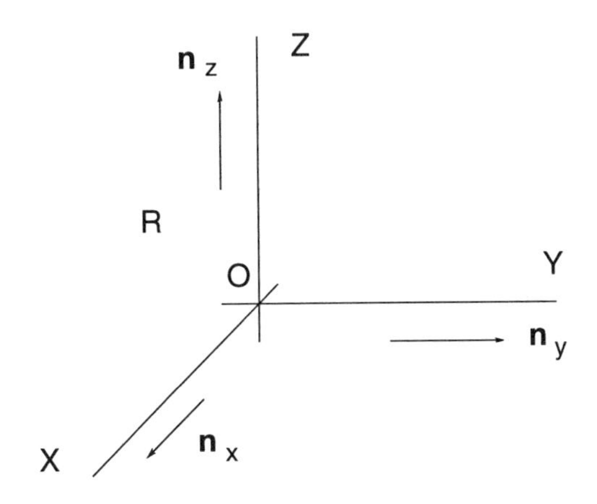

Fig. 3.9.1 A Fixed Reference Frame R

Similarly, let $\hat{X}, \hat{Y}, \hat{Z}$ be axes of a moving Cartesian reference frame with origin $\hat{O}$ as in Figure 3.9.2, where $\hat{\mathbf{n}}_x$, $\hat{\mathbf{n}}_y$, and $\hat{\mathbf{n}}_z$ are parallel to $\hat{X}, \hat{Y}$, and $\hat{Z}$ as shown.

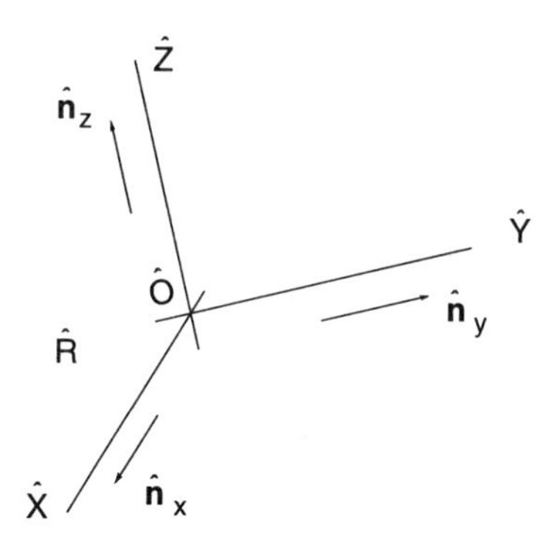

Fig. 3.9.2 A Moving Reference Frame $\hat{R}$

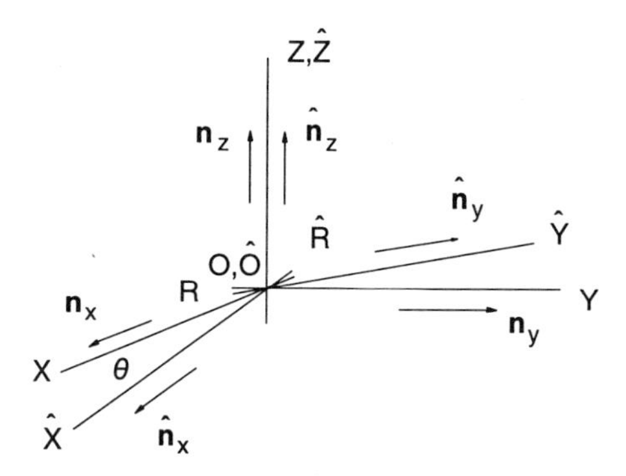

Fig. 3.9.3 Reference Frames R (X,Y,Z-fixed) and $\hat{R}$ ($\hat{X},\hat{Y},\hat{Z}$-rotating)

Let the motion of $\hat{R}$ be restricted such that its origin $\hat{O}$ is coincident with the origin O of R, and also such that the $\hat{Z}$ axis is coincident with the Z axis of R as in Figure 3.9.3 Let θ be the angle between X and $\hat{X}$ (and hence, also between Y and $\hat{Y}$). Then the unit vectors $\hat{\mathbf{n}}_x$ and $\hat{\mathbf{n}}_y$ move parallel to the X-Y plane and from Equation (1) of Example 2.17.6 (Section 2.17) their derivatives may be expressed as:

$$d\hat{\mathbf{n}}_x/dt = \dot{\theta}\,\mathbf{n}_z \times \hat{\mathbf{n}}_x \quad \text{and} \quad d\hat{\mathbf{n}}_y/dt = \dot{\theta}\mathbf{n}_z \times \hat{\mathbf{n}}_y \tag{3.9.1}$$

Observe that it is also true that the derivative of $\hat{\mathbf{n}}_z$ (which is zero) may be expressed in the same form. That is,

$$d\hat{\mathbf{n}}_z/dt = \dot{\theta}\,\mathbf{n}_z \times \hat{\mathbf{n}}_z \tag{3.9.2}$$

In Equations (3.9.1) and (3.9.2) the term $\dot{\theta}\,\mathbf{n}_z$ is sometimes called "the angular velocity of $\hat{R}$ in R" with θ being the "turning angle." In this context, $\dot{\theta}\,\mathbf{n}_z$ is often written as ${}^{R}\boldsymbol{\omega}^{\hat{R}}$, $\boldsymbol{\omega}^{\hat{R}}$, or simply as $\boldsymbol{\omega}$. Equations (3.9.1) and (3.9.2) then become

$$d\hat{\mathbf{n}}_x/dt = \boldsymbol{\omega} \times \hat{\mathbf{n}}_x\,, \quad d\hat{\mathbf{n}}_y/dt = \boldsymbol{\omega} \times \hat{\mathbf{n}}_y\,, \quad d\hat{\mathbf{n}}_z/dt = \boldsymbol{\omega} \times \hat{\mathbf{n}}_z \tag{3.9.3}$$

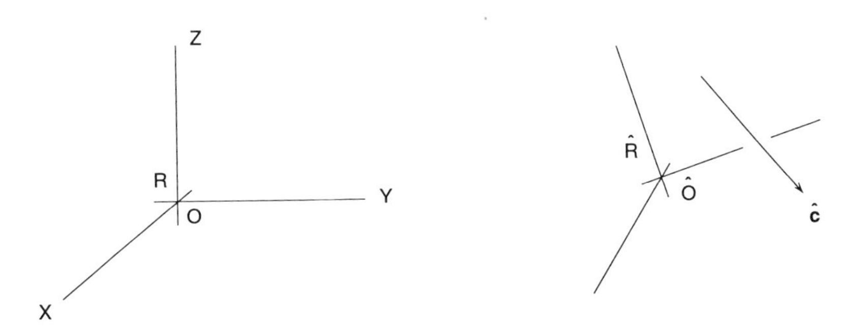

Fig. 3.9.4 Reference Frame $\hat{R}$ with Fixed Vector $\hat{\mathbf{c}}$ Moving in Reference Frame R

As a generalization of Equation (3.9.3) let $\hat{R}$ now move in an arbitrary, or unrestricted, manner relative to R. Let $\hat{\mathbf{c}}$ be any vector fixed in $\hat{R}$, as in Figure 3.9.4. Then $\hat{\mathbf{c}}$ has a constant magnitude and a fixed direction in $\hat{R}$, but a varying direction in R. It can be shown that the derivative of $\hat{\mathbf{c}}$ in R has the form

$$d\hat{\mathbf{c}}/dt = \boldsymbol{\omega} \times \hat{\mathbf{c}} \tag{3.9.4}$$

where $\boldsymbol{\omega}$ is now a general angular velocity of $\hat{R}$ in R. Equation (3.9.4) is sometimes written in the more explicit form [3.1 to 3.4]:

$$^{R}d\hat{c}/dt = {}^{R}\boldsymbol{\omega}^{\hat{R}} \times \hat{c} \qquad (3.9.5)$$

In the context of general angular velocity, as in Equations (3.9.4) and (3.9.5), the more restricted angular velocity of Equation (3.9.1) is sometimes called "simple angular velocity."

Equation (3.9.4) is useful for finding velocities of particles of rigid bodies. Another useful expression is the "addition formula" for angular velocity. If R_1, R_2, and R_3 are reference frames, then it may be seen that

$$^{R_1}\boldsymbol{\omega}^{R_3} = {}^{R_1}\boldsymbol{\omega}^{R_2} + {}^{R_2}\boldsymbol{\omega}^{R_3} \qquad (3.9.6)$$

Finally, one other useful expression for developing the kinematics of particles is a differentiation algorithm relating the derivatives of vector quantities in two different reference frames:

$$^{R_1}d(\;\;)/dt = {}^{R_1}d(\;\;)/dt + {}^{R_1}\boldsymbol{\omega}^{R_2} \times (\;\;) \qquad (3.9.7)$$

where ( ) is any vector quantity [3.1 to 3.4].

3.10 Angular Acceleration

The angular acceleration of a body B in a reference R is defined as the derivative in R of the angular velocity of B in R. That is:

$$^{R}\boldsymbol{\alpha}^{B} = {}^{R}d\,{}^{R}\boldsymbol{\omega}^{B}/dt \qquad (3.10.1)$$

or when the reference frame is understood, as:

$$\boldsymbol{\alpha}^{B} = d\boldsymbol{\omega}^{B}/dt \qquad (3.10.2)$$

3.11 Rigid Bodies and Reference Frames

From the perspective of kinematics, with a focus upon motion, as opposed to forces, and as a consequence where mass is unimportant, there is no distinction between a rigid body and a reference frame. A rigid body is defined as a collection of particles whose distances from one another remains constant. Let B be a rigid body and let P and Q be particles of B, as in Figure 3.11.1. Let **r** be a position vector locating P relative to Q. Then since P and Q are fixed in B, and since the distance between P and Q is constant, **r** has constant magnitude and is also fixed relative to B. With P and Q being particles, their size is regarded as "small." Then with the masses of P and Q also being unimportant in a kinematic analysis, we may let P and Q be represented by points also called P and Q. In this latter regard, let XYZ be a Cartesian Axis system, called Reference Frame R, also fixed in B as in Figure 3.11.2.

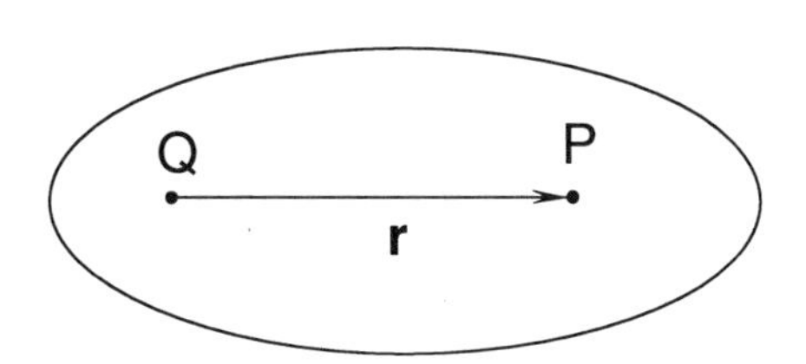

Fig. 3.11.1 A Rigid Body B and Typical Particles P and Q

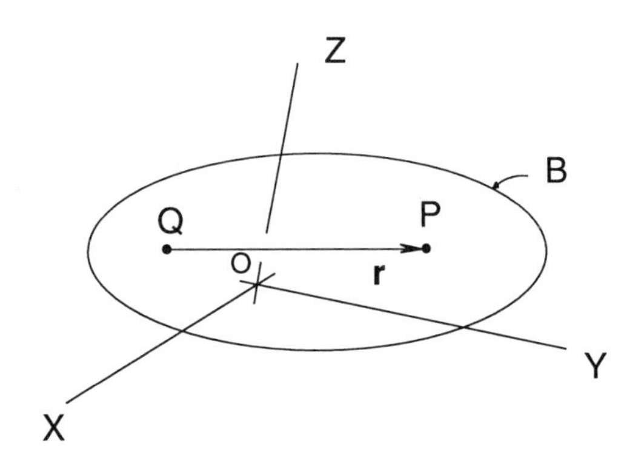

Fig. 3.11.2 A Rigid Body B with an Imbedded (Internally Fixed) Axis System XYZ

Then as B moves so also does R. Hence, P and Q may be regarded as points of R, and **r** is then a vector fixed in R.

3.12 Relative Velocity

Let P and Q be two particles moving in a reference frame R as in Figure 3.12.1. The *relative velocity* of P and Q in R, or alternatively, "the velocity of P relative to Q in R," written as ${}^{R}\mathbf{V}^{P/Q}$, is defined as the difference of the velocities of P and Q in R. That is,

$$ {}^{R}\mathbf{V}^{P/Q} = {}^{R}\mathbf{V}^{P} - {}^{R}\mathbf{V}^{Q} \tag{3.12.1} $$

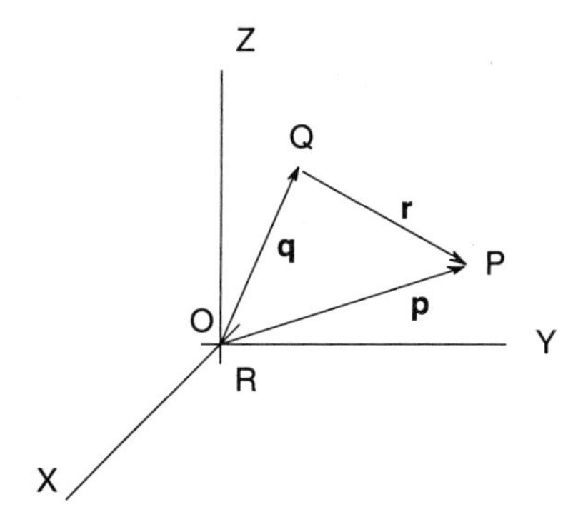

Fig. 3.12.1 Particles P and Q Moving in a Reference Frame R

If the reference frame is understood, then Equation (3.12.1) may be written in the simpler form:

$$ \mathbf{V}^{P/Q} = \mathbf{V}^{P} - \mathbf{V}^{Q} \tag{3.12.2} $$

or as

$$\mathbf{V}^{P} = \mathbf{V}^{Q} + \mathbf{V}^{P/Q} \tag{3.12.3}$$

In this context, $\mathbf{V}^P$ and $\mathbf{V}^Q$ are sometimes called the "absolute" velocities of P and Q.

From Figure 3.12.1, we see that the position vectors **p**, **q**, and **r** locating P and Q relative to the origin O, and relative to each other, are related by

$$\mathbf{p} = \mathbf{q} + \mathbf{r} \tag{3.12.4}$$

Then by differentiating and comparing the results with Equations (3.6.1) and (3.12.3) we have

$$\mathbf{V}^{P/Q} = d\mathbf{r}/dt \tag{3.12.5}$$

3.13 Relative Acceleration

The relative acceleration of two particles is defined as the derivative of the relative velocity of the particles. That is,

$${}^{R}\mathbf{a}^{P/Q} = {}^{R}d\,{}^{R}\mathbf{V}^{P/Q}/dt \tag{3.13.1}$$

or when the reference frame is understood, as

$$\mathbf{a}^{P/G} = d\mathbf{V}^{P/Q}/dt \tag{3.13.2}$$

Then from Equation (3.12.3) we have

$$\mathbf{a}^{P/Q} = \mathbf{a}^P - \mathbf{a}^Q \tag{3.13.3}$$

and

$$\mathbf{a}^P = \mathbf{a}^Q + \mathbf{a}^{P/Q} \tag{3.13.4}$$

where $\mathbf{a}^P$ and $\mathbf{a}^Q$ are now the "absolute" accelerations of P and Q.

3.14 Relative Velocity of Two Particles of a Rigid Body

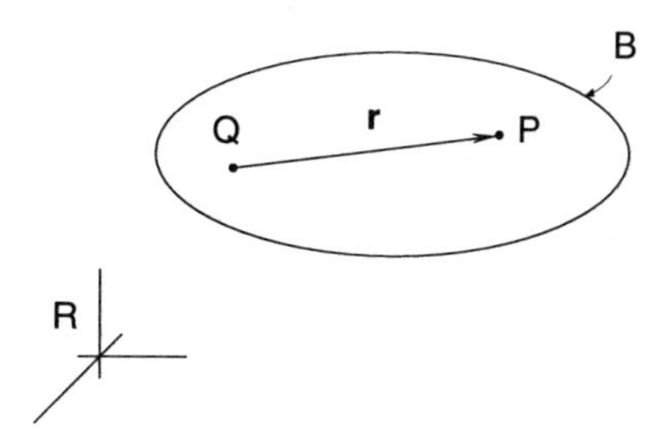

Fig. 3.14.1 Particles P and Q of Body B, Moving in Reference Frame R

Consider a rigid body B moving in a reference frame R as in Figure 3.14.1. let P and Q be particles of B and let $\mathbf{r}$ be the position vector locating P relative to Q. Then P, Q, and $\mathbf{r}$ are fixed in B. From Equations (3.9.5) and (3.12.5) the velocity of P relative to Q in R may be expressed as

$$^R\mathbf{V}^{P/Q} = {}^R\boldsymbol{\omega}^B \times \mathbf{r} \tag{3.14.1}$$

where $^R\boldsymbol{\omega}^B$ is the angular velocity of B in R. From Equation (3.12.3) we then also have the expression

$$\mathbf{V}^P = \mathbf{V}^Q + \boldsymbol{\omega}^B \times \mathbf{r} \tag{3.14.2}$$

where $\mathbf{V}^P$, $\mathbf{V}^Q$, and $\boldsymbol{\omega}^B$ are all evaluated in R.

Equation (3.14.2) shows that the velocity of any particle P may be interpreted as the velocity of any reference point Q plus a rotation about Q. The equation also shows that if the velocity of any point (say Q) of B is known, and if the angular velocity of B is known, then the velocity of any and all other points of B can be found.

3.15 Relative Acceleration of Two Particles of a Rigid Body

Consider again particles P and Q, fixed in body B, as in Figure 3.14.1. Then by differentiating in Equations (3.14.1) and (3.14.2) the relative acceleration of P and Q and the acceleration of P itself may be expressed as:

$$^R\mathbf{a}^{P/Q} = {}^R\boldsymbol{\alpha}^B \times \mathbf{r} + {}^R\boldsymbol{\omega}^B \times ({}^R\boldsymbol{\omega}^B \times \mathbf{r}) \tag{3.15.1}$$

and

$$\mathbf{a}^P = \mathbf{a}^Q + \boldsymbol{\alpha}^B \times \mathbf{r} + \boldsymbol{\omega}^B \times (\boldsymbol{\omega}^B \times \mathbf{r}) \tag{3.15.2}$$

where $^R\boldsymbol{\alpha}^B$ is the acceleration of B in R and where $\mathbf{a}^P$, $\mathbf{a}^Q$, $\boldsymbol{\alpha}^B$, and $\boldsymbol{\omega}^B$ are all evaluated in R.

Equation (3.15.2) shows that if the acceleration of any point (say Q) of B is known, and if the angular velocity and angular acceleration of B are known, then the acceleration of any and all other points of B can be found.

3.16 Velocity of a Particle Moving Relative to a Moving Body

Consider a rigid body B moving in a reference frame R and a particle P moving relative to B as in Figure 3.16.1. Let Q be a point of B and let $\mathbf{r}$ locate P relative to Q. Then the velocity of P in R is

$$ {}^{R}\mathbf{V}^{P} = {}^{R}\mathbf{V}^{Q} + {}^{R}d\mathbf{r}/dt \tag{3.16.1} $$

From Equation (3.9.7) ${}^{R}d\mathbf{r}/dt$ may be written as:

$$ \begin{aligned} {}^{R}d\mathbf{r}/dt &= {}^{B}d\mathbf{r}/dt + {}^{R}\boldsymbol{\omega}^{B} \times \mathbf{r} \\ &= {}^{B}\mathbf{V}^{P} + {}^{R}\boldsymbol{\omega}^{B} \times \mathbf{r} \end{aligned} \tag{3.16.2} $$

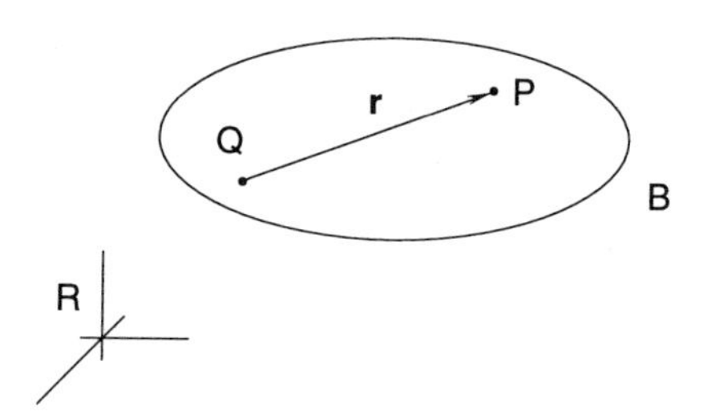

Fig. 3.16.1 A Rigid Body B with Particle P Moving Relative to B

Then the velocity of P in R is

$$ {}^{R}\mathbf{V}^{P} = {}^{R}\mathbf{V}^{Q} + {}^{B}\mathbf{V}^{P} + {}^{R}\boldsymbol{\omega}^{P} \times \mathbf{r} \tag{3.16.3} $$

Finally, if at an instant of interest Q is chosen as that point P^* of B which coincides with P, then $\mathbf{r}$ is zero and ${}^{R}\mathbf{V}^{P}$ has the form:

$$ {}^{R}\mathbf{V}^{P} = {}^{R}\mathbf{V}^{P^*} + {}^{B}\mathbf{V}^{P} \tag{3.16.4} $$

3.17 Acceleration of a Particle Moving Relative to a Moving Body

Consider again a rigid body B moving in a reference frame R and a particle P moving relative to B as in Figure 3.16.1. Then with the same notation as in Section 3.16, the acceleration of P in R ${}^R\mathbf{a}^P$ may be obtained by differentiating in Equation (3.16.1):

$$ {}^R\mathbf{a}^P = {}^R\mathbf{a}^Q + {}^R d^2\mathbf{r}/dt^2 \tag{3.17.1} $$

From Equations (3.9.7) and (3.16.2) ${}^R d^2\mathbf{r}/dt^2$ may be developed as follows:

$$ \begin{aligned} {}^R d^2\mathbf{r}/dt^2 &= {}^R d({}^R d\mathbf{r}/dt)/dt \\ &= {}^R d({}^B d\mathbf{r}/dt)dt + {}^R d({}^R\boldsymbol{\omega}^B \times \mathbf{r})/dt \\ &= {}^B d({}^B d\mathbf{r}/dt)/dt + {}^R\boldsymbol{\omega}^B \times {}^B d\mathbf{r}/dt + {}^R\boldsymbol{\alpha}^B \times \mathbf{r} \\ &\quad + {}^R\boldsymbol{\omega}^B \times {}^R d\mathbf{r}/dt \\ &= {}^B\mathbf{a}^P + {}^R\boldsymbol{\omega}^B \times {}^B\mathbf{V}^P + {}^R\boldsymbol{\alpha}^B \times \mathbf{r} \\ &\quad + {}^R\boldsymbol{\omega}^B \times ({}^B\mathbf{V}^P + {}^R\boldsymbol{\omega}^B \times \mathbf{r}) \\ &= {}^B\mathbf{a}^P + 2{}^R\boldsymbol{\omega}^B \times {}^B\mathbf{V}^P + {}^R\boldsymbol{\alpha}^B \times \mathbf{r} \\ &\quad + {}^R\boldsymbol{\omega}^B \times ({}^R\boldsymbol{\omega}^B \times \mathbf{r}) \end{aligned} \tag{3.17.2} $$

Then the acceleration of P in R is:

$$ \begin{aligned} {}^R\mathbf{a}^P = {}^B\mathbf{a}^P + {}^R\mathbf{a}^Q + 2{}^R\boldsymbol{\omega}^B \times {}^B\mathbf{V}^P + {}^R\boldsymbol{\alpha}^B \times \mathbf{r} \\ + {}^R\boldsymbol{\omega}^B \times ({}^R\boldsymbol{\omega}^B \times \mathbf{r}) \end{aligned} \tag{3.17.3} $$

Finally, if at an instant of interest Q is chosen as that point P^* of B which coincides with P, then $\mathbf{r}$ is zero and ${}^R\mathbf{a}^P$ has the form:

$$ {}^R\mathbf{a}^P = {}^B\mathbf{a}^P + {}^R\mathbf{a}^Q + 2{}^R\boldsymbol{\omega}^B \times {}^B\mathbf{V}^P \tag{3.17.4} $$

3.18 Summary of Particle Kinematic Formulas

Table 3.18.1 provides a summary listing of the principal equations of the immediate foregoing sections. The notation is the same as that in the referenced equations. This table may be viewed as a supplement to Table 3.8.1 which provides a summary of particle kinematics in component form in various coordinate systems.

Table 3.18.1 Summary of Selected Vector Particle Kinematic Formulas

Configuration	Equation	Reference Equation
1. Relative Velocity and Acceleration	$\mathbf{V}^{P/Q} = \mathbf{V}^P - \mathbf{V}^Q$ $\mathbf{a}^{P/Q} = \mathbf{a}^P - \mathbf{a}^Q$	(3.12.2) (3.13.3)
2. Relative Velocity and Acceleration of Two Particles of a Rigid Body	$\mathbf{V}^{P/Q} = \boldsymbol{\omega}^B \times \mathbf{r}$ $\mathbf{a}^{P/Q} = \boldsymbol{\alpha}^B \times \mathbf{r} + \boldsymbol{\omega}^B \times (\boldsymbol{\omega}^B \times \mathbf{r})$	(3.14.1) (3.15.2)
3. Velocity and Acceleration of a Particle Moving Relative to a Moving Body	${}^R\mathbf{V}^P = {}^B\mathbf{V}^P + {}^R\mathbf{V}^{P^*}$ ${}^R\mathbf{a}^P = {}^B\mathbf{a}^P + {}^R\mathbf{a}^{P^*} + 2\,{}^R\boldsymbol{\omega}^B \times {}^B\mathbf{V}^P$	(3.16.4) (3.17.4)

3.19 Application: Motion of a Particle in a Straight Line

The simplest motion a particle can have (aside from rest) is movement in a straight line ("rectilinear motion"). Let a particle P move along the X-axis as in Figure 3.19.1, where O is the origin and the position of P is measured by the coordinate x(t). Let P have coordinate x_0 with speed v_0 when t is zero.

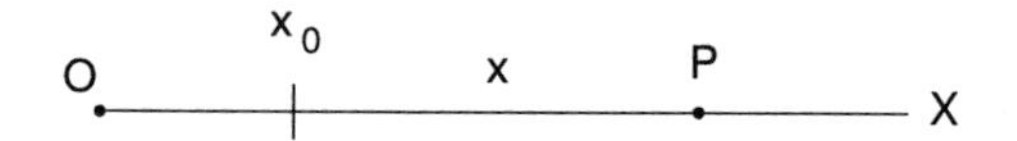

Fig. 3.19.1 Particle P Moving Along the X-Axis

If x(t) is known, then the velocity and acceleration may be obtained by differentiation. Alternatively, if the acceleration of P is known, the velocity and position may be obtained by integration. Integration, however, is not generally as simple as differentiation. Integration constants need to be determined. Also the known acceleration may not be a function of time, but instead a function of either the velocity or the position. Nevertheless, depending upon the acceleration function, the integration can generally be performed. Table 3.19.1 outlines the formal procedures of these integrations, depending upon the form of the acceleration function.

Table 3.19.1 Motion of a Particle, Given the Acceleration (Rectilinear Motion)

Case	Description	Motion
1. $a = f(t)$	Acceleration is given as a function of time	$v = v_0 + \int_0^t f(t)dt$ $x = x_0 + \int_{v_0}^{v} \frac{dv}{f(v)}$
2. $a = f(v)$	Acceleration is given as a unction of velocity	$t = \int_{v_0}^{v} \frac{dv}{f(v)}$
3. $a = f(x)$	Acceleration is given as a function of displacement	$v^2 = v_0^2 + 2\int_{x_0}^{x} f(x)dx$ $x = x_0 + \int_0^t v(t)dt$
4. $a = a_0$	Acceleration is given as a constant	$v = v_0 + a_0 t$ $x = x_0 + v_0 t + \frac{1}{2}at^2$ $v^2 = v_0^2 + 2a_0(x - x_0)$

3.20 Application: Motion of a Particle in a Circle

Among the most common motions of a particle is movement in a circle. Analysis of such motion is a direct application of the formulas of Table 3.8.1 for Plane Polar Coordinates. Let a particle P, represented by a point P, move on a circle with radius r as in Figure 3.20.1, where O is the center of the circle and the angular position of P is measured by the angle θ. Then by letting r be a constant in the Plane Polar representation of Table 3.8.1, the velocity and acceleration of P may be expressed in component form as shown in Figure 3.20.2. These components may be expressed in terms of r and θ as:

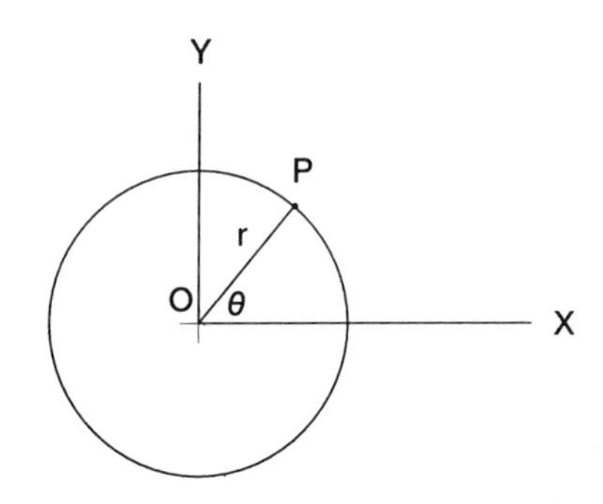

Fig. 3.20.1 A Particle Moving in a Circle

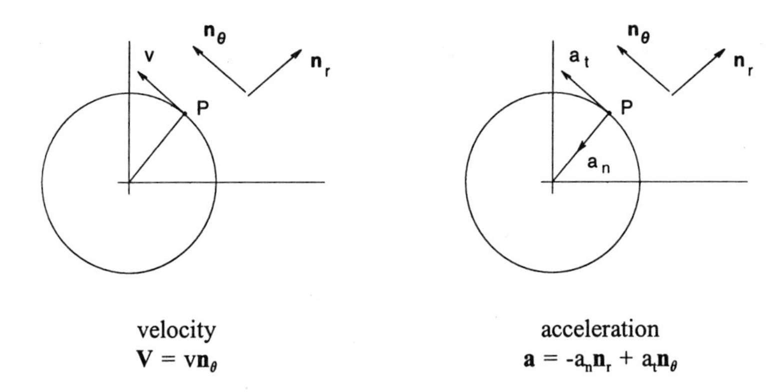

velocity
$\mathbf{V} = v\mathbf{n}_\theta$

acceleration
$\mathbf{a} = -a_n\mathbf{n}_r + a_t\mathbf{n}_\theta$

Fig. 3.20.2 Velocity and Acceleration of a Particle Moving in a Circle

$$\mathbf{V} = r\dot{\theta} = r\omega \tag{3.20.1}$$

$$a_n = r\dot{\theta}^2 = r\omega^2 = v^2/r \tag{3.20.2}$$

$$a_t = r\ddot{\theta} = r\dot{\omega} = r\alpha \tag{3.20.3}$$

where ω and α are defined by inspection as:

$$\omega = \dot{\theta} \qquad \alpha = \dot{\omega} = \ddot{\theta} \tag{3.20.4}$$

3.21 Application: Projectile Motion

A third major application of particle kinematic analysis is in the study of projectile motion. Consider a projectile P moving in a fixed (inertial) reference frame R as depicted in Figure 3.21.1. Let X,Y,Z be a Cartesian axis system fixed in R as shown. Then from an elementary free-body diagram of P, the differential equations governing the motion of P are seen to be

$$\ddot{x} = 0 \tag{3.21.1}$$
$$\ddot{z} = -g \tag{3.21.3}$$
$$\ddot{y} = 0 \tag{3.21.2}$$

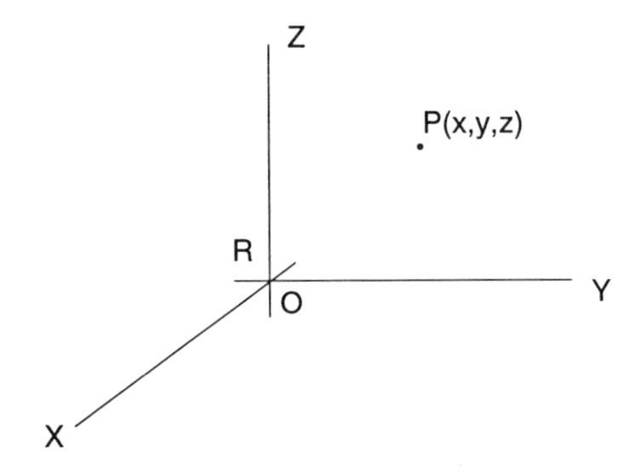

Fig. 3.21.1 A Projectile Moving in a Fixed (Inertial) Reference Frame.

where, as before, the overdot designates time differentiation. Also, in Equation (3.21.1), (3.21.2), and (3.21.3) effects of air resistance are neglected.

Integrating in Equations (3.21.1), (3.21.2), and (3.21.3) then leads to the X, Y and Z velocity components of P as:

$$\dot{x} = \dot{x}_0 = v_{0x} \tag{3.21.4}$$

$$\dot{y} = \dot{y}_0 = v_{0y} \tag{3.21.5}$$

$$\dot{z} = -gt + \dot{z}_0 = -gt + v_{0z} \tag{3.21.6}$$

where $\dot{x}_0$, $\dot{y}_0$, and $\dot{z}_0$ (also defined as v_{0x}, v_{0y}, and v_{0z}) represent the X, Y, and Z velocity components of P at time t = 0.

Integrating Equations (3.21.4), (3.21.5), and (3.21.6) we obtain the position coordinates of P as:

$$x = v_{0x}t + x_0 \tag{3.21.7}$$

$$y = v_{0y}t + y_0 \tag{3.21.8}$$

$$z = -gt^2/2 + v_{0z}t + z_0 \tag{3.21.9}$$

where x_0, y_0, and z_0 are the X, Y and Z coordinates of P at t = 0.

An examination of Equations (3.21.7) and (3.21.8) shows that the movement of P in both the X and Y directions is linear in t. This means that P moves in a vertical plane. Hence, without loss of generality, let us assume P is moving in the X-Z plane, or equivalently let $y_0 = v_{0y} = 0$. For simplicity, let us further assume that P is launched from the origin O, so that $x_0 = z_0 = 0$, as in Figure 3.21.2.

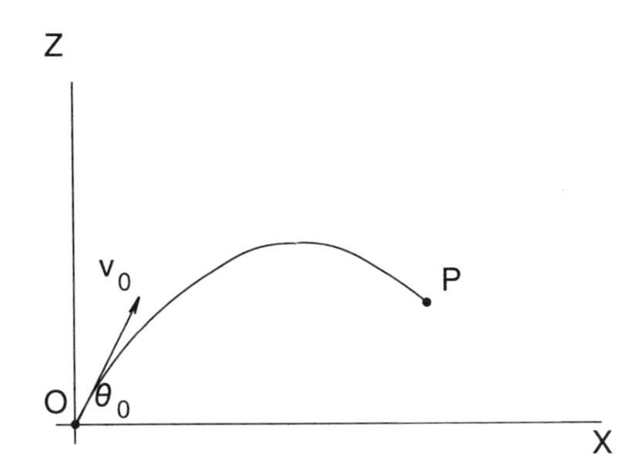

Fig. 3.21.2 Projectile Launched from the Origin in the X-Z Plane

The position Equations (3.21.7), (3.21.8), and (3.21.9) then become:

$$x = v_{0x}t = (v_0 \cos\theta_0)t \tag{3.21.10}$$

$$y = 0 \tag{3.21.11}$$

$$z = -gt^2/2 + v_{0z}t = -gt^2/2 + (v_0 \sin\theta_0)t \tag{3.21.12}$$

where v_0 is the initial speed of P and θ_0 is the launch angle as shown in Figure 3.21.2. By eliminating time t between Equations (3.21.10) and (3.21.12) we obtain the equation describing the path of P as:

$$z = -\left(\frac{g}{2v_0^2 \cos^2\theta_0}\right)x^2 + (\tan\theta_0)x \tag{3.21.13}$$

Thus we see that P moves on a parabola.

In the following paragraphs are special cases of interest.

Case 1. Vertical Projection

Suppose P is projected vertically along the Z-axis. That is, let $\theta_0 = 90°$. Then the position equations are

$$x = y = 0 \tag{3.21.14}$$

and

$$z = -gt^2/2 + v_0 t \tag{3.21.15}$$

The vertical speed of P is then

$$\dot{z} = -gt + v_0 \tag{3.21.16}$$

Observe that the vertical speed of P is zero when t is v_0/g. Observe further that when the vertical speed is zero P has stopped ascending and is about to descend. At this point P has reached its maximum altitude z_{max}. Hence, from Equation (3.21.15) we see that the maximum height h reached by P is

$$h = z_{max} = v_0^2/2g \tag{3.21.17}$$

Case 2. Projection from and Onto a Horizontal Surface

Consider the case of a projectile launched from the origin and returning again to the horizontal as in Figure 3.21.3. Equation (3.21.3) shows that the projectile range x_d is:

$$x_d = (g/2v_0)^2 \sin\theta_0 \cos\theta_0 = (g/v_0^2) \sin 2\theta_0 \tag{3.21.18}$$

(Note that for a horizontal projection surface the range is determined when the elevation z is zero.)

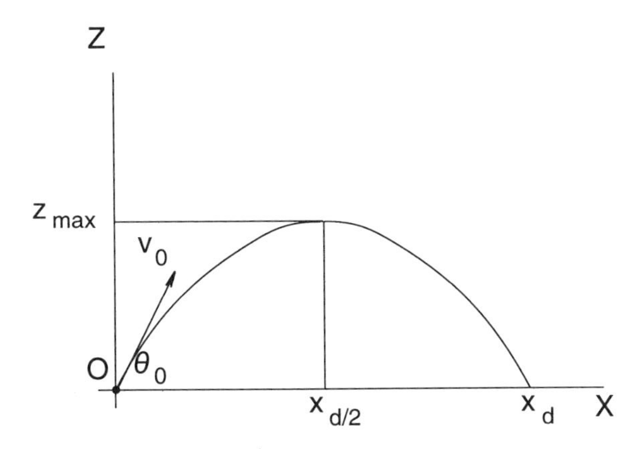

Fig. 3.21.3 Projection Onto a Horizontal Surface

From Equation (3.21.6) the time t_{max} for the vertical component of the velocity $\dot{z}$ to vanish, and hence, for the elevation z to become a maximum is:

$$t_{max} = (v_0 \sin\theta_0)/g \tag{3.21.19}$$

Then from Equation (3.21.12) the maximum elevation z_{max} is:

$$z_{max} = (v_0^2 \sin^2\theta_0)/2g \tag{3.21.20}$$

Finally, observe that from the symmetry of the parabolic curve, we see that the time t_d to reach the full range is

$$t_d = 2t_{max} = (2v_0 \sin\theta_0/g) \tag{3.21.21}$$

Also, the half range distance is:

$$x_d/2 = (g/2v_0^2) \sin 2\theta_0 \tag{3.21.22}$$

Case 3. Maximum Horizontal Range for a Given Launch Speed

A classical problem in projectile motion is to determine the launch angle θ_0 to obtain the maximum projectile range x_{max} for a given launch speed v_0 from a horizontal projection surface. An equivalent problem is to determine θ_0 for a given range x_{max} so as to minimize the launch speed v_0. (Minimizing v_0 minimizes the kinetic energy required to launch the projectile.)

From Equation (3.21.18) we see that the maximum range x_{max} will occur when θ is 45 degrees, or $\pi/4$ radians, and that x_{max} is

$$x_{max} = g/v_0^2 \tag{3.21.23}$$

Figure 3.21.4 shows these results.

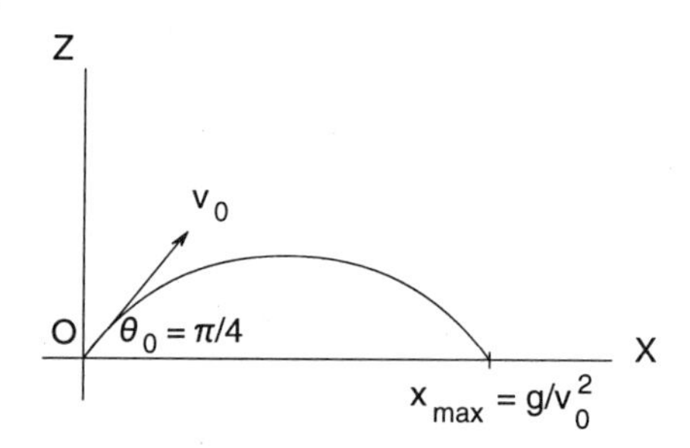

Fig. 3.21.4 Projection Angle for Maximum Horizontal Range

Case 4. Minimum Launch Speed to Reach a Given Target

A related problem is to determine the minimum launch speed v_0 to have P reach a target at a desired point with coordinates $(\hat{x},\hat{z})$. From Equations (3.21.10) and (3.21.12) we see that by eliminating t between the equations and by setting x and z to the target values $(\hat{x},\hat{z})$ the launch speed v_0 is given by the expression

$$v_0^2 = \frac{g\hat{x}^2/2}{\hat{x}\sin\theta_0\cos\theta_0 - \hat{z}\cos^2\theta_0} \tag{3.21.24}$$

Setting the derivative of v_0^2 equal to zero (to obtain the minimum v_0^2) and solving for θ_0 produces

$$\tan 2\theta_0 = -\hat{x}/\hat{z} \tag{3.21.25}$$

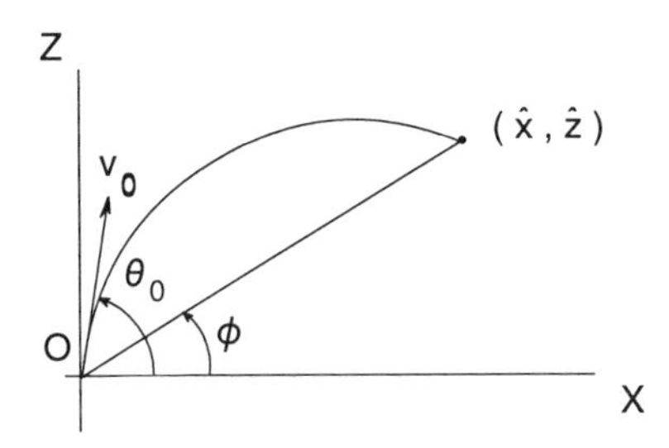

Fig. 3.21.5 Launch Angle to Reach a Given Target

Figure 3.21.5 shows the launch angle and target point. Observe that the results of Equation (3.21.25) may be expressed in terms of the angle ϕ of Figure 3.21.5 as:

$$\tan 2\theta_0 = -\cot\phi \tag{3.21.26}$$

Using the trigonometric identity: $\cot\phi \equiv -\tan(\pi/2 + \phi)$ we have $2\theta_0 = \pi/2 + \phi$ or

$$\theta_0 = \pi/4 + \phi/2 \tag{3.21.27}$$

References

3.1 T. R. Kane, *Analytical Elements of Mechanics*, Vol. 2, Academic Press, New York, 1961.

3.2 T. R. Kane, *Dynamics*, Holt, Rinehart and Winston, New York, 1968.

3.3 T. R. Kane and D. A. Levinson, *Dynamics: Theory and Applications*, McGraw Hill, New York, 1985.

3.4 R. L. Huston, *Multibody Dynamics*, Butterworth-Heinemann, Boston, 1990.

Chapter 4

PARTICLE KINETICS

4.1 Introduction

Whereas *kinematics* is a study of motion without regard to the forces causing the motion, *kinetics* is a study of forces producing the motion. The relations between the forces and the motion (that is, between the kinetics and the kinematics) is the essence of dynamics. These relations, referred to as: "equations of motion," are presented and discussed in the next chapter.

In this chapter we will review the notion of forces and various other concepts useful for developing the equations of motion.

We begin with a review of the most fundamental of these concepts.

4.2 Fundamental Concepts

Table 4.2.1 provides a listing of four fundamental quantities needed in the study of kinematics, kinetics, and dynamics together with somewhat intuitive descriptions of these quantities.

Table 4.2.1 Fundamental Quantities

Quantity	Description	Typical Units	Typical Symbol
Time	Time is difficult to describe, perhaps impossible to define. Intuitively, time is a measure of change generally associated with a uniform physical change such as the rotation of the earth. It is measured on a positive (increasing) linear scale.	Seconds (sec) Minutes (min) Hours (hr)	t

Continued

Table 4.2.1 Continued

Quantity	Description	Typical Units	Symbol
Length	Length is usually defined as the measure of a distance function. Intuitively, the distance between two particles is the magnitude of the difference of position vectors locating the particles.	Inches (in) Feet (ft) Miles (mi) Millimeters (mm) Centimeters (cm) Meters (m) Kilometers (k)	ℓ
Mass	Mass is a measure of the "gravitational strength" of a particle. Near the earth, the mass of a particle is proportional to its weight or physical attraction to the earth.	Slug Pound Mass (lb m) Gram (g) Kilogram (kg)	m
Force	Intuitively, force is a "push" or a "pull" in a given direction. Forces are characterized by vectors and are usually associated with a line of action or point of application	Pounds (lb) Ounces (oz) Newtons (N) Dynes (D)	F

Tables 4.2.2 to 4.2.4 provide conversion factors between the various systems of units for length, mass, and force. [The conversions of time units are generally known (1 hr = 60 min, 1 min = 60 sec).]

Table 4.2.2 Length Conversion Multiplying Factors

To From	Inches	Feet	Miles	Milli-meters	Centi-meters	Meters	Kilo-meters
Inches	1.0	0.0833	1.578×10^{-5}	25.4	2.54	0.0254	2.54×10^{-5}
Feet	12.0	1.0	1.893×10^{-4}	3.045×10^{2}	30.48	0.3048	3.048×10^{-4}
Miles	6.336×10^{4}	5280	1.0	1.609×10^{6}	1.609×10^{5}	1.609×10^{3}	1.609
Millimeters	3.937×10^{-2}	3.28×10^{-3}	6.21×10^{-7}	1.0	0.1	10^{-3}	10^{-6}
Centimeters	0.3937	3.28×10^{-2}	6.21×10^{-6}	10.0	1.0	10^{-2}	10^{-4}
Meters	39.37	3.28	6.21×10^{-4}	10^{3}	10^{2}	1.0	10^{-2}
Kilometers	3.937×10^{4}	3.28×10^{3}	0.621	10^{6}	10^{5}	10^{3}	1.0

Table 4.2.3 Mass Conversion Multiplying Factors

To / From	Pound Mass	Slug	Kilogram	Gram
Pound Mass	1.0	3.105×10^{-2}	0.453	4.53×10^{2}
Slug	32.2	1.0	14.61	1.461×10^{4}
Kilogram	2.204	6.846×10^{-2}	1.0	10^{3}
Gram	2.204×10^{-3}	6.846×10^{-5}	10^{-3}	1.0

Table 4.2.4 Force Conversion Multiplying Factors

To / From	Pounds	Ounces	Dynes	Newtons
Pounds	1.0	16.0	4.448×10^{5}	4.448
Ounces	0.0625	1.0	2.78×10^{4}	0.278
Dynes	2.248×10^{-4}	3.597×10^{-5}	1.0	10^{-5}
Newtons	0.2248	3.597	10^{5}	1.0

The units of Tables 4.2.2, 4.2.3, and 4.2.4 are also related through the units of time. That is

$$\text{force} = \text{mass length}/(\text{time})^2 \tag{4.2.1}$$

Specifically, for the English and metric systems we have

$$1.0 \text{ pounds} = 1.0 \text{ slug feet}/(\text{second})^2 \tag{4.2.2}$$
$$1.0 \text{ Newton} = 1.0 \text{ kilogram meter}/(\text{second})^2 \tag{4.2.3}$$
$$1.0 \text{ Dyne} = 1.0 \text{ gram centimeter}/(\text{second})^2 \tag{4.2.4}$$

See also Chapter 1, Sections 1.5, 1.6, and 1.7.

4.3 Applied (Active) Forces

Forces may be divided into two major categories: Applied (or "active") forces and Inertia (or "passive") forces. Here we consider the applied forces. In the next section we will consider the inertia forces.

Applied forces on a particle arise from entities external to the particle in the form of gravity, contact, and electromagnetic forces.

4.3.1 Gravity Forces

Gravity forces exerted on a particle are forces of attraction, proportional to the mass of the particle, and directed toward the attracting particle or body. To illustrate this consider first two particles P_1 and P_2 with masses m_1 and m_2 and separated by a distance d as in Figure 4.3.1. Then Newton's gravitational theory states that the particles are attracted to one another with a force F given by [see Equation (1.2.2)]:

$$F = G\frac{m_1 m_2}{d^2} \tag{4.3.1}$$

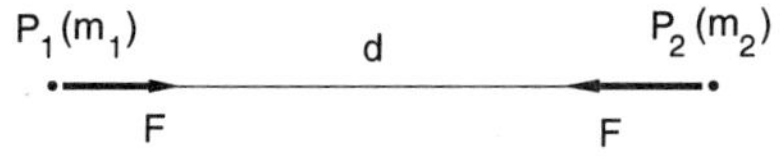

$P_1(m_1)$ d $P_2(m_2)$

F F

Fig. 4.3.1 Two Gravitational Attracting Particles

where G is a universal gravity constant given by

$$G = 6.673 \times 10^{-11}\ m^3/kg\ s^2 = 3.438 \times 10^{-8}\ ft^2/slug\ s^2 \tag{4.3.2}$$

Observe in Equation (4.3.1) that the gravitational force decreases as the square of the distance separating the particles.

Next consider a particle P with mass m in the vicinity of a body B as in Figure 4.3.2. Then the gravitational attraction (or force) exerted on P by B may be determined using Equation (4.3.1) where B is regarded as a set particles occupying differential elements of the volume (or region) occupied by B. Specifically, let **p** be a position vector locating a point of a typical differential element of B as in Figure 4.3.3. Then the gravitational force **F** exerted on P by B may be expressed as:

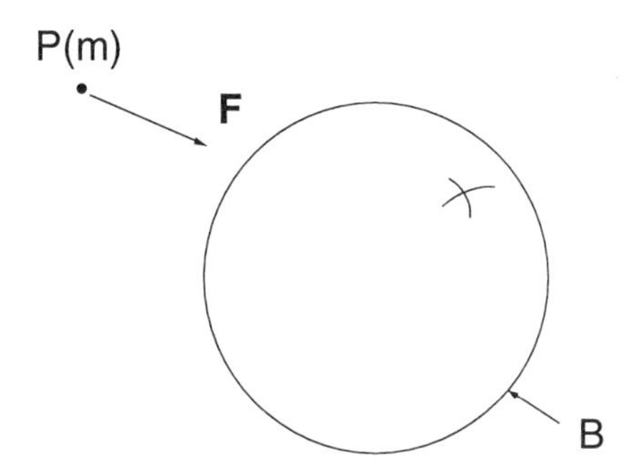

Fig. 4.3.2 A Particle P with Mass m Gravitationally Attracted to a Body B.

$$\mathbf{F} = Gm \int_V \frac{\mathbf{p}\delta dV}{|\mathbf{p}|^3} \tag{4.3.3}$$

where δ is the mass density of B and V represents the region occupied by B.

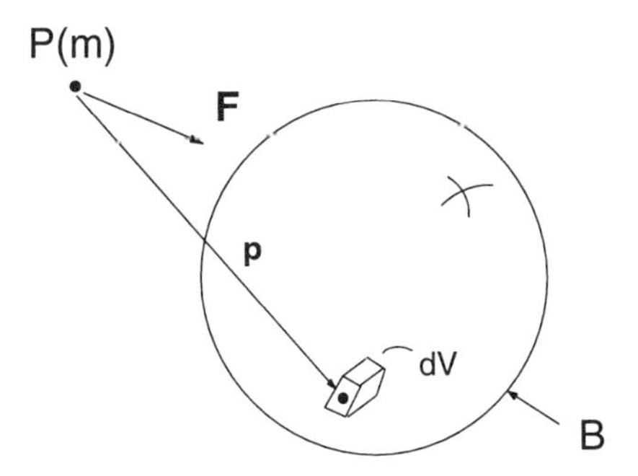

Fig. 4.3.3 Position Vector **p** Locating a Point of B Relative to Particle P

Example 4.3.1 Attraction of a Sphere

To illustrate the application of Equation (4.3.3), consider a particle P with mass m being attracted by a solid homogeneous sphere as depicted in Figure 4.3.4. Let the sphere S have radius a and let P be located at a distance b away from the center O of S. Let b be greater than a so that P is outside S. Let a Cartesian axis system XYZ be located with its origin at O and let P be located on the Z-axis as shown in Figure 4.3.4. Finally, let Q be a point of a differential element of S a distance r from O and positioned relative to X, Y, and Z axes by the orientation angles θ and ϕ, and let ψ be the angle between PQ and the Z-axis. Then due to symmetry the X and Y components of the gravitational force on P are zero and from Equation (4.3.3) the Z component is seen to be:

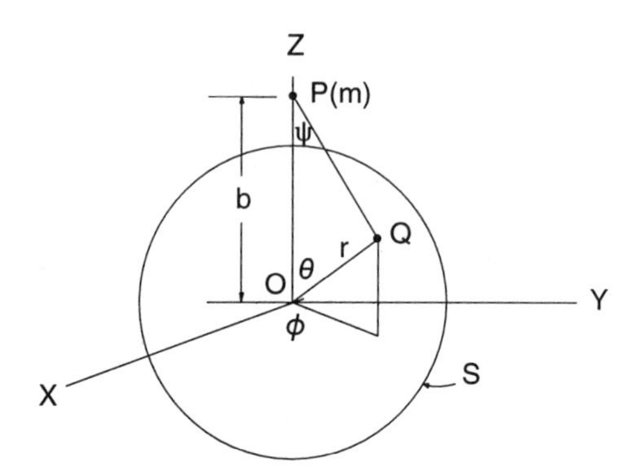

Fig. 4.3.4 A Solid Homogeneous Sphere Attracting a Particle P

$$F_z = -Gm \int_S \frac{\delta \cos \psi}{p^2} dV \tag{4.3.4}$$

where δ is the mass density of S, dV is a differential volume element of S, and p is the distance PQ.

From Figure 4.3.4, p^2, $\cos \psi$, and dV may be expressed as:

$$p^2 = b^2 + r^2 - 2br \cos \theta \tag{4.3.5}$$

$$\cos\psi = (b - r\cos\theta)/p \tag{4.3.6}$$

$$dV = r^2 \sin\theta \, dr \, d\theta \, d\phi \tag{4.3.7}$$

Hence F_z becomes:

$$F_z = -Gm \int_0^{2\pi} \left\{ \int_0^a \left[\int_0^\pi \frac{\delta \gamma^2 (b - \gamma \cos\theta) \sin\theta \, d\theta}{(b^2 + r^2 - 2br\cos\theta)^{3/2}} \right] dr \right\} d\phi \tag{4.3.8}$$

Upon integrating, F_z becomes

$$F_z = -\frac{Gm}{b^2} \delta \frac{4}{3} \pi a^2 = -\frac{GmM}{b^2} \tag{4.3.9}$$

where M is the total mass of S.

This example and the resulting Equation (4.3.9) show that a sphere with a spherically symmetric mass distribution attracts as if it were a point with concentrated mass at its center.

Example 4.3.2 Attraction of the Earth

As an application of Equation (4.3.9), let P be a particle near the surface of the earth. Let the earth have radius R. Then the gravitational attraction W on P by the earth (the "weight" of P) is:

$$W = -\frac{GmM}{R^2} \tag{4.3.10}$$

The earth mass M is approximately 5.976×10^{24} kg, or 4.096×10^{23} slug, and its radius R is approximately 6.371×10^{6} m, or 3960 miles [see Equations (1.3.3) and (1.3.4)]. By substituting into Equation (4.3.10) W is seen to be

$$W = -mg \tag{4.3.11}$$

where g is the gravitational acceleration given by

$$g = 9.81 \text{ m/s}^2 = 32.2 \text{ ft/sec}^2 \tag{4.3.12}$$

Equation (4.3.3) may be applied with other common shapes. Table 4.3.1 provides a summary listing of these results (see References 4.1 and 4.2).

Table 4.3.1 Gravitational Force on a Particle by Bodies of Various Shapes

	Gravitational Force	Figure
Uniform Rod	$F_z = -\frac{2Gm\rho}{h}\sin\frac{\alpha+\beta}{2}\sin\frac{\alpha-\beta}{2}$ $F_y = -\frac{2Gm\rho}{h}\sin\frac{\alpha-\beta}{2}\cos\frac{\alpha+\beta}{2}$	
Uniform Rod	$\mathbf{F}_x = -\frac{GMm}{a(a+\ell)}$	

Continued

Table 4.3.1 Continued

	Gravitational Force	Figure
Semicircular Wire	$\mathbf{F}_y = -\dfrac{2GMm}{\pi R^2}$	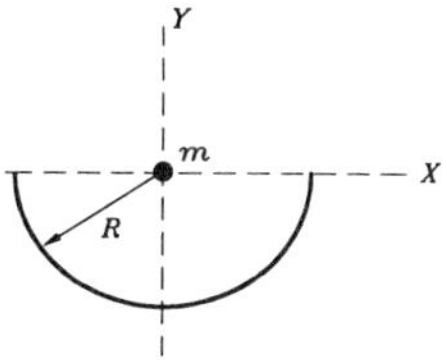
Uniform Circular Wire	$\mathbf{F}_z = -\dfrac{GMmh}{(R^2 + h^2)^{3/2}}$	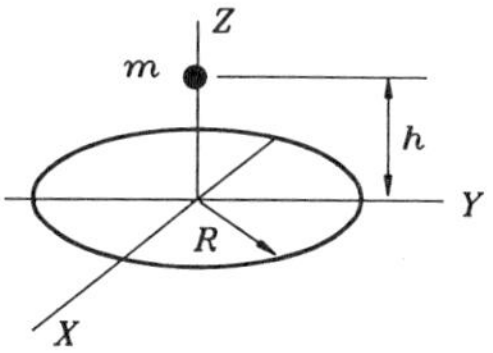
Disk	$\mathbf{F}_z = -2\pi G m \rho \left[1 - \dfrac{h}{\sqrt{R^2 + h^2}}\right]$	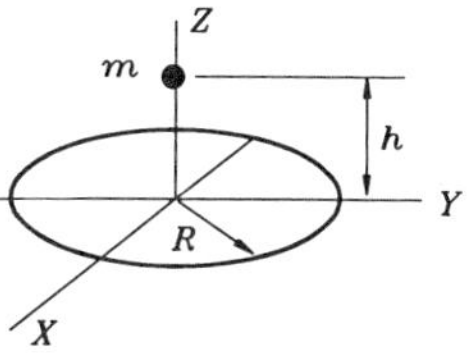
Hemisphere Shell	$\mathbf{F}_z = -\dfrac{GMm}{2R^2}$	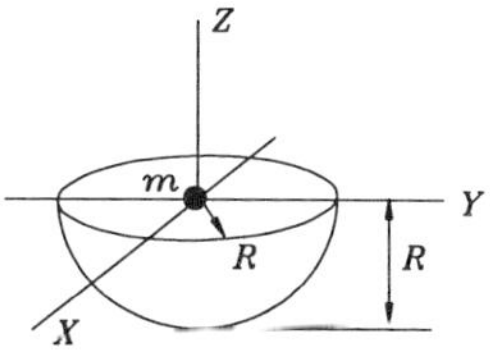
Solid Hemisphere	$\mathbf{F}_z = -\dfrac{GMm}{R^2}(\sqrt{2} - 1)$	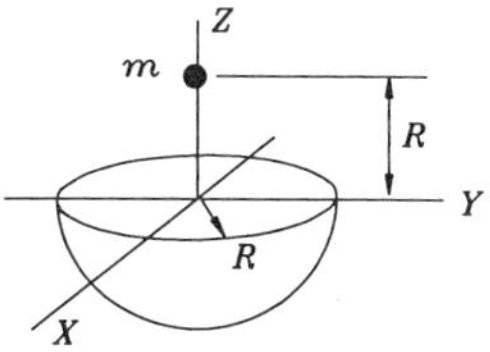

Continued

Table 4.3.1 Continued

	Gravitational Force	Figure
Solid Hemisphere	$\mathbf{F}_z = -\frac{3}{2}\frac{GMm}{R^2}$	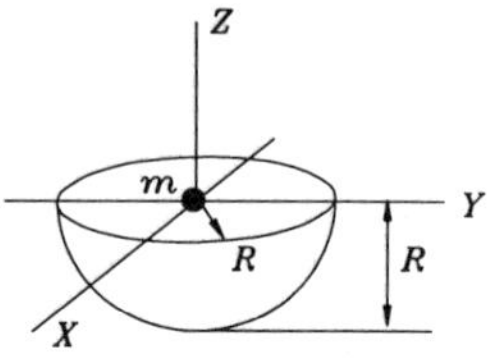
Solid Sphere	$\mathbf{F}_z = -\frac{GMm}{b^2} \quad b \geq R$	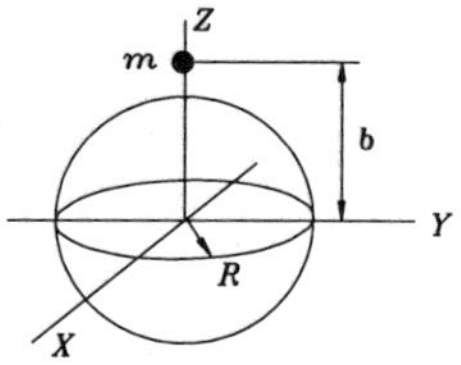
Solid Sphere	$\mathbf{F}_z = -\frac{GMm}{R^2}\left(\frac{h}{R}\right) \quad h < R$	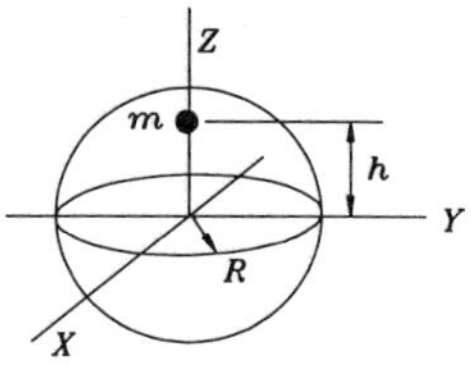
Solid Right Circular Cone	$\mathbf{F}_z = -2\pi Gm\rho h\left[1 - \frac{h}{\sqrt{R^2 + h^2}}\right]$	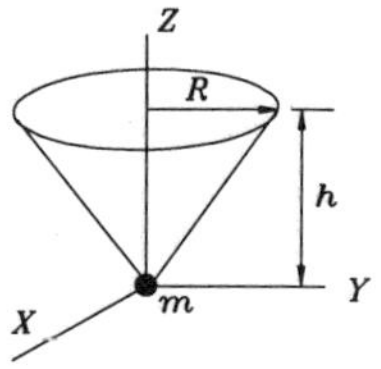
Uniform Open-Ended Cylindrical Shell	$\mathbf{F}_z = -\frac{GMm}{h}\left[\frac{1}{\sqrt{R^2 + b^2}} - \frac{1}{\sqrt{R^2 + (h - b)^2}}\right]$	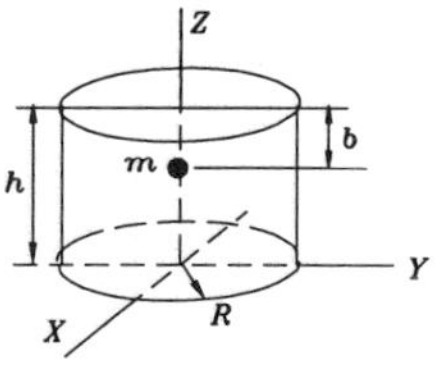

Continued

Table 4.3.1 Continued

	Gravitational Force	Figure
Cylindrical Shell	$F_z = -2\pi G m \rho \left[b - a - \sqrt{b^2 + h^2} + \sqrt{a^2 + h^2} \right]$	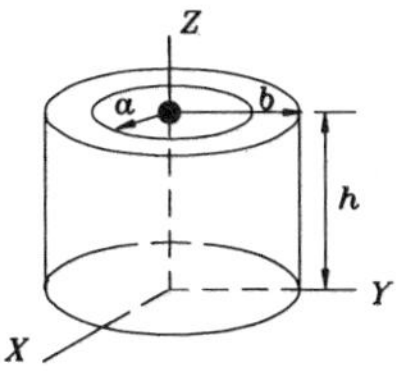
Solid Cylinder	$F_z = -2\pi G M \rho \left[R^2 - \sqrt{R^2 + h^2} + h \right]$	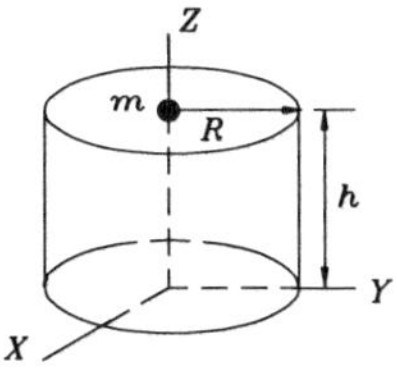
Solid Cylinder	$F_z = -2\pi G M \rho \left[h - \sqrt{(h + a)^2 + R^2} + \sqrt{a^2 + R^2} \right]$	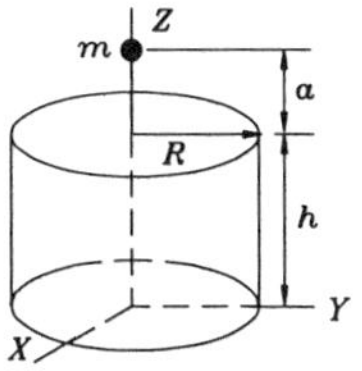
Spherical Shell	$F_z = -\dfrac{GMm}{b^2} \qquad b > R$	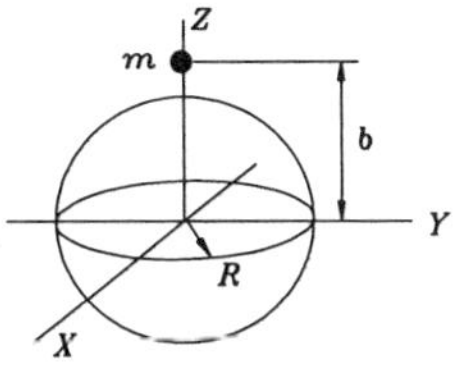
Spherical Shell	$F = 0 \qquad h < R$	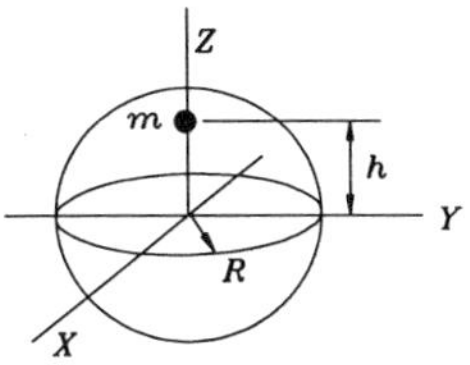

Example 4.3.3 Gravitational Force Magnitudes

For particles and objects near the earth's surface, the dominant gravitational force is due to the earth itself. That is, compared to the earth's attraction, the attraction from other particles and objects is negligible. To illustrate this consider two uniform spheres, each weighing 1000 lb with their centers separated by 2.0 ft as in Figure 4.3.5.

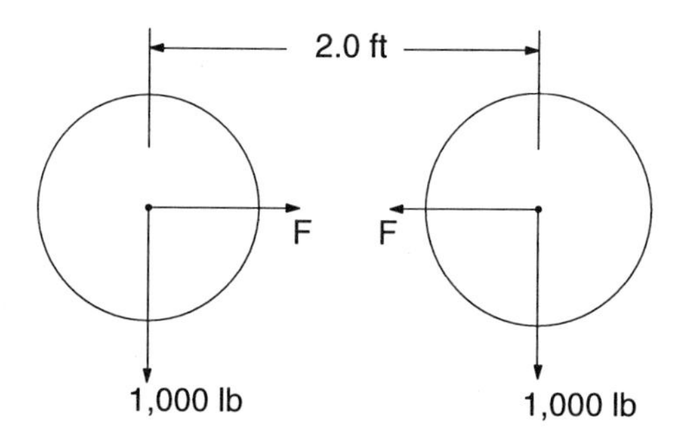

Fig. 4.3.5 Gravitation Attraction of Two Spheres

Then the gravitational attraction F may be determined using Equation (4.3.1) by noting that spheres attract as point masses at their centers (see Example 4.3.2). Specifically, with weights of 1000 lb each sphere has a mass of 1000/32.2 or 31.06 slug. Hence the gravitational force F is

$$F = G\frac{m^2}{d^2} = (3.438 \times 10^{-8})(31.06)^2/(2)^2$$

$$= 8.29 \times 10^{-6} \text{ pounds} \qquad (4.3.13)$$

4.3.2 Contact Forces

Next to gravitational forces, contact forces are the most common of applied forces. Unlike gravity forces, which are constant in most applications, contact forces are directly dependent upon the application.

Consider a particle P in contact with a surface S as in Figure 4.3.6. Let P move along S. Then S will exert forces on P depending upon the physical properties of P and S. These forces can usually be represented by a single force **C** passing through P. [That is, there is usually no appreciable couple torque since P has no significant dimension (see Example 2.10.8)]. **C** is sometimes called the "traction" force.

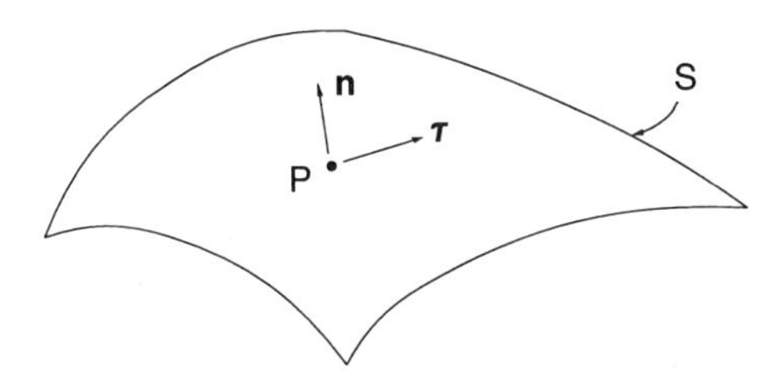

Fig. 4.3.6 A Particle P Moving on a Surface S

It is usually convenient to express **C** in terms of components tangent and normal to S. Specifically, let **n** and $\boldsymbol{\tau}$ be unit vectors normal and tangent to S at P. Then **C** may be expressed as:

$$\mathbf{C} = N\mathbf{n} + F\boldsymbol{\tau} \tag{4.3.14}$$

where N is called the "normal" force and F is called the "friction" force.

If P is sliding on S, F is approximately proportional to N and may be expressed as:

$$F = \mu N \tag{4.3.15}$$

where μ is called the "coefficient of friction" or "drag factor," with values between zero and one. For "smooth" surfaces μ is zero and then the contact force **C** is normal to S.

Spring forces are a major source of contact forces in mechanical systems — particularly models of mechanical systems. Springs are used to model flexibility effects. Usually these are linear springs where the force exerted is proportional to the extension or compression of the spring. Specifically, consider a spring exerting a force on a particle P as in Figure 4.3.7. Let the X-axis be parallel to the spring

Fig. 4.3.7 Spring Exerting a Force on a Particle P

and let x represent the displacement of P away from its equilibrium position. Then for a linear spring the force F exerted on P may be expressed as:

$$F = kx \tag{4.3.16}$$

where k is a constant, sometimes called the "spring modulus."

For nonlinear springs Equation (4.3.16) may be expressed as

$$F = f(x) \tag{4.3.17}$$

where f(x) describes the spring non-linearity. Typically, f(x) may have the form

$$f(x) = k_1 x - k_2 x^3 \tag{4.3.18}$$

where k_1 and k_2 are constants.

Damper forces are also a major source of contact forces in models of mechanical systems. Damping forces are related to the speed of a particle. Consider the damper model of Figure 4.3.8. Again, let the X-axis be parallel to the damper

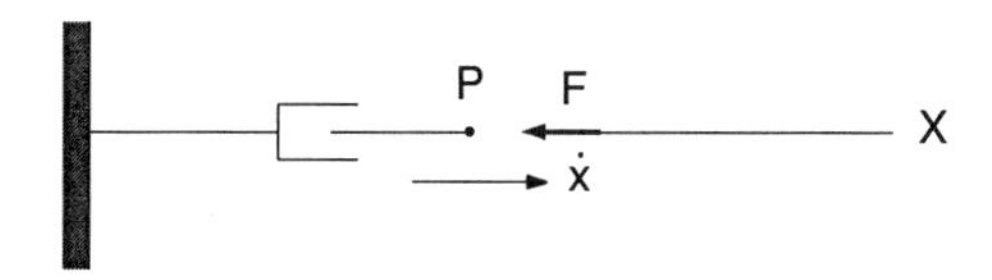

Fig. 4.3.8 Damper Exerting a Force on a Particle P

and let $\dot{x}$ measure the speed of a particle P in line with the damper. Then the force F exerted on P by the damper is typically represented as

$$F = -c\dot{x} \tag{4.3.19}$$

where c is a constant.

The linear damping represented by Equation (4.3.19) is often called "coulomb damping." Unfortunately, however, the physical damping occurring in mechanical systems is usually non-linear requiring a more general representation such as

$$F = g(x,\dot{x}) \tag{4.3.20}$$

4.4 Inertia (Passive) Forces

Inertia (or "passive") forces occur when particles with masses are accelerated in an inertial reference frame. An "inertial reference frame" is defined as a fixed, or non-moving, reference frame, attached to "fixed stars."

There are philosophical and cosmological questions as to whether inertial reference frames actually exist, since the entire universe appears to be evolving and moving. Hence, there may be no "fixed stars."

To avoid these difficulties, analysts have defined inertial reference frames as reference frames in which Newton's laws are valid. Inertial reference frames are thus sometimes called Newtonian reference frames.

Whether absolute inertial reference frames exist or not, for most problems of practical importance, the earth may be regarded as an inertial reference frame. Experimental evidence has shown Newton's laws to be nearly valid for mechanical systems near the surface of the earth.

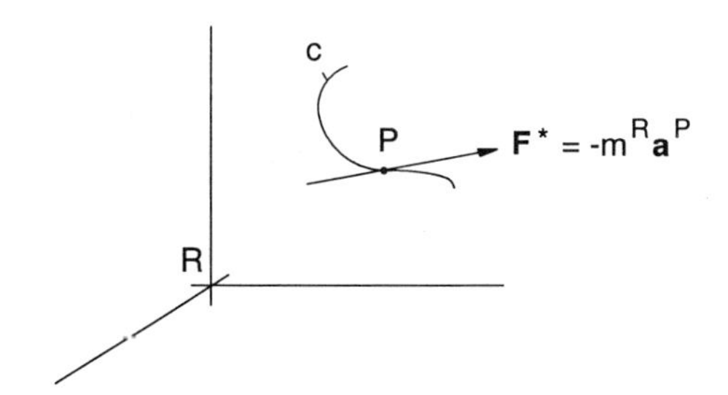

Fig. 4.4.1 Inertia Force Acting on a Particle in an Inertial Reference Frame R

Let R be an inertial reference frame and let P be a particle with mass m moving on a curve C in R as in Figure 4.4.1. Then the inertia force $\mathbf{F}^*$ acting on P is defined as

$$\mathbf{F}^* = -m\,{}^R\mathbf{a}^P \tag{4.4.1}$$

where ${}^R\mathbf{a}^P$ is the acceleration of P in R.

Inertia forces are similar to gravitational forces and are sometimes identified as such in cosmological theories.

4.5 Generalized Forces — Kinematic Preliminaries

It is convenient in the analysis of mechanical systems to be able to neglect some of the components of forces and still answer the questions stimulating the analysis. Expressed another way, it is often desired to focus only upon some of the components of the forces, particularly the projection of the forces in the directions of motion of a system. To obtain these components and projections it is convenient to introduce and define the concept of "partial velocity vectors," as in the subsequent paragraphs.

4.5.1 Coordinates

A coordinate is a parameter, variable, or mathematical entity, used to define the location of points, or alternatively to define the configuration of a mechanical system. Coordinates are not unique. Consider, for example, the point P moving in a plane as in Figure 4.5.1. The position of P may be defined by the coordinates (x,y) or (r,θ). These coordinates are related by the expressions:

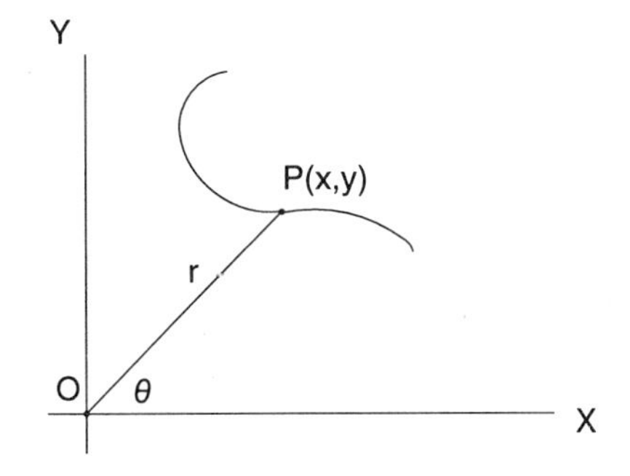

Figure 4.5.1 A Point P Moving in a Plane

$$x = r\cos\theta \qquad y = r\sin\theta \tag{4.5.1}$$

and

$$r = \sqrt{x^2+y^2} \qquad \theta = \tan^{-1} y/x \tag{4.5.2}$$

4.5.2 Constraints

A restriction in the movement of a particle or system is called a "constraint." For example, if a particle or point is restricted to move in a plane as in Figure 4.5.1, then the constraint might be expressed as

$$z = 0 \tag{4.5.3}$$

Constraints are often classified as being geometric (or "holonomic") or kinematic (or "non-holonomic"). Equation (4.5.3) is an example of a geometric constraint. An example of a kinematic or non-holonomic constraint is "rolling." That is, a body B is said to be rolling on a surface S if the contact point C of B on S has zero velocity (see Figure 4.5.2). Then the rolling condition is

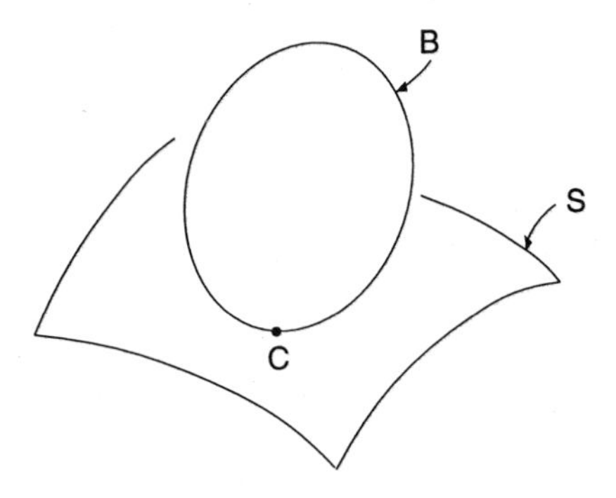

Fig. 4.5.2 A Body B Rolling on a Surface S

$$ {}^{S}\mathbf{v}^{C} = 0 \tag{4.5.4} $$

Equation (4.5.4) is a kinematic, or non-holonomic, constraint.

Suppose a mechanical system has n coordinates q_r $(r = 1, \ldots, n)$. then geometric, or holonomic, constraints have the form:

$$ f(q_r, t) = 0 \tag{4.5.5} $$

Kinematic, or non-holonomic, constraints have the form:

$$ f(q_r, \dot{q}_r, \ddot{q}_r, \ldots, t) = 0 \tag{4.5.6} $$

If a non-holonomic constraint does not involve derivatives of second or higher order and if it is linear in the $\dot{q}_r$, it is called a "simple non-holonomic constraint" and the constrained mechanical system is then called a "simple non-holonomic system."

If a mechanical system is not constrained or if it has only geometric, or holonomic, constraints, it is called a "holonomic system."

4.5.3 Degrees of Freedom

Suppose a mechanical system has n coordinates and m constraint equations, then the difference $n - m$ is the "number of degrees of freedom" of the system.

4.5.4 Partial Velocity Vectors

Suppose a mechanical system S has n coordinates q_r $(r = 1, \ldots, n)$. Let **p** be a position vector locating a particle P of S in an inertial reference frame R (see Figure 4.5.3). Then in general **p** will depend upon the q_r and t having the form

$$ \mathbf{p} = \mathbf{p}(q_r, t) \tag{4.5.7} $$

The velocity of P in R may then be obtained by differentiating in Equation (4.5.7) leading to the expression

$$ {}^{R}\mathbf{v}^{P} = \frac{\partial \mathbf{p}}{\partial t} + \sum_{r=1}^{n} \frac{\partial \mathbf{p}}{\partial q_r} \dot{q}_r \tag{4.5.8} $$

The terms $\partial \mathbf{p}/\partial t$ and $\partial \mathbf{p}/\partial q_r$ are called "partial velocity vectors" relative to t and q_r respectively. They are analogous to base vectors employed in continuum mechanics. They play a central role in the development of generalized forces and generalized mechanics.

In Equation (4.5.8) it is convenient to introduce the following notation:

$$ \mathbf{v}_{\dot{q}_r} = \frac{\partial \mathbf{p}}{\partial q_r} = \frac{\partial \mathbf{v}}{\partial \dot{q}_r} \qquad \mathbf{v}_t = \frac{\partial \mathbf{p}}{\partial t} \tag{4.5.9} $$

Equation (4.5.8) then becomes

$$ \mathbf{v} = \mathbf{v}_t + \sum_{r=1}^{n} \mathbf{v}_{\dot{q}_r} \dot{q}_r \tag{4.5.10} $$

Observe further that by differentiating with respect to q_s in Equation (4.5.8) and by interchanging the order of mixed partial derivatives we have

$$ \frac{\partial \mathbf{v}}{\partial q_s} = \frac{d}{dt}\left(\frac{\partial \mathbf{p}}{\partial q_s}\right) = \frac{d}{dt}(\mathbf{v}_{\dot{q}_s}) \tag{4.5.11} $$

Finally, observe that by computing the derivative of the projection of the velocity on the partial velocity $(\mathbf{v} \bullet \mathbf{v}_{\dot{q}_r})$ we obtain:

$$\frac{d}{dt}(\mathbf{v} \bullet \mathbf{v}_{\dot{q}_r}) = \frac{d\mathbf{v}}{dt} \bullet \mathbf{v}_{\dot{q}_r} + \mathbf{v} \bullet \frac{d}{dt}\mathbf{v}_{\dot{q}_r}$$
$$= \mathbf{a} \bullet \mathbf{v}_{\dot{q}_r} + \mathbf{v} \bullet \frac{\partial \mathbf{v}}{\partial q_r} \qquad (4.5.12)$$

where as before **a** is the acceleration of P in R.

Equation (4.5.12) may readily be rewritten and rearranged into the following form:

$$\mathbf{a} \bullet \mathbf{v}_{\dot{q}_r} = \frac{d}{dt}\left[\frac{\partial}{\partial \dot{q}_r}(v^2/2)\right] - \frac{\partial}{\partial q_r}(v^2/2) \qquad (4.5.13)$$

Table 4.5.1 provides a summary listing of the foregoing results.

Table 4.5.1 Formulas Related to Partial Velocity Vectors

Name	Expression
Partial Velocity with Respect to q_r	$\frac{\partial \mathbf{p}}{\partial q_r} = \frac{\partial \mathbf{v}}{\partial \dot{q}_r} = \mathbf{v}_{\dot{q}_r}$
Derivative of Partial Velocity Vector	$\frac{d\mathbf{v}_{\dot{q}_r}}{dt} = \frac{\partial \mathbf{v}}{\partial q_r}$
Projection of Acceleration along the Partial Velocity Vectors	$\mathbf{a} \bullet \mathbf{v}_{\dot{q}_r} = \frac{d}{dt}\left[\frac{\partial}{\partial \dot{q}_r}(v^2/2)\right] - \frac{\partial}{\partial q_r}(v^2/2)$

4.6 Generalized Applied (or Active) Forces

Consider a force **F** applied to a particle P of a mechanical S in an inertial frame R as in Figure 4.6.1.

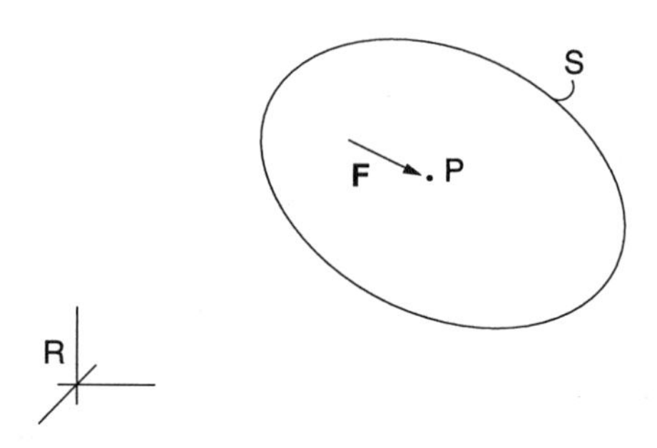

Fig. 4.6.1 A Force **F** on a Particle P of a Mechanical System S

Let S have n degrees of freedom represented by the coordinates q_r (r = 1,...,n). Then the generalized applied force F_{q_r} for the coordinate q_r is defined as

$$F_{q_r} = \mathbf{F} \bullet \mathbf{v}_{\dot{q}_r} \tag{4.6.1}$$

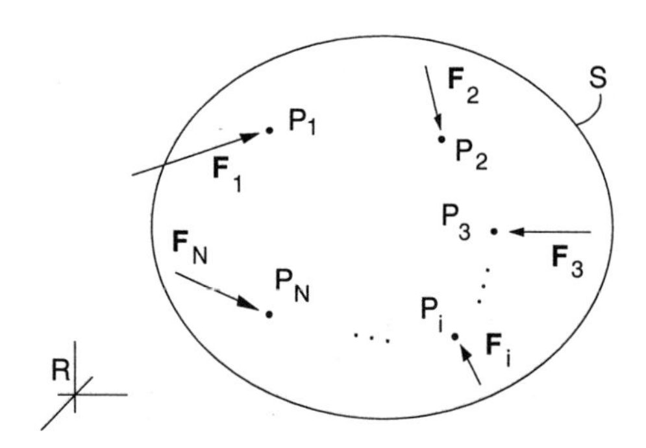

Fig. 4.6.2 A Set of Forces Acting on a Set of Particles of a Mechanical System

Next, consider a set of particles P_i $(i = 1,\ldots,N)$ each experiencing an applied force as in Figure 4.6.2. Then the generalized applied force F_{q_r} is defined as the sum of the generalized forces for each force. That is,

$$F_{q_r} = \sum_{i=1}^{N} \mathbf{F}_i \cdot \mathbf{v}_{\dot{q}_r}^{P_i} \tag{4.6.2}$$

Observe in Equation (4.6.1) that the generalized force F_{q_r} may be zero even though the force $\mathbf{F}$ is not zero. This can happen if $\mathbf{F}$ is perpendicular to $\mathbf{v}_{\dot{q}_r}$ or if $\mathbf{v}_{\dot{q}_r}$ is zero. When this occurs, $\mathbf{F}$ is said to be "non-contributing" or "non-working."

If non-contributing or non-working forces can be identified in dynamical analyses, then those forces may be neglected in the analysis, thus simplifying the analytical effort. Table 4.6.1 provides a listing of conditions where forces are non-contributing or non-working [4.3, 4.4].

Table 4.6.1 Conditions Producing Non-Contributing Forces for Generalized Active Forces (i.e., $F_{q_r} = 0$)

1. Force $\mathbf{F}$ is zero
2. Force $\mathbf{F}$ is perpendicular to partial velocity $\mathbf{v}_{\dot{q}_r}$. (This can occur with forces exerted over smooth surfaces internal to a mechanical system or by smooth surfaces whose motion is a prescribed function of time [4.3].)
3. Partial velocity $\mathbf{v}_{\dot{q}_r}$ is zero. (This can occur if the particle velocity is zero, as the contact point of a rolling body.)

Finally, for two common applied forces, namely gravity forces and spring forces, the contribution to the generalized forces may be developed in greater detail. The following two examples illustrate this.

Example 4.6.1 <u>Contribution of Gravity Forces to Generalized Applied Forces</u>

Consider a particle P with mass m moving on a curve C in a gravitational field as depicted in Figure 4.6.3. Let P be part of a mechanical system S with n degrees of freedom represented by coordinates q_r $(r = 1, \ldots, n)$.

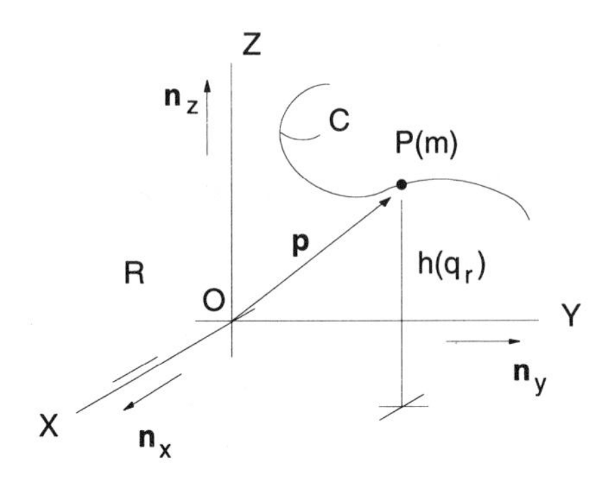

Fig. 4.6.3 A Particle P Moving in a Gravitational Field

Let the Z-axis be vertical and let h measure the elevation of P above a fixed level such as the X-Y plane as shown in Figure 4.6.3. Then the position of P in R may be expressed as

$$\mathbf{p} = x\mathbf{n}_x + y\mathbf{n}_y + z\mathbf{n}_z \tag{4.6.3}$$

where z is identified with h which in turn is a function of the system coordinates q_r. By differentiating, the velocity of P is

$$\mathbf{v} = \dot{x}\mathbf{n}_x + \dot{y}\mathbf{n}_y + \dot{z}\mathbf{n}_z \tag{4.6.4}$$

Then the partial velocity of P relative to q_r is

$$\mathbf{v}_{\dot{q}_r} = \frac{\partial \dot{x}}{\partial \dot{q}_r}\mathbf{n}_x + \frac{\partial \dot{y}}{\partial \dot{q}_r}\mathbf{n}_y + \frac{\partial \dot{z}}{\partial \dot{q}_r}\mathbf{n}_z \tag{4.6.5}$$

The weight (or gravity) force $\mathbf{w}$ on P is

$$\mathbf{w} = -mg\mathbf{n}_z \tag{4.6.6}$$

From Equations (4.6.1), (4.6.5), and (4.6.6) the generalized applied force F_{q_r} associated with the coordinate q_r is:

$$F_{q_r} = \mathbf{w} \bullet \mathbf{v}_{\dot{q}} = -mg\frac{\partial \dot{z}}{\partial \dot{q}_r} \tag{4.6.7}$$

However, since z is $h(q_r)$, $\dot{z}$ may be expressed as

$$\dot{z} = \frac{dh}{dt} = \sum_{r=1}^{n} \frac{\partial h}{\partial q_r}\dot{q}_r \tag{4.6.8}$$

Hence, $\partial \dot{z}/\partial \dot{q}_r$ is $\partial h/\partial q_r$ and F_{q_r} becomes

$$F_{q_r} = -mg\,\partial h/\partial q_r \tag{4.6.9}$$

Example 4.6.2 Contribution of Spring Forces to Generalized Applied Forces

Consider a spring connecting particles P_1 and P_2 of a mechanical system S as depicted in Figure 4.6.4. Let S have n degrees of freedom represented by the coordinates q_r $(r = 1,\dots,n)$. Let the spring have natural length ℓ and extension x as shown and let **n** be a unit vector along the axis of the spring. Then in general x will depend upon the q_r that is, $x = x(q_r)]$.

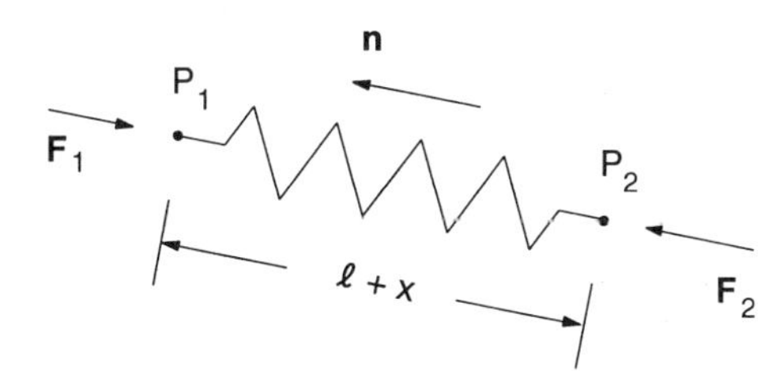

Fig. 4.6.4 A Spring Connecting Particles of $\mathbf{F}_1$ in a Mechanical System

The spring will exert forces $\mathbf{F}_1$ and $\mathbf{F}_2$ on P_1 and P_2 where $\mathbf{F}_1$ and $\mathbf{F}_2$ may be expressed as:

$$\mathbf{F}_1 = -\mathbf{F}_2 = -f(x)\mathbf{n} \tag{4.6.10}$$

where f(x) represents the spring modulus.

Observe that the velocities of P_1 and P_2 in an inertial frame R may be related by the expression

$$\mathbf{v}^{P_1} = \mathbf{v}^{P_2} + \mathbf{v}^{P_1/P_2} \quad (4.6.11)$$

where the relative velocity $\mathbf{v}^{P_1/P_2}$ may be expressed as

$$\mathbf{v}^{P_1/P_2} = \dot{x}\mathbf{n} + (\quad)\mathbf{n}_\perp \quad (4.6.12)$$

where $(\quad)\mathbf{n}_\perp$ is the projection of $\mathbf{v}^{P_1/P_2}$ perpendicular to **n**. By differentiating with respect to $\dot{q}_r$ in Equation (4.6.11) we obtain

$$\mathbf{v}_{\dot{q}_r}^{P_1} = \mathbf{v}_{\dot{q}_r}^{P_2} + \mathbf{v}_{\dot{q}_r}^{P_1/P_2} \quad (4.6.13)$$

where from Equation (4.6.12) $\mathbf{v}_{\dot{q}_r}^{P_1/P_2}$ is

$$\mathbf{v}_{\dot{q}_r}^{P_1/P_2} = \frac{\partial \dot{x}}{\partial \dot{q}_r}\mathbf{n} + \frac{\partial(\quad)}{\partial \dot{q}_r}\mathbf{n}_\perp \quad (4.6.14)$$

Since x is a function of the q_r, $\dot{x}$ may be expressed as

$$\dot{x} = \sum_{r=1}^{n} \frac{\partial x}{\partial q_r}\dot{q}_r \quad (4.6.15)$$

Then $\partial\dot{x}/\partial\dot{q}_r$ is

$$\frac{\partial \dot{x}}{\partial \dot{q}_r} = \frac{\partial x}{\partial q_r} \quad (4.6.16)$$

From Equation (4.6.2), the contribution of the spring forces $\mathbf{F}_1$ and $\mathbf{F}_2$ to the generalized active force F_{q_r} is

$$F_{q_r} = \mathbf{F}_1 \bullet \mathbf{v}_{\dot{q}_r}^{P_1} + \mathbf{F}_2 \bullet \mathbf{v}_{\dot{q}_r}^{P_2} \tag{4.6.17}$$

By substituting from Equation (4.6.10), (4.6.13), and (4.6.14) into (4.6.17) we have

$$\begin{aligned} F_{q_r} &= -f(x)\mathbf{n} \bullet (\mathbf{v}_{\dot{q}_r}^{P_2} + \mathbf{v}_{\dot{q}_r}^{P_1/P_2}) + f(x)\mathbf{n} \bullet \mathbf{v}_{\dot{q}_r}^{P_2} \\ &= -f(x)\mathbf{n} \bullet \mathbf{v}_{\dot{q}_r}^{P_1/P_2} \\ &= -f(x)\frac{\partial \dot{x}}{\partial \dot{q}_r} \end{aligned} \tag{4.6.18}$$

Then from Equation (4.6.16) F_{q_r} becomes:

$$F_{q_r} = -f(x)\,\partial x/\partial q_r \tag{4.6.19}$$

Observe that if the spring is linear such that f(x) is kx (k being the spring constant) then F_{q_r} is

$$F_{q_r} = -kx\,\partial x/\partial q_r \tag{4.6.20}$$

4.7 Generalized Inertia (or Passive) Forces

The definition of generalized inertia forces is similar to and directly analogous to that of generalized active forces. Consider a particle P with mass m which is part of a mechanical system S with n degrees of freedom represented by the coordinates q_r $(r = 1, \ldots, n)$. Let P be moving in an inertial reference frame R as in Figure 4.7.1. Then if P has acceleration **a** in an inertial reference frame R, P

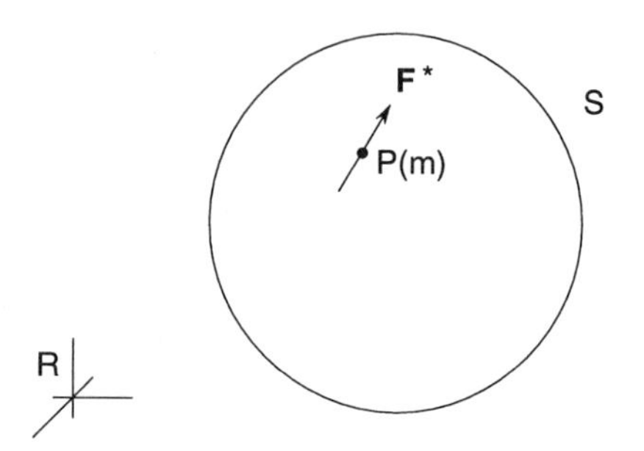

Fig. 4.7.1 A Particle P Moving in an Inertial Frame R with Inertia Force **F***

will experience an inertia force **F** where **F** is $-m\mathbf{a}$ [see Equation (4.4.1)]. The generalized inertia force $F^*_{q_r}$ for the coordinate q_r is then defined as:

$$F^*_{q_r} = \mathbf{F}^* \bullet \mathbf{v}_{\dot{q}_r} \tag{4.7.1}$$

In view of Equation (4.4.1) $F^*_{q_r}$ may also be written as

$$F^*_{q_r} = -m\mathbf{a} \bullet \mathbf{v}_{\dot{q}_r} \tag{4.7.2}$$

Next, consider a set of particles P_i $(i = 1,\ldots,N)$ of a mechanical system S with n degrees of freedom with coordinates q_r $(r = 1,\ldots,n)$ moving in an inertial frame R, as in Figure 4.7.2. Let the particles have masses m_i $(i = 1,\ldots,N)$ respectively. Then if a typical particle P_i has an acceleration $\mathbf{a}_i$ in R, it will experience an inertia force $\mathbf{F}^*_i$, where $\mathbf{F}^*_i$ is $-m\mathbf{a}_i$. The generalized inertia force $F^*_{q_r}$ for the coordinate q_r is then the sum of the generalized inertia force for each of the particles. That is,

$$F^*_{q_r} = \sum_{i=1}^{N} \mathbf{F}^*_i \bullet \mathbf{v}^{P_i}_{\dot{q}_r} \tag{4.7.3}$$

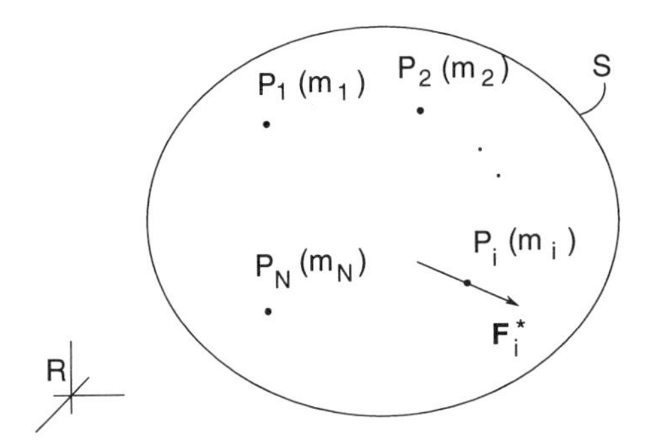

Fig. 4.7.2 Particles P_i Moving in an Inertial Frame R with Inertia Forces

In view of Equation (4.4.1) $F^*_{q_r}$ may also be written as:

$$F^*_{q_r} = -\sum_{i=1}^{N} m_i \mathbf{a}_i \bullet \mathbf{v}^{P_i}_{\dot{q}_r} \tag{4.7.4}$$

4.8 Associated Applied (or Active) Kinetic Quantities

There are several associated applied kinetic quantities — namely impulse, potential energy, and work — which are useful in dynamical analyses. These are defined in this section. Associated inertial kinetic quantities are discussed in the following section.

4.8.1 Impulse

Let **F** be an applied force. The "impulse" **I** of **F** is defined as the time integral of **F**:

$$\mathbf{I} = \int_{t_1}^{t_2} \mathbf{F}\,dt \tag{4.8.1}$$

where the time interval (t_1, t_2) is arbitrary and may be selected for convenience in a given analysis.

Impulses are primarily useful in working with "impulsive forces" which in turn are relatively large forces acting over a short time interval. Such forces typically occur in impact, as in collision of particles and bodies.

Example 4.8.1 Impulsive Forces

Figure 4.8.1 shows the time profile of the magnitude of a typical impulsive force **F**, acting in the direction of a unit vector **n**. Then the impulse of **F** is:

$$\mathbf{I} = \int_0^{\hat{t}} \mathbf{F}\,dt = (1/2)F_0\hat{t}\,\mathbf{n} \tag{4.8.2}$$

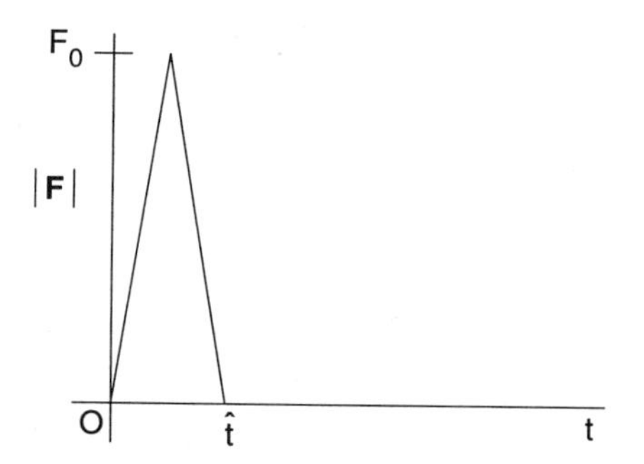

Fig. 4.8.1 Time Profile of the Magnitude of an Impulsive Force

4.8.2 Potential Energy

Potential energy is a second useful applied kinetic quantity. Let F_{q_r} be a generalized applied force of a mechanical system S having n degrees of freedom with coordinates q_r $(r = 1, \ldots, n)$. Then a potential energy P of S is defined as a function $P(q_r)$ such that

$$F_{q_r} = -\frac{\partial P}{\partial q_r} \quad (r = 1,\ldots,n) \tag{4.8.3}$$

For any given system S, a potential energy function may not exist. If it does exist, however, the system is said to be "conservative."

Potential energy is not unique. From the definition of Equation (4.8.3) we see that if P is a potential energy then so also is $P + c$ where c is a constant.

Potential energy is often constructed as the sum of potential functions which in turn are associated with particular classes of forces. Specifically, for gravity and spring forces, potential functions are readily developed, as in the following examples.

Example 4.8.2 Potential Function for Gravity Forces

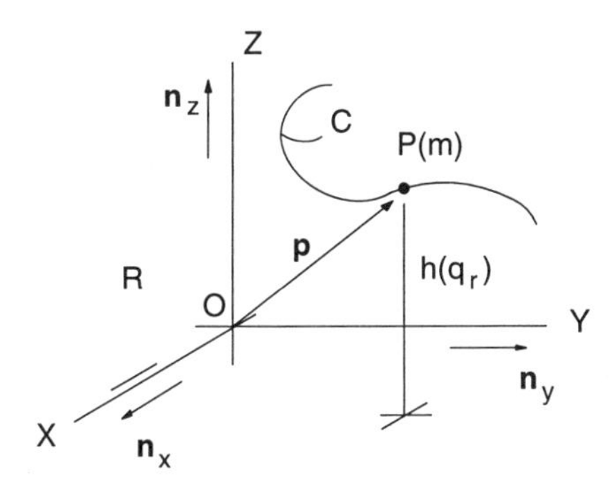

Fig. 4.8.2 A Particle P Moving in a Gravitational Field

Let P be a particle with mass m belonging to a mechanical system S having n degrees of freedom and coordinates q_r $(r = 1,\ldots,n)$ as in Example 4.6.1. Let P move in a gravity field as in Figure 4.6.3 and shown again in Figure 4.8.2. Then from Equation (4.6.9) the generalized applied forces F_{q_r} $(r = 1,\ldots,n)$ were seen to be

$$F_{q_r} = -mg\,\partial h/\partial q_r \tag{4.8.4}$$

where h is the elevation of P above an arbitrary reference plane as shown in Figure 4.8.2. Hence, by comparing Equations (4.8.3) and (4.8.4) we see that a gravitational potential function is

$$P = mg\,h \tag{4.8.5}$$

Example 4.8.3 Potential Function for Spring Forces

Consider a spring connecting particles P_1 and P_2 of a mechanical system S as in Example 4.6.2, as shown in Figure 4.6.4, and shown again in Figure 4.8.3. As before, let S have n degrees of freedom represented by coordinates q_r $(r = 1,\ldots,n)$.

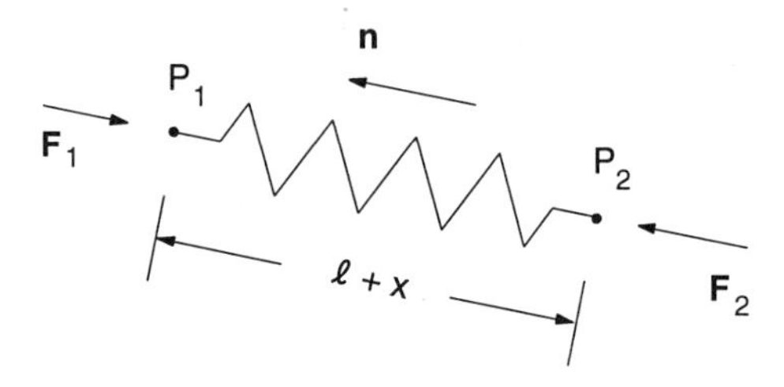

Fig. 4.8.3 A Spring Connecting Particles in a Mechanical System

Let the spring have natural length ℓ and extension $x(q_r)$ as shown and let the forces $\mathbf{F}_1$ and $\mathbf{F}_2$ exerted on the particles be

$$\mathbf{F}_1 = -\mathbf{F}_2 = -f(x)\,\mathbf{n} \tag{4.8.6}$$

where f(x) is the spring modulus and **n** is a unit vector along the spring axis as in Figure 4.8.3. Then from Equation (4.6.19) the generalized forces F_{q_r} $(r = 1,\ldots,n)$

are

$$F_{q_r} = -f(x)\,\partial x/\partial q_r \tag{4.8.7}$$

By comparing Equations (4.8.3) and (4.8.7) we see that potential function for the spring is

$$P = \int_0^x f(\xi)\,d\xi \tag{4.8.8}$$

If the spring is linear such that f(x) is kx then the potential function becomes

$$P = (1/2)kx^2 \tag{4.8.9}$$

Observe that P is positive, independent of whether x is positive (extension) or negative (compression).

4.8.3 Work

The work of a force on a particle is defined as the product of the force magnitude and the distance moved by the particle in the direction of the force. Consider a particle P of a mechanical system S having n degrees of freedom with coordinates q_r $(r = 1,\ldots,n)$. Let P move on a curve C as depicted in Figure 4.8.4. Let C be defined by parametric equations of the form:

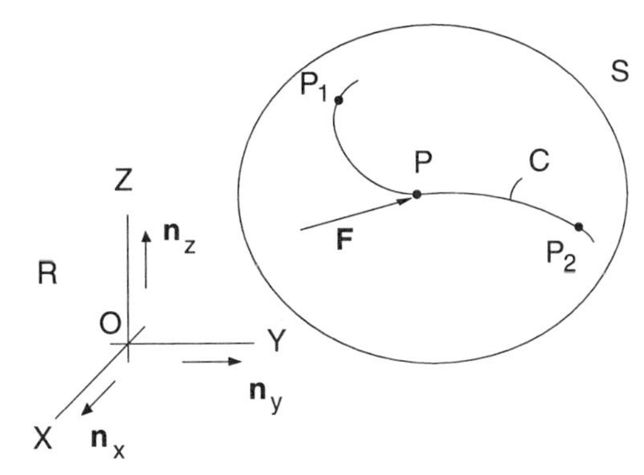

Fig. 4.8.4 A Particle P of a Mechanical System S, Moving on a Curve C, and Acted Upon by a Force **F**

$$x = x(q_r) \qquad y = y(q_r) \qquad z = z(q_r) \tag{4.8.10}$$

where x, y, and z are the coordinates of P relative to a Cartesian axis system of an inertial frame R, and where the q_r are themselves functions of t $[q_r = q_r(t)]$. Let P be acted upon by a force **F** as in Figure 4.8.4. Then the work W done by **F** on P as P moves along C from P_1 to P_2 is

$$W = \int_{P_1}^{P_2} \mathbf{F} \cdot \mathbf{ds} \tag{4.8.11}$$

where the integral is a line integral along C and where **ds** is a differential arc element vector tangent to C at P. In terms of unit vectors $\mathbf{n}_x$, $\mathbf{n}_y$, and $\mathbf{n}_z$ along X, Y, and Z, **ds** may be expressed as

$$\begin{aligned} \mathbf{ds} &= dx\,\mathbf{n}_x + dy\,\mathbf{n}_y + dz\,\mathbf{n}_z \\ &= \sum_{r=1}^{n} \left(\frac{\partial x}{\partial q_r}\mathbf{n}_x + \frac{\partial y}{\partial q_r}\mathbf{n}_y + \frac{\partial z}{\partial q_r}\mathbf{n}_z \right) dq_r \\ &= \sum_{r=1}^{n} \frac{\partial \mathbf{p}}{\partial q_r} dq_r \end{aligned} \tag{4.8.12}$$

where **p** is the position vector locating P relative to the origin O of the Cartesian frame.

Observe from Equation (4.5.9) that $\partial \mathbf{p}/\partial q_r$ is a partial velocity vector and therefore in view of Equation (4.6.1), the work done on P may be expressed as

$$W = \int_{P_1}^{P_2} \sum_{r=1}^{n} F_{q_r} dq_r \tag{4.8.13}$$

Finally, suppose that S is a conservative system such that F_{q_r} may be expressed in terms of a potential energy function P as $-\partial P/\partial q_r$ $(r = 1, \ldots, n)$. [See

Equation (4.8.3).] Then the work W of **F** may be expressed as

$$W = -\int_{P_1}^{P_2} \sum_{r=1}^{n} \frac{\partial P}{\partial q_r} dq_r = -\int_{P_1}^{P_2} dP$$
$$= P_1 - P_2 \tag{4.8.14}$$

where P_1 and P_2 are the values of P at points P_1 and P_2.

Example 4.8.4 Work Done by Gravity

Consider a particle P with mass m moving in a gravity field on a curve C between points P_1 and P_2 as depicted in Figure 4.8.5. Determine the work done by the gravitational or weight force on P.

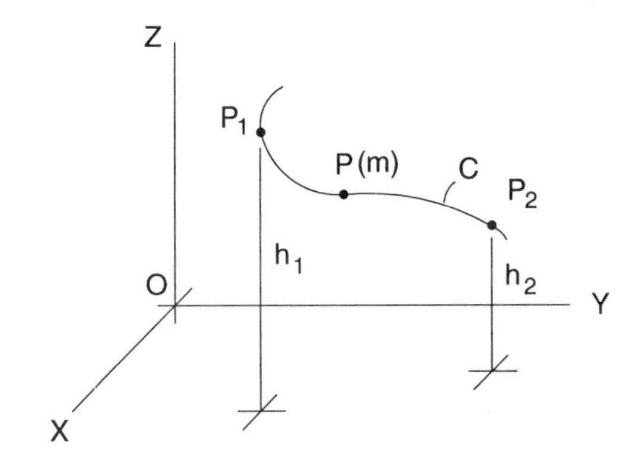

Fig. 4.8.5 A Particle P Moving in a Gravity Field

From Equations (4.8.5) and (4.8.4) we immediately see that the work from P_1 to P_2, written as ${}_1W_2$, is

$${}_1W_2 = mg(h_1 - h_2) = mg\Delta h \tag{4.8.15}$$

where h_1 and h_2 are the elevations of P_1 and P_2 above a reference plane and by inspection Δh is the difference in elevations.

Observe that the gravity work is independent of the curve C and depends only upon the weight mg and the elevation change Δh. Observe further that the work may be positive or negative depending upon the sign of Δh.

Example 4.8.5 Work Done by a Spring

Consider a particle P acted upon by a spring as represented in Figure 4.8.6. Let the spring move P from point P_1 to P_2. Let the natural length of the spring be ℓ and let the spring extension (or compression) at P_1 and P_2 be x_1 and x_2. Let the spring modulus be f(x) where x is a measure of the extension (positive) or compression (negative). Determine the work done by the spring on the particle.

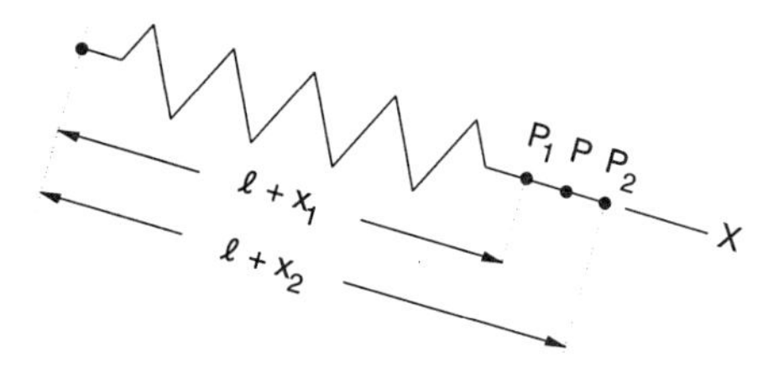

Fig. 4.8.6 A Particle Acted Upon by a Spring

From Equations (4.8.8) and (4.8.14) we immediately see that the work done on P from P_1 to P_2 is

$$ {}_1W_2 = \int_0^{x_1} f(\xi)d\xi - \int_0^{x_2} f(\xi)d\xi = -\int_{x_1}^{x_2} f(\xi)d\xi \qquad (4.8.16) $$

For a linear spring where f(x) is kx, ${}_1W_2$ is

$$ {}_1W_2 = (1/2)kx_1^2 - (1/2)kx_2^2 \tag{4.8.17} $$

4.9 Associated Inertia (or Passive) Kinetic Quantities

Analogous to the associated applied kinetic quantities there are several associated inertia kinetic quantities namely linear momentum, angular momentum, and kinetic energy, which are useful in dynamic analyses. These are defined in this section.

4.9.1 Linear Momentum

Consider a particle P having mass m and moving in an inertial reference frame R. Then the linear momentum ${}^R\mathbf{L}^P$ of P in R is defined as:

$$ {}^R\mathbf{L}^P = m\,{}^R\mathbf{V}^P \tag{4.9.1} $$

or simply as:

$$ \mathbf{L} = m\mathbf{V} \tag{4.9.2} $$

where ${}^R\mathbf{V}^P$ is the velocity of P in R.

4.9.2 Angular Momentum

Consider again a particle P having a mass m and moving in an inertial reference frame R. Let O be an arbitrary point of R, as in Figure 4.9.1. Then the angular momentum of P relative to O in R ${}^R\mathbf{A}^{P/O}$ is defined as:

$$ {}^R\mathbf{A}^{P/O} = \mathbf{p} \times m\,{}^R\mathbf{V}^P \tag{4.9.3} $$

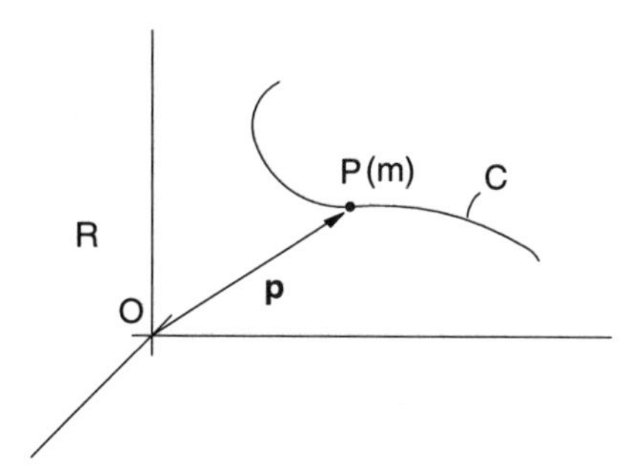

Fig. 4.9.1 Particle P Moving in an Inertial Reference Frame R

or simply as

$$\mathbf{A}_O = \mathbf{p} \times m\mathbf{V} \tag{4.9.4}$$

where **P** is the position vector locating P relative to O as in Figure 4.9.1, and **V** is the velocity of P in R.

4.9.3 Kinetic Energy

Consider a particle P having mass m moving in an inertial reference frame R. Then the kinetic energy K of P in R is defined as

$$K = (1/2)m\mathbf{V}^2 \tag{4.9.5}$$

where **V** is the velocity of P in R.

Kinetic energy is useful in determining generalized inertia forces as illustrated in the following example.

Example 4.9.1 Determining Generalized Inertia Forces

Consider a particle P of a holonomic mechanical system S (see Section 4.5) having n degrees of freedom represented by coordinates q_r $(r = 1,\dots,n)$. Then from Equation (4.5.13) we see that the projection of the acceleration **a** of P along the partial velocity $\mathbf{V}_{\dot{q}_r}$ is

$$\mathbf{a} \bullet \mathbf{V}_{\dot{q}_r} = \frac{d}{dt}\left[\frac{\partial}{\partial \dot{q}_r}\left(\frac{\mathbf{V}^2}{2}\right)\right] - \frac{\partial}{\partial q_r}\left(\frac{\mathbf{V}^2}{2}\right) \tag{4.9.6}$$

where **V** is the velocity of P and where all kinematic quantities are computed in an inertial reference frame R.

Let P have mass m. Then by multiplying by -m we may express Equation (4.9.6) in the form:

$$-m\mathbf{a} \bullet \mathbf{V}_{\dot{q}_r} = \frac{\partial}{\partial q_r}\left(\frac{m\mathbf{V}^2}{2}\right) - \frac{d}{dt}\left[\frac{\partial}{\partial \dot{q}_1}\left(\frac{m\mathbf{V}^2}{2}\right)\right] \tag{4.9.7}$$

From Equation (4.7.2) we see that the left side of Equation (4.9.7) is the generalized inertia force $F^*_{q_r}$, and thus from the definition of kinetic energy of Equation (4.9.5) we have the result:

$$F^*_{q_r} = \frac{\partial K}{\partial q_r} - \frac{d}{dt}\left(\frac{\partial K}{\partial \dot{q}_r}\right) \quad (r = 1,\dots,n) \tag{4.9.8}$$

Observe that Equation (4.9.8) provides a means for determining generalized inertia forces without using acceleration vectors.

4.10 Summary of Formulas for Associated Applied (Active) and Inertia (Passive) Force Quantities

Table 4.10.1 provides a summary of formulas for applied force quantities together with source equations.

Table 4.10.1 Formulas for Applied (Active) Force Quantities

Quantity	Symbol	Formula	Source Equation
Impulse	$\mathbf{I}$	$\mathbf{I} = \int_{t_1}^{t_2} \mathbf{F}dt$	(4.8.1)
Potential for Gravity Forces	P	$P = mgh$	(4.8.5)
Potential for Spring Forces	P	$(1/2)kx^2$	(4.8.9)
Generalized Applied Forces	F_{q_r}	$F_{q_r} = \partial P/\partial q_r \;\; (r = 1,...,n)$	(4.8.3)
Work	W	${}_1W_2 = \int_{P_1}^{P_2} \mathbf{F} \bullet d\mathbf{s}$	(4.8.11)
Work done by Gravity Force	${}_1W_2$	${}_1W_2 = mg(h_1 - h_2)$	(4.8.15)
Work done by Linear Spring Force	${}_1W_2$	${}_1W_2 = (1/2)k(x_1^2 - x_2^2)$	(4.8.17)

Table 4.10.2 provides a summary of formulas for inertia force quantities together with source equations.

Table 4.10.2 Formulas for Inertia (Passive) Force Quantities

Quantity	Symbol	Formula	Source Equation
Linear Momentum	$\mathbf{L}$	$\mathbf{L} = m\mathbf{V}$	(4.9.2)
Angular Momentum	$\mathbf{A}$	$\mathbf{A} = \mathbf{p} \times m\mathbf{V}$	(4.9.4)
Kinetic Energy	K	$K = (1/2)m\mathbf{V}^2$	(4.9.5)
Generalized Inertia Forces (Holonomic Systems)	$F^*_{q_r}$	$F^*_{q_r} = \frac{\partial K}{\partial q_r} - \frac{d}{dt}\left(\frac{\partial K}{\partial \dot{q}_r}\right)$ $(r = 1,...,n)$	(4.9.8)

References

4.1 M. Kline, *Calculus*, John Wiley and Sons, Inc., New York, 1967.

4.2 T. R. Kane, P. W. Likins, and D. A. Levinson, *Spacecraft Dynamics*, McGraw Hill, New York, 1983.

4.3 T. R. Kane, *Dynamics*, Holt, Rinehart and Winston, New York, 1968.

4.4 R. L. Huston, *Multibody Dynamics*, Butterworth, Stoneham, MA, 1990.

Chapter 5

PARTICLE DYNAMICS

5.1 Introduction

In this chapter we list commonly used formulas for dynamic analyses of particles. These formulas arise from the basic principles of dynamics and the corresponding laws of motion. In later chapters we review additional, more advanced procedures for application with rigid bodies and mechanical systems.

Initially, we will review the dynamics principles themselves. We then consider a few elementary applications, followed by a listing of formulas for impact/collision analysis. We conclude with a comparison of the relative advantages and disadvantages of the various methods.

5.2 Principles of Dynamics/Laws of Motion

Most principles of dynamics and the resulting laws of motion have their roots in Newton's laws of motion. In this sense they are all equivalent and can be developed one from another. Table 5.2.1 provides a listing of these principles and laws commonly used in particle dynamics analyses.

Table 5.2.1 Dynamics Principles/Laws of Motion for Particle Dynamics

Name	Principle	Formulas/Equations
1. Newton's Laws	First Law: A particle at rest or in uniform motion remains at rest or in uniform motion unless acted upon by a force.	If $\mathbf{F} = 0$ $\mathbf{v} = \mathbf{c}$ (a constant) (5.2.1)

Name	Principle	Formulas/Equations
1. Newton's Laws	Second Law: The acceleration of a particle is proportional to the force applied to the particle and inversely proportional to the mass of the particle. Third Law: If a particle P exerts a force on a particle Q, then Q exerts an equal and opposite force on P (Action-Reaction).	$\mathbf{F} = m\mathbf{a}$ (5.2.2) $\mathbf{F}_{P/Q} = -\mathbf{F}_{Q/P}$ (5.2.3)
2. d'Alembert's Principle	A particle with mass m and acceleration **a** in an inertial frame R experiences an inertial force $\mathbf{F}^*$ proportional to m and **a** and directed opposite to **a**. The sum of the applied and inertia forces is zero.	$\mathbf{F}^* = -m\mathbf{a}$ (5.2.4) $\mathbf{F} + \mathbf{F}^* = 0$ (5.2.5)
3. Work-Energy	The work done on a particle in moving it from position P_1 to position P_2 is equal to the change in kinetic energies of the particle between P_1 and P_2.	${}_1W_2 = K_2 - K_1$ (5.2.6)
4. Impulse-Momentum	The impulse on a particle between times t_i and t_2 is equal to the change in momentum of the particle between t_1 and t_2.	${}_1\mathbf{I}_2 = \mathbf{L}_2 - \mathbf{L}_1 = m(\mathbf{v}_2 - \mathbf{v}_1)$ (5.2.7)
5. Kane's Equations	Let P be a particle belonging to a mechanical system S having n degrees of freedom represented by coordinates q_r $(r = 1,\ldots,n)$. Then the sum of the generalized applied (active) and inertia (passive) forces is zero.	$F_{q_r} + F^*_{q_r} = 0$ $(r = 1,\ldots,n)$ (5.2.8)

Table 5.2.1 Continued

Name	Principle	Formulas/Equations
6. Lagranges Equations	(See 5.) Let S be holonomic. Then Equation (4.9.8) produces the generalized inertia forces, and Kane's equation have the form of Equation (5.2.9). Alternatively, the time integral of the Lagrangian L (kinetic minus potential energy) is a minimum (Hamilton's principle). The calculus of variations then leads to Equations (5.2.10).	$\frac{d}{dt}\left(\frac{\partial K}{\partial \dot{q}_r}\right) - \frac{\partial K}{\partial q_r} = F_{q_r}$ $(r = 1, \ldots, n)$ (5.2.9) $\frac{d}{dt}\left(\frac{\partial L}{\partial \dot{q}_r}\right) - \frac{\partial L}{\partial q_r} = 0$ $(r = 1, \ldots, n)$ (5.2.10)

5.3 Application: Dynamics of a Simple Pendulum

To illustrate these principles and laws of motion consider the simple pendulum consisting of a particle P with mass m supported by a light flexible cable with length ℓ and moving in a vertical plane as depicted in Figure 5.3.1. Let θ measure the inclination and movement of the pendulum as shown. The objective in the following examples is to determine the governing equations of motion of the system.

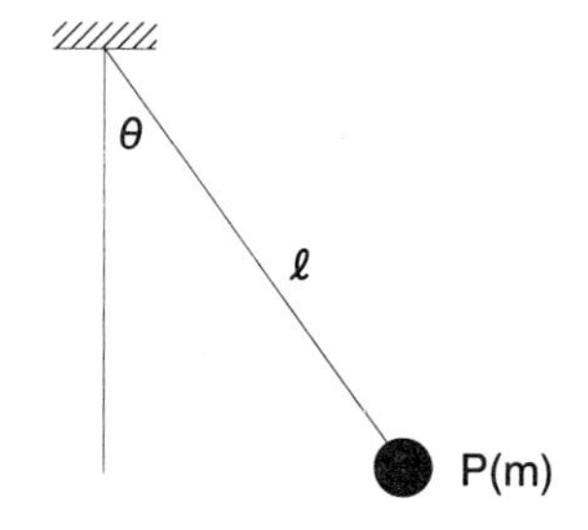

Fig. 5.3.1 A Simple Pendulum

In preparation, it is helpful to first develop the kinematics. In the vertical plane P moves on a circle with radius ℓ. Hence, it is convenient to express the movement of P in terms of radial and tangential unit vectors as in Figure 5.3.2.

Then the velocity and acceleration of P are

$$\mathbf{V} = \ell\dot{\theta}\,\mathbf{n}_t \tag{5.3.1}$$

and

$$\mathbf{a} = \ell\dot{\theta}^2\mathbf{n}_r + \ell\ddot{\theta}\,\mathbf{n}_t \tag{5.3.2}$$

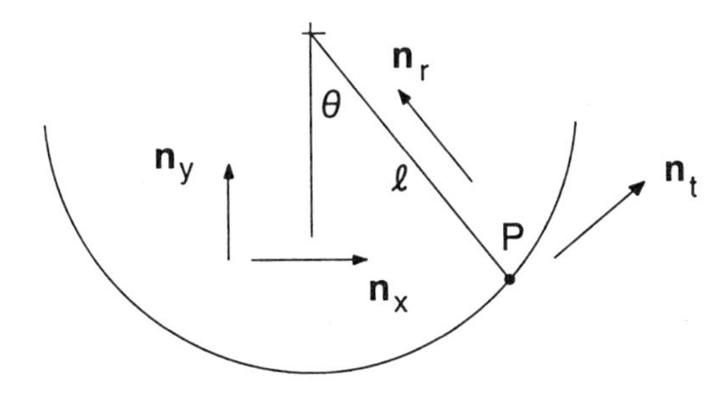

Fig. 5.3.2 Pendulum Bob Moving on a Circle

Example 5.3.1 <u>Simple Pendulum with Newton's Laws</u>

Consider a free-body diagram of the pendulum bob P of Figure 5.3.1, as in Figure 5.3.3, where T is the cable tension and the weight **W** may be expressed as:

$$\mathbf{W} = -mg\,\mathbf{n}_y \tag{5.3.3}$$

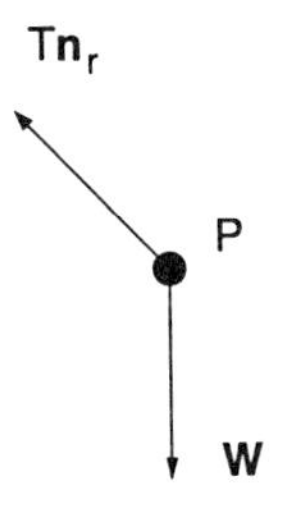

Fig. 5.3.3 Free-Body Diagram of P

with $\mathbf{n}_y$ being the vertical unit vector as in Figure 5.3.2.

From Newton's second law, the equation of motion is simply

$$T\mathbf{n}_r + \mathbf{W} = m\mathbf{a} \tag{5.3.4}$$

By substituting from Equation (5.3.2) and by expressing $\mathbf{W}$ in terms of $\mathbf{n}_r$ and $\mathbf{n}_t$ Equation (5.3.4) may be expressed as:

$$\begin{aligned} &T\mathbf{n}_r - mg\cos\theta\,\mathbf{n}_r - mg\sin\theta\,\mathbf{n}_t \\ &\quad = m\ell\dot{\theta}^2\mathbf{n}_r + m\ell\ddot{\theta}\,\mathbf{n}_t \end{aligned} \tag{5.3.5}$$

Then by equating like components we have:

$$T - mg\cos\theta = m\ell\dot{\theta}^2 \tag{5.3.6}$$

and

$$-mg\sin\theta = m\ell\ddot{\theta} \tag{5.3.7}$$

Finally, by solving for $\ddot{\theta}$ and T, the desired equations are:

$$\ddot{\theta} + (g/\ell)\sin\theta \tag{5.3.8}$$

and

$$T = m\ell\dot{\theta}^2 + mg\cos\theta \tag{5.3.9}$$

Example 5.3.2 Simple Pendulum with d'Alembert's Principle

The use of d'Alembert's principle is very similar to the use of Newton's second law. The principle difference is the introduction of an inertia force $\mathbf{F}^*$ which has the effect of moving the $m\mathbf{a}$ term from the right side to the left side of the equation. To see this consider the free body diagram of the pendulum bob P as in Figure 5.3.4 where $\mathbf{F}^*$ is the inertia force defined as

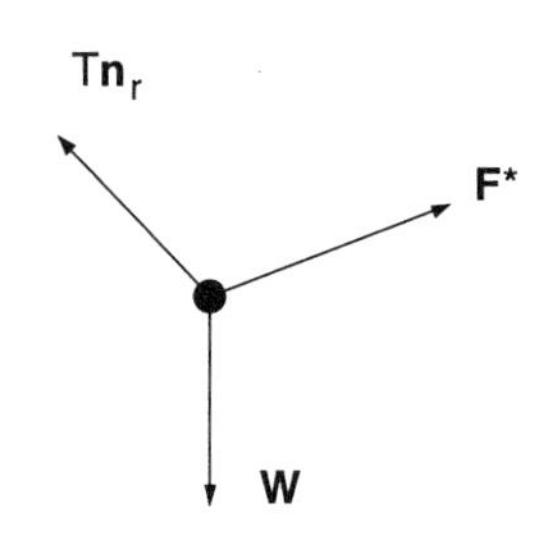

Fig. 5.3.4 Free-Body Diagram of P

$$\mathbf{F}^* = -m\mathbf{a} \tag{5.3.10}$$

The equation of motion is then

$$T\mathbf{n}_r + \mathbf{W} + \mathbf{F}^* = 0 \tag{5.3.11}$$

By substituting from Equations (5.3.2), (5.3.3), and (5.3.10) we obtain

$$\begin{aligned} T\mathbf{n}_r - mg\cos\theta\,\mathbf{n}_r - mg\sin\theta\,\mathbf{n}_t - m\ell\dot{\theta}^2\mathbf{n}_r \\ -m\ell\ddot{\theta}\,\mathbf{n}_t = 0 \end{aligned} \tag{5.3.12}$$

By setting the $\mathbf{n}_r$ and $\mathbf{n}_t$ components equal to zero we have

$$T - mg\cos\theta - m\ell\dot{\theta}^2 = 0 \tag{5.3.13}$$

and

$$-mg\sin\theta - m\ell\ddot{\theta} = 0 \tag{5.3.14}$$

from which we immediately obtain the desired equation:

$$\ddot{\theta} + (g/\ell)\sin\theta \tag{5.3.15}$$

and

$$T = m\ell\dot{\theta}^2 + mg\cos\theta \tag{5.3.16}$$

identical to Equations (5.3.8) and (5.3.9).

Example 5.3.3 Simple Pendulum Using the Work-Energy Principle

From Equations (4.8.15) the work done on P as it moves from position P_1 to P_2 is:

$${}_1W_2 = mg(h_2 - h_1) \tag{5.3.17}$$

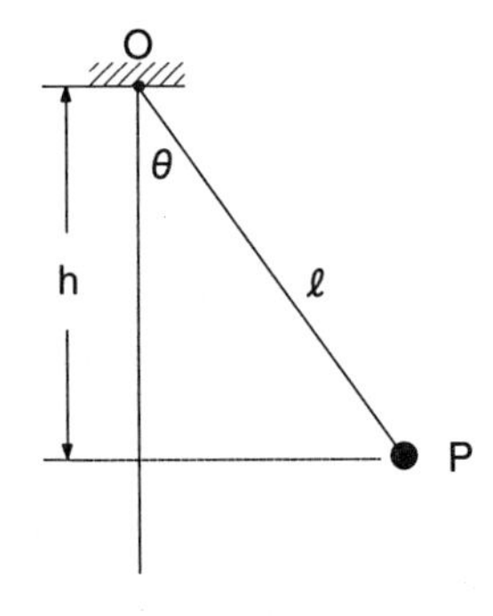

Fig. 5.3.5 Pendulum Bob Elevation h

where h is the position elevation relative to an arbitrary reference. Suppose we measure h from the pendulum support O as in Figure 5.3.5. Then h is

$$h = -\ell \cos\theta \tag{5.3.18}$$

Then from Equation (5.3.17) the work ${}_1W_2$ between P_1 and P_2 is:

$${}_1W_2 = mg\ell(\cos\theta_2 - \cos\theta_1) \tag{5.3.19}$$

where θ_1 and θ_2 are the pendulum orientations when P is at P_1 and P_2.

Observe next that as P moves on its circle its motion is perpendicular to the direction of the cable tension. Therefore, the tension does no work on P.

From Equations (4.9.5) and (5.3.1) the kinetic energy of P is

$$K = \frac{1}{2}mv^2 = \frac{1}{2}\ell^2\dot{\theta}^2 \tag{5.3.20}$$

Hence, the change in kinetic energy between positions P_1 and P_2 is

$$K_2 - K_1 = \frac{1}{2}m\ell^2(\dot{\theta}_2^2 - \dot{\theta}_1^2) \tag{5.3.21}$$

Therefore, from Equations (5.2.6), (5.3.19), and (5.3.21) the governing equation is

$$mg\ell(\cos\theta_2 - \cos\theta_1) = \left(\frac{1}{2}\right)m\ell^2(\dot{\theta}_2^2 - \dot{\theta}_1^2) \tag{5.3.22}$$

or

$$\dot{\theta}_2^2 = \dot{\theta}_1^2 + 2(g/\ell)(\cos\theta_2 - \cos\theta_1) \tag{5.3.23}$$

Observe that only a single equation [Equation (5.3.23)] is produced with the work-energy principle as opposed to the two equations [Equations (5.3.8) and (5.3.9) or Equations (5.3.15) and (5.3.16)] produced by Newton's laws or d'Alembert's principle. Observe further that the cable tension T does not appear in Equation (5.3.23). (This is advantageous if there is no interest in T.) Finally, observe that if θ_2 is regarded as a variable in Equation (5.3.23) then Equation (5.3.23) may be differentiated to obtain Equation (5.3.8) or (5.3.15). Alternatively, Equation (5.3.23) may be viewed as being a "first integral" of Equation (5.3.9) or (5.3.15).

Example 5.3.4 Simple Pendulum with Kane's Equations

The pendulum has one degree of freedom represented by the coordinate θ. From Equation (5.3.1) the partial velocity of the pendulum bob P relative to θ is

$$\mathbf{v}_{\dot{\theta}} = \ell\,\mathbf{n}_t \tag{5.3.24}$$

From Figures 5.3.1 and 5.3.2 the applied forces are the weight **W** and the cable tension **T** given by

$$\mathbf{W} = -mg\,\mathbf{n}_y \quad \text{and} \quad \mathbf{T} = T\,\mathbf{n}_r \tag{5.3.25}$$

Then from Equation (4.6.1) the generalized applied force for the coordinate θ is

$$F_\theta = \mathbf{v}_{\dot{\theta}} \cdot \mathbf{W} + \mathbf{v}_{\dot{\theta}} \cdot \mathbf{T}$$

$$= \ell\mathbf{n}_t \cdot (-mg\,\mathbf{n}_y) + \ell\,\mathbf{n}_t \cdot T\,\mathbf{n}_r \tag{5.3.26}$$

or

$$F_\theta = -mg\ell\sin\theta$$

Similarly, the generalized inertia force for θ is:

$$F_\theta^* = \mathbf{v}_{\dot\theta} \bullet \mathbf{F}^* \tag{5.3.27}$$

Then from Equations (5.3.10) and (5.3.2) we have

$$F_\theta^* = \ell\mathbf{n}_t \bullet (-m)(\ell\dot\theta^2\mathbf{n}_r + \ell\ddot\theta\mathbf{n}_t)$$

or

$$F_\theta^* = -m\ell^2\ddot\theta \tag{5.3.28}$$

Finally, Kane's equations [Equations (5.2.8)] are

$$F_\theta + F_\theta^* = 0 \tag{5.3.29}$$

or

$$\ddot\theta + (g/\ell)\sin\theta = 0 \tag{5.3.30}$$

Observe that Equation (5.3.30) is identical to Equations (5.3.8) and (5.3.15) as obtained by Newton's laws and d'Alembert's principle. Observe further, however, that like the work-energy principle only one equation is obtained. The unknown cable tension T is eliminated from the analysis by the projection of the cable tension along the partial velocity vector. This is advantageous if T is not of interest, and, of course, disadvantageous if T is of interest.

Example 5.3.5 Simple Pendulum with Lagrange's Equations

From Equations (4.9.5) and (5.3.1) the kinetic energy of the pendulum is:

$$K = \frac{1}{2}mv^2 = \frac{1}{2}m\ell^2\dot\theta^2 \tag{5.3.31}$$

From Equation (5.3.26) the generalized applied force is

$$F_\theta = -mg\ell \sin\theta \tag{5.3.32}$$

From Equation (5.2.9) Lagrange's equations are

$$\frac{d}{dt}\left(\frac{\partial K}{\partial \dot{\theta}}\right) - \frac{\partial K}{\partial \theta} = F_\theta \tag{5.3.33}$$

Then by substitution from Equations (5.3.31) and (5.3.32) we obtain

$$\ddot{\theta} + (g/\ell)\sin\theta = 0 \tag{5.3.34}$$

Observe that Equation (5.3.34) is identical to Equations (5.3.8), (5.3.15), and (5.3.30) obtained by Newton's laws, d'Alembert's principle, and Kane's equations. Observe further that like Kane's equations only one equation is obtained with the unknown cable tension being eliminated from the analysis.

In the following section we will briefly illustrate how the cable tension, and constraint forces in general, can be determined using Kane's equations and Lagrange's equations.

5.4 Determination of Unknown Constraint Force or Moment Components

As observed in the foregoing examples, the use of generalized applied forces as with Kane's equations and Lagrange's equations leads to an automatic elimination of the constraint force components. That is, the cable tension which restrains the pendulum bob P is directed perpendicular to the partial velocity of P and thus does not enter into the expression for the generalized force. While this is an advantage in that the resulting analysis is simplified, it is a disadvantage if the values of the restraining force components (the cable tension) are of interest.

It is relatively easy to determine the unknown constraining force components by temporarily introducing a degree of freedom for the system in the direction of the unknown and desired force component. This will allow the desired force component to appear in the generalized force expressions. Then once the resulting governing equations are determined, the coordinate corresponding to the introduced degree of freedom may then be set back to its prelease value (usually zero).

To illustrate this procedure consider again the simple pendulum of Figure 5.3.1 and redrawn again in Figure 5.4.1. Let the pendulum have a temporary degree of freedom in the radial direction as shown. Then the velocity and acceleration of P are (see Table 3.8.1)

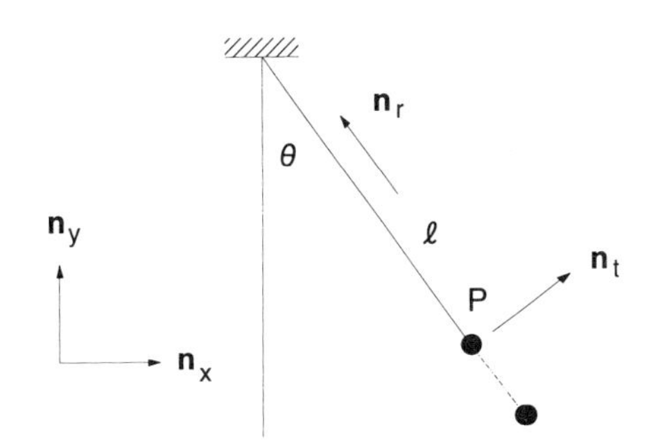

Fig. 5.4.1 Simple Pendulum with a Temporary Release in the Radial Direction

$$\mathbf{v} = -\dot{\ell}\,\mathbf{n}_r + \ell\dot{\theta}\,\mathbf{n}_t \qquad (5.4.1)$$

and

$$\mathbf{a} = (-\ddot{\ell} + \ell\dot{\theta}^2)\mathbf{n}_r + (\ell\ddot{\theta} + 2\dot{\ell}\,\dot{\theta})\mathbf{n}_t \qquad (5.4.2)$$

There are now two degrees of freedom represented by the coordinates θ and ℓ. The corresponding partial velocity vectors are

$$\mathbf{v}_{\dot{\theta}} = \ell\,\mathbf{n}_t \quad \text{and} \quad \mathbf{v}_{\dot{\ell}} = -\mathbf{n}_r \qquad (5.4.3)$$

Recall from Figure 5.3.3 that the applied forces are the weight and cable tension given by

$$\mathbf{W} = -mg\mathbf{n}_y \quad \text{and} \quad \mathbf{T} = T\mathbf{n}_r \qquad (5.4.4)$$

Then from Equation (5.4.4) the generalized applied forces for θ and ℓ are

$$F_\theta = \mathbf{v}_{\dot{\theta}} \cdot (\mathbf{W} + \mathbf{T}) = -mg\ell \sin\theta \quad (5.4.5)$$

and

$$F_\ell = \mathbf{v}_{\dot{\ell}} \cdot (\mathbf{W} + \mathbf{T}) = mg \cos\theta - T \quad (5.4.6)$$

From Equation (5.4.2) the inertia force on P is

$$\mathbf{F}^* = -m(-\ddot{\ell} + \ell\dot{\theta}^2)\mathbf{n}_r - m(\ell\ddot{\theta} + 2\dot{\ell}\dot{\theta})\mathbf{n}_t \quad (5.4.7)$$

Then from Equation (5.4.4) the generalized inertia forces for θ and ℓ are

$$F_\theta^* = \mathbf{v}_{\dot{\theta}} \cdot \mathbf{F}^* = -m\ell(\ell\ddot{\theta} + 2\dot{\ell}\dot{\theta}) \quad (5.4.8)$$

and

$$F_\ell^* = \mathbf{v}_{\dot{\ell}} \cdot \mathbf{F}^* = m(-\ddot{\ell} + \ell\dot{\theta}^2) \quad (5.4.9)$$

Kane's equations then become

$$F_\theta + F_\theta^* = 0 \quad \text{or} \quad \ell\ddot{\theta} + 2\dot{\ell}\dot{\theta} + g\sin\theta = 0 \quad (5.4.10)$$

and

$$F_\ell + F_\ell^* = 0 \quad \text{or} \quad m(-\ddot{\ell} + \ell\dot{\theta}^2) + mg\cos\theta - T = 0 \tag{5.4.11}$$

If $\dot{\ell}$ and $\ddot{\ell}$ are set equal to zero, these equations may be expressed as:

$$\ddot{\theta} + (g/\ell)\sin\theta = 0 \tag{5.4.12}$$

and

$$T = m\ell\dot{\theta}^2 + mg\cos\theta \tag{5.4.13}$$

These are identical to Equations (5.3.8) and (5.3.9) obtained with Newton's laws and Equations (5.3.15) and (5.3.16) obtained with d'Alembert's principle. Thus by solving Equation (5.4.12) for θ the cable tension T may be determined from Equation (5.4.13).

Next, for Lagrange's equations, from Equation (5.4.1) the kinetic energy of P is

$$K = (1/2)m(\dot{\ell}^2 + \ell^2\dot{\theta}^2) \tag{5.4.14}$$

Then Lagrange's equations become:

$$\frac{d}{dt}\left(\frac{\partial K}{\partial \dot{\theta}}\right) - \frac{\partial K}{\partial \theta} = F_\theta \quad \text{or} \quad m\ell^2\ddot{\theta} + 2m\dot{\ell}\dot{\theta} = mg\ell\sin\theta \tag{5.4.15}$$

and

$$\frac{d}{dt}\left(\frac{\partial K}{\partial \dot{\ell}}\right) - \frac{\partial K}{\partial \ell} = F_\ell \quad \text{or} \quad m\ddot{\ell} - m\ell\dot{\theta}^2 = mg\cos\theta - T \tag{5.4.16}$$

If $\dot{\ell}$ and $\ddot{\ell}$ are set equal to zero these equations may be expressed in the forms:

$$\ddot{\theta} + (g/\ell)\sin\theta = 0 \tag{5.4.17}$$

and

$$T = m\ell\dot{\theta}^2 + mg\cos\theta \tag{5.4.18}$$

These equations are identical to Equations (5.4.12) and (5.4.13) and to those obtained using Newton's laws and d'Alembert's principle. Thus again, once Equation (5.4.17) is solved for θ, Equation (5.4.18) may be used to find T.

5.5 Application: The Linear Oscillator (Mass-Spring-Damper System)

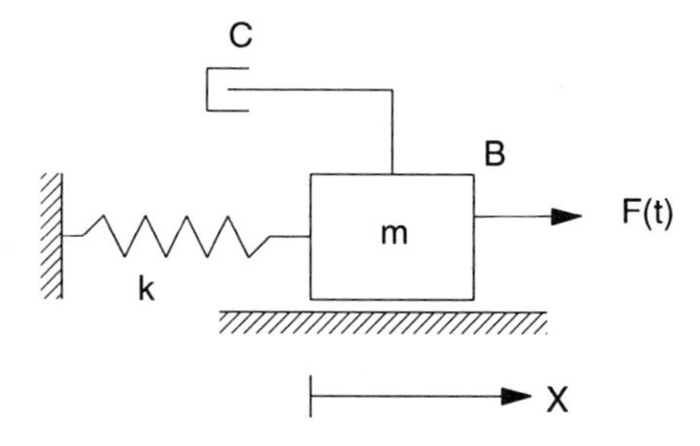

Fig. 5.5.1 Mass-Spring-Damper System

As a second application consider the mass-spring-damper system depicted in Figure 5.5.1. Let the spring consist of a rigid block B having mass m moving on a smooth horizontal surface and with its motion governed by a linear spring with stiffness k, a Coulomb damper with damping coefficient C, and horizontally applied force F(t). Let N measure the movement away from a static equilibrium position [when F(t) is zero].

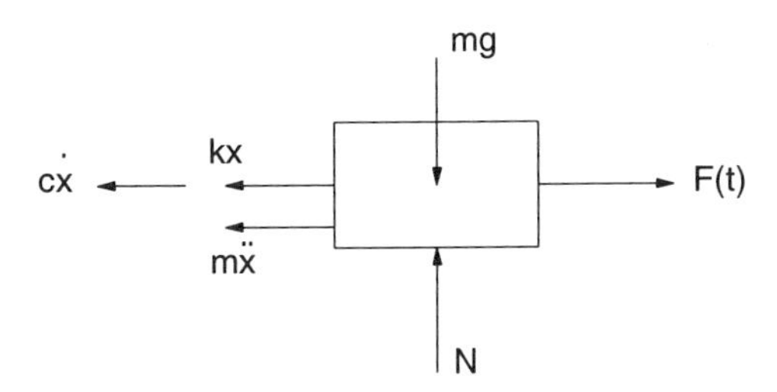

Fig. 5.5.2 Free-Body Diagram of Block

Consider a free-body diagram of B as in Figure 5.5.2 where kx and $c\dot{x}$ represent the spring and damper forces, and $m\ddot{x}$ is the inertia force. Then from d'Alembert's principle [Equation (5.2.5)] the governing equations are

$$N = mg \tag{5.5.1}$$

$$m\ddot{x} + c\dot{x} + kx = F(t) \tag{5.5.2}$$

From solution procedures for differential equations [see References 5.1 to 5.4] we see that the solution of Equation (5.5.2) may be written in the form

$$x = x_h + x_p \tag{5.5.3}$$

where x_h is the solution of the homogeneous equation with the right side F(t) being zero and where x_p is any solution of the equation itself (the "particular solution"). Then

$$m\ddot{x}_h + c\dot{x}_h + kx_h = 0 \tag{5.5.4}$$

and

$$m\ddot{x}_p + c\dot{x}_p + kx_p = F(t) \tag{5.5.5}$$

The solution of Equation (5.5.4) depends upon the relative values of m, c, and k, and specifically, the relative values of 4km and c^2. That is

$$x_h = e^{-\mu t}[A \cos \omega t + B \sin \omega t] \quad \text{for } 4km > c^2 \tag{5.5.6}$$

and

$$x_h = Ae^{-(\mu+\nu)t} + Be^{-(\mu-\nu)t} \quad \text{for } 4km < c^2 \tag{5.5.7}$$

and

$$x_h = e^{-\mu t}(A + Bt) \quad \text{for } 4km = c^2 \tag{5.5.8}$$

where A and B are constants to be determined from initial conditions and where μ, ω and ν are defined as:

$$\mu = c/2m \tag{5.5.9}$$

$$\omega = \left[\frac{k}{m} - \frac{c^2}{4m^2}\right]^{1/2} \tag{5.5.10}$$

and

$$\nu = \left[\frac{c^2}{4m^2} - \frac{k}{m}\right]^{1/2} \tag{5.5.11}$$

Equations (5.5.6), (5.5.7), and (5.5.8) are called the "underdamped," "overdamped," and "critically damped" cases respectively.

Typically, the "forcing function" F(t) has the form

$$F(t) = F_0 \cos pt \qquad (5.5.12)$$

where F_0 and p are constants. Then the particular solution x_p of Equation (5.5.5) may be written in the form:

$$x_p = (F_0/\Delta)[(k - mp^2) \cos pt + cp \sin pt] \qquad (5.5.13)$$

where Δ is defined as:

$$\Delta = (k - mp^2)^2 + c^2p^2 \qquad (5.5.14)$$

In the case when c and F are zero in Equation (5.5.2), we have undamped free oscillations, and Equation (5.5.2) may be expressed as:

$$\ddot{x} + \omega^2 x = 0 \qquad (5.5.15)$$

where as before ω is $\sqrt{k/m}$. Then from Equation (5.5.6) the solution is

$$x = A \cos \omega t + B \sin \omega t \qquad (5.5.16)$$

The solution may also be expressed in the form:

$$x = \hat{A} \cos(\omega t + \phi) \qquad (5.5.17)$$

where $\hat{A}$ and ϕ may be expressed in terms of A and B as

$$\hat{A} = \sqrt{A^2+B^2} \quad \text{and} \quad \phi = \tan^{-1}(-B/A) \tag{5.5.18}$$

In Equation (5.5.17) $\hat{A}$ is called the "amplitude," ϕ is called the "phase," and ω is called the "circular frequency," which may be related to the frequency f and period T as

$$\omega = 2\pi f = 2\pi/T \tag{5.5.19}$$

If the initial conditions are such that x is x_0 and $\dot{x}$ is $\dot{x}_0$ when t is zero, the undamped free oscillator solution [Equation (5.5.16)] becomes

$$x = x_0 \cos \omega t + (\dot{x}_0/\omega) \sin \omega t \tag{5.5.20}$$

Finally, observe that with the simple pendulum, if the angle θ is small so that the sine of θ is nearly equal to θ, the governing equation [Equation (5.3.8)] takes in the form:

$$\ddot{\theta} + (g/\ell)\theta = 0 \tag{5.5.21}$$

which is identical in form to Equation (5.5.15).

5.6 Application: Projectile Motion

A third immediate application of the dynamics principles is to establish the equations of motion of a projectile as used in Section 3.21. To this end, consider a projectile P with mass m, moving in the air, relative to a fixed Cartesian inertial frame R as represented in Figure 5.6.1.

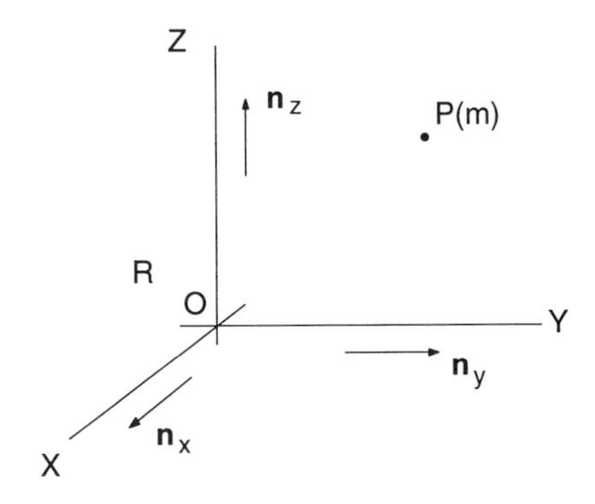

Fig. 5.6.1 A Particle P Moving in an Inertial Reference Frame R

Consider a free-body diagram of P as in Figure 5.6.2. The applied forces are simply the weight **W** and possibly air resistance **D**, and $\mathbf{F}^*$ represents the inertia force. Let the air resistance be proportional to the velocity **v** of P, but opposite to the direction of **v**. Then **W**, **D**, and $\mathbf{F}^*$ may be expressed as

$$\mathbf{W} = -mg\,\mathbf{n}_z \tag{5.6.1}$$

$$\mathbf{D} = -\mu\mathbf{v} \tag{5.6.2}$$

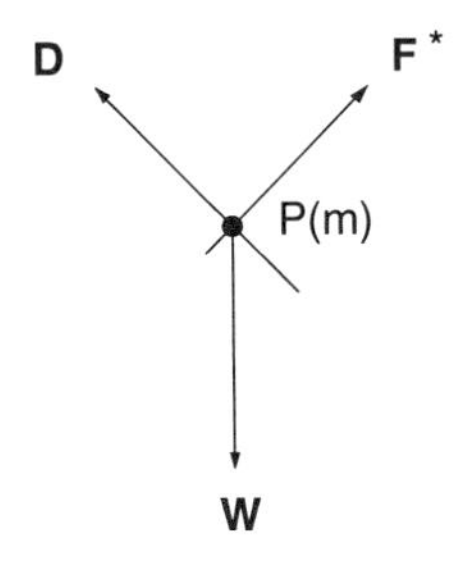

Fig. 5.6.2 A Free-body Diagram of P

and

$$\mathbf{F}^* = -m\mathbf{a} \tag{5.6.3}$$

where μ is the "drag" coefficient and where the velocity and acceleration of P may be expressed as

$$\mathbf{v} = \dot{x}\mathbf{n}_x + \dot{y}\mathbf{n}_y + \dot{z}\mathbf{n}_z \tag{5.6.4}$$

and

$$\mathbf{a} = \ddot{x}\mathbf{n}_x + \ddot{y}\mathbf{n}_y + \ddot{z}\mathbf{n}_z \tag{5.6.5}$$

where (x,y,z) are the X,Y,Z coordinates of P (see Figure 5.6.1) and where $\mathbf{n}_x$, $\mathbf{n}_y$, and $\mathbf{n}_z$ are unit vectors parallel to X, Y, and Z as shown. Then d'Alembert's principle leads to:

$$\mathbf{W} + \mathbf{D} + \mathbf{F}^* = 0 \tag{5.6.6}$$

or

$$\begin{aligned} -mg\mathbf{n}_z - \mu\dot{x}\mathbf{n}_x - \mu\dot{y}\mathbf{n}_y - \mu\dot{z}\mathbf{n}_z - m\ddot{x}\mathbf{n}_x \\ -m\ddot{y}\mathbf{n}_y - m\ddot{z}\mathbf{n}_z = 0 \end{aligned} \tag{5.6.7}$$

This in turn leads to the three scalar equations:

$$\ddot{x} = -(\mu/m)\dot{x} \tag{5.6.8}$$

$$\ddot{y} = -(\mu/m)\dot{y} \tag{5.6.9}$$

and

$$\ddot{z} = -g - (\mu/m)\dot{z} \tag{5.6.10}$$

If the air resistance is small so that μ is negligible, then Equations (5.6.8), (5.6.9), and (5.6.10) are identical to Equations (3.21.1), (3.21.2), and (3.21.3). The solutions of the equations are then provided in Section 3.21. The principal results are contained in Equations (3.21.7), (3.21.8), and (3.21.9):

$$\begin{aligned} x &= v_{0x}t + x_0 \\ y &= v_{0y}t + y_0 \\ z &= -gt^2/2 + v_{0z}t + z_0 \end{aligned} \tag{5.6.11}$$

where (x_0, y_0, z_0) and (v_{0x}, v_{0y}, v_{0z}) are the X,Y,Z components of the initial position and velocity. It was also found that the projectile moves on a parabola [see Equation (3.21.13)].

If the air resistance is not negligible, Equations (5.6.8), (5.6.9), and (5.6.10) may be integrated leading to:

$$\dot{x} = v_{0x}e^{-(\mu/m)t} \tag{5.6.12}$$

$$\dot{y} = v_{0y}e^{-(\mu/m)t} \tag{5.6.13}$$

and

$$\dot{z} = -v_{0z}e^{-(\mu/m)t} - (gm/\mu)\left[1 - e^{-(\mu/m)t}\right] \tag{5.6.14}$$

and to

$$x = (mv_{0x}/\mu)\left[1 - e^{-(\mu/m)t}\right] + x_0 \tag{5.6.15}$$

$$y = (mv_{0y}/\mu)\left[1 - e^{-(\mu/m)t}\right] + y_0 \tag{5.6.16}$$

and

$$z = [(mv_{0z}/\mu) + (gm^2/\mu^2)]\left[1 - e^{-(\mu/m)t}\right] - (mg/\mu)t + z_0 \tag{5.6.17}$$

where as before (x_0, y_0, z_0) and (v_{0x}, v_{0y}, v_{0z}) are initial X, Y, Z position and velocity components.

Observe in Equations (5.6.12), (5.6.13), and (5.6.14) that with increasing time, the velocity components approach the terminal values.

$$\dot{x} \to \dot{x}_\infty = 0 \tag{5.6.18}$$

$$\dot{y} \to \dot{y}_\infty = 0 \tag{5.6.19}$$

$$\dot{z} \to \dot{z}_\infty = -gm/\mu \tag{5.6.20}$$

Similarly, the position components approach the terminal values:

$$x \to x_\infty = (mv_{0x}/\mu) + x_0 \tag{5.6.21}$$

$$y \to y_\infty = (mv_{0y}/\mu) + y_0 \tag{5.6.22}$$

$$z \rightarrow z_\infty = [(mv_{0z}/\mu) + (gm^2/\mu^2)] - (mg/\mu)t + z_0 \qquad (5.6.23)$$

Thus the motion approaches a vertical asymptote.

5.7 Application: Impact

The foregoing applications illustrate the dynamics principles of Table 5.2.1 except for the principle of impulse and momentum. This principle is not conveniently used in routine classical problems such as the pendulum, mass-spring system, and projectiles. However, the impulse-momentum principle is ideally suited for application with problems involving impact. Impact is defined as a collision between two (or more) particles or bodies occurring over a relatively short time interval, but with large forces exerted between the colliding particles or bodies during that time interval.

The forces exerted during impact (so-called "impact forces") are usually so large that they dominate non-impact forces (such as gravity forces). Thus, in comparison, non-impact forces are relatively unimportant and may be neglected in impact analyses. Similarly, the impact time interval is so short that during impact the relative positions of the colliding particles remain essentially the same. The velocities, however, can change substantially.

The collision of two particles may be classified as being either *direct* or *oblique*. Direct impact occurs when the pre-impact velocities of the particles are colinear as in Figure 5.7.1. Oblique impact occurs when the pre-impact velocities are not colinear as in Figure 5.7.2. Most collisions are oblique.

Although a "particle" does not generally have size or dimension, the collision of particles is often modeled as the collision of small spheres as in Figure 5.7.3 [see Reference 5.5].

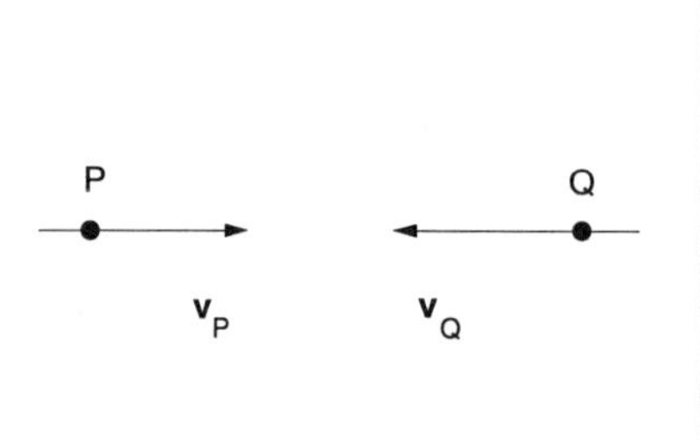

Fig. 5.7.1 Particles with Colinear Velocities Prior to Impact

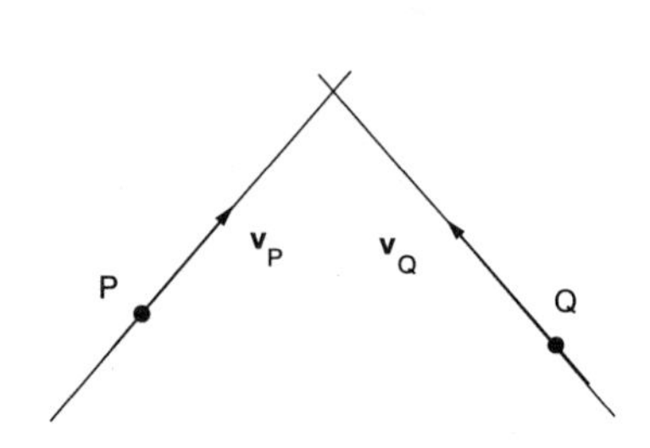

Fig. 5.7.2 Particles with Non-linear Velocities Prior to Impact

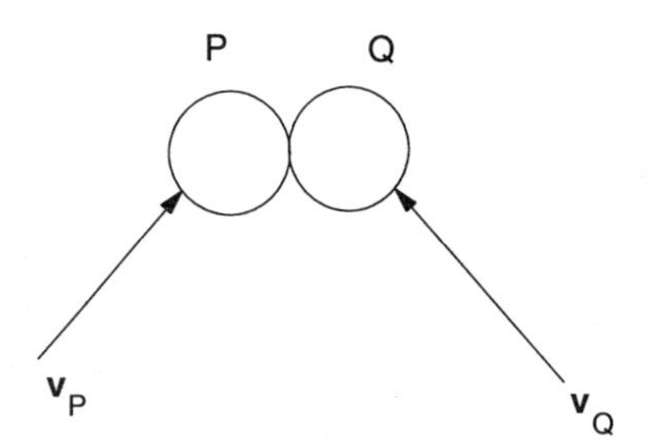

Fig. 5.7.3 Colliding Particles Modeled as Spheres

Collision between bodies or between particles and bodies is somewhat more complex, but a discussion of such problems may be found in Reference 5.6.

When particles collide they either remain together (a "plastic" collision) or they bounce apart (a "semi-elastic" collision). If they bounce apart, or rebound, the ratio of separation speed (post-impact) to the approach speed (pre-impact) is called the *coefficient of restitution.* Specifically, consider two colliding particles, represented as colliding spheres, as in Figure 5.7.3. Let n and t represent the directions normal and tangent to the contacting surfaces. Then the coefficient of restitution e is defined as

$$e = \frac{\hat{v}_{Qn} - \hat{v}_{Pn}}{v_{Pn} - v_{Qn}} \tag{5.7.1}$$

where v_{Pn} and v_{Qn} are the normal (n) components of the velocities of P and Q before impact, and $\hat{v}_{Pn}$ and $\hat{v}_{Qn}$ are the normal (n) components of the velocities of P and Q after impact. (In the sequel the "overhat" ($\hat{}$) is used to designate post-impact velocities.)

The coefficient of restitution e ranges in value between zero and one. Table 5.7.1 lists three characterizations of the impact depending upon the values of e.

Table 5.7.1 Impact Characterizations

Coefficient of Restitution	Characterization
$e = 0$	Plastic
$0 < e < 1$	Partially Elastic
$e = 1$	Fully Elastic

5.8 Application: Direct Impact

Direct impact occurs when the colliding particles have colinear motion as in Figure 5.8.1 Although direct impact will seldom occur in an exact sense, it is nevertheless representative of many physical occurrences. Perhaps the most common and of greatest importance of these are head-on and rear-end automobile collisions.

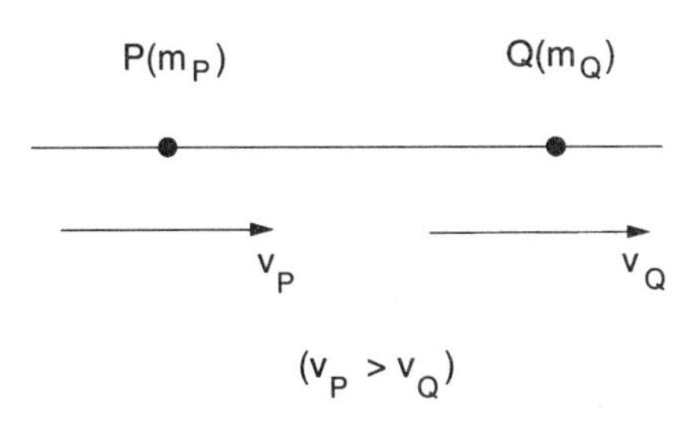

Fig. 5.8.1 Colliding Particles with Colinear Motion: Direct Impact

Referring to Figure 5.8.1 let P and Q be the colliding particles having masses m_P and m_Q. Since the particles move on the same straight line, it is not necessary to use vector notation to describe their movements. To that end let v_P and v_Q represent the velocities of P and Q (positive to the right) before collision and let $\hat{v}_P$ and $\hat{v}_Q$ represent the velocities of P and Q after collision. Then collision will occur only if v_P is greater than v_Q.

The principal problem in direct impact analysis is: Given the pre-impact parameters, that is, m_P, m_Q, v_P, v_Q, and the restitution coefficient e, determine the post-impact speeds $\hat{v}_P$ and $\hat{v}_Q$. Two equations are needed to determine these two unknowns. The first of these may be determined by observing that if P and Q are taken together as a system, then the impulsive forces of impact are internal to the system. This means that the net impulse on the system is zero and hence the overall momentum of the system is unchanged during the impact. That is, the system momentum is conserved leading to the equation:

$$m_P v_P + m_Q v_Q = m_P \hat{v}_P + m_Q \hat{v}_Q \tag{5.8.1}$$

The second equation may be obtained by observing that during impact the coefficient provides a relation between the relative pre-impact and post-impact velocities as in Equation (5.7.1). That is,

$$e(v_P - v_Q) = \hat{v}_Q - \hat{v}_P \tag{5.8.2}$$

Solving Equations (5.8.1) and (5.8.2) for $\hat{v}_P$ and $\hat{v}_Q$ gives:

$$\hat{v}_P = [(m_P - em_Q)v_P + (1 + e)m_Q v_Q]/(m_P + m_Q) \tag{5.8.3}$$

and

$$\hat{v}_Q = [(m_Q - em_P)v_Q + (1 + e)m_P v_P]/(m_P + m_Q) \tag{5.8.4}$$

Although the momentum is conserved during impact, the energy is not conserved. Equations (5.8.3) and (5.8.4) may be used to determine the loss of energy: That is,

$$\Delta K = K - \hat{K}$$

$$= \left[(1/2)m_P v_P^2 + (1/2)m_Q v_Q^2\right] \tag{5.8.5}$$

or

$$\Delta K = (1 - e^2)m_P m_Q (v_Q - v_P)^2/2(m_P + m_Q)$$

Example 5.8.1 Elastic Impact of Identical Particles

To illustrate the use of Equations (5.8.3) and (5.8.4) consider the perfectly elastic ($e = 1$) collision of identical particles ($m_P = m_Q$). In this case Equations (5.8.3) and (5.8.4) take the form:

$$\hat{v}_P = v_Q \tag{5.8.6}$$

and

$$\hat{v}_Q = v_P \tag{5.8.7}$$

Observe that there is a complete momentum exchange between the particles. Observe further from Equation (5.8.5) that there is no energy loss during the collision.

Example 5.8.2 Elastic Impact of Vastly Dissimilar Particles

Consider next an elastic collision where there is great disparity of the masses, that is, say m_P greatly exceeds m_Q. In this case Equation (5.8.3) and (5.8.4) become:

$$\hat{v}_P = v_P \tag{5.8.8}$$

and

$$\hat{v}_Q = -v_Q + 2v_P \tag{5.8.9}$$

Observe here that if the initial speed of Q is zero, then Q may be projected at twice the speed of the striking particle.

Example 5.8.3 Plastic Impact

Consider a plastic impact ($e = 0$) where the colliding particles stick together after the impact. In this case Equations (5.8.3) and (5.8.4) become

$$\hat{v}_P = (m_P v_P + m_Q v_Q)/(m_P + m_Q) \tag{5.8.10}$$

and

$$\hat{v}_Q = (m_P v_P + m_Q v_Q)/(m_P + v_Q) \tag{5.8.11}$$

Observe that in this plastic case the post-impact speeds of the particles are identical. Observe further that if the initial speed of Q (the "struck" particle) is zero, then the post-impact speeds are

$$\hat{v}_P = \hat{v}_Q = m_P v_P/(m_P + m_Q) \tag{5.8.12}$$

Observe that the energy loss is not zero. That is, from Equation (5.8.5) we see that the kinetic energy loss is:

$$\Delta K = m_P m_Q (v_P - v_Q)^2/2(m_P + m_Q) \tag{5.8.13}$$

If the particles have identical masses and if the initial speed of Q is zero, the kinetic energy loss is

$$\Delta K = (1/4) m v_P^2 \tag{5.8.14}$$

That is, the energy is reduced by one half.

These results and others are summarized in Table 5.8.1.

Table 5.8.1 Typical Results for Direct Impact of Colliding Particles

Conditions	Pre-Impact Speeds		Post-Impact Speeds	
	v_P	v_Q	$\hat{v}_P$	$\hat{v}_Q$
1. Elastic Impact (e = 1)	v	0	$(m_P - m_Q)v/(m_P + m_Q)$	$2m_P v/(m_P + M_Q)$
2. Elastic Impact (e = 1) and Industrial particles ($m_P = m_Q = m$)	v	0	0	v
3. Elastic Impact (e = 1) and P much more massive than Q ($m_P \gg m_Q$)	v_P	v_Q	v_P	$-v_Q + 2v_P$
4. Elastic Impact (e = 1) and P much more massive than Q ($m_P \gg m_Q$)	v	0	v	2v
5. Plastic Impact (e = 0)	v	0	$m_P v/(m_P + m_Q)$	$m_P v/(m_P + m_Q)$

Continued

Table 5.8.1 Continued

Conditions	Pre-Impact Speeds		Post-Impact Speeds	
	v_P	v_Q	$\hat{v}_P$	$\hat{v}_Q$
6. Plastic Impact ($e = 1$) and Identical Particles ($m_P = m_Q = m$)	v	0	v/2	v/2

Example 5.8.4 Determination of Coefficient of Restitution

Let a particle P be dropped onto a fixed horizontal surface from rest at a distance h above the surface. If P is observed to bounce back to a height $\hat{h}$, find the coefficient of restitution e.

Solution: From the work-energy principle, if P is dropped from a height h, its speed v at impact is $-\sqrt{2gh}$. Similarly, if P rebounds to a height $\hat{h}$, its rebound speed $\hat{v}$ is $\sqrt{2g\hat{h}}$. Since the coefficient of restitution e is the ratio of $\hat{v}$ to v [see Equation (5.7.1)] we see that e is

$$e = \sqrt{\hat{h}/h} \tag{5.8.15}$$

5.9 Application: Oblique Impact

When particles collide they usually do not have colinear motion prior to the impact. That is, direct impact generally does not occur. Instead, colliding particles generally approach each other on paths inclined to one another as depicted in Figure 5.9.1 — so called "oblique impact." As with direct impact, the typical problem is to determine the post-impact conditions when given the pre-impact conditions. That is, given the masses of the colliding particles and their immediate pre-impact velocities, the objective is to determine the immediate post-impact velocities.

Observe that the approach paths, as in Figure 5.9.1, determine a plane where the impact occurs. This impact plane, however, is not necessarily a horizontal or vertical plane or even a convenient reference plane. Nevertheless, the impact analysis can be conducted in that plane, thus allowing the analysis to be two-dimensional.

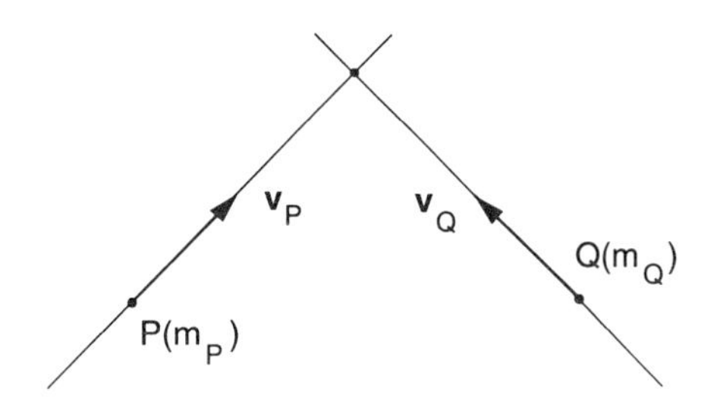

Fig. 5.9.1 Paths of Approach for Particles with Oblique Impact

In the plane of impact each velocity vector will have two components. This means that for the two particles, there will be four pre-impact velocity components and four post-impact velocity components. Then if the four post-impact velocity components are to be determined, we will need four governing equations.

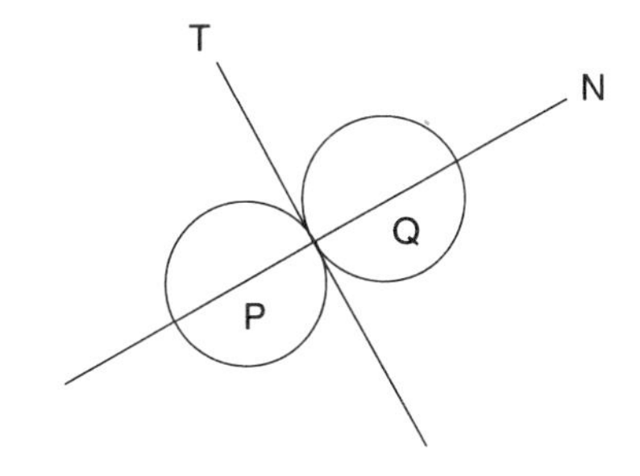

Fig. 5.9.2 Directions Normal and Tangent to the Plane of Impact

To develop these governing equations it is convenient, as before, to consider the colliding particles as colliding spheres and to conduct the analysis in directions normal and tangent to the contact plane as in Figure 5.9.2. The movement of the particles in the normal direction may be analyzed in exactly the same way as with direct impact. That is, the overall momentum in the normal direction is conserved and the ratio of the separation speed to the approach speed for the particles, in the normal direction, is the coefficient of restitution. Specifically, the momentum conservation equation is [see Equation (5.8.1)]:

$$m_P v_{Pn} + m_Q v_{Qn} = m_P \hat{v}_{Pn} + m_Q \hat{v}_{Qn} \tag{5.9.1}$$

and the impact equation is [see Equation (5.8.2)]:

$$e(v_{Pn} - v_{Qn}) = \hat{v}_{Qn} - \hat{v}_{Pn} \tag{5.9.2}$$

where the subscript n designates the velocity component in the normal direction.

The movement of the particles in the tangential direction may be analyzed by assuming that there is no friction between the colliding spheres which would exert forces in the tangential direction on the particles during impact.* Then with the absence of tangential forces the momenta of the individual particles in the tangential direction are conserved. This leads to the equations:

$$m_P v_{Pt} = m_P \hat{v}_{Pt} \tag{5.9.3}$$

and

$$m_Q v_{Qt} = m_Q \hat{v}_{Qt} \tag{5.9.4}$$

where the subscript t designates the velocity component in the tangential direction.

Equations (5.9.1), (5.9.2), (5.9.3), and (5.9.4) may be solved for the post-impact velocity components. The results are

$$\hat{v}_{Pn} = [(m_P - em_Q)v_{Pn} + (1 + e)m_Q v_{Qn}]/(m_P + m_Q) \tag{5.9.5}$$

$$\hat{v}_{Qn} = [(m_Q - em_P)v_{Qn} + (1 + e)m_P v_{Pn}]/(m_P + m_Q) \tag{5.9.6}$$

---

* This is a reasonable assumption for small particles with little, if any, rotational inertia.

$$\hat{v}_{Pt} = v_{Pt} \tag{5.9.7}$$

$$\hat{v}_{Qt} = v_{Qt} \tag{5.9.8}$$

The kinetic energy lost during impact is

$$\Delta K = (1 - e^2) m_P m_Q (v_{Qn} - v_{Pn})^2 / 2(m_P + m_Q) \tag{5.9.9}$$

Example 5.9.1 Typical Oblique Collision

To illustrate the use of Equation (5.9.5) to (5.9.8), consider the problem of finding the post-impact velocities of two colliding particles P and Q as in Figure 5.9.3. For simplicity, let the particles have equal masses and let the collision of the particles (represented as small spheres) occur in the X-Y plane as shown, and let the collision itself occur on surfaces normal to the X-direction as in Figure 5.9.4.

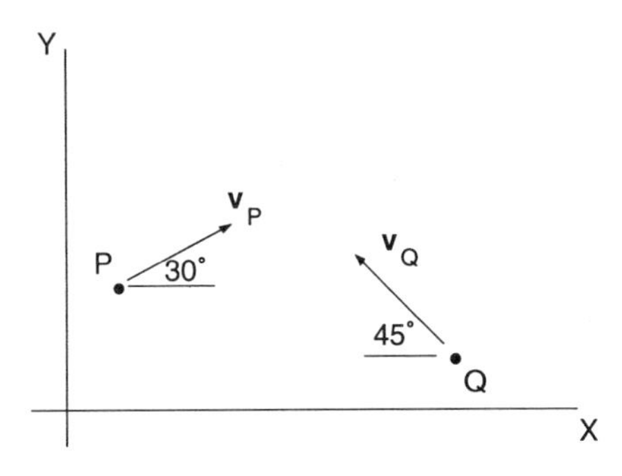

Fig. 5.9.3 Typical Oblique Impact

Finally, for specificity, let the magnitudes of the particle velocities, $\mathbf{v}_P$ and $\mathbf{v}_Q$, be 10 and 15 m/s respectively and let the restitution coefficient e have the values: 0.0, 0.5, and 1.0 representing plastic, semi-elastic,

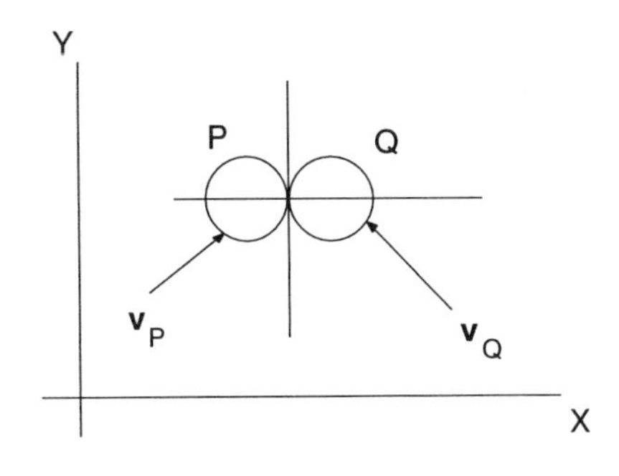

Fig. 5.9.4 Representation of the Particle Impact

and elastic collisions.

Solution: From Figure 5.9.3 and with $|\mathbf{v}_P|$ and $|\mathbf{v}_Q|$ having the values of 10 and 15 m/s respectively we see that the pre-impact velocity components are:

$$v_{Pn} = 8.66 \qquad v_{Pt} = 5 \qquad v_{Qn} = -10.61 \qquad v_{Qt} = 10.61 \text{ m/s}$$

Then from Equations (5.9.5) to (5.9.8) we see that the post-impact velocity components are as listed in Table 5.9.1.

Table 5.9.1 Post-Impact Velocity Components (m/s) and Directions [θ (Deg)] for Example 5.9.1

Condition	$\hat{v}_{Pn}$	$\hat{v}_{Pt}$	$\hat{\theta}_P$	$\hat{v}_{Qn}$	$\hat{v}_{Qt}$	$\hat{\theta}_Q$
Plastic Impact (e = 0.0)	-0.973	5	101	-0.973	10.61	95.2
Semi-Elastic Impact (e = 0.5)	-5.79	5	139	3.84	10.61	70.1
Elastic (e = 1.0)	-10.61	5	115	8.66	10.61	51

Observe that for the plastic impact, the post-impact tangential velocity components are different for P and Q, whereas the normal components are the same. The difference in the tangent components is due to the assumption of frictionless impact.

5.10 Summary, Comparison of Methods/Formulas

The listing of dynamics principles and laws in Table 5.2.1 gives the analyst a menu of methods to use for any given problem. The method to choose will, of course, depend upon the problem itself. There are, however, several guidelines which might be helpful, based upon the relative advantages and disadvantages of the various methods. Table 5.10.1 provides a listing of these advantages and disadvantages.

Table 5.10.1 Advantages and Disadvantages of Dynamics Principles/Laws/Methods

Method Name	Advantages	Disadvantages
1. Newton's Second Law	Simple formulation; no restrictions on applica-tions; well-known and well-understood	Can be cumbersome in working with large numbers of particles; requires knowledge of accelerations
2. d'Alembert's Principle	Similar to Newton's Second Law; use of inertia force gives intuitive understanding in many problems	Can be cumbersome in working with large numbers of particles; requires knowledge of acceleration; some analysts believe inertia forces to be fictitious
3. Work-Energy	Uses energy functions and thus avoids the need to calculate accelerations	Provides only one equation
4. Impulse-Momentum	Very useful for problems involving collisions and impact	Can be cumbersome in problems which do not involve collision or impact
5. Kane's Equations	Automatically eliminates non-working constraint forces from the analysis. No kinematic restrictions on the range of applications.	Less intuitive than Newton's Second Law or d'Alembert's Principle; requires knowledge of acceleration

Continued

Table 5.10.1 Continued

Method Name	Advantages	Disadvantages
6. Lagrange's Equations	Automatically eliminates non-working constraint forces from the analysis. Uses energy functions and thus avoids the need to calculate accelerations. Unlike work-energy, Lagrange's equations provide an equation for each degree of freedom	Not readily applicable with non-holonomic systems. Differentiation of kinetic energy functions can be tedious for large systems.

References

5.1 W. E. Boyce and R. C. DiPrima, *Elementary Differential Equations*, Wiley, New York, NY, 1965.

5.2 E. A. Coddington, *An Introduction to Ordinary Differential Equations*, Prentice Hall, Englewood Cliffs, NJ, 1961.

5.3 M. Golomb and M. Shanks, *Elements of Ordinary Differential Equations*, McGraw Hill, New York, NY, 1965.

5.4 G. M. Murphy, *Ordinary Differential Equations and Their Solutions*, D. Van Nostrand, New York, 1960.

5.5 F. P. Beer and E. R. Johnston, Jr., *Vector Mechanics for Engineers*, Fourth Edition, McGraw Hill, New York, NY, 1984, pp. 562-581.

5.6 R. M. Brach, *Mechanical Impact Dynamics—Rigid Body Collisions*, John Wiley & Sons, New York, NY, 1991.

Chapter 6

KINEMATICS OF BODIES

6.1 Introduction

A body may be regarded as a set of particles bonded together with adjoining particles maintaining their positions relative to one another. If the distance between all pairs of particles remains constant, the body is said to be "rigid." In this chapter we review and summarize formulas for the kinematics of bodies with a focus upon rigid bodies.

For a rigid body there are four basic quantities which determine a complete description of the kinematics of the body. These are: angular velocity, angular acceleration, mass center velocity and mass center acceleration. Of these four, the most fundamental is angular velocity. Indeed, the angular velocity may often be used to determine the other three quantities. Hence, we will begin our review with angular velocity.

Since angular velocity may be regarded as the time rate of change of orientation, it is helpful to first consider concepts of orientation.

6.2 Orientation of Bodies

Angular velocity, regarded as the time rate of change of orientation, is analogous to velocity which is the time rate of change of position. Unlike position, however, orientation is somewhat more difficult to define. Orientation might be described as inclination, cantering, or angular positioning; but whereas particle position is conveniently defined by a position vector, there is no corresponding vector to define body orientation.

Consider a body B moving in a reference frame R as in Figure 6.2.1. Let B be rigid and be regarded as being composed of physical particles — as a sand stone is made up of sand crystals. Let $\hat{B}$ be a reference frame embedded in B and fixed relative to the particles of B as indicated in Figure 6.2.1. A reference frame may be regarded as a set of axes or as the set of points in the space of the axes. If a reference frame is regarded in terms of points these points in turn may be identified with the particles of B, if the particles are considered to be "small," or without significant size. In this way we may identify the body B with the reference frame $\hat{B}$ and conversely $\hat{B}$ with B. That is, in a kinematic analysis where mass is not important, there is no difference between a rigid body and a reference frame.

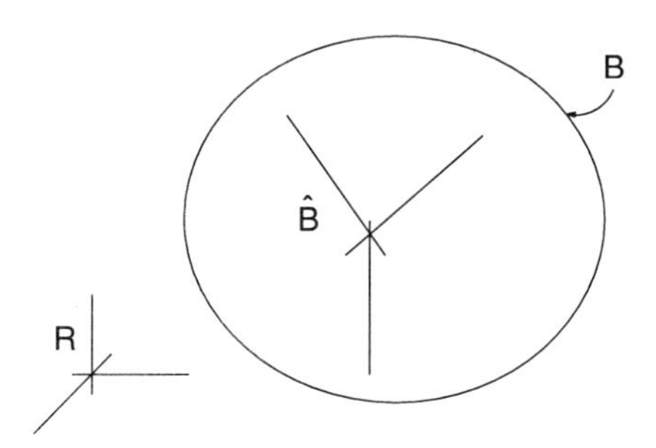

Fig. 6.2.1 A Rigid Body B Moving in a Reference Frame R

By identifying a body B with a reference frame $\hat{B}$ the orientation of B in a reference frame R may then be defined in terms of the orientation of $\hat{B}$ in R as in Figure 6.2.2. The orientation of $\hat{B}$ in R may be defined in terms of the angles between the respective axes. Specifically, let unit vectors $\mathbf{N}_i$ and $\mathbf{n}_i$ $(i = 1,2,3)$ be unit vectors parallel to the axes of R and $\hat{B}$ as shown. (For convenience, let the axes be mutually perpendicular.) Then the angles describing the inclination of the axes of $\hat{B}$ relative to those of R may be represented by their cosines determined by the products: $\mathbf{n}_i \cdot \mathbf{n}_j$ $(i,j = 1,2,3)$. Then we can form an array of these cosines,

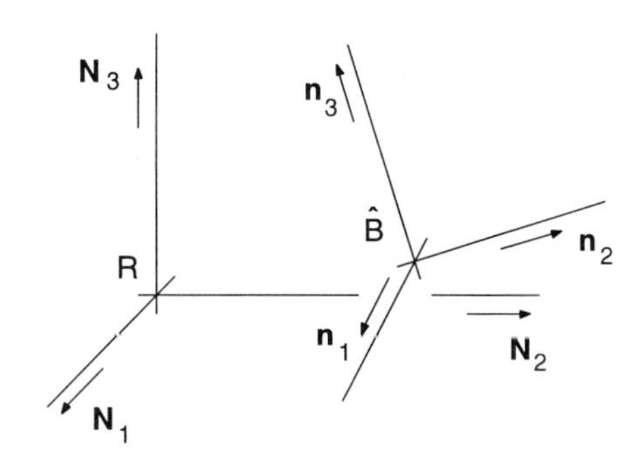

Fig. 6.2.2 Reference Frames $\hat{B}$ and R and Associated Unit Vectors

called the "direction cosine array" or "orientation array" S defined as

$$S = [S_{ij}] = [\mathbf{N}_i \bullet \mathbf{n}_j] \tag{6.2.1}$$

It is readily seen that S is non-singular and orthogonal. That is, the inverse is the transpose

$$S^{-1} = S^T \tag{6.2.2}$$

S may also be regarded as a "transformation matrix" relating the components of a vector relative to $\mathbf{N}_i$ and $\mathbf{n}_j$ (see Section 2.20). That is, if a vector $\mathbf{v}$ is expressed as:

$$\mathbf{V} = V_i \mathbf{N}_i = v_j \mathbf{n}_j \tag{6.2.3}$$

then the V_i and v_j are related by the expressions

$$V_i = S_{ij} v_j \quad \text{and} \quad v_j = S_{ij} V_i \tag{6.2.4}$$

where as before there is a sum over the range of the repeated index.

S may also be used to relate the unit vectors themselves to one another. That is,

$$\mathbf{N}_i = S_{ij} \mathbf{n}_j \quad \text{and} \quad \mathbf{n}_j = S_{ij} \mathbf{N}_i \tag{6.2.5}$$

[Observe the similarity of Equations (6.2.4) and (6.2.5).]

S may be conveniently constructed and expressed in terms of orientation angles as follows: Let the axes of $\hat{B}$ be initially aligned with those of R. Let $\hat{B}$ then be brought to a general orientation relative to R by three successive dextral rotations of $\hat{B}$ about $\mathbf{n}_1$, $\mathbf{n}_2$, and $\mathbf{n}_3$ through angles α, β, and γ. Then S may be expressed as

$$S = ABC = \begin{bmatrix} 1 & 0 & 0 \\ 0 & c_\alpha & -s_\alpha \\ 0 & s_\alpha & c_\alpha \end{bmatrix} \begin{bmatrix} c_\beta & 0 & s_\beta \\ 0 & 1 & 0 \\ -s_\beta & 0 & c_\beta \end{bmatrix} \begin{bmatrix} c_\gamma & -s_\gamma & 0 \\ s_\gamma & c_\gamma & 0 \\ 0 & 0 & 1 \end{bmatrix} \tag{6.2.6}$$

where s_α and c_α are abbreviations for $\sin\alpha$ and $\cos\alpha$, etc., and where the matrices A, B, and C are defined by inspection. [Observe that (like S) A, B, and C are also orthogonal.]

By carrying out the multiplication of Equation (6.2.6) S may be expressed as:

$$S = \begin{bmatrix} c_\beta c_\gamma & -c_\beta s_\gamma & s_\beta \\ (c_\alpha s_\gamma + s_\alpha s_\beta c_\gamma) & (c_\alpha c_\gamma - s_\alpha s_\beta s_\gamma) & -s_\alpha c_\beta \\ (s_\alpha s_\gamma - c_\alpha s_\beta c_\gamma) & (s_\alpha c_\gamma + c_\alpha s_\beta s_\gamma) & c_\alpha c_\beta \end{bmatrix} \tag{6.2.7}$$

The following section presents a discussion of the development of Equation (6.2.6).

6.3 Configuration Graphs

In addition to defining the orientation of a body B in a reference frame R, the orientation matrix (or transformation matrix) S provides a means for expressing unit vectors fixed in B in terms of those fixed in R and conversely. To develop algorithms for this, as in Equation (6.2.6), consider a pair of unit vector sets $\mathbf{N}_i$ and $\hat{\mathbf{N}}_i$ $(i = 1,2,3)$ as in Figure 6.3.1 where the unit vectors $\mathbf{N}_1$ and $\hat{\mathbf{N}}_i$ are coincident and where the unit vector pairs $\mathbf{N}_2, \hat{\mathbf{N}}_2$ and $\mathbf{N}_3, \hat{\mathbf{N}}_3$ are inclined relative to each other by an angle α as shown. Observe that for this configuration it is relatively easy to express the unit vectors in terms of one another, leading to the equations:

$$
\begin{aligned}
\mathbf{N}_1 &= \hat{\mathbf{N}}_1 & \hat{\mathbf{N}}_1 &= \mathbf{N}_1 \\
\mathbf{N}_2 &= c_\alpha \hat{\mathbf{N}}_2 - s_\alpha \hat{\mathbf{N}}_3 \quad \text{and} & \hat{\mathbf{N}}_2 &= c_\alpha \mathbf{N}_2 + s_\alpha \mathbf{N}_3 \\
\mathbf{N}_3 &= s_\alpha \hat{\mathbf{N}}_2 + c_\alpha \hat{\mathbf{N}}_3 & \hat{\mathbf{N}}_3 &= -s_\alpha \mathbf{N}_2 + c_\alpha \mathbf{N}_3
\end{aligned}
\tag{6.3.1}
$$

where as before s_α and c_α are abbreviations for $\sin\alpha$ and $\cos\alpha$.

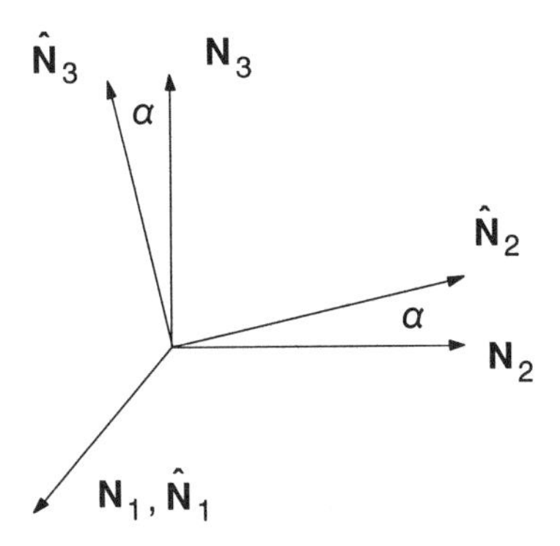

Fig. 6.3.1 Unit Vector Sets $\mathbf{N}_i$ and $\hat{\mathbf{N}}_i$ (i = 1,2,3)

In matrix form Equations (6.3.1) may be written as

$$
\begin{bmatrix} \mathbf{N}_1 \\ \mathbf{N}_2 \\ \mathbf{N}_3 \end{bmatrix} = \begin{bmatrix} 1 & 0 & 0 \\ 0 & c_\alpha & -s_\alpha \\ 0 & s_\alpha & c_\alpha \end{bmatrix} \begin{bmatrix} \hat{\mathbf{N}}_1 \\ \hat{\mathbf{N}}_2 \\ \hat{\mathbf{N}}_3 \end{bmatrix} \quad \text{and} \quad \begin{bmatrix} \hat{\mathbf{N}}_1 \\ \hat{\mathbf{N}}_2 \\ \hat{\mathbf{N}}_3 \end{bmatrix} = \begin{bmatrix} 1 & 0 & 0 \\ 0 & c_\alpha & s_\alpha \\ 0 & -s_\alpha & c_\alpha \end{bmatrix} \begin{bmatrix} \mathbf{N}_1 \\ \mathbf{N}_2 \\ \mathbf{N}_3 \end{bmatrix}
\tag{6.3.2}
$$

or as

$$
\mathrm{N} = \mathrm{A}\hat{\mathrm{N}} \quad \text{and} \quad \hat{\mathrm{N}} = \mathrm{A}^{\mathrm{T}}\mathrm{N}
\tag{6.3.3}
$$

where N and $\hat{\mathrm{N}}$ are the column arrays of Equation (6.3.2), and where A is the transformation matrix which by inspection is seen to be orthogonal. That is,

$$A^T = A^{-1} \tag{6.3.4}$$

Equations (6.3.1) and as a result, Equations (6.3.2) to (6.3.4) are readily developed by observing that when the inclination angle α is zero, the unit vector sets are mutually aligned. This in turn means that when α is not zero, the unit vectors with like subscripts ($\mathbf{N}_2$, $\hat{\mathbf{N}}_2$ and $\mathbf{N}_3$, $\hat{\mathbf{N}}_3$) are related by $\cos\alpha$. [Recall that $\cos 0 = 1$ and $\cos(\pi/2) = 0$.] Similarly, unit vectors with unlike subscripts are related by $\pm\sin\alpha$. [Recall that $\sin 0 = 0$ and $\sin(\pi/2) = 1$.] Then the only decision in Equations (6.3.1) is where to place the minus signs. In the plane of the vectors $\mathbf{N}_2$, $\hat{\mathbf{N}}_2$, $\mathbf{N}_3$, and $\hat{\mathbf{N}}_3$ as in Figure 6.3.2, we see that two of the vectors, namely $\hat{\mathbf{N}}_2$ and $\mathbf{N}_3$, are "inside" the vectors $\mathbf{N}_2$ and $\hat{\mathbf{N}}_3$. Alternatively, $\mathbf{N}_2$ and $\hat{\mathbf{N}}_3$ are "outside" of $\hat{\mathbf{N}}_2$ and $\mathbf{N}_3$. By inspection (imagine α being $\pi/2$, or 90°) we see that the outside vectors require the negative signs for the $\sin\alpha$ terms.

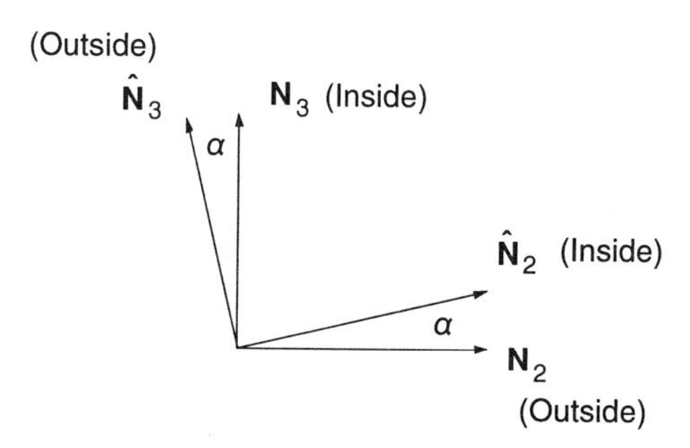

Fig. 6.3.2 Plane of Unit Vector Show Inside and Outside Vectors

The information contained in Equations (6.3.1) to (6.3.3) and the description of the foregoing paragraph may be condensed into a simple diagram called a "configuration graph" [6.1] as in Figure 6.3.3. In this diagram there are six dots, or nodes, each representing one of the six unit vectors of Figure 6.3.2 and as indicated in Figure 6.3.3. The horizontal line then designated equality ($\mathbf{N}_1 = \hat{\mathbf{N}}_1$). The inclined line connects nodes associated with the "inside" unit vectors ($\mathbf{N}_3$ and $\hat{\mathbf{N}}_2$). Consequently, the unconnected nodes are associated with the "outside" unit vectors

($\mathbf{N}_2$ and $\hat{\mathbf{N}}_3$). These observations enable us to immediately use the diagram of Figure 6.3.3 to produce Equations (6.3.1) and (6.3.2). Indeed, we could even develop an algorithm for equation development as in Figure 6.3.4.

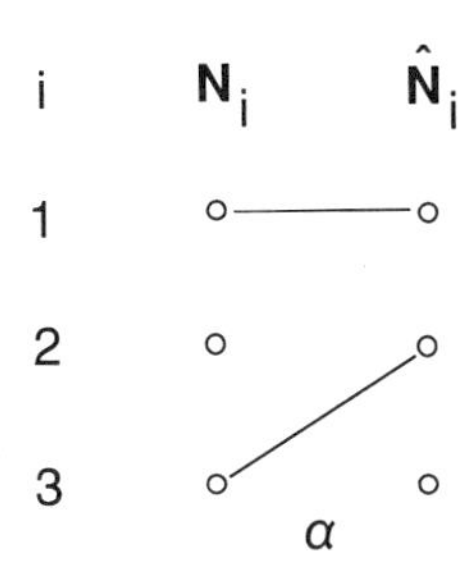

Fig. 6.3.3 Configuration Graph for the Unit Vector Sets of Figure 6.3.1

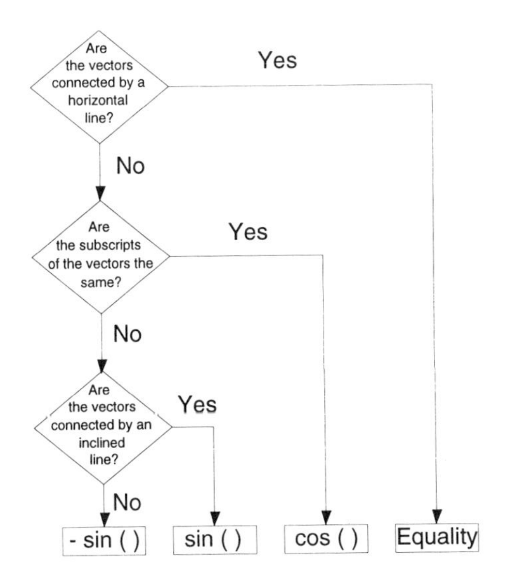

Fig. 6.3.4 Equation Coefficients and Signs Obtained from a Configuration Graph

Example 6.3.1 Inclinations About the 2- and 3-Axes

To illustrate the use of configuration graphs consider a pair of mutually perpendicular unit vector sets $\hat{\mathbf{N}}_i$ and $\hat{\mathbf{n}}_i$ (i = 1,2,3) inclined relative to each other by the angle β as in Figure 6.3.5. By observation in the figure we see that the vectors $\hat{\mathbf{N}}_1$ and $\hat{\mathbf{n}}_2$ are "inside" vectors, and that $\hat{\mathbf{n}}_1$ and $\hat{\mathbf{N}}_3$ are "outside" vectors. The configuration graph relating $\hat{\mathbf{N}}_i$ and $\hat{\mathbf{n}}_i$ is then seen to be as shown in Figure 6.3.6.

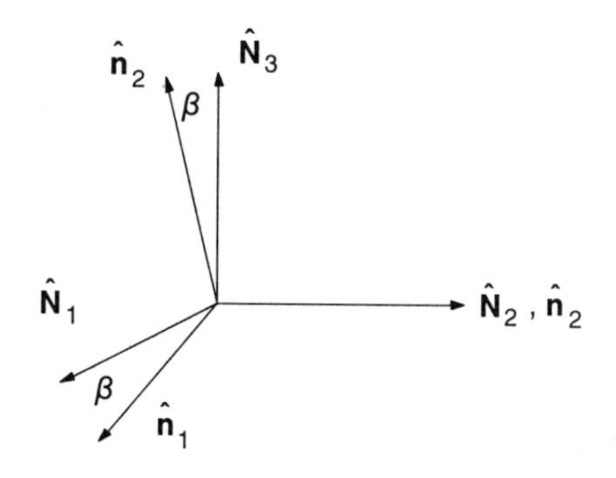

Fig. 6.3.5 Unit Vector Sets $\hat{\mathbf{N}}_i$ and $\hat{\mathbf{n}}_i$ (i = 1,2,3)

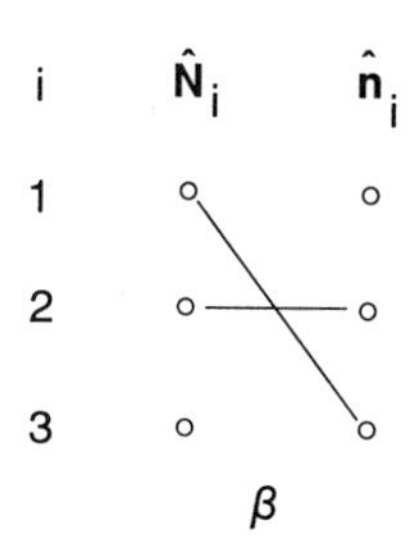

Fig. 6.3.6 Configuration Graph for the Unit Vector Scts of Figure 6.3.5

From this graph we immediately obtain the expressions:

$$\begin{aligned} \hat{\mathbf{N}}_1 &= c_\beta \hat{\mathbf{n}}_1 + s_\beta \hat{\mathbf{n}}_3 & \qquad \hat{\mathbf{n}}_1 &= c_\beta \hat{\mathbf{N}}_1 - s_\beta \hat{\mathbf{N}}_3 \\ \hat{\mathbf{N}}_2 &= \hat{\mathbf{n}}_2 & \text{and} \quad \hat{\mathbf{n}}_2 &= \hat{\mathbf{N}}_2 \\ \hat{\mathbf{N}}_3 &= -s_\beta \hat{\mathbf{n}}_1 + c_\beta \hat{\mathbf{n}}_3 & \qquad \hat{\mathbf{n}}_3 &= s_\beta \hat{\mathbf{N}}_1 + c_\beta \hat{\mathbf{N}}_3 \end{aligned} \tag{6.3.5}$$

and

$$\begin{bmatrix} \hat{\mathbf{N}}_1 \\ \hat{\mathbf{N}}_2 \\ \hat{\mathbf{N}}_3 \end{bmatrix} = \begin{bmatrix} c_\beta & 0 & s_\beta \\ 0 & 1 & 0 \\ -s_\beta & 0 & c_\beta \end{bmatrix} \begin{bmatrix} \hat{\mathbf{n}}_1 \\ \hat{\mathbf{n}}_2 \\ \hat{\mathbf{n}}_3 \end{bmatrix} \quad \text{and} \quad \begin{bmatrix} \hat{\mathbf{n}}_1 \\ \hat{\mathbf{n}}_2 \\ \hat{\mathbf{n}}_3 \end{bmatrix} = \begin{bmatrix} c_\beta & 0 & -s_\beta \\ 0 & 1 & 0 \\ s_\beta & 0 & c_\beta \end{bmatrix} \begin{bmatrix} \hat{\mathbf{N}}_1 \\ \hat{\mathbf{N}}_2 \\ \hat{\mathbf{N}}_3 \end{bmatrix} \tag{6.3.6}$$

or

$$\hat{\mathbf{N}} = \mathrm{B}\hat{\mathbf{n}} \quad \text{and} \quad \hat{\mathbf{n}} = \mathrm{B}^{\mathrm{T}}\hat{\mathbf{N}} \tag{6.3.7}$$

where the arrays $\hat{\mathbf{N}}$, $\hat{\mathbf{n}}$ and B are defined by comparing Equations (6.3.6) and (6.3.7).

Similarly, let unit vector sets $\hat{\mathbf{n}}_i$ and $\mathbf{n}_i$ be inclined relative to each other by the angle γ as in Figure 6.3.7. By observation in this figure we see that the vectors $\hat{\mathbf{n}}_3$ and $\mathbf{n}_3$ are equal, that $\mathbf{n}_1$ and $\hat{\mathbf{n}}_2$ are "inside" vectors, and that $\hat{\mathbf{n}}_1$ and $\mathbf{n}_2$ are "outside" vectors. Then the configuration graph relating $\hat{\mathbf{n}}_i$ and $\mathbf{n}_i$ is as shown in Figure 6.3.8, and expressions relating the unit vector sets are:

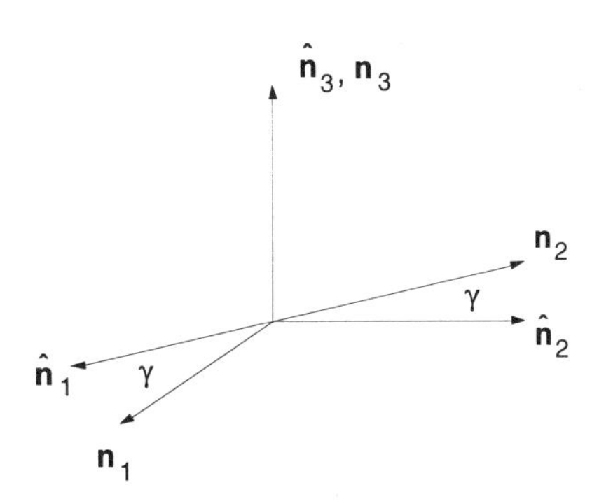

Fig. 6.3.7 Unit Vector Sets $\hat{\mathbf{n}}_i$ and $\mathbf{n}_i$ (i = 1,2,3)

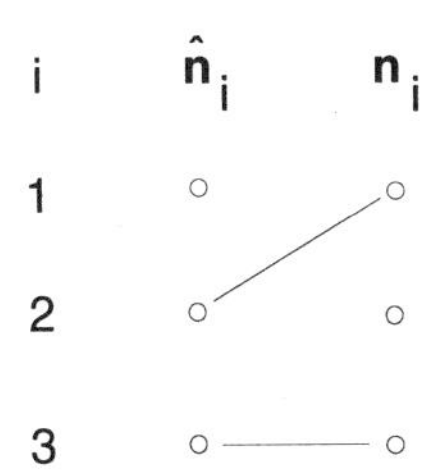

Fig. 6.3.8 Configuration Graph for the Unit Vector Sets of Figure 6.3.7

$$\begin{aligned} \hat{\mathbf{n}}_1 &= c_\gamma \mathbf{n}_1 - s_\gamma \mathbf{n}_2 \\ \hat{\mathbf{n}}_2 &= s_\gamma \mathbf{n}_1 + c_\gamma \mathbf{n}_2 \\ \hat{\mathbf{n}}_3 &= \mathbf{n}_3 \end{aligned} \quad \text{and} \quad \begin{aligned} \mathbf{n}_1 &= c_\gamma \hat{\mathbf{n}}_1 + s_\gamma \hat{\mathbf{n}}_2 \\ \mathbf{n}_2 &= -s_\gamma \hat{\mathbf{n}}_1 + c_\gamma \hat{\mathbf{n}}_2 \\ \mathbf{n}_3 &= \hat{\mathbf{n}}_3 \end{aligned} \tag{6.3.8}$$

and

$$\begin{bmatrix} \hat{\mathbf{n}}_1 \\ \hat{\mathbf{n}}_2 \\ \hat{\mathbf{n}}_3 \end{bmatrix} = \begin{bmatrix} c_\gamma & -s_\gamma & 0 \\ s_\gamma & c_\gamma & 0 \\ 0 & 0 & 1 \end{bmatrix} \begin{bmatrix} \mathbf{n}_1 \\ \mathbf{n}_2 \\ \mathbf{n}_3 \end{bmatrix} \quad \text{and} \quad \begin{bmatrix} \mathbf{n}_1 \\ \mathbf{n}_2 \\ \mathbf{n}_3 \end{bmatrix} = \begin{bmatrix} c_\gamma & s_\gamma & 0 \\ -s_\gamma & c_\gamma & 0 \\ 0 & 0 & 1 \end{bmatrix} \begin{bmatrix} \hat{\mathbf{n}}_1 \\ \hat{\mathbf{n}}_2 \\ \hat{\mathbf{n}}_3 \end{bmatrix} \tag{6.3.9}$$

or

$$\hat{\mathbf{n}} = c\mathbf{n} \quad \text{and} \quad \mathbf{n} = c\hat{\mathbf{n}} \tag{6.3.10}$$

where the arrays $\hat{\mathbf{n}}$, $\mathbf{n}$, and c are defined by comparing Equations (6.3.9) and (6.3.10).

Several additional comments might be helpful as an aid to using the configuration graphs and these expressions. First, observe that in Figures 6.3.1, 6.3.5, and 6.3.7 we have simulated rotations about the 1, 2, and 3-directions (through the angles α, β, and γ). By appropriately choosing α, β, and γ and by linking these rotations together we can bring a body B into any desired orientation relative to a reference frame R, as briefly outlined at the end of Section 6.2. Specifically, if the $\mathbf{N}_i$ $(i = 1,2,3)$ are fixed in R and if the $\mathbf{n}_i$ $(i = 1,2,3)$ are fixed in B, then the relations between the $\mathbf{N}_i$ and the $\mathbf{n}_i$ are conveniently obtained by adjoining the respective configuration graphs of Figures 6.3.3, 6.3.6, and 6.3.8. That is, by observing the common intermediate unit vector sets, namely the $\hat{\mathbf{N}}_i$ and $\hat{\mathbf{n}}_i$ $(i = 1,2,3)$ we obtain the general, or "three-rotation" graph of Figure 6.3.9.

Next, observe that by comparing the forms of the configuration graphs with the forms of the associated transformation matrices, we can recognize a pattern

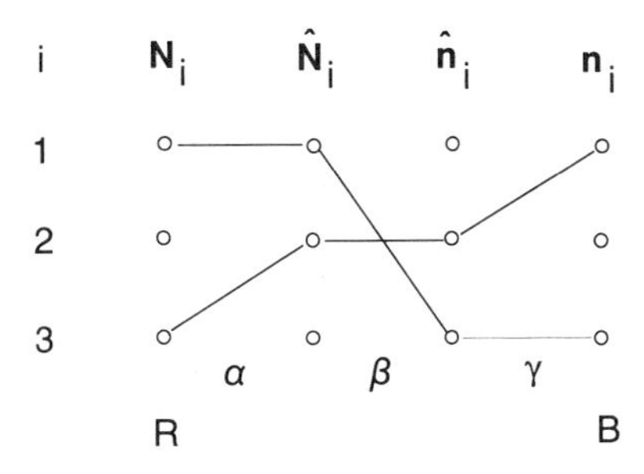

Fig. 6.3.9 Configuration Graph to Relate Unit Vectors Fixed in Body B and Reference Frame R

enabling us to immediately obtain the transformation matrix, once the configuration graph is constructed. Specifically, consider the graphs of Figures 6.3.3, 6.3.6, and 6.3.8 and the corresponding matrices A, B, and C of Equations (6.3.3), (6.3.6), and (6.3.9) placed atop one another as in Figure 6.3.10.

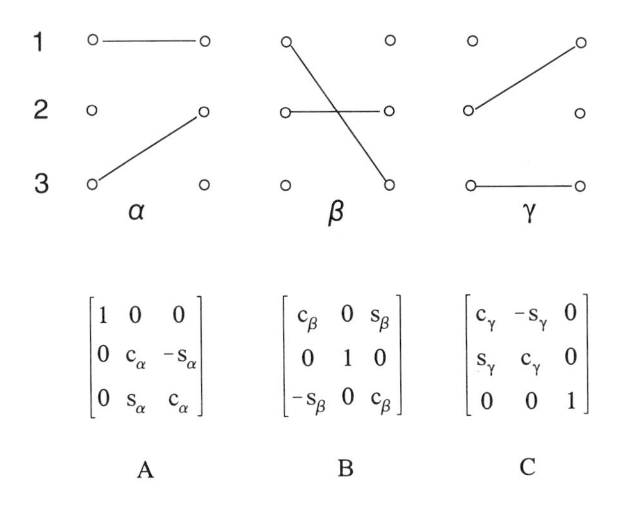

$$\begin{bmatrix} 1 & 0 & 0 \\ 0 & c_\alpha & -s_\alpha \\ 0 & s_\alpha & c_\alpha \end{bmatrix} \qquad \begin{bmatrix} c_\beta & 0 & s_\beta \\ 0 & 1 & 0 \\ -s_\beta & 0 & c_\beta \end{bmatrix} \qquad \begin{bmatrix} c_\gamma & -s_\gamma & 0 \\ s_\gamma & c_\gamma & 0 \\ 0 & 0 & 1 \end{bmatrix}$$

Fig. 6.3.10 Superposition and Juxtaposition of Configuration Graphs and Transformation Matrices

Then the position of the 1's and 0's in the matrices corresponds to the row of the horizontal line of the graphs. The cosines are then placed on the diagonals and the sines off-diagonal. The minus sign (which is always associated with one of the sine terms) is placed with the sine term in the *upper* row of the matrix if the slope of the inclined line in the graph is *positive* (that is upward). Alternatively, when the inclined line slope of the graph is *negative* (that is, downward), the minus sign is placed with the sine term in the *lower* row of the matrix.

Finally, observe that in view of the direct relation between the configuration graphs and the transformation matrices as represented in Figure 6.3.10, we see from the three-rotation graph of Figure 6.3.9 that the desired relations between the $\mathbf{N}_i$ and $\mathbf{n}_i$ unit vectors may be obtained from the expression

$$S = ABC \tag{6.3.11}$$

where from Equation (6.2.1), the elements S_{ij} of S are $\mathbf{N}_i \bullet \mathbf{n}_j$, and S itself is given by Equation (6.2.7).

The development of S in the foregoing paragraphs is based upon an envisioned three-axis rotation of B about the 1, 2, and 3-directions successively through the angles α, β, and γ. The angles as defined are positive when B has a "dextral" or "right-hand" rotation about the respective axes. Hence, α, β, and γ as defined are called "dextral" or "Bryan angles."

B may also be brought into a general orientation relative to R by different rotation sequences. For example, we could rotate B about the 1, 3, and 1 directions as developed in the following paragraphs:

Example 6.3.2 Alternative Rotation Sequences: 1,3,1-Rotations (Euler Angles)

To illustrate an alternative sequence let the unit vectors $\mathbf{n}_i$ of B be initially aligned with the $\mathbf{N}_i$ of R. Then let B be successively rotated about $\mathbf{n}_1$, $\mathbf{n}_3$, and then $\mathbf{n}_1$ again through the angles θ_1, θ_2, and θ_3 respectively. Then from Figures 6.3.9

and 6.3.10 the corresponding three-rotation configuration graph is as shown in Figure 6.3.11. The transformation matrix S relating the $\mathbf{N}_i$ and $\mathbf{n}_i$ may then be expressed as

$$S = [S_1(\theta_1)]\,[S_3(\theta_2)]\,[S_1(\theta_3)]$$

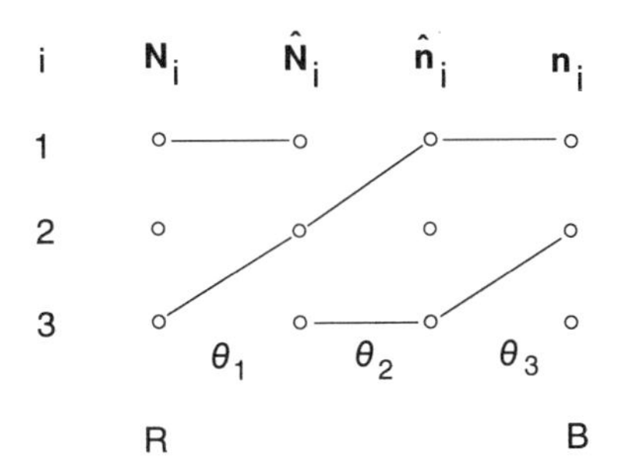

Fig. 6.3.11 Configuration Graph for a 1,3,1-Rotation Sequence

where from Figure 6.3.10, $S_1(\theta_1)$, $S_3(\theta_2)$ and $S_1(\theta_3)$ are

$$S_1(\theta_1) = \begin{bmatrix} 1 & 0 & 0 \\ 0 & c_1 & -s_1 \\ 0 & s_1 & c_1 \end{bmatrix} \quad S_3(\theta_2) = \begin{bmatrix} c_2 & -s_2 & 0 \\ s_2 & c_2 & 0 \\ 0 & 0 & 1 \end{bmatrix} \quad S_1(\theta_3) = \begin{bmatrix} 1 & 0 & 0 \\ 0 & c_3 & -s_3 \\ 0 & s_3 & c_3 \end{bmatrix} \tag{6.3.12}$$

where s_i and c_i are $\sin\theta_i$ and $\cos\theta_i$ $(i = 1,2,3)$. By carrying out the indicated multiplication, S becomes

$$S = \begin{bmatrix} c_2 & -s_2c_3 & s_2s_3 \\ c_1s_2 & (c_1c_2c_3 - s_1s_3) & (-c_1c_2s_3 - s_1c_3) \\ s_1s_2 & (s_1c_2c_3 + c_1s_3) & (-s_1c_2s_3 + c_1c_3) \end{bmatrix} \tag{6.3.13}$$

Still another way of orienting a body B in a reference frame R is to initially align the unit vectors $\mathbf{n}_i$ of B with the $\mathbf{N}_i$ of R and then perform successive rotations of B about the $\mathbf{N}_i$ axes. Such rotations may be called "space-fixed" rotations. (Correspondingly, rotation of B about the $\mathbf{n}_i$ are called "body-fixed" rotations [6.2].) The following example illustrates a space-fixed rotation about $\mathbf{N}_1$, $\mathbf{N}_2$, and $\mathbf{N}_3$.

Example 6.3.3 Space-Fixed X,Y,Z-Rotation

To illustrate a space-fixed rotation sequence consider a body B with unit vectors and axes aligned with those of a reference frame R, as in Figure 6.3.12. Next, let B be rotated about the X-axis (or equivalently, about $\mathbf{N}_1$) through the angle α as in Figure 6.3.13. A configuration graph relating the $\mathbf{N}_i$ and the $\mathbf{n}_i$ is then shown in Figure 6.3.14.

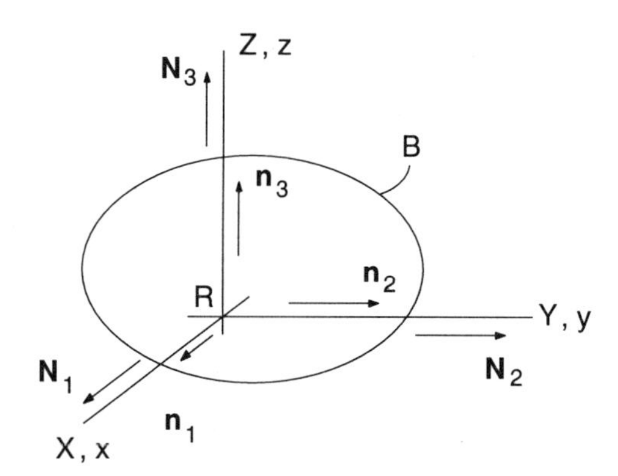

Fig. 6.3.12 Body B and Reference Frame R with Mutually Aligned Axes.

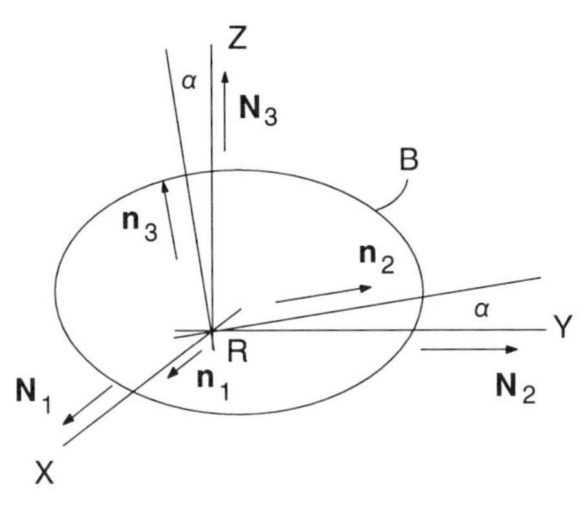

Fig. 6.3.13 Rotation of Body B about the X-Axis of Reference Frame R

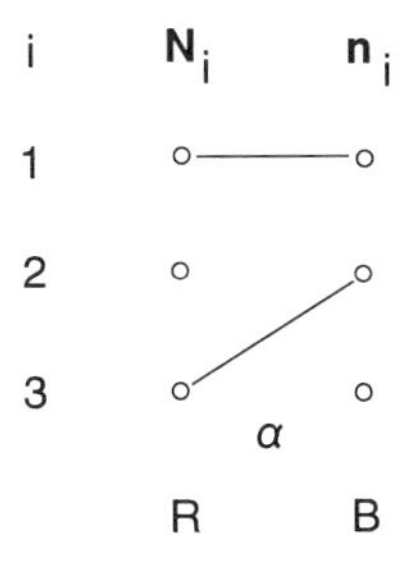

Fig. 6.3.14 Configuration Graph for the X-Axis Rotation of Figure 6.3.13

Next, let B be rotated about the Y-axis (or equivalently about $\mathbf{N}_2$) through the angle β as in Figure 6.3.15 where the $\hat{x}$, $\hat{y}$, and $\hat{z}$-axes and the $\hat{\mathbf{n}}_i$ $(i = 1,2,3)$ are axes and unit vectors of B which are aligned with the X, Y, and Z-axes and the $\mathbf{N}_i$ $(i = 1,2,3)$ of R immediately after the rotation of B about the X-axis. Figure 6.3.16 shows the configuration graph corresponding to the Y-rotation of Figure 6.3.15.

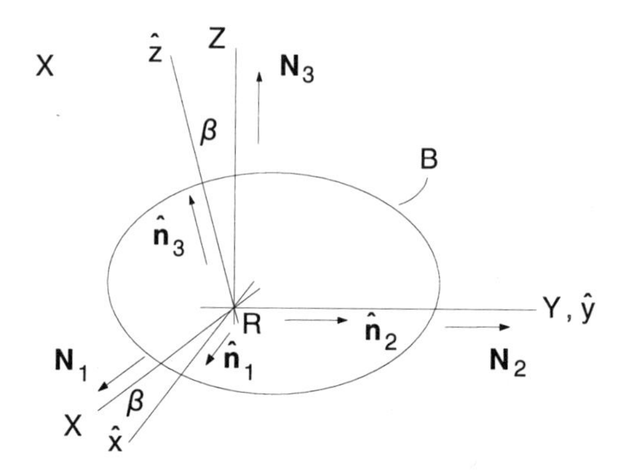

Fig. 6.3.15 Rotation of B about the Y-Axis of R

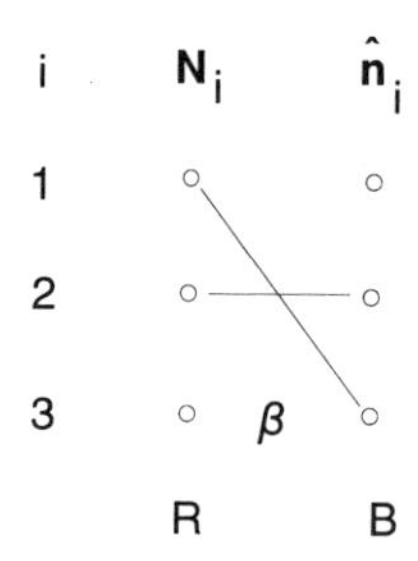

Fig. 6.3.16 Configuration Graph for the Y-Axis Rotation of Figure 6.3.15

Finally, let B be rotated about the Z-Axis (or equivalently about $\mathbf{N}_3$) through the angle γ as in Figure 6.3.17 where the $\hat{X}$, $\hat{Y}$, and $\hat{Z}$-axes and the $\hat{\mathbf{N}}_i$ $(i = 1,2,3)$ are axes and unit vectors of B which are aligned with the X, Y, and Z-axes and the $\mathbf{N}_i$ $(i = 1,2,3)$ of R immediately after the rotation of B about the Y-axis. Figure 6.3.18 shows the configuration graph corresponding to the Z-rotation of Figure 6.3.17.

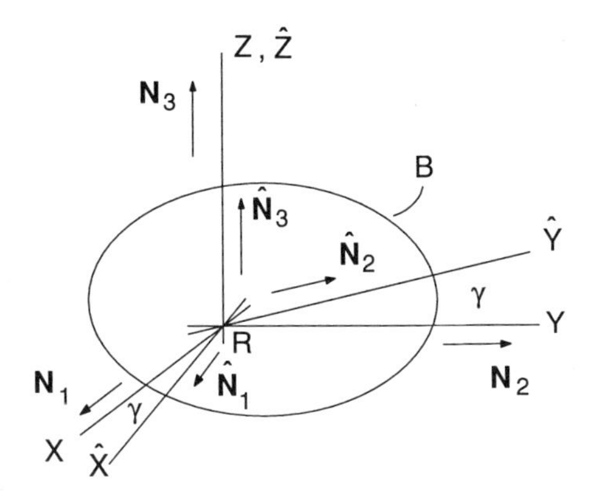

Fig. 6.3.17 Rotation of B about the Z-Axis of R

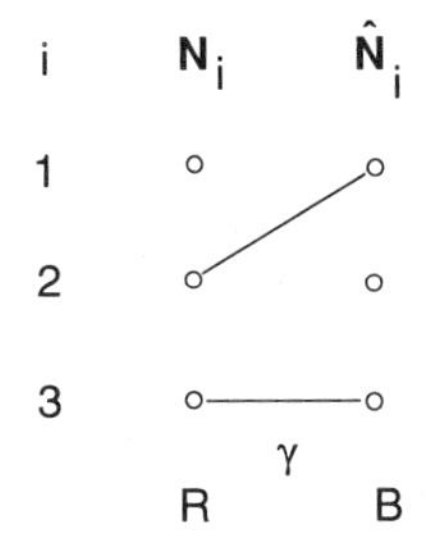

Fig. 6.3.18 Configuration Graph for the Z-Axis Rotation of Figure 6.3.17

Finally, by combining the configuration graphs of Figures 6.3.14, 6.3.16, and 6.3.18, we obtain the three-rotation graph of Figure 6.3.19. The corresponding transformation matrix T between the $\mathbf{N}_i$ and the $\mathbf{n}_i$ is then

$$T = CBA \tag{6.3.14}$$

where by inspection of the configuration graph C, B, and A are

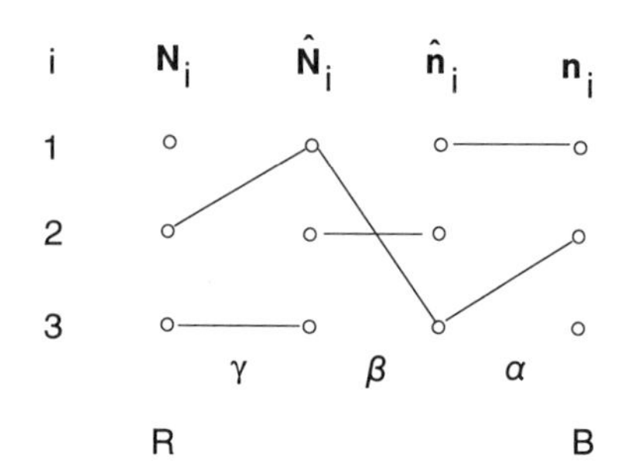

Fig. 6.3.19 Configuration Graph for the X,Y,Z Space Rotation of B in R.

$$C = \begin{bmatrix} c_\gamma & -s_\gamma & 0 \\ s_\gamma & c_\gamma & 0 \\ 0 & 0 & 1 \end{bmatrix} \quad B = \begin{bmatrix} c_\beta & 0 & s_\beta \\ 0 & 1 & 0 \\ -s_\beta & 0 & c_\beta \end{bmatrix} \quad A = \begin{bmatrix} 1 & 0 & 0 \\ 0 & c_\alpha & -s_\alpha \\ 0 & s_\alpha & c_\alpha \end{bmatrix} \tag{6.3.15}$$

By carrying out the multiplication of Equation (6.3.16), S becomes

$$T = \begin{bmatrix} c_\beta c_\gamma & (s_\alpha s_\beta c_\gamma - c_\alpha s_\gamma) & (s_\alpha s_\gamma + c_\alpha s_\beta c_\gamma) \\ c_\beta s_\gamma & (s_\alpha s_\beta s_\gamma + c_\alpha c_\gamma) & (c_\alpha s_\beta s_\gamma - s_\alpha c_\gamma) \\ -s_\beta & s_\alpha c_\beta & c_\alpha c_\beta \end{bmatrix} \tag{6.3.16}$$

Observe in Equation (6.3.14) that T is **not** the inverse of S of Equation (6.3.11). That is, although rotations of a body about body-fixed and space-fixed axes are similar, they are nevertheless distinct and independent.

6.4 Transformation Matrices for Various Rotation Sequences

Reference [6.2] provides a comprehensive listing of transformation matrices obtained from both body-fixed axis and space-fixed axis rotation sequences. Tables

6.4.1 and 6.4.2 contain these listings together with the associated configuration graphs.

In using these tables recall that $\mathbf{N} = S\mathbf{n}$ and $\mathbf{n} = S^T\mathbf{N}$, where $\mathbf{N}$ is the column array of unit vectors $\mathbf{N}_i$ $(i = 1,2,3)$ fixed in a stationary reference frame R, and where $\mathbf{n}$ is the column array of unit vectors $\mathbf{n}_i$ $(i = 1,2,3)$ fixed in the rotating body B. Note further that s_i and c_i represent $\sin\theta_i$ and $\cos\theta_i$ $(i = 1,2,3)$. Finally, the matrices S_1, S_2, S_3, T_1, T_2, and T_3 are defined as:

$$S_1(\theta_i) = T_1(\theta_i) = \begin{bmatrix} 1 & 0 & 0 \\ 0 & c_i & -s_i \\ 0 & s_i & c_i \end{bmatrix} \tag{6.4.1}$$

$$S_2(\theta_i) = T_2(\theta_i) = \begin{bmatrix} c_i & 0 & s_i \\ 0 & 1 & 0 \\ -s_i & 0 & c_i \end{bmatrix} \tag{6.4.2}$$

and

$$S_3(\theta_i) = T_3(\theta_i) = \begin{bmatrix} c_i & -s_i & 0 \\ s_i & c_i & 0 \\ 0 & 0 & 1 \end{bmatrix} \tag{6.4.3}$$

Table 6.4.1 Transformation Matrices and Configuration Graphs for Body-Fixed Axis Rotation

Rotation Sequence	Configuration Graph	Transformation Matrix
1. 1-2-3 or x-y-z (Dextral or Bryan Angles) (Body Axes)	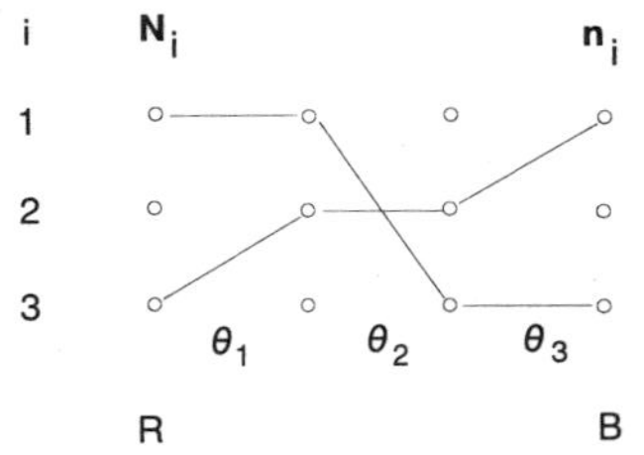 i, N_i, n_i; 1, 2, 3; θ_1, θ_2, θ_3; R, B	$S = S_1(\theta_1)S_2(\theta_2)S_3(\theta_3)$ $\begin{bmatrix} c_2c_3 & -c_2s_3 & s_2 \\ (s_1s_2c_3+c_1s_3) & (-s_1s_2s_3+c_1c_3) & -s_1c_2 \\ (-c_1s_2c_3+s_1s_3) & (c_1s_2s_3+s_1c_3) & c_1c_2 \end{bmatrix}$
2. 2-3-1 or y-z-x (Body Axes)	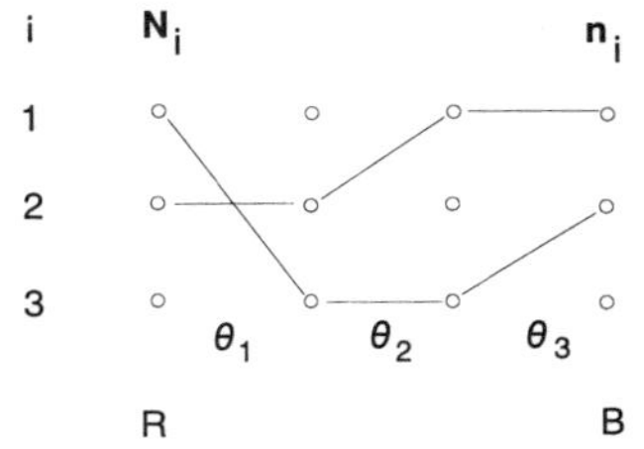 i, N_i, n_i; 1, 2, 3; θ_1, θ_2, θ_3; R, B	$S = S_2(\theta_1)S_3(\theta_2)S_1(\theta_3)$ $\begin{bmatrix} c_1c_2 & (-c_1s_2c_3+s_1s_3) & (c_1s_2s_3+s_1c_3) \\ s_2 & c_2c_3 & -c_2s_3 \\ -s_1c_2 & (s_1s_2c_3+c_1s_3) & (-s_1s_2s_3+c_1c_3) \end{bmatrix}$

Rotation Sequence	Configuration Graph	Transformation Matrix
3. 3-1-2 or z-x-y (Body Axes)	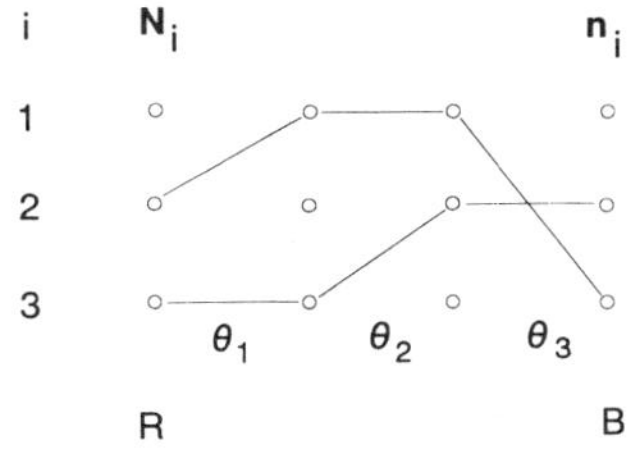	$S = S_3(\theta_1)S_1(\theta_2)S_2(\theta_3)$ $\begin{bmatrix} (-s_1s_2s_3+c_1c_3) & -s_1c_2 & (s_1s_2c_3+c_1s_3) \\ (c_1s_2s_3+s_1c_3) & c_1c_2 & (-c_1s_2c_3+s_1s_3) \\ -c_2s_3 & s_2 & c_2c_3 \end{bmatrix}$
4. 1-3-2 or x-z-y (Body Axes)	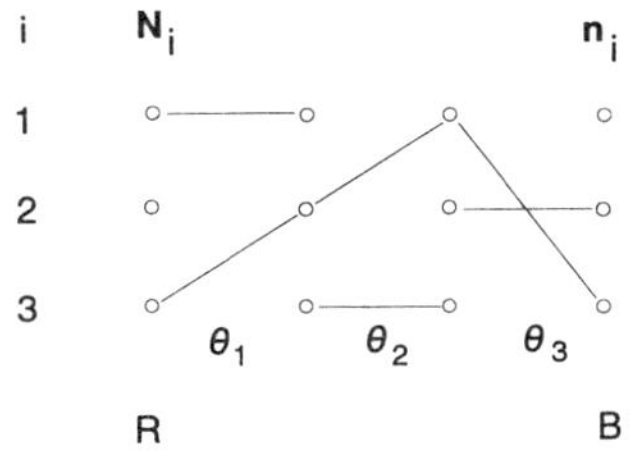	$S = S_1(\theta_1)S_3(\theta_2)S_2(\theta_3)$ $\begin{bmatrix} c_2c_3 & -s_2 & c_2s_3 \\ (c_1s_2c_3+s_1s_3) & c_1c_2 & (c_1s_2s_3-c_3s_1) \\ (s_1s_2c_3-c_1s_3) & s_1c_2 & (s_1s_2s_3+c_1c_3) \end{bmatrix}$

Rotation Sequence	Configuration Graph	Transformation Matrix
5. 2-1-3 or y-x-z (Body Axes)	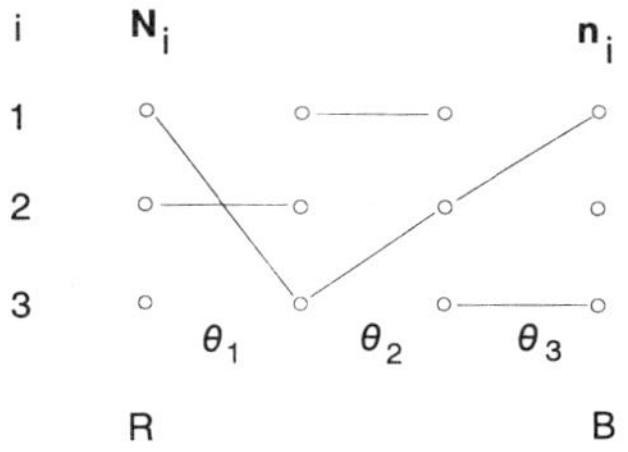 i, N_i, n_i; 1, 2, 3; θ_1 θ_2 θ_3; R, B	$S = S_2(\theta_1)\,S_1(\theta_2)\,S_3(\theta_3)$ $\begin{bmatrix} (s_1s_2s_3+c_1c_3) & (s_1s_2c_3-c_1s_3) & s_1c_2 \\ c_2s_3 & c_2c_3 & -s_2 \\ (c_1s_2s_3-s_1c_3) & (c_1s_2c_3+s_1s_3) & c_1c_2 \end{bmatrix}$
6. 3-2-1 or z-y-x (Body Axes)	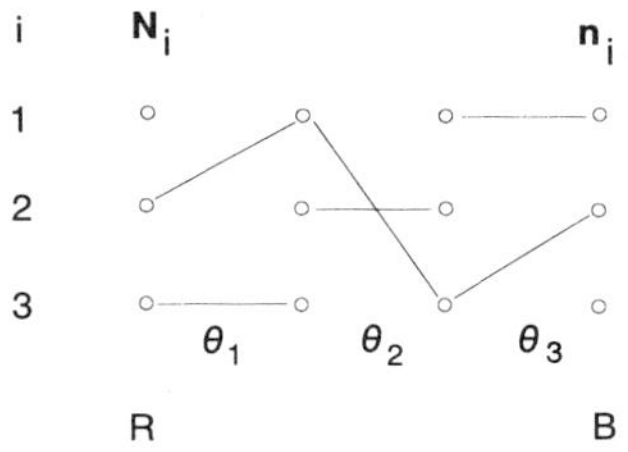 i, N_i, n_i; 1, 2, 3; θ_1 θ_2 θ_3; R, B	$S = S_3(\theta_1)\,S_2(\theta_2)\,S_1(\theta_3)$ $\begin{bmatrix} c_1c_2 & (c_1s_2s_3-s_1c_3) & (c_1s_2c_3+s_1s_3) \\ s_1c_2 & (s_1s_2s_3+c_1c_3) & (s_1s_2c_3-c_1s_3) \\ -s_2 & c_2s_3 & c_2c_3 \end{bmatrix}$

Rotation Sequence	Configuration Graph	Transformation Matrix
7. 1-2-1 or x-y-x (Body Axes)	i, $\mathbf{N}_i$, $\mathbf{n}_i$; 1, 2, 3; θ_1 θ_2 θ_3; R B	$S = S_1(\theta_1)s_2(\theta_2)s_1(\theta_3)$ $\begin{bmatrix} c_2 & s_2s_3 & s_2c_3 \\ s_1s_2 & (-s_1c_2s_3+c_1c_3) & (-s_1c_2c_3-c_1s_3) \\ -c_1s_2 & (c_1c_2s_3+s_1c_3) & (c_1c_2c_3-s_1s_3) \end{bmatrix}$
8. 1-3-1 or x-z-x (Euler Angles) (Body Axes)	i, $\mathbf{N}_i$, $\mathbf{n}_i$; 1, 2, 3; θ_1 θ_2 θ_3; R B	$S = s_1(\theta_1)s_3(\theta_2)s_1(\theta_3)$ $\begin{bmatrix} c_2 & -s_2c_3 & s_2s_3 \\ c_1s_2 & (c_1c_2c_3-s_1s_3) & (-c_1c_2s_3-s_1c_3) \\ s_1s_2 & (s_1c_2c_3+c_1s_3) & (-s_1c_2s_3+c_1c_3) \end{bmatrix}$

Rotation Sequence	Configuration Graph	Transformation Matrix
9. 2-1-2 or y-x-y (Body Axes)	i, N_i, n_i; 1, 2, 3; θ_1 θ_2 θ_3; R B	$S = S_2(\theta_1)S_1(\theta_2)S_2(\theta_3)$ $\begin{bmatrix} (-s_1c_2s_3+c_1c_3) & s_1s_2 & (s_1c_2c_3+c_1s_3) \\ s_2s_3 & c_2 & -s_2c_3 \\ (-c_1c_2s_3-s_1c_3) & c_1s_2 & (c_1c_2c_3-s_1s_3) \end{bmatrix}$
10. 2-3-2 or y-z-y (Body Axes)	i, N_i, n_i; 1, 2, 3; θ_1 θ_2 θ_3; R B	$S = S_2(\theta_1)S_3(\theta_2)S_2(\theta_3)$ $\begin{bmatrix} (c_1c_2c_3-s_1s_3) & -c_1s_2 & (c_1c_2s_3+s_1c_3) \\ s_2c_3 & c_2 & s_2s_3 \\ (-s_1c_2c_3-c_1s_3) & s_1s_2 & (-s_1c_2s_3+c_1c_3) \end{bmatrix}$

Rotation Sequence	Configuration Graph	Transformation Matrix
11. 3-1-3 or z-x-z (Euler Angles) (Body Axes)	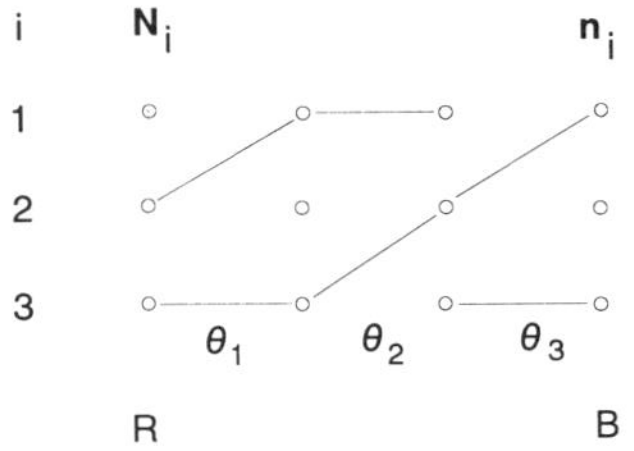 i, N_i, n_i; 1, 2, 3; θ_1, θ_2, θ_3; R, B	$S = S_3(\theta_1)S_1(\theta_2)S_3(\theta_3)$ $\begin{bmatrix} (-s_1c_2s_3+c_1c_3) & (-s_1c_2c_3-c_1s_3) & s_1s_2 \\ (c_1c_2s_3+s_1c_3) & (c_1c_2c_3-s_1s_3) & -c_1s_2 \\ s_2s_3 & s_2c_3 & c_2 \end{bmatrix}$
12. 3-2-3 or z-y-z (Body Axes)	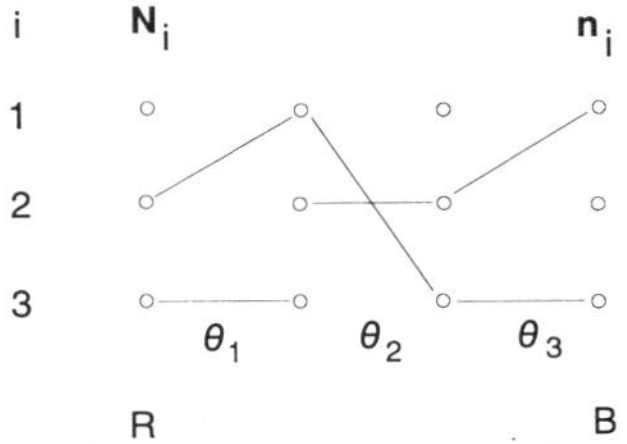 i, N_i, n_i; 1, 2, 3; θ_1, θ_2, θ_3; R, B	$S = S_3(\theta_1)S_2(\theta_2)S_3(\theta_3)$ $\begin{bmatrix} (c_1c_2c_3-s_1s_3) & (-c_1c_2s_3-s_1c_2) & c_1s_2 \\ (s_1c_2c_3+c_1s_3) & (-s_1c_2s_3+c_1c_3) & s_1s_2 \\ -s_2c_3 & s_2s_3 & c_2 \end{bmatrix}$

Table 6.4.2 Transformation Matrices and Configuration Graphs for Space-Fixed Axis Rotation

Rotation Sequence	Configuration Graph	Transformation Matrix
1. 1-2-3 or X-Y-Z (Dextral or Bryan Angles) (Space Axes)	i, N_i, n_i; 1, 2, 3; θ_3 θ_2 θ_1; R B	$T = T_3(\theta_3)T_2(\theta_2)T_1(\theta_1)$ $\begin{bmatrix} c_2c_3 & (s_1s_2c_3 - c_1s_3) & (c_1s_2c_3 + s_1s_3) \\ c_2s_3 & (s_1s_2s_3 + c_1c_3) & (c_1s_2s_3 - s_1c_3) \\ -s_2 & s_1c_2 & c_1c_2 \end{bmatrix}$
2. 2-3-1 or Y-Z-X (Space Axes)	i, N_i, n_i; 1, 2, 3; θ_3 θ_2 θ_1; R B	$T = T_1(\theta_3)T_3(\theta_2)T_2(\theta_1)$ $\begin{bmatrix} c_1c_2 & -s_2 & s_1c_2 \\ (c_1s_2c_3 + s_1s_3) & c_2c_3 & (s_1s_2c_3 - c_1s_3) \\ (c_1s_2s_3 - s_1c_3) & c_2s_3 & (s_1s_2s_3 + c_1c_3) \end{bmatrix}$

Rotation Sequence	Configuration Graph	Transformation Matrix
3. 3-1-2 or Z-X-Y (Space Axes)	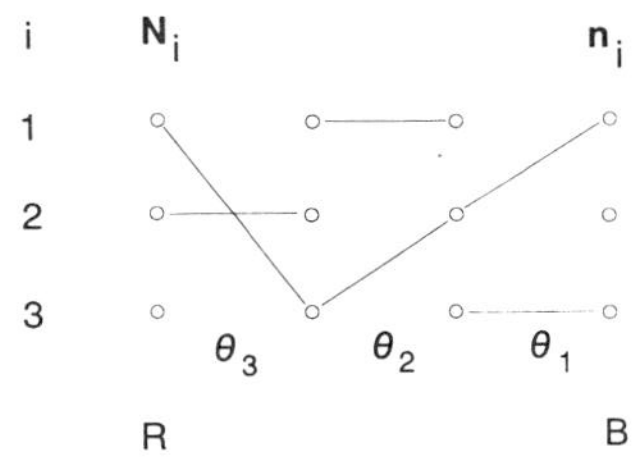 i, N_i, n_i; 1, 2, 3; θ_3, θ_2, θ_1; R, B	$T = T_2(\theta_3)T_1(\theta_2)T_3(\theta_1)$ $\begin{bmatrix} (s_1s_2s_3+c_1c_3) & (c_1s_2s_3-s_1c_3) & c_2s_3 \\ s_1c_2 & c_1c_2 & -s_2 \\ (s_1s_2c_3-c_1s_3) & (c_1s_2c_3+s_1s_3) & c_2c_3 \end{bmatrix}$
4. 1-3-2 or X-Z-Y (Space Axes)	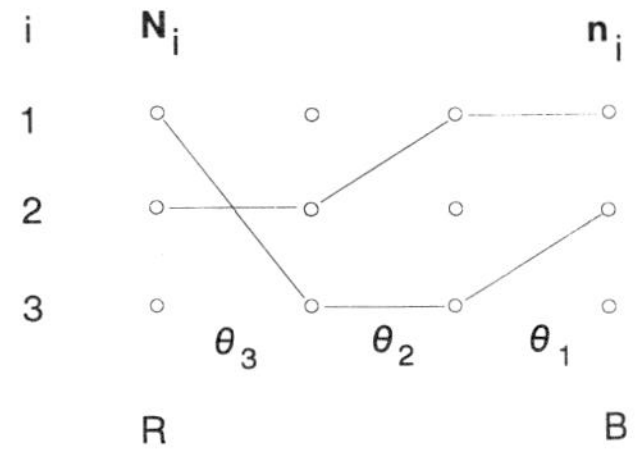 i, N_i, n_i; 1, 2, 3; θ_3, θ_2, θ_1; R, B	$T = T_2(\theta_3)T_3(\theta_2)T_1(\theta_1)$ $\begin{bmatrix} c_2c_3 & (-c_1s_2c_3+s_1s_3) & (s_1s_2c_3+c_1s_3) \\ s_2 & c_1c_2 & -s_1c_2 \\ -c_2s_3 & (s_1c_3+c_1s_2s_3) & (-s_1s_2s_3+c_1c_3) \end{bmatrix}$

Rotation Sequence	Configuration Graph	Transformation Matrix
5. 2-1-3 or Y-X-Z (Space Axes)	i, N_i, n_i; 1, 2, 3; θ_3 θ_2 θ_1; R B	$T = T_2(\theta_3)T_1(\theta_2)T_3(\theta_1)$ $\begin{bmatrix} (-s_1s_2s_3+c_1c_3) & -c_2s_3 & (c_1s_2s_3+s_1c_3) \\ (s_1s_2c_3+c_1s_3) & c_2c_3 & (-c_1s_2c_3+s_1s_3) \\ -s_1c_2 & s_2 & c_1c_2 \end{bmatrix}$
6. 3-2-1 or Z-Y-X (Space Axes)	i, N_i, n_i; 1, 2, 3; θ_3 θ_2 θ_1; R B	$T = T_1(\theta_3)T_2(\theta_2)T_1(\theta_1)$ $\begin{bmatrix} c_1c_2 & -s_1c_2 & s_2 \\ (c_1s_2s_3+s_1c_3) & (-s_1s_2s_3+c_1c_3) & -c_2s_3 \\ (-c_1s_2c_3+s_1s_3) & (s_1s_2c_3+c_1s_3) & c_2c_3 \end{bmatrix}$

Rotation Sequence	Configuration Graph	Transformation Matrix
7. 1-2-1 or X-Y-X (Space Axes)	i, N_i, n_i; 1, 2, 3; θ_3, θ_2, θ_1; R, B	$T = T_1(\theta_3)T_2(\theta_2)T_1(\theta_1)$ $\begin{bmatrix} c_2 & s_1s_2 & c_1s_2 \\ s_2s_3 & (-s_1c_2s_3+c_1c_3) & (-c_1c_2s_3-s_1c_3) \\ -s_2c_3 & (s_1c_2c_3+c_1s_3) & (c_1c_2c_3-s_1s_3) \end{bmatrix}$
8. 1-3-1 or X-Z-X (Euler Angles) (Space Axes)	i, N_i, n_i; 1, 2, 3; θ_3, θ_2, θ_1; R, B	$T = T_1(\theta_3)T_3(\theta_2)T_1(\theta_1)$ $\begin{bmatrix} c_2 & -c_1s_2 & s_1s_2 \\ s_2c_3 & (c_1c_2c_3-s_1s_3) & (-s_1c_2c_3-c_1s_3) \\ s_2s_3 & (c_1c_2s_3+s_1c_3) & (-s_1c_2s_3+c_1c_3) \end{bmatrix}$

Rotation Sequence	Configuration Graph	Transformation Matrix
9. 2-1-2 or Y-X-Y (Space Axes)	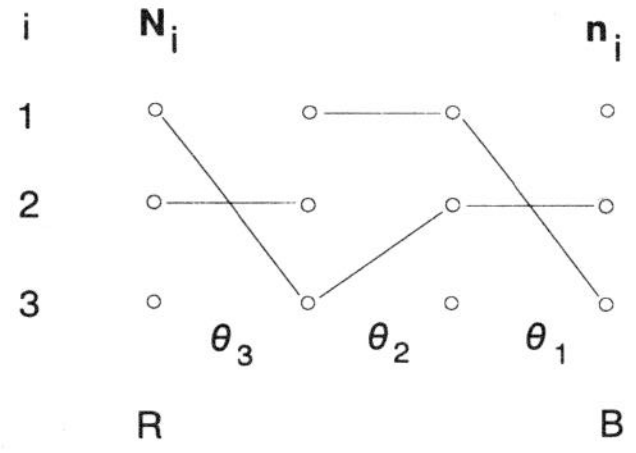	$T = T_2(\theta_3)T_1(\theta_2)T_2(\theta_1)$ $\begin{bmatrix} (-s_1c_2s_3+c_1c_3) & s_2s_3 & (c_1c_2s_3+s_1c_3) \\ s_1s_2 & c_2 & -c_1s_2 \\ (-s_1c_2c_3-c_1s_3) & s_2c_3 & (c_1c_2c_3-s_1s_3) \end{bmatrix}$
10. 2-3-2 or Y-Z-Y (Space Axes)	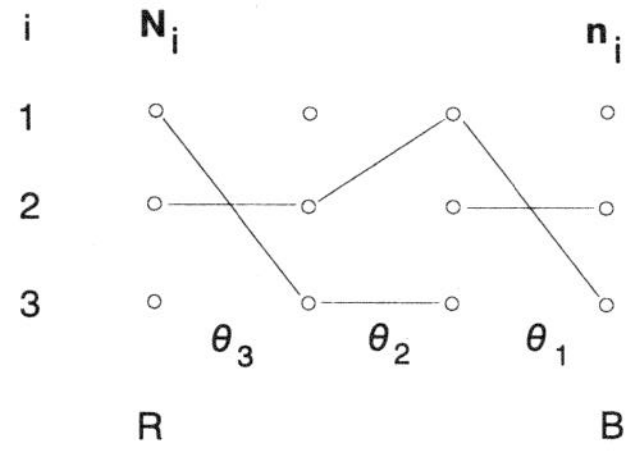	$T = T_2(\theta_3)T_3(\theta_2)T_2(\theta_1)$ $\begin{bmatrix} (c_1c_2c_3-s_1s_3) & -s_2c_3 & (s_1c_2c_3+c_1s_3) \\ c_1s_2 & c_2 & s_1s_2 \\ (-c_1c_2s_3-s_1c_3) & s_2s_3 & (-s_1c_2s_3+c_1c_3) \end{bmatrix}$

Rotation Sequence	Configuration Graph	Transformation Matrix
11. 3-1-3 or Z-X-Z (Euler Angles) (Space Axes)	i; N_i; n_i; 1; 2; 3; θ_3; θ_2; θ_1; R; B	$T = T_3(\theta_3)T_1(\theta_2)T_3(\theta_1)$ $\begin{bmatrix} (-s_1c_2s_3+c_1c_3) & (-c_1c_2s_3-s_1c_3) & s_2s_3 \\ (s_1c_2c_3+c_1s_3) & (c_1c_2c_3-s_1s_3) & -s_2c_3 \\ s_1s_2 & c_1s_2 & c_2 \end{bmatrix}$
12. 3-2-3 or Z-Y-Z (Space Axes)	i; N_i; n_i; 1; 2; 3; θ_3; θ_2; θ_1; R; B	$T = T_3(\theta_3)T_2(\theta_2)T_3(\theta_1)$ $\begin{bmatrix} (c_1c_2c_3-s_1s_3) & (-s_1c_2c_3-c_1s_3) & s_2c_3 \\ (c_1c_2s_3+s_1c_3) & (-s_1c_2s_3+c_1c_3) & s_2s_3 \\ -c_1s_2 & s_1s_2 & c_2 \end{bmatrix}$

6.5 Angular Velocity

6.5.1 Definitions

Consider a body B moving in a reference frame R as in Figure 6.5.1. The angular velocity of B in R is defined as the time rate of change of orientation of B in R.

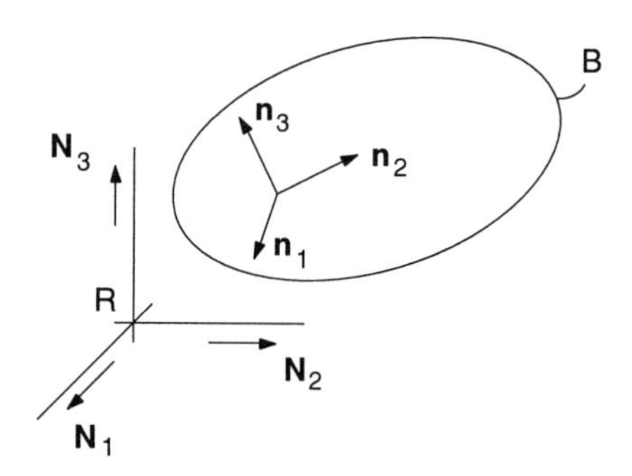

Fig. 6.5.1 A Body B Moving in a Reference Frame R

This definition is analogous to the definition for the velocity of a particle P in R as the time rate of change of the position vector **P** locating P in R. Unlike position, however, orientation is not conveniently represented as a vector. Instead, orientation is represented in terms of the relative inclination of unit vectors $\mathbf{n}_i$ and $\mathbf{N}_i$ (i = 1,2,3) fixed in B and R (see Figure 6.5.1). From Section 6.2 we see that the relative unit vector inclinations may be defined in terms of the direction cosine matrix (or transformation matrix) S whose elements S_{ij} are

$$S_{ij} = \mathbf{N}_i \cdot \mathbf{n}_j \qquad (6.5.1)$$

Then the time rate of change of the elements are

$$dS_{ij}/dt = d(\mathbf{N}_i \bullet \mathbf{n}_i)/dt = \mathbf{N}_i \bullet d\mathbf{n}_i/dt \tag{6.5.2}$$

where relative to R, the $\mathbf{N}_i$ are fixed and thus their derivative in R is zero. The derivatives of the $\mathbf{n}_j$, however, are not zero and thus may be used to define the angular velocity vector. specifically, the angular velocity of B in R is defined as the vector ${}^R\boldsymbol{\omega}^B$ such that

$$d\mathbf{n}_j/dt = {}^R\boldsymbol{\omega}^B \times \mathbf{n}_j \quad (j = 1,2,3) \tag{6.5.3}$$

Equations (6.5.3) show that ${}^R\boldsymbol{\omega}^B$ may be expressed in the form

$${}^R\boldsymbol{\omega}^B = \left(\frac{d\mathbf{n}_2}{dt} \bullet \mathbf{n}_3\right)\mathbf{n}_1 + \left(\frac{d\mathbf{n}_3}{dt} \bullet \mathbf{n}_1\right)\mathbf{n}_2 + \left(\frac{d\mathbf{n}_1}{dt} \bullet \mathbf{n}_2\right)\mathbf{n}_3 \tag{6.5.4}$$

To see this, observe that ${}^R\boldsymbol{\omega}^B$ may be expressed identically as

$${}^R\boldsymbol{\omega}^B \equiv \left({}^R\boldsymbol{\omega}^B \bullet \mathbf{n}_1\right)\mathbf{n}_1 + \left({}^R\boldsymbol{\omega}^B \bullet \mathbf{n}_2\right)\mathbf{n}_2 + \left({}^R\boldsymbol{\omega}^B \bullet \mathbf{n}_3\right)\mathbf{n}_3 \tag{6.5.5}$$

Then since the $\mathbf{n}_i$ $(i = 1,2,3)$ form a mutually perpendicular set of unit vectors, we have

$$\begin{aligned}
{}^R\boldsymbol{\omega}^B &= ({}^R\boldsymbol{\omega}^B \bullet \mathbf{n}_2 \times \mathbf{n}_3)\mathbf{n}_1 + ({}^R\boldsymbol{\omega}^B \bullet \mathbf{n}_3 \times \mathbf{n}_1)\mathbf{n}_2 + ({}^R\boldsymbol{\omega}^B \bullet \mathbf{n}_1 \times \mathbf{n}_2)\mathbf{n}_3 \\
&= ({}^R\boldsymbol{\omega}^B \times \mathbf{n}_2 \bullet \mathbf{n}_3)\mathbf{n}_1 + ({}^R\boldsymbol{\omega}^B \times \mathbf{n}_3 \bullet \mathbf{n}_1)\mathbf{n}_2 + ({}^R\boldsymbol{\omega}^B \times \mathbf{n}_1 \bullet \mathbf{n}_2)\mathbf{n}_3 \\
&= \left(\frac{d\mathbf{n}_2}{dt} \bullet \mathbf{n}_3\right)\mathbf{n}_1 + \left(\frac{d\mathbf{n}_3}{dt} \bullet \mathbf{n}_2\right)\mathbf{n}_2 + \left(\frac{d\mathbf{n}_1}{dt} \bullet \mathbf{n}_2\right)\mathbf{n}_3
\end{aligned} \tag{6.5.6}$$

where the last equality follows from Equations (6.5.3).

6.5.2 Remarks

Several questions arise in view of the form of Equation (6.5.4) for the angular velocity vector ${}^R\boldsymbol{\omega}^B$. First, even though the form of Equation (6.5.4) satisfies the requirements of Equations (6.5.3), is it the only solution? That is, is angular velocity as represented in Equation (6.5.4) unique? Next, how is the representation of Equation (6.5.4) consistent with angular velocity as discussed in elementary courses in mechanics and physics for bodies rotating about fixed areas? Finally, how is Equation (6.5.4) to be useful as a formula for dynamic analysis? Indeed, in comparing Equations (6.5.3) and (6.5.4), if a knowledge of unit vector derivatives in Equation (6.5.4) is needed to obtain unit vector derivatives as in Equations (6.5.3), why is the angular velocity vector needed at all?

These questions are discussed and answered in the following paragraphs.

6.5.3 Uniqueness of Angular Velocity

To address the first of the these questions, consider again a body B moving in a reference frame R as in Figure 6.5.2. Let **c** be a vector fixed in B as shown.

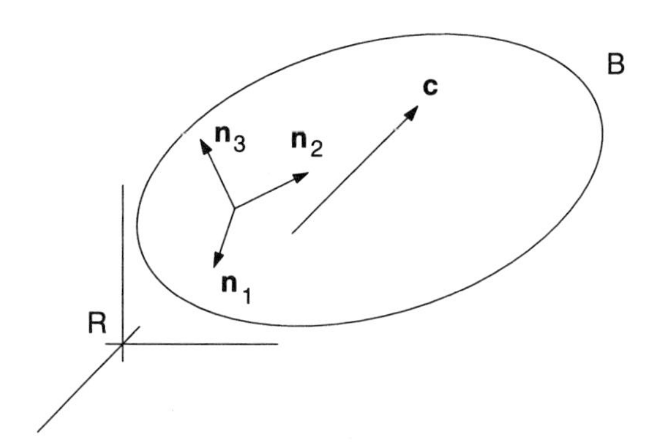

Fig. 6.5.2 A Body B with Fixed Vector **c** Moving in a Reference Frame R

Then if $\mathbf{n}_1$, $\mathbf{n}_2$ and $\mathbf{n}_3$ are mutually perpendicular unit vectors fixed in B, $\mathbf{c}$ may be expressed in the form

$$\mathbf{c} = c_1\mathbf{n}_1 + c_2\mathbf{n}_2 + c_3\mathbf{n}_3 = c_i\mathbf{n}_i \tag{6.5.7}$$

where the c_i $(i = 1,2,3)$ are constants. The derivative of $\mathbf{c}$ in R is then:

$$\frac{d\mathbf{c}}{dt} = c_1\frac{d\mathbf{n}_1}{dt} + c_2\frac{d\mathbf{n}_2}{dt} + c_3\frac{d\mathbf{n}_3}{dt} = c_i\frac{d\mathbf{n}_i}{dt} \tag{6.5.8}$$

By substituting from Equations (6.5.3) we then have

$$\frac{d\mathbf{c}}{dt} = c_i\,{}^R\boldsymbol{\omega}^B \times \mathbf{n}_i = {}^R\boldsymbol{\omega}^B \times (c_i\mathbf{n}_i)$$

or

$$\frac{d\mathbf{c}}{dt} = {}^R\boldsymbol{\omega}^B \times \mathbf{c} \tag{6.5.9}$$

Equation (6.5.9) may be used as an alternative to Equation (6.5.3) for defining angular velocity. Specifically, the angular velocity of B in R is defined as the vector ${}^R\boldsymbol{\omega}^B$ satisfying Equation (6.5.9) for all vectors $\mathbf{c}$ fixed in B.

In this context it is relatively easy to see that ${}^R\boldsymbol{\omega}^B$ is unique: Suppose ${}^R\hat{\boldsymbol{\omega}}^B$ is a second vector satisfying Equation (6.5.9). That is,

$$\frac{d\mathbf{c}}{dt} = {}^R\hat{\boldsymbol{\omega}}^B \times \mathbf{c} \tag{6.5.10}$$

Then by subtracting Equations (6.5.9) and (6.5.10) we have

$$0 = ({}^R\boldsymbol{\omega}^B - {}^R\hat{\boldsymbol{\omega}}^B) \times \mathbf{c} \tag{6.5.11}$$

This expression is satisfied for all $\mathbf{c}$ only if ${}^R\boldsymbol{\omega}^B - {}^R\hat{\boldsymbol{\omega}}^B$ is zero. That is

$${}^R\hat{\boldsymbol{\omega}}^B = {}^R\boldsymbol{\omega}^B \tag{6.5.12}$$

6.5.4 Alternative Definition and Forms for Angular Velocity

Consider again a body B moving in a reference frame R as in Figure 6.5.3. Let $\mathbf{c}$ be an arbitrary vector fixed in B and let $\mathbf{a}$ and $\mathbf{b}$ be non-zero, non-parallel

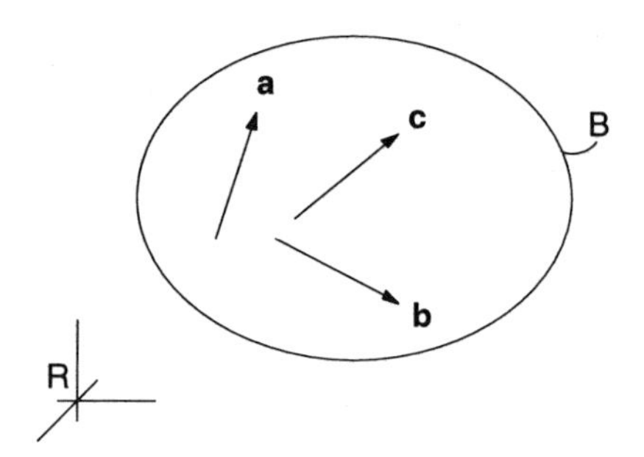

Figure 6.5.3 A Body B Moving in a Reference Frame R with Vectors **a**, **b** and **c** Fixed in B

vectors also fixed in B. Then it is readily seen that the angular velocity vector ${}^R\boldsymbol{\omega}^B$ satisfying Equation (6.5.9) for all vectors $\mathbf{c}$ may be written in the form:

$${}^R\boldsymbol{\omega}^B = \frac{\dfrac{d\mathbf{a}}{dt} \times \dfrac{d\mathbf{b}}{dt}}{\dfrac{d\mathbf{a}}{dt} \bullet \mathbf{b}} \tag{6.5.13}$$

This expression has an advantage over Equation (6.5.4) in that is involves only two vectors which need not be either perpendicular nor unit vectors, whereas Equation (6.5.4) requires three mutually perpendicular unit vectors. The disadvantage of Equation (6.5.13) is its rather awkward appearance and its non-symmetry. The non-symmetry may appear to be a problem in that there is no distinction in the definition of **a** and **b**, but they do not appear symmetrically in Equation (6.5.13). The apparent non-symmetry, however, is misleading in that **a** and **b** can be interchanged. To see this, observe that since **a** and **b** are fixed in B the products $\mathbf{a} \bullet \mathbf{a}$, $\mathbf{a} \bullet \mathbf{b}$, and $\mathbf{b} \bullet \mathbf{b}$ are all scalar constants. That is,

$$\mathbf{a} \bullet \mathbf{a} = k_1 \qquad \mathbf{a} \bullet \mathbf{b} = k_2 \qquad \mathbf{b} \bullet \mathbf{b} = k_3 \tag{6.5.14}$$

Then by differentiation we have

$$\mathbf{a} \bullet \frac{d\mathbf{a}}{dt} = 0 \qquad \frac{d\mathbf{a}}{dt} \bullet \mathbf{b} + \mathbf{a} \bullet \frac{d\mathbf{b}}{dt} = 0 \qquad \mathbf{b} \bullet \frac{d\mathbf{b}}{dt} = 0 \tag{6.5.15}$$

Then using the second of these we can rewrite Equation (6.5.13) as

$$^{R}\boldsymbol{\omega}^{B} = \frac{\dfrac{d\mathbf{a}}{dt} \times \dfrac{d\mathbf{b}}{dt}}{\dfrac{d\mathbf{a}}{dt} \bullet \mathbf{b}} = \frac{\dfrac{d\mathbf{a}}{dt} \times \dfrac{d\mathbf{b}}{dt}}{-\mathbf{a} \bullet \dfrac{d\mathbf{b}}{dt}} = \frac{\dfrac{d\mathbf{b}}{dt} \times \dfrac{d\mathbf{a}}{dt}}{\dfrac{d\mathbf{b}}{dt} \bullet \mathbf{a}} \tag{6.5.16}$$

In view of Equations (6.5.13) and (6.5.16) we can express $^{R}\boldsymbol{\omega}^{B}$ in the form

$$^{R}\boldsymbol{\omega}^{B} = \frac{1}{2}\left[\frac{\dfrac{d\mathbf{a}}{dt} \times \dfrac{d\mathbf{b}}{dt}}{\dfrac{d\mathbf{a}}{dt} \bullet \mathbf{b}} + \frac{\dfrac{d\mathbf{b}}{dt} \times \dfrac{d\mathbf{a}}{dt}}{\dfrac{d\mathbf{b}}{dt} \bullet \mathbf{a}}\right] \tag{6.5.17}$$

To see that $^{R}\boldsymbol{\omega}^{B}$ in either of the forms of Equations (6.5.13), (6.5.16), or (6.5.17) satisfies Equation (6.5.9), observe that since **a**, **b**, and **c** are fixed in B and that **a** and **b** are non-zero and non-parallel, there exist constant scalars α, β, and γ such that

$$\mathbf{c} = \alpha\mathbf{a} + \beta\mathbf{b} + \gamma\mathbf{a}\times\mathbf{b} \tag{6.5.18}$$

Consider the expression ${}^{R}\boldsymbol{\omega}^{B} \times \mathbf{a}$: From Equation (6.5.13) we have

$$\begin{aligned} {}^{R}\boldsymbol{\omega}^{B}\times\mathbf{a} &= \left(\frac{\dfrac{d\mathbf{a}}{dt}\times\dfrac{d\mathbf{b}}{dt}}{\dfrac{d\mathbf{a}}{dt}\bullet\mathbf{b}}\right)\times\mathbf{a} = \left[\left(\overset{0}{\cancel{\frac{d\mathbf{a}}{dt}\bullet\mathbf{a}}}\right)\frac{d\mathbf{b}}{dt} - \left(\frac{d\mathbf{b}}{dt}\bullet\mathbf{a}\right)\frac{d\mathbf{a}}{dt}\right]\Big/\frac{d\mathbf{a}}{dt}\bullet\mathbf{b} \\ &= \frac{d\mathbf{a}}{dt} \end{aligned} \tag{6.5.19}$$

where Equation (2.13.2) is used to expand the triple vector product and where the final equality follows from Equations (6.5.15).

In a similar manner we see that

$$\frac{d\mathbf{b}}{dt} = {}^{R}\boldsymbol{\omega}^{B}\times\mathbf{b} \tag{6.5.20}$$

Consider next $d(\mathbf{a}\times\mathbf{b})/dt$: From Equations (6.5.19) and (6.5.20) we have

$$\begin{aligned} \frac{d}{dt}(\mathbf{a}\times\mathbf{b}) &= \frac{d\mathbf{a}}{dt}\times\mathbf{b} + \mathbf{a}\times\frac{d\mathbf{b}}{dt} = ({}^{R}\boldsymbol{\omega}^{B}\times\mathbf{a})\times\mathbf{b} + \mathbf{a}\times({}^{R}\boldsymbol{\omega}^{B}\times\mathbf{b}) \\ &= ({}^{R}\boldsymbol{\omega}^{B}\bullet\mathbf{b})\mathbf{a} - (\mathbf{a}\bullet\mathbf{b}){}^{R}\boldsymbol{\omega}^{B} + (\mathbf{a}\bullet\mathbf{b}){}^{R}\boldsymbol{\omega}^{B} - (\mathbf{a}\bullet{}^{R}\boldsymbol{\omega}^{B})\mathbf{b} \\ &= ({}^{R}\boldsymbol{\omega}^{B}\bullet\mathbf{b})\mathbf{a} - ({}^{R}\boldsymbol{\omega}^{B}\bullet\mathbf{a})\mathbf{b} \\ &= {}^{R}\boldsymbol{\omega}^{B}\times(\mathbf{a}\times\mathbf{b}) \end{aligned} \tag{6.5.21}$$

where, again, Equations (2.13.2) and (2.13.3) have been used to expand the triple vector products.

In view of the results of Equations (6.5.19), (6.5.20), and (6.5.21) we see that the derivative of **c** in Equation (6.5.18) is:

$$\begin{aligned}\frac{d\mathbf{c}}{dt} &= \alpha\frac{d\mathbf{a}}{dt} + \beta\frac{d\mathbf{b}}{dt} + \gamma\frac{d}{dt}(\mathbf{a}\times\mathbf{b}) \\ &= \alpha\,{}^{R}\boldsymbol{\omega}^{B}\times\mathbf{a} + \beta\,{}^{R}\boldsymbol{\omega}^{B}\times\mathbf{b} + \gamma\,{}^{R}\boldsymbol{\omega}^{B}\times(\mathbf{a}\times\mathbf{b} \\ &= {}^{R}\boldsymbol{\omega}^{B}\times[\alpha\mathbf{a} + \beta\mathbf{b} + \gamma(\mathbf{a}\times\mathbf{b})] \\ &= {}^{R}\boldsymbol{\omega}^{B}\times\mathbf{c} \end{aligned} \tag{6.5.22}$$

thus verifying that the forms of Equations (6.5.13), (6.5.16), and (6.5.17) satisfy Equation (6.5.9).

6.5.5 Simple Angular Velocity

A question raised earlier is: How does the angular velocity as expressed in Equation (6.5.4) and now also in Equation (6.5.13), relate to angular velocity as used in elementary mechanics to describe the movement of rotating bodies? To answer this question consider the body B rotating about a fixed axis A-A in a reference frame R as in Figure 6.5.4. That is, let A-A be fixed in both B and R. Let $\mathbf{n}_1$, $\mathbf{n}_2$, and $\mathbf{n}_3$ be mutually perpendicular unit vectors fixed in B with $\mathbf{n}_3$ being parallel to A-A. Let θ measure the angle of rotation of B in R, as shown.

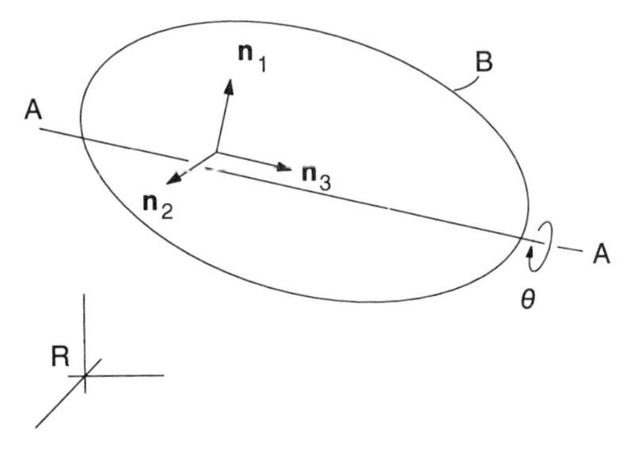

Fig. 6.5.4 A Body Rotating About a Fixed Axis in Reference Frame R

Then from elementary mechanics, the angular velocity of B in R is

$$ {}^{R}\boldsymbol{\omega}^{B} = \dot{\theta}\,\mathbf{n}_3 \tag{6.5.23} $$

Recall from Example 2.16.6 of Chapter 2 that the derivative of a unit vector $\mathbf{n}$ which remains perpendicular to a fixed line is

$$ \frac{d\mathbf{n}}{dt} = \dot{\theta}\,\mathbf{n}_3 \times \mathbf{n} \tag{6.5.24} $$

where $\mathbf{n}_3$ is parallel to the line and θ defines the orientation of $\mathbf{n}$. In Figure 6.5.4 $\mathbf{n}_1$ and $\mathbf{n}_2$ are unit vectors satisfying the conditions for $\mathbf{n}$. Therefore, their derivatives are

$$ \frac{d\mathbf{n}_1}{dt} = \dot{\theta}\,\mathbf{n}_3 \times \mathbf{n}_1 = \dot{\theta}\,\mathbf{n}_2 \quad \text{and} \quad \frac{d\mathbf{n}_2}{dt} = \dot{\theta}\mathbf{n}_3 \times \mathbf{n}_2 = -\dot{\theta}\,\mathbf{n}_1 \tag{6.5.25} $$

Since $\mathbf{n}_3$ remains parallel to A-A, its derivative is zero. Therefore, from Equation (6.5.4) the angular velocity of B in R is:

$$ \begin{aligned} {}^{R}\boldsymbol{\omega}^{B} &= \left(\frac{d\mathbf{n}_2}{dt} \bullet \mathbf{n}_3\right)\mathbf{n}_1 + \left(\frac{d\mathbf{n}_3}{dt} \bullet \mathbf{n}_1\right)\mathbf{n}_2 + \left(\frac{d\mathbf{n}_1}{dt} \bullet \mathbf{n}_2\right)\mathbf{n}_3 \\ &= (-\dot{\theta}\,\mathbf{n}_1 \bullet \mathbf{n}_3)\mathbf{n}_1 + 0 + (\dot{\theta}\,\mathbf{n}_2 \bullet \mathbf{n}_2)\mathbf{n}_3 \\ &= \dot{\theta}\,\mathbf{n}_3 \end{aligned} \tag{6.5.26} $$

which is identical to Equation (6.5.23) used in elementary mechanics.

Finally, since angular velocity is unique and since Equation (6.5.4) can be developed from Equation (6.5.13) the relation between Equations (6.5.13) and (6.5.23) is also established.

The question of the utility of Equations (6.5.4) and (6.5.13) still remains. This question is addressed in the following sections.

6.5.6 Summary

Table 6.5.1 provides a summary listing of the principal formulas related to the definition of angular velocity for a body B in a reference frame R.

Table 6.5.1 Summary of Formulas/Definitions for Angular Velocity of a Body B in a Reference Frame R (${}^R\boldsymbol{\omega}^B$)

Condition	Formula/Equations	Reference Equation
1. $\mathbf{n}_i$ (i = 1,2,3) are mutually perpendicular unit vectors fixed in B	$\dfrac{d\mathbf{n}_i}{dt} = {}^R\boldsymbol{\omega}^B \times \mathbf{n}_i \ (i = 1,2,3)$	(6.5.3)
2. **c** is a vector fixed in B	$\dfrac{d\mathbf{c}}{dt} = {}^R\boldsymbol{\omega}^B \times \mathbf{c}$	(6.5.9)
3. $\mathbf{n}_i$ (i = 1,2,3) are mutually perpendicular unit vectors fixed in B	${}^R\boldsymbol{\omega}^B = \left(\dfrac{d\mathbf{n}_2}{dt} \bullet \mathbf{n}_3\right)\mathbf{n}_1 + \left(\dfrac{d\mathbf{n}_3}{dt} \bullet \mathbf{n}_1\right)\mathbf{n}_2 + \left(\dfrac{d\mathbf{n}_1}{dt} \bullet \mathbf{n}_2\right)\mathbf{n}_3$	(6.5.4)
4. **a** and **b** are non-zero, non-parallel vectors fixed in B	${}^R\boldsymbol{\omega}^B = \dfrac{\dfrac{d\mathbf{a}}{dt} \times \dfrac{d\mathbf{b}}{dt}}{\dfrac{d\mathbf{a}}{dt} \bullet \mathbf{b}}$	(6.5.13)
5. **a** and **b** are non-zero, non-parallel vectors fixed in B	${}^R\boldsymbol{\omega}^B = \dfrac{1}{2}\left(\dfrac{\dfrac{d\mathbf{a}}{dt} \times \dfrac{d\mathbf{b}}{dt}}{\dfrac{d\mathbf{a}}{dt} \bullet \mathbf{b}} + \dfrac{\dfrac{d\mathbf{b}}{dt} \times \dfrac{d\mathbf{a}}{dt}}{\dfrac{d\mathbf{b}}{dt} \bullet \mathbf{a}}\right)$	(6.5.17)

6.6 Differentiation Algorithms

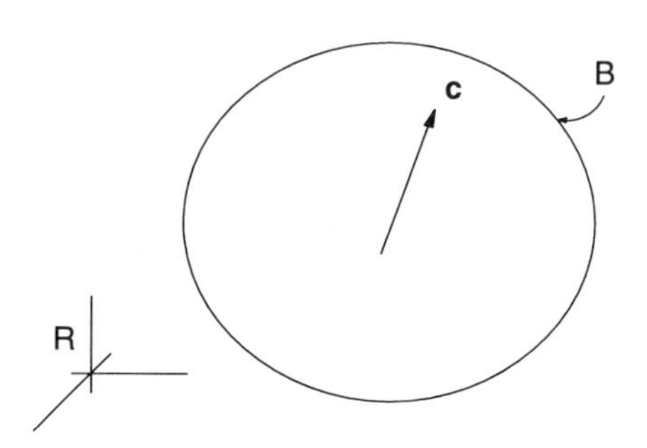

Fig. 6.6.1 A Body B with Fixed Vector **c** Moving in a Reference Frame R

Consider again a body B moving in a reference frame R and let **c** be a vector fixed in B as in Figure 6.6.1. Then from Equation (6.5.9) the derivative of **c** relative to an observer in R is

$$\frac{{}^{R}d\mathbf{c}}{dt} = {}^{R}\boldsymbol{\omega}^{B} \times \mathbf{c} \tag{6.6.1}$$

The superscript R on the left side of this equation designates that the derivative is computed in R. This distinction is important since the derivative of **c** relative to an observer on B is zero. That is,

$$\frac{{}^{R}d\mathbf{c}}{dt} = 0 \tag{6.6.2}$$

Observe further that Equation (6.6.1) is a potentially useful algorithm for computing the derivative of **c**. That is, the derivative is computed by a multiplication operation — a procedure ideally suited for numerical methods.

Observe still further that since in the absence of mass there is no distinction between a rigid body and a reference frame, Equation (6.6.1) could be reformulated using a reference frame $\hat{R}$ instead of B. Specifically, let $\hat{R}$ be a reference frame moving relative to R and let **c** now be a vector fixed in $\hat{R}$ as in Figure 6.6.2. Then Equation (6.6.1) takes the form

$$\frac{{}^{R}d\mathbf{c}}{dt} = {}^{R}\boldsymbol{\omega}^{\hat{R}} \times \mathbf{c} \tag{6.6.3}$$

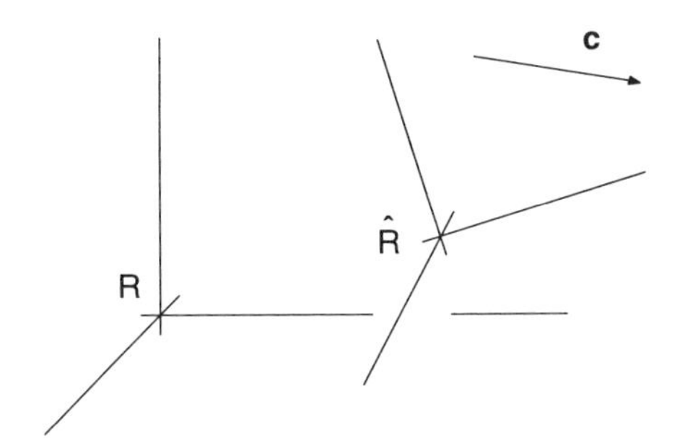

Fig. 6.6.2 A Reference Frame $\hat{R}$ with Fixed Vector **c** Moving in a Reference Frame R

As a generalization, consider an arbitrary vector **V** and reference frames R and $\hat{R}$, where **V** is not fixed in either R or $\hat{R}$. Let $\hat{R}$ move relative to R. Let $\hat{n}_i$ $(i = 1,2,3)$ be mutually perpendicular unit vectors fixed in $\hat{R}$ as in Figure 6.6.3. Then **V** may be expressed in terms of the $\hat{n}_i$ as

$$\mathbf{V} = v_1(t)\hat{\mathbf{n}}_1 + v_2(t)\hat{\mathbf{n}}_2 + v_3(t)\hat{\mathbf{n}}_3 = v_i(t)\hat{\mathbf{n}}_i \tag{6.6.4}$$

where, as indicated, the scalar components v_i $(i = 1,2,3)$ are now functions of time. The derivatives of **V** in R and $\hat{R}$ are then

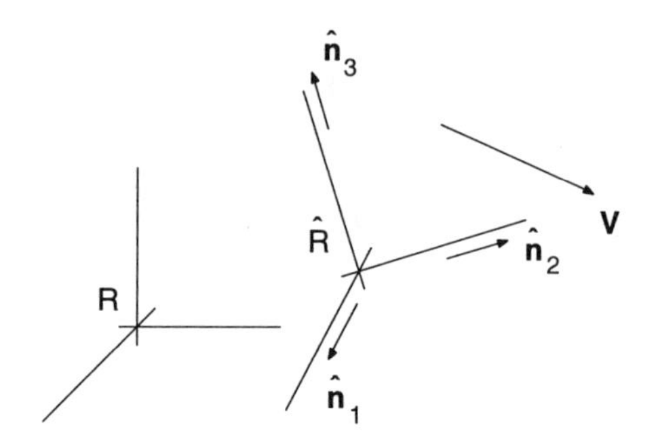

Fig. 6.6.3 An Arbitrary Vector **V** and Reference Frames R and $\hat{R}$

$$\frac{{}^{R}d\mathbf{V}}{dt} = \dot{v}_1\hat{\mathbf{n}}_1 + v_1\frac{{}^{R}d\hat{\mathbf{n}}_1}{dt} + \dot{v}_2\hat{\mathbf{n}}_2 + v_2\frac{{}^{R}d\hat{\mathbf{n}}_2}{dt}$$

$$+ \dot{v}_3\hat{\mathbf{n}}_3 + v_3\frac{{}^{R}d\hat{\mathbf{n}}_2}{dt}$$

$$= \dot{v}_i\hat{\mathbf{n}}_i + v_i\frac{{}^{R}d\hat{\mathbf{n}}_i}{dt}$$

$$= \dot{v}_i\hat{\mathbf{n}}_i + v_i\,{}^{R}\boldsymbol{\omega}^{\hat{R}} \times \hat{\mathbf{n}}_i$$

$$= \dot{v}_i\hat{\mathbf{n}}_i + {}^{R}\boldsymbol{\omega}^{\hat{R}} \times \mathbf{V} \tag{6.6.5}$$

and

$$\frac{{}^{\hat{R}}d\mathbf{V}}{dt} = \dot{v}_1\hat{\mathbf{n}}_1 + \dot{v}_2\hat{\mathbf{n}}_2 + \dot{v}_3\hat{\mathbf{n}}_3 = \dot{v}_i\hat{\mathbf{n}}_i \tag{6.6.6}$$

By substituting from Equation (6.6.6) into (6.6.5) we obtain

$$\frac{{}^{R}d\mathbf{V}}{dt} = \frac{{}^{\hat{R}}d\mathbf{V}}{dt} + {}^{R}\boldsymbol{\omega}^{\hat{R}} \times \mathbf{V} \tag{6.6.7}$$

Since **V** is arbitrary, we can express this result as

$$\frac{{}^{R}d(\ )}{dt} = \frac{{}^{\hat{R}}d(\ )}{dt} + {}^{R}\boldsymbol{\omega}^{\hat{R}} \times (\ ) \tag{6.6.8}$$

where ( ) is any vector quantity.

Observe that as with Equation (6.6.1), Equation (6.6.8) is an algorithm, relating derivatives for observers in different reference frames. This expression is useful in the sequel for developing addition theorems for angular velocity and for developing expressions for Coriolis acceleration.

6.7 Addition Theorem for Angular Velocity

A remaining question in view of the definition for angular velocity as in Equations (6.5.4) and (6.5.13) is: What is the utility of these definitions? That is, how do they lead to formulas for the solution of dynamics problems. The addition theorem for angular velocity as presented in this section, together with the configuration graphs of Section 6.3, provide the answers to these questions. The addition theorem itself is a direct consequence of the differentiation algorithm of Equation (6.6.8).

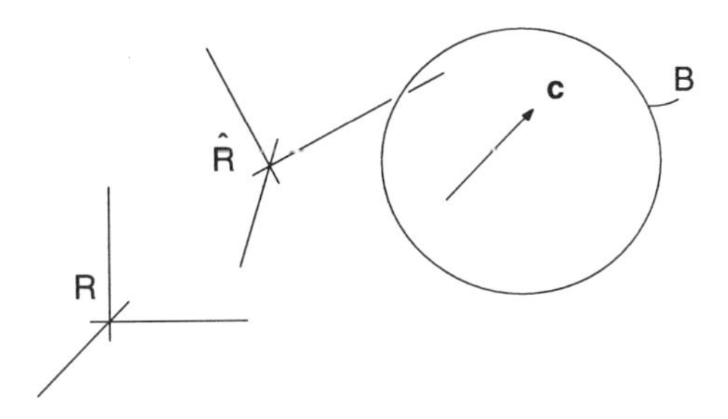

Fig. 6.7.1 A Body B with Fixed Vector **c** Moving in Reference Frames R and $\hat{R}$

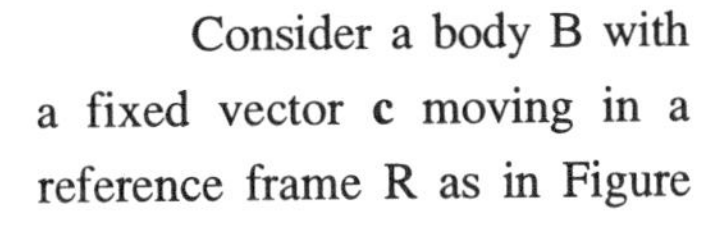

Consider a body B with a fixed vector **c** moving in a reference frame R as in Figure

6.7.1. Let $\hat{R}$ be a second reference frame in which B moves and which in turn moves relative to R. Then from Equation (6.5.9) we have

$$\frac{{}^{R}d\mathbf{c}}{dt} = {}^{R}\boldsymbol{\omega}^{B} \times \mathbf{c} \quad \text{and} \quad \frac{{}^{\hat{R}}d\mathbf{c}}{dt} = {}^{\hat{R}}\boldsymbol{\omega}^{B} \times \mathbf{c} \tag{6.7.1}$$

However, from Equation (6.6.7) we also have

$$\frac{{}^{R}d\mathbf{c}}{dt} = \frac{{}^{\hat{R}}d\mathbf{c}}{dt} + {}^{R}\boldsymbol{\omega}^{\hat{R}} \times \mathbf{c} \tag{6.7.2}$$

By substituting from Equation (6.7.1) to (6.7.2) we have

$${}^{R}\boldsymbol{\omega}^{B} \times \mathbf{c} = {}^{\hat{R}}\boldsymbol{\omega}^{B} \times \mathbf{c} + {}^{R}\boldsymbol{\omega}^{\hat{R}} \times \mathbf{c} = ({}^{\hat{R}}\boldsymbol{\omega}^{B} + {}^{R}\boldsymbol{\omega}^{\hat{R}}) \times \mathbf{c} \tag{6.7.3}$$

Then since $\mathbf{c}$ is an arbitrary vector fixed in B, we have

$${}^{R}\boldsymbol{\omega}^{B} = {}^{\hat{R}}\boldsymbol{\omega}^{B} + {}^{R}\boldsymbol{\omega}^{\hat{R}} \tag{6.7.4}$$

That is, the angular velocity of B in R may be obtained as the sum of the angular velocity of B in an intermediate frame $\hat{R}$ and the angular velocity of $\hat{R}$ in R. This statement, and the corresponding Equation (6.7.4), is the "addition theorem for angular velocity."

The addition theorem is readily generalized to include more than one intermediate reference frame. Consider again a body B moving in a reference frame R and let R_1, R_2, ..., R_n be other reference frames moving relative to each other and relative to R as depicted in Figure 6.7.2. Then by repeated use of Equation (6.7.4) and by recognizing that B itself is a reference frame, we have:

$${}^{R}\boldsymbol{\omega}^{B} = {}^{R}\boldsymbol{\omega}^{R_1} + {}^{R_1}\boldsymbol{\omega}^{R_2} + {}^{R_2}\boldsymbol{\omega}^{R_3} + \dots + {}^{R_{n-1}}\boldsymbol{\omega}^{R_n} + {}^{R_n}\boldsymbol{\omega}^{B} \tag{6.7.5}$$

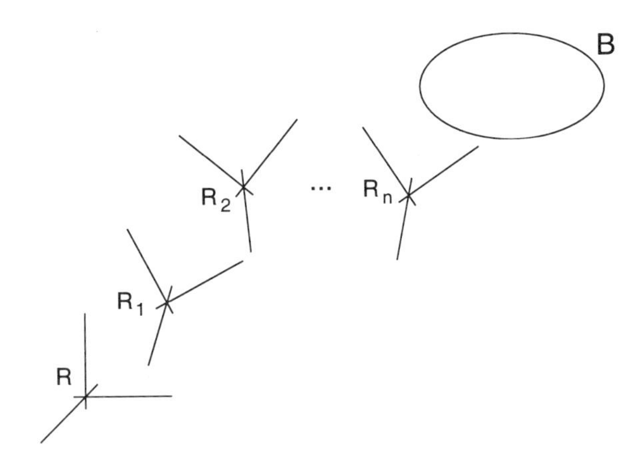

Fig. 6.7.2 A Body B, Reference Frame R, and Intermediate Reference Frames $R_1, \ldots, R_n$

Example 6.7.1 Use of Angular Velocity Addition Theorem and Configuration Graphs

Consider a body B moving in a reference frame R as in Figure 6.7.3. Let $\mathbf{N}_i$ and $\mathbf{n}_i$ be unit vectors fixed in B and R as shown. Let the orientation of B in R be described by dextral orientation angles α, β, and γ: Let B initially be oriented so that the unit vectors $\mathbf{n}_i$ and $\mathbf{N}_i$ $(i = 1,2,3)$ are respectively aligned. Then B can be brought to any given orientation by successive rotations about axes parallel to $\mathbf{n}_1$, $\mathbf{n}_2$, and $\mathbf{n}_3$ through the angles α, β, and γ, regarded as positive for dextral or right-hand rotations. Let $\hat{R}$ represent a reference frame whose axes are respectively

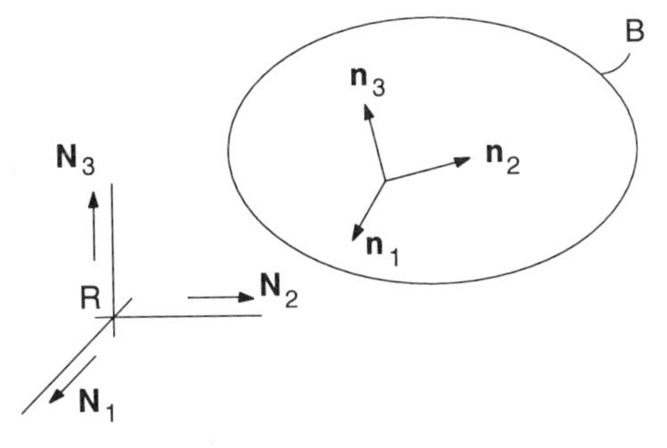

Fig. 6.7.3 A Body B, a Reference Frame R and Respective Unit Vector Sets $\mathbf{n}_i$ and $\mathbf{N}_i$ $(i = 1,2,3)$

parallel to the $\mathbf{n}_i$ after the first (α) rotation and similarly let $\hat{\mathrm{B}}$ be a reference frame with axes respectively parallel to the $\mathbf{n}_i$ after the second (β) rotation. We can then conveniently represent this procedure by a configuration graph as in Figure 6.7.4 where $\hat{\mathbf{N}}_i$ and $\hat{\mathbf{n}}_i$ $(i = 1, 2, 3)$ are unit vectors parallel to the axes of $\hat{\mathrm{R}}$ and $\hat{\mathrm{B}}$ respectively.

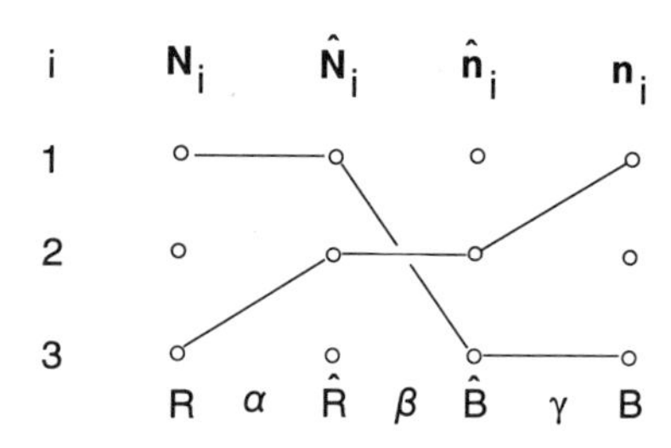

Fig. 6.7.4 Configuration Graph for the Orientation of Body B in Reference Frame R using Dextral Orientation Angles

The movements of B in $\hat{\mathrm{B}}$, of $\hat{\mathrm{B}}$ in $\hat{\mathrm{R}}$, and of $\hat{\mathrm{R}}$ in R are each rotations about axes fixed in the corresponding reference frame pairs. Then the rotation rates of B in $\hat{\mathrm{B}}$, of $\hat{\mathrm{B}}$ and $\hat{\mathrm{R}}$, and of $\hat{\mathrm{R}}$ in R are each "simple angular velocities" as in Section 6.5.5. The corresponding unit vectors and rotation angles are immediately seen in the configuration graph of Figure 6.7.4. Specifically, from the graph and Equation (6.5.23) we have

$$^{R}\boldsymbol{\omega}^{\hat{R}} = \dot{\alpha}\mathbf{N}_1 = \dot{\alpha}\hat{\mathbf{N}}_1 \tag{6.7.6}$$

$$^{\hat{R}}\boldsymbol{\omega}^{\hat{B}} = \dot{\beta}\hat{\mathbf{N}}_2 = \dot{\beta}\hat{\mathbf{n}}_2 \tag{6.7.7}$$

and

$$^{\hat{B}}\boldsymbol{\omega}^{B} = \dot{\gamma}\hat{\mathbf{n}}_3 = \dot{\gamma}\mathbf{n}_3 \tag{6.7.8}$$

The addition theorem [Equation (6.7.5)] then gives us

$$ {}^{R}\boldsymbol{\omega}^{B} = {}^{R}\boldsymbol{\omega}^{\hat{R}} + {}^{R}\boldsymbol{\omega}^{\hat{B}} + {}^{\hat{B}}\boldsymbol{\omega}^{B} \qquad (6.7.9) $$

By substituting from Equations (6.7.6), (6.7.7), and (6.7.8) into (6.7.9), and by again using the configuration graph we can express ${}^{R}\boldsymbol{\omega}^{B}$ in terms of either the $\mathbf{N}_i$ or the $\mathbf{n}_i$. Specifically,

$$ \begin{aligned} {}^{R}\boldsymbol{\omega}^{B} &= \dot{\alpha}\mathbf{N}_1 + \dot{\beta}\hat{\mathbf{N}}_2 + \dot{\gamma}\hat{\mathbf{n}}_3 \\ &= \dot{\alpha}\mathbf{N}_1 + \dot{\beta}(c_\alpha \mathbf{N}_2 + s_\alpha \mathbf{N}_3) + \dot{\gamma}(c_\beta \hat{\mathbf{N}}_3 + s_\beta \hat{\mathbf{N}}_1) \\ &= \dot{\alpha}\mathbf{N}_1 + \dot{\beta}c_2\mathbf{N}_2 + \dot{\beta}s_2\mathbf{N}_3 + \dot{\gamma}c_\beta(c_2\mathbf{N}_3 - s_\alpha\mathbf{N}_2) + \dot{\gamma}s_\beta\mathbf{N}_1 \end{aligned} $$

or

$$ {}^{R}\boldsymbol{\omega}^{B} = (\dot{\alpha} + \dot{\gamma}s_\beta)\mathbf{N}_1 + (\dot{\beta}c_\alpha - \dot{\gamma}c_\beta s_\alpha)\mathbf{N}_2 + (\dot{\beta}s_\alpha + \dot{\gamma}c_\beta c_\alpha)\mathbf{N}_3 \qquad (6.7.10) $$

and

$$ \begin{aligned} {}^{R}\boldsymbol{\omega}^{B} &= \dot{\alpha}\hat{\mathbf{N}}_1 + \dot{\beta}\hat{\mathbf{n}}_2 + \dot{\gamma}\mathbf{n}_3 \\ &= \dot{\alpha}(c_\beta\hat{\mathbf{n}}_1 + s_\beta\hat{\mathbf{n}}_3) + \dot{\beta}(c_\gamma\mathbf{n}_2 + s_\gamma\mathbf{n}_1) + \dot{\gamma}\mathbf{n}_3 \\ &= \dot{\alpha}c_\beta(c_\gamma\mathbf{n}_1 - s_\beta\mathbf{n}_2) + \dot{\alpha}s_\beta\mathbf{n}_3 + \dot{\beta}c_\gamma\mathbf{n}_2 + \dot{\beta}s_\gamma\mathbf{n}_1 + \dot{\gamma}\mathbf{n}_3 \end{aligned} $$

or

$$ {}^{R}\boldsymbol{\omega}^{B} = (\dot{\alpha}c_\beta c_\gamma + \dot{\beta}s_\beta)\mathbf{n}_1 + (-\dot{\gamma}c_\beta s_\gamma + \dot{\beta}c_\gamma)\mathbf{n}_2 + (\dot{\alpha}s_\beta + \dot{\gamma})\mathbf{n}_3 \qquad (6.7.11) $$

Equations (6.7.10) and (6.7.11) may be expressed in the component forms as:

$$ {}^{R}\boldsymbol{\omega}^{B} = \Omega_1\mathbf{N}_1 + \Omega_2\mathbf{N}_2 + \Omega_3\mathbf{N}_3 = \Omega_i\mathbf{N}_i \qquad (6.7.12) $$

and

$$ {}^{R}\boldsymbol{\omega}^{B} = \omega_1\mathbf{n}_1 + \omega_2\mathbf{n}_2 + \omega_3\mathbf{n}_3 = \omega_i\mathbf{n}_i \qquad (6.7.13) $$

where the components Ω_i and ω_i $(i = 1,2,3)$ are

$$\begin{aligned} \Omega_1 &= \dot{\alpha} + \dot{\gamma} s_\beta \\ \Omega_2 &= \dot{\beta} c_\alpha - \dot{\gamma} c_\beta s_\alpha \\ \Omega_3 &= \dot{\beta} s_\alpha + \dot{\gamma} c_\beta c_\alpha \end{aligned} \quad \text{and} \quad \begin{aligned} \omega_1 &= \dot{\alpha} c_\beta c_\gamma + \dot{\beta} s_\gamma \\ \omega_2 &= -\dot{\alpha} c_\beta s_\gamma + \dot{\beta} c_\gamma \\ \omega_3 &= \dot{\alpha} s_\beta + \dot{\gamma} \end{aligned} \tag{6.7.14}$$

This example demonstrates the convenience of using the addition theorem, simple angular velocity, and a configuration graph to obtain angular velocity components for general rotation of a body in a reference frame. Since the definitions of angular velocity as in Equations (6.5.4) and (6.5.13) are consistent with the concept of simple angular velocity (see Section 6.5.5), the utility of the definitions is established.

Finally, observe that the components of Equation (6.7.14) are also related by the transformation matrix S with elements $S_{ij} = \mathbf{N}_i \bullet \mathbf{n}_j$. That is

$$\Omega_i = S_{ij}\omega_j \quad \text{and} \quad \omega_j = S_{ij}\Omega_i \tag{6.7.15}$$

[See Equation (6.2.4).]

The following section presents analogous results for other rotation sequences.

6.8 Angular Velocity Components for Various Rotation Sequences

Table 6.8.1 presents a listing of angular velocity vector components for a body B moving in a reference frame R for various orientation angle definitions, for body-fixed axis rotation sequences. (See Table 6.4.1 and Reference [6.2].) The components are developed for unit vectors fixed in both R and B.

Table 6.8.2 provides similar listings for orientation angles developed from space-fixed axis rotation sequences (see Table 6.4.2 and Reference [6.2].)

Table 6.8.1 Angular Velocity Components for Orientation Angles Defined by Body-Fixed Axis Rotation Sequences

Rotation Sequence	Configuration Graph	Angular Velocity Components
1. 1-2-3 or x-y-z (Dextral or Bryan Angles) (Body Axes)	i, N_i, n_i; 1, 2, 3; θ_1 θ_2 θ_3; R, B	${}^R\boldsymbol{\omega}^B = \Omega_i \mathbf{N}_i = \omega_i \mathbf{n}_i$ $\mathbf{N}_i$: $\Omega_1 = \dot{\theta}_1 + \dot{\theta}_3 s_2$ $\Omega_2 = \dot{\theta}_2 c_1 - \dot{\theta}_3 c_2 s_1$ $\Omega_3 = \dot{\theta}_2 s_1 + \dot{\theta}_3 c_2 c_1$ $\mathbf{n}_i$: $\omega_1 = \dot{\theta}_1 c_2 c_3 + \dot{\theta}_2 s_3$ $\omega_2 = -\dot{\theta}_1 c_2 s_3 + \dot{\theta}_2 c_3$ $\omega_3 = \dot{\theta}_1 s_2 + \dot{\theta}_3$
2. 2-3-1 or y-z-x (Body Axes)	i, N_i, n_i; 1, 2, 3; θ_1 θ_2 θ_3; R, B	$\mathbf{N}_i$: $\Omega_1 = \dot{\theta}_2 s_1 + \dot{\theta}_3 c_1 c_2$ $\Omega_2 = \dot{\theta}_1 + \dot{\theta}_3 s_2$ $\Omega_3 = \dot{\theta}_2 c_1 - \dot{\theta}_3 s_1 c_2$ $\mathbf{n}_i$: $\omega_1 = \dot{\theta}_1 s_2 + \dot{\theta}_3$ $\omega_2 = \dot{\theta}_1 c_2 c_3 + \dot{\theta}_2 s_3$ $\omega_3 = -\dot{\theta}_1 c_2 s_3 + \dot{\theta}_2 c_3$

Rotation Sequence	Configuration Graph	Angular Velocity Components
3. 3-1-2 or z-y-x (Body Axes)	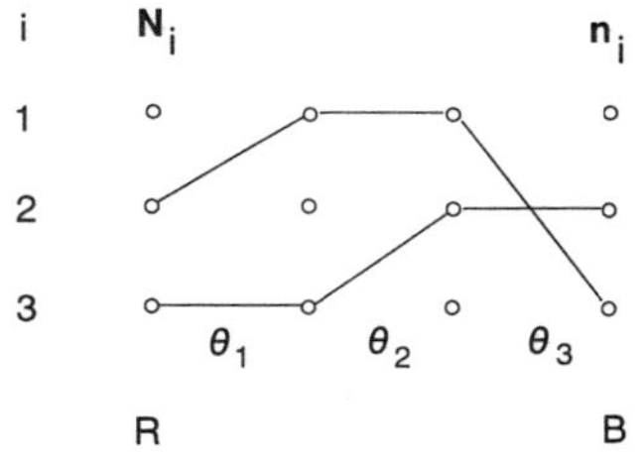 i: 1, 2, 3; N_i, n_i; θ_1, θ_2, θ_3; R, B	N_i: $\Omega_1 = \dot{\theta}_2 c_1 - \dot{\theta}_3 s_1 c_2$ $\Omega_2 = \dot{\theta}_2 s_1 + \dot{\theta}_3 c_1 c_2$ $\Omega_3 = \dot{\theta}_1 + \dot{\theta}_3 s_2$ n_i: $\omega_1 = -\dot{\theta}_1 c_2 s_3 + \dot{\theta}_2 c_3$ $\omega_2 = \dot{\theta}_1 s_2 + \dot{\theta}_3$ $\omega_3 = \dot{\theta}_1 c_2 c_3 + \dot{\theta}_2 s_3$
4. 1-3-2 or x-z-y (Body Axes)	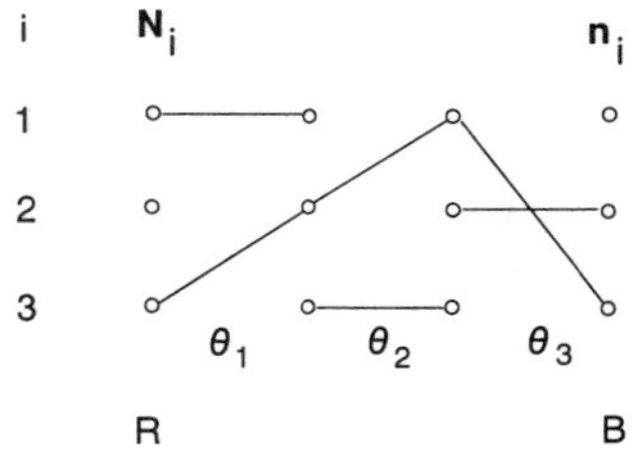 i: 1, 2, 3; N_i, n_i; θ_1, θ_2, θ_3; R, B	N_i: $\Omega_1 = \dot{\theta}_1 - \dot{\theta}_3 s_2$ $\Omega_2 = -\dot{\theta}_2 s_1 + \dot{\theta}_3 c_1 c_2$ $\Omega_3 = \dot{\theta}_2 c_1 + \dot{\theta}_3 s_1 c_2$ n_i: $\omega_1 = \dot{\theta}_1 c_2 c_3 - \dot{\theta}_2 s_3$ $\omega_2 = -\dot{\theta}_1 s_2 + \dot{\theta}_3$ $\omega_3 = \dot{\theta}_1 c_2 s_3 + \dot{\theta}_2 c_3$

Rotation Sequence	Configuration Graph	Angular Velocity Components
5. 2-1-3 or y-x-z (Body Axes)	i $\mathbf{N}_i$ $\mathbf{n}_i$ 1 2 3 θ_1 θ_2 θ_3 R B	$\mathbf{N}_i$: $\Omega_1 = \dot{\theta}_2 c_1 + \dot{\theta}_3 s_1 c_2$ $\Omega_2 = \dot{\theta}_1 - \dot{\theta}_3 s_2$ $\Omega_3 = -\dot{\theta}_2 s_1 + \dot{\theta}_3 c_1 c_2$ $\mathbf{n}_i$: $\omega_1 = \dot{\theta}_1 c_2 s_3 + \dot{\theta}_2 c_3$ $\omega_2 = \dot{\theta}_1 c_2 c_3 - \dot{\theta}_2 s_3$ $\omega_3 = -\dot{\theta}_1 s_2 + \dot{\theta}_3$
6. 3-2-1 or z-y-x (Body Axes)	i $\mathbf{N}_i$ $\mathbf{n}_i$ 1 2 3 θ_1 θ_2 θ_3 R B	$\mathbf{N}_i$: $\Omega_1 = -\dot{\theta}_2 s_1 + \dot{\theta}_3 c_1 c_2$ $\Omega_2 = \dot{\theta}_2 c_1 + \dot{\theta}_3 s_1 c_2$ $\Omega_3 = \dot{\theta}_1 - \dot{\theta}_3 s_2$ $\mathbf{n}_i$: $\omega_1 = -\dot{\theta}_1 s_2 + \dot{\theta}_3$ $\omega_2 = \dot{\theta}_1 c_2 s_3 + \dot{\theta}_2 c_3$ $\omega_3 = \dot{\theta}_1 c_2 c_3 - \dot{\theta}_2 s_3$

Rotation Sequence	Configuration Graph	Angular Velocity Components
7. 1-2-1 or X-Y-X (Body Axes)	i, $\mathbf{N}_i$, $\mathbf{n}_i$; 1, 2, 3; θ_1, θ_2, θ_3; R, B	$\mathbf{N}_i$: $\Omega_1 = \dot{\theta}_1 + \dot{\theta}_3 c_2$ $\Omega_2 = \dot{\theta}_2 c_1 + \dot{\theta}_3 s_1 s_2$ $\Omega_3 = \dot{\theta}_2 s_1 - \dot{\theta}_3 c_1 s_2$ $\mathbf{n}_i$: $\omega_1 = \dot{\theta}_1 c_2 + \dot{\theta}_3$ $\omega_2 = \dot{\theta}_1 s_2 s_3 + \dot{\theta}_2 c_3$ $\omega_3 = \dot{\theta}_1 s_2 c_3 - \dot{\theta}_2 s_3$
8. 1-3-1 or x-z-x (Euler Angles) (Body Axes)	i, $\mathbf{N}_i$, $\mathbf{n}_i$; 1, 2, 3; θ_1, θ_2, θ_3; R, B	$\mathbf{N}_i$: $\Omega_1 = \dot{\theta}_1 + \dot{\theta}_3 c_2$ $\Omega_2 = -\dot{\theta}_2 s_1 + \dot{\theta}_3 c_1 s_2$ $\Omega_3 = -\dot{\theta}_2 c_1 + \dot{\theta}_3 s_1 s_2$ $\mathbf{n}_i$: $\omega_1 = \dot{\theta}_1 c_2 + \dot{\theta}_3$ $\omega_2 = -\dot{\theta}_1 s_2 c_3 + \dot{\theta}_2 s_3$ $\omega_3 = \dot{\theta}_1 s_2 s_3 + \dot{\theta}_2 c_3$

Rotation Sequence	Configuration Graph	Angular Velocity Components
9. 2-1-2 or y-x-y (Body Axes)	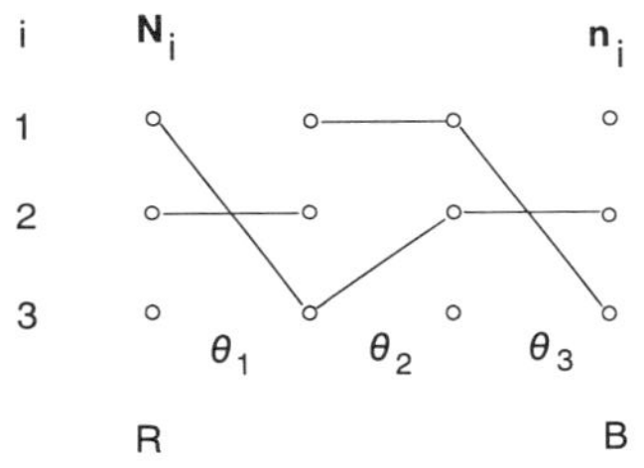 i, $\mathbf{N}_i$, $\mathbf{n}_i$; 1, 2, 3; θ_1 θ_2 θ_3; R, B	$\mathbf{N}_i$: $\Omega_1 = \dot{\theta}_2 c_1 + \dot{\theta}_3 s_1 s_2$ $\Omega_2 = \dot{\theta}_1 + \dot{\theta}_3 c_2$ $\Omega_3 = -\dot{\theta}_2 s_1 + \dot{\theta}_3 c_1 s_2$ $\mathbf{n}_i$: $\omega_1 = \dot{\theta}_1 s_2 s_3 + \dot{\theta}_2 c_3$ $\omega_2 = \dot{\theta}_1 c_2 + \dot{\theta}_3$ $\omega_3 = -\dot{\theta}_1 s_2 c_3 + \dot{\theta}_2 s_3$
10. 2-3-2 or y-z-y (Body Axes)	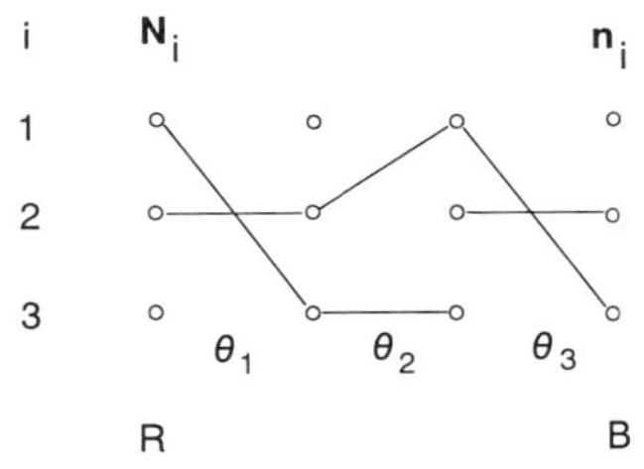 i, $\mathbf{N}_i$, $\mathbf{n}_i$; 1, 2, 3; θ_1 θ_2 θ_3; R, B	$\mathbf{N}_i$: $\Omega_1 = \dot{\theta}_2 s_1 - \dot{\theta}_3 c_1 s_2$ $\Omega_2 = \dot{\theta}_1 + \dot{\theta}_3 c_2$ $\Omega_3 = \dot{\theta}_2 c_1 + \dot{\theta}_3 s_1 s_2$ $\mathbf{n}_i$: $\omega_1 = \dot{\theta}_1 s_2 c_3 - \dot{\theta}_2 s_3$ $\omega_2 = \dot{\theta}_1 c_2 + \dot{\theta}_3$ $\omega_3 = \dot{\theta}_1 s_2 s_3 + \dot{\theta}_2 c_3$

Rotation Sequence	Configuration Graph	Angular Velocity Components
11. 3-1-3 or z-x-z (Euler Angles) (Body Axes)	i, N_i, n_i; 1, 2, 3; θ_1 θ_2 θ_3; R B	$\mathbf{N}_i$: $\Omega_1 = \dot{\theta}_2 c_1 + \dot{\theta}_3 s_1 s_2$ $\Omega_2 = \dot{\theta}_2 s_1 - \dot{\theta}_3 c_1 s_2$ $\Omega_3 = \dot{\theta}_1 + \dot{\theta}_3 c_2$ $\mathbf{n}_i$: $\omega_1 = \dot{\theta}_1 s_2 s_3 + \dot{\theta}_2 c_3$ $\omega_2 = \dot{\theta}_1 s_2 c_3 - \dot{\theta}_2 s_3$ $\omega_3 = \dot{\theta}_3 + \dot{\theta}_1 c_2$
12. 3-2-3 or z-y-z (Body Axes)	i, N_i, n_i; 1, 2, 3; θ_1 θ_2 θ_3; R B	$\mathbf{N}_i$: $\Omega_1 = -\dot{\theta}_2 s_1 + \dot{\theta}_3 c_1 s_2$ $\Omega_2 = \dot{\theta}_2 c_1 + \dot{\theta}_3 s_2 c_2$ $\Omega_3 = \dot{\theta}_1 + \dot{\theta}_3 c_2$ $\mathbf{n}_i$: $\omega_1 = -\dot{\theta}_1 s_2 c_3 + \dot{\theta}_2 s_3$ $\omega_2 = \dot{\theta}_1 s_2 s_3 + \dot{\theta}_2 c_3$ $\omega_3 = \dot{\theta}_3 + \dot{\theta}_1 c_2$

Table 6.8.2 Angular Velocity Components for Orientation Angles Defined by Space-Fixed Axis Rotation Sequences

Rotation Sequence	Configuration Graph	Angular Velocity Components
1. 1-2-3 or X-Y-Z (Dextral or Bryan Angles) (Body Axes) (Space Axes)	i, $\mathbf{N}_i$, $\mathbf{n}_i$; 1, 2, 3; θ_3, θ_2, θ_1; R, B	${}^R\boldsymbol{\omega}^B = \Omega_i \mathbf{N}_i = \omega_i \mathbf{n}_i$ $\mathbf{N}_i$: $\Omega_1 = \dot{\theta}_1 c_2 c_3 - \dot{\theta}_2 s_3$ $\Omega_2 = \dot{\theta}_1 c_2 s_3 + \dot{\theta}_2 c_3$ $\Omega_3 = -\dot{\theta}_1 s_2 + \dot{\theta}_3$ $\mathbf{n}_i$: $\omega_1 = \dot{\theta}_1 - s_2 \dot{\theta}_3$ $\omega_2 = \dot{\theta}_2 c_1 + \dot{\theta}_3 s_1 c_2$ $\omega_3 = -\dot{\theta}_2 s_1 + \dot{\theta}_3 c_1 c_2$
2. 2-3-1 or Y-Z-X (Space Axes)	i, $\mathbf{N}_i$, $\mathbf{n}_i$; 1, 2, 3; θ_3, θ_2, θ_1; R, B	$\mathbf{N}_i$: $\Omega_1 = -\dot{\theta}_1 s_2 + \dot{\theta}_3$ $\Omega_2 = \dot{\theta}_1 c_2 c_3 - \dot{\theta}_2 s_3$ $\Omega_3 = \dot{\theta}_1 c_2 s_3 + \dot{\theta}_2 c_3$ $\mathbf{n}_i$: $\omega_1 = -\dot{\theta}_2 s_1 + \dot{\theta}_3 c_1 c_2$ $\omega_2 = \dot{\theta}_1 - \dot{\theta}_3 s_2$ $\omega_3 = \dot{\theta}_2 c_1 + \dot{\theta}_3 s_1 c_2$

Rotation Sequence	Configuration Graph	Angular Velocity Components
3. 3-1-2 or Z-Y-X (Space Axes)	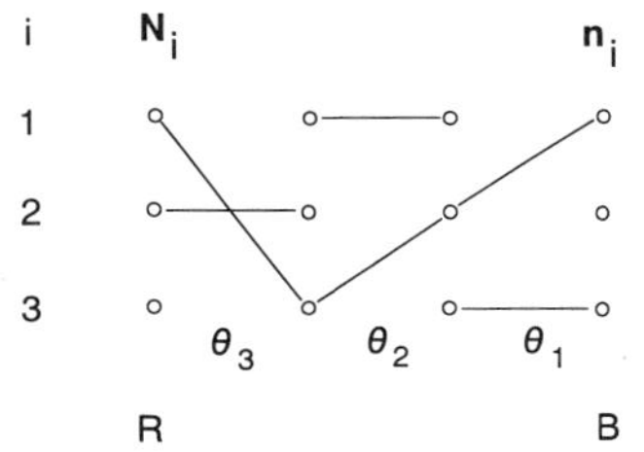 i, N_i, n_i; 1, 2, 3; θ_3 θ_2 θ_1; R, B	$\mathbf{N}_i$: $\Omega_1 = \dot{\theta}_1 c_2 s_3 + \dot{\theta}_2 c_3$ $\Omega_2 = -\dot{\theta}_1 s_2 + \dot{\theta}_3$ $\Omega_3 = \dot{\theta}_1 c_2 c_3 - \dot{\theta}_2 s_3$ $\mathbf{n}_i$: $\omega_1 = \dot{\theta}_2 c_1 + \dot{\theta}_3 s_1 c_2$ $\omega_2 = -\dot{\theta}_2 s_1 + \dot{\theta}_3 c_1 c_2$ $\omega_3 = \dot{\theta}_1 - \dot{\theta}_3 s_2$
4. 1-3-2 or X-Z-Y (Space Axes)	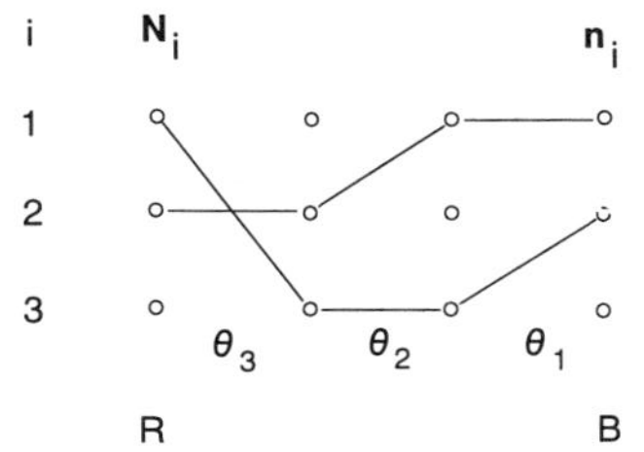 i, N_i, n_i; 1, 2, 3; θ_3 θ_2 θ_1; R, B	$\mathbf{N}_i$: $\Omega_1 = \dot{\theta}_1 c_2 c_3 + \dot{\theta}_2 s_3$ $\Omega_2 = -\dot{\theta}_1 s_2 + \dot{\theta}_3$ $\Omega_3 = -\dot{\theta}_1 c_2 s_3 + \dot{\theta}_2 c_3$ $\mathbf{n}_i$: $\omega_1 = \dot{\theta}_1 + \dot{\theta}_3 s_2$ $\omega_2 = \dot{\theta}_2 s_1 + \dot{\theta}_3 c_1 c_2$ $\omega_3 = \dot{\theta}_2 c_1 - \dot{\theta}_3 s_1 c_2$

Rotation Sequence	Configuration Graph	Angular Velocity Components
5. 2-1-3 or Y-X-Z (Space Axes)	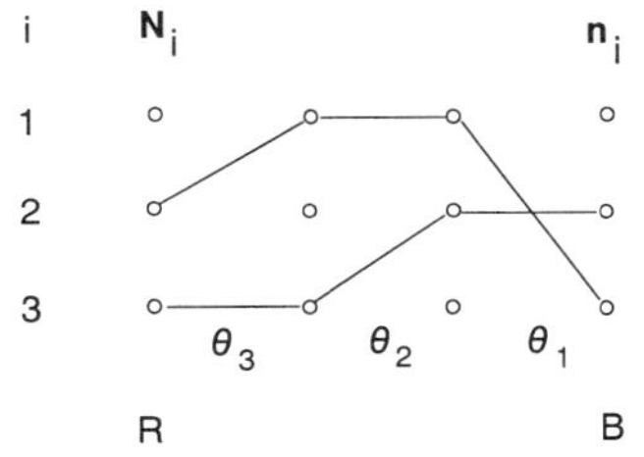 i, N_i, n_i; 1, 2, 3; θ_3 θ_2 θ_1; R, B	$\mathbf{N}_i$: $\Omega_1 = -\dot{\theta}_1 c_2 s_3 + \dot{\theta}_2 c_3$ $\Omega_2 = \dot{\theta}_1 c_2 c_3 + \dot{\theta}_2 s_3$ $\Omega_3 = \dot{\theta}_1 s_2 + \dot{\theta}_3$ $\mathbf{n}_i$: $\omega_1 = \dot{\theta}_2 c_1 - \dot{\theta}_3 s_1 c_2$ $\omega_2 = \dot{\theta}_1 + \dot{\theta}_3 s_2$ $\omega_3 = \dot{\theta}_2 s_1 + \dot{\theta}_3 c_1 c_2$
6. 3-2-1 or Z-Y-X (Space Axes)	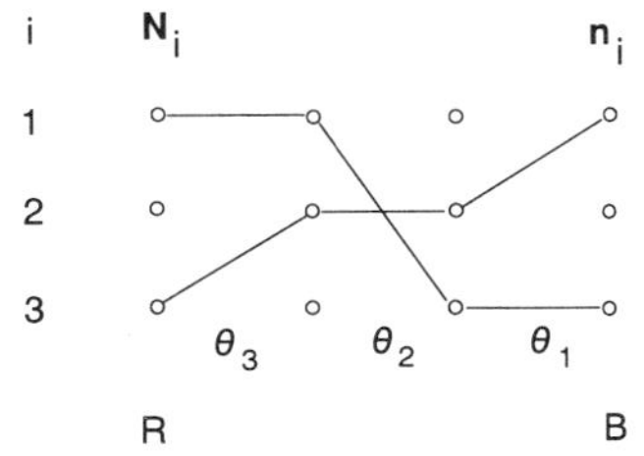 i, N_i, n_i; 1, 2, 3; θ_3 θ_2 θ_1; R, B	$\mathbf{N}_i$: $\Omega_1 = \dot{\theta}_1 s_2 + \dot{\theta}_3$ $\Omega_2 = -\dot{\theta}_1 c_2 s_3 + \dot{\theta}_2 c_3$ $\Omega_3 = \dot{\theta}_1 c_2 c_3 + \dot{\theta}_2 s_3$ $\mathbf{n}_i$: $\omega_1 = \dot{\theta}_2 s_1 + \dot{\theta}_3 c_1 c_2$ $\omega_2 = \dot{\theta}_2 c_1 - \dot{\theta}_3 s_1 c_2$ $\omega_3 = \dot{\theta}_1 + \dot{\theta}_3 s_2$

Rotation Sequence	Configuration Graph	Angular Velocity Components
7. 1-2-1 or X-Y-X (Space Axes)	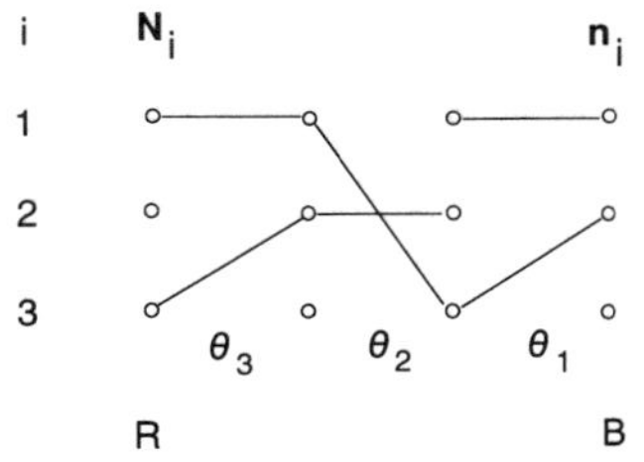 i, N_i, n_i; 1, 2, 3; θ_3, θ_2, θ_1; R, B	N_i: $\Omega_1 = \dot\theta_1 c_2 + \dot\theta_3$ $\Omega_2 = \dot\theta_1 s_2 s_3 + \dot\theta_2 c_3$ $\Omega_3 = -\dot\theta_1 s_2 c_3 + \dot\theta_2 s_3$ n_i: $\omega_1 = \dot\theta_1 + \dot\theta_3 c_2$ $\omega_2 = \dot\theta_2 c_1 + \dot\theta_3 s_1 s_2$ $\omega_3 = -\dot\theta_2 s_1 + \dot\theta_3 c_1 s_2$
8. 1-3-1 or X-Z-X (Euler Angles) (Space Axes)	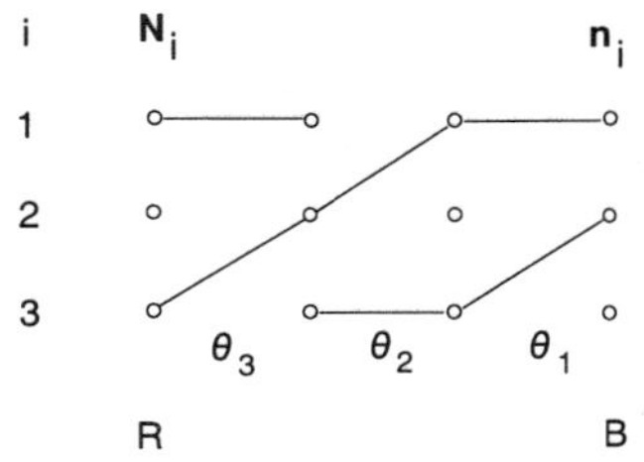 i, N_i, n_i; 1, 2, 3; θ_3, θ_2, θ_1; R, B	N_i: $\Omega_1 = \dot\theta_1 c_2 + \dot\theta_3$ $\Omega_2 = \dot\theta_1 s_2 c_3 - \dot\theta_2 s_3$ $\Omega_3 = \dot\theta_1 s_2 s_3 + \dot\theta_2 c_3$ n_i: $\omega_1 = \dot\theta_1 + \dot\theta_3 c_2$ $\omega_2 = \dot\theta_2 s_1 - \dot\theta_3 c_1 s_2$ $\omega_3 = \dot\theta_2 c_1 + \dot\theta_3 s_1 s_2$

Rotation Sequence	Configuration Graph	Angular Velocity Components
9. 2-1-2 or Y-X-Y (Space Axes)	i, N_i, n_i; 1, 2, 3; θ_3 θ_2 θ_1; R B	$\mathbf{N}_i$: $\Omega_1 = \dot{\theta}_1 s_2 s_3 + \dot{\theta}_2 c_3$ $\Omega_2 = \dot{\theta}_1 c_2 + \dot{\theta}_3$ $\Omega_3 = \dot{\theta}_1 s_2 c_3 - \dot{\theta}_2 s_3$ $\mathbf{n}_i$: $\omega_1 = \dot{\theta}_2 c_1 + \dot{\theta}_3 s_1 s_2$ $\omega_2 = \dot{\theta}_1 + \dot{\theta}_3 c_2$ $\omega_3 = \dot{\theta}_2 s_1 - \dot{\theta}_3 s_2 c_1$
10. 2-3-2 or Y-Z-Y (Space Axes)	i, N_i, n_i; 1, 2, 3; θ_3 θ_2 θ_1; R B	$\mathbf{N}_i$: $\Omega_1 = -\dot{\theta}_1 s_2 c_3 + \dot{\theta}_2 s_3$ $\Omega_2 = \dot{\theta}_1 c_2 + \dot{\theta}_3$ $\Omega_3 = \dot{\theta}_1 s_2 s_3 + \dot{\theta}_2 c_3$ $\mathbf{n}_i$: $\omega_1 = -\dot{\theta}_2 s_1 + \dot{\theta}_3 c_1 s_2$ $\omega_2 = \dot{\theta}_1 + \dot{\theta}_3 c_2$ $\omega_3 = \dot{\theta}_2 c_1 - \dot{\theta}_3 s_1 s_2$

Rotation Sequence	Configuration Graph	Angular Velocity Components
11. 3-1-3 or z-x-z (Euler Angles) (Space Axes)	i, N_i, n_i; 1, 2, 3; θ_3, θ_2, θ_1; R, B	N_i: $\Omega_1 = \dot{\theta}_1 s_2 s_3 + \dot{\theta}_2 c_3$ $\Omega_2 = -\dot{\theta}_2 s_2 c_3 + \dot{\theta}_2 s_3$ $\Omega_3 = \dot{\theta}_1 c_2 + \dot{\theta}_3$ n_i: $\omega_1 = \dot{\theta}_2 c_1 + \dot{\theta}_3 s_1 s_2$ $\omega_2 = -\dot{\theta}_2 s_1 + \dot{\theta}_3 c_1 s_2$ $\omega_3 = \dot{\theta}_1 + \dot{\theta}_3 c_2$
12. 3-2-3 or Z-Y-Z (Space Axes)	i, N_i, n_i; 1, 2, 3; θ_3, θ_2, θ_1; R, B	N_i: $\Omega_1 = \dot{\theta}_1 s_2 c_3 - \dot{\theta}_2 s_3$ $\Omega_2 = \dot{\theta}_1 s_2 s_3 + \dot{\theta}_2 c_3$ $\Omega_3 = \dot{\theta}_1 c_2 + \dot{\theta}_3$ n_i: $\omega_1 = \dot{\theta}_2 s_1 - \dot{\theta}_3 c_1 s_2$ $\omega_2 = \dot{\theta}_2 c_1 + \dot{\theta}_3 s_1 s_2$ $\omega_3 = \dot{\theta}_1 + \dot{\theta}_3 c_2$

6.9 Angular Acceleration

6.9.1 Definition

Consider a body B moving in a reference frame R. The angular velocity of B in R, ${}^{R}\alpha^{B}$, is defined as the derivative of the angular velocity of B in R:

$$ {}^{R}\boldsymbol{\alpha}^{B} = {}^{R}d\,{}^{R}\boldsymbol{\omega}^{B}/dt \tag{6.9.1} $$

6.9.2 Addition Theorem

Recall that for a body B moving in a reference frame R, with intermediate reference frames R_1, R_2, ..., R_n we have the addition theorem for angular velocity [Equation (6.7.5)]:

$$ {}^{R}\boldsymbol{\omega}^{B} = {}^{R}\boldsymbol{\omega}^{R_1} + {}^{R_1}\boldsymbol{\omega}^{R_2} + {}^{R_2}\boldsymbol{\omega}^{R_3} + \dots + {}^{R_{n-1}}\boldsymbol{\omega}^{R_n} + {}^{R_n}\boldsymbol{\omega}^{B} \tag{6.9.2} $$

This equation, however, does not generalize to angular acceleration. That is

$$ {}^{R}\boldsymbol{\alpha}^{B} \neq {}^{R}\boldsymbol{\alpha}^{R_1} + {}^{R_1}\boldsymbol{\alpha}^{R_2} + {}^{R_2}\boldsymbol{\alpha}^{R_3} + \dots + {}^{R_{n-1}}\boldsymbol{\alpha}^{R_n} + {}^{R_n}\boldsymbol{\alpha}^{B} \tag{6.9.3} $$

To see the essence of the inequality, consider a body B moving in a reference frame R with a single intermediate reference frame $\hat{R}$. Then we have

$$ {}^{R}\boldsymbol{\omega}^{B} = {}^{R}\boldsymbol{\omega}^{\hat{R}} + {}^{\hat{R}}\boldsymbol{\omega}^{B} \tag{6.9.4} $$

By differentiating, and using Equation (6.6.7), we have

$$^{R}d\,^{R}\boldsymbol{\omega}^{B}/dt = {}^{R}d\,^{R}\boldsymbol{\omega}^{\hat{R}}/dt + {}^{R}d\,^{\hat{R}}\boldsymbol{\omega}^{B}/dt$$

$$= {}^{R}\boldsymbol{\alpha}^{\hat{R}} + {}^{B}d\,^{\hat{R}}\boldsymbol{\omega}^{B}/dt + {}^{R}\boldsymbol{\omega}^{B} \times {}^{\hat{R}}\boldsymbol{\omega}^{B}$$

or

$$^{R}\boldsymbol{\alpha}^{B} = {}^{R}\boldsymbol{\alpha}^{\hat{R}} + {}^{\hat{R}}\boldsymbol{\alpha}^{B} + {}^{R}\boldsymbol{\omega}^{B} \times {}^{\hat{R}}\boldsymbol{\omega}^{B} \tag{6.9.5}$$

The presence of the terms: $^{R}\boldsymbol{\omega}^{B} \times {}^{\hat{R}}\boldsymbol{\omega}^{B}$ gives rise to the inequality in (6.9.3). If $^{R}\boldsymbol{\omega}^{B} \times {}^{\hat{R}}\boldsymbol{\omega}^{B}$ is zero, we have an angular acceleration addition theorem analogous to Equation (6.9.4). This can occur in the trivial cases with either $^{R}\boldsymbol{\omega}^{B}$ or $^{\hat{R}}\boldsymbol{\omega}^{B}$ being parallel, as with parallel simple angular velocities.

6.9.3 Computation Algorithms

From the definition of Equation (6.9.1) it is seen that the components of the angular acceleration vector are different from the simple derivatives of the components of the angular vector. That is, $\mathbf{n}_i$ $(i = 1,2,3)$ are mutually perpendicular unit vectors and if $^{R}\boldsymbol{\omega}^{B}$ and $^{R}\boldsymbol{\alpha}^{B}$ are expressed as:

$$^{R}\boldsymbol{\omega}^{B} = \omega_i \mathbf{n}_i \quad \text{and} \quad {}^{R}\boldsymbol{\alpha}^{B} = \alpha_i \mathbf{n}_i \tag{6.9.6}$$

then in general,

$$\alpha_i \neq \dot{\omega}_i \tag{6.9.7}$$

However, if the $\mathbf{n}_i$ are fixed in either B or R then

$$\alpha_i = \dot{\omega}_i \tag{6.9.8}$$

[If the $\mathbf{n}_i$ are fixed in R their derivatives are zero, and if they are fixed in B, then $^{R}d\,^{R}\boldsymbol{\omega}^{B}/dt = {}^{B}d\,^{R}\boldsymbol{\omega}^{B}/dt$ due to Equation (6.6.7) since then we have $^{R}\boldsymbol{\omega}^{B} \times {}^{R}\boldsymbol{\omega}^{B} = 0$.]

6.10 Velocity of Particles, or Points, of a Body

6.10.1 Introduction

Consider a body B moving in a reference frame R as in Figure 6.10.1. Let B be modeled as a set of particles fixed relative to one another and let P and Q be typical particles of B. Then if **r** locates P relative to Q, the magnitude of **r** is a constant, and **r** is fixed in B. Recall that if we regard the particles of B as being

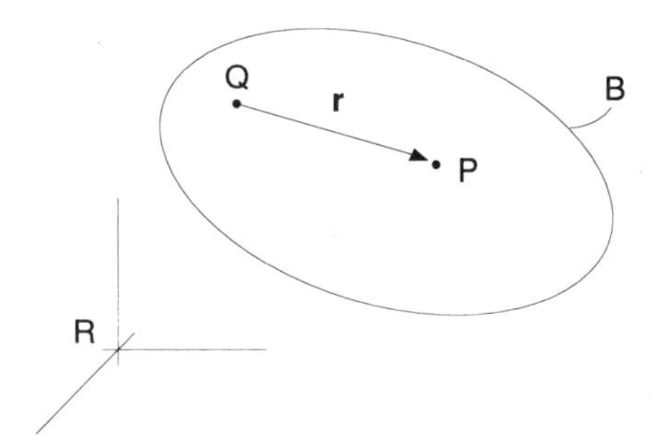

Fig. 6.10.1 A Body B, with Particles P and Q Moving in a Reference Frame R

infinitesimally small, and if we are unconcerned about the masses of the particles, then there is no difference between a particle P and a point P located at the particle (see Section 3.1). The objective in this section is to review the formulas for the velocity of particle or points (such as P and Q) of a body.

6.10.2 Relative Velocity of Two Points of a Body (see also Section 3.14)

Let P and Q be typical points of a body B as in Figure 6.10.1 with **r** locating P relative to Q. Then the velocity of P relative to Q in R is simply the derivative of **r** in R. That is:

$$ {}^{R}\mathbf{V}^{P/Q} = {}^{R}\mathbf{V}^{P} - {}^{R}\mathbf{V}^{Q} = {}^{R}d\mathbf{r}/dt \qquad (6.10.1) $$

Then since $\mathbf{r}$ is fixed in B we have from Equation (6.5.9)

$$^{R}d\mathbf{r}/dt = {}^{R}\boldsymbol{\omega}^{B} \times \mathbf{r} \tag{6.10.2}$$

Hence, we also have

$$^{R}\mathbf{V}^{P} = {}^{R}\mathbf{V}^{Q} + {}^{R}\boldsymbol{\omega}^{B} \times \mathbf{r} \tag{6.10.3}$$

Equation (6.10.3) may be interpreted as showing that if the velocity of a point Q of B is known and if the angular velocity of B in R is known, then the velocity of any and all other points of B can be found.

6.10.3 Motion Classification

The velocities of the points of a body may be used to classify the movement of the body itself:

1) If all points have the same velocity, the body is in *translation.*

2) If the points of a body in translation move in a straight line, the body has *rectilinear translation.*

3) If the points on a line L fixed in a body B all have zero velocity and if the other points of B have non-zero velocity, moving in circles about L, then B has *simple rotation* and L is called the *axis of rotation.*

4) If all the points of a body move parallel to a plane, the body has *plane motion.*

5) Rectilinear translation and simple rotation are spinal cases of plane motion. If a body has plane motion but is neither in rectilinear translation nor simple rotation, then the body is said to have *general plane motion.*

6) If the points of a body B on a line L move along L while B itself is rotating about L, B has *screw motion* about L.

7) A body with neither translation, plane motion nor screw motion is said to have *general motion.*

6.10.4 Center of Rotation

If a body has simple rotation or screw motion then points on the axis of rotation or on the screw axis are called "centers of rotation."

If a body has general motion it is possible to regard the body, at an instant, as rotating about an axis or center of rotation. For bodies having plane motion it is possible to graphically locate the instantaneous center of rotation: Consider such a body as represented in Figure 6.10.2. Let P and Q be points of B and let $\mathbf{V}^P$ and $\mathbf{V}^Q$ be the velocities of P and Q. Then lines drawn through P and Q, perpendicular to $\mathbf{V}^P$ and $\mathbf{V}^Q$, will intersect at the instant center of rotation C. The magnitude of the instantaneous angular speed ω of B about C may then be determined by dividing the magnitude of the velocity of a point by its distance from C. That is,

$$\omega = |\mathbf{V}^P|/PC = |\mathbf{V}^Q|/QC \tag{6.10.4}$$

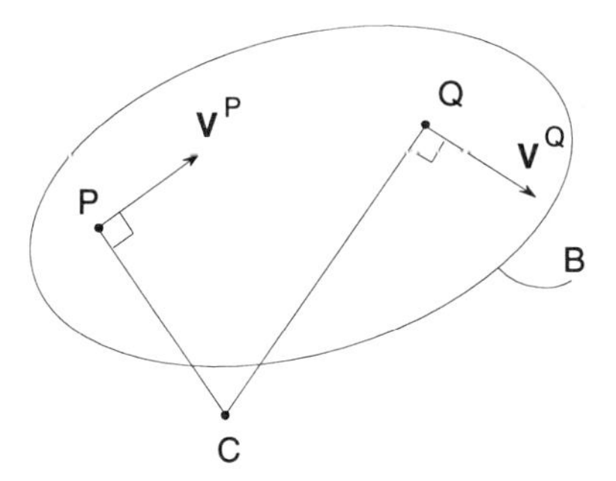

Fig. 6.10.2 A Body B with General Plane Motion and with Instant Center of Rotation C

If a body has general plane motion and if its instant center C and angular speed ω. The direction of the velocity of P is perpendicular to PC as depicted in Figure 6.10.3.

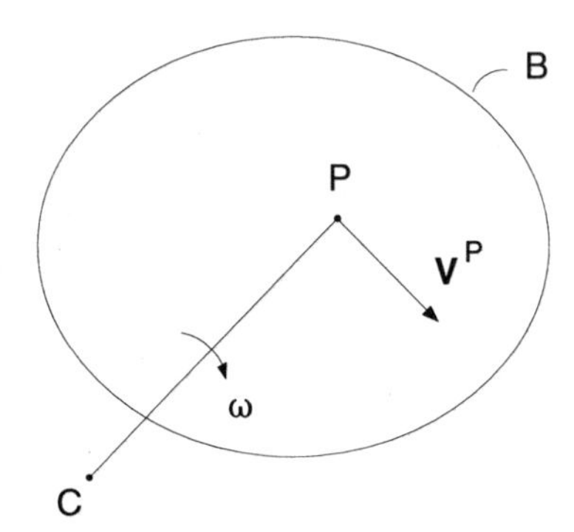

Fig. 6.10.3 Determining the Velocity of a Point P of a Body in General Plane Motion when the Instant Center C and Angular speed ω are known

Finally, if a body B moves in translation its center of rotation is regarded as being infinitely far away in a direction perpendicular to the velocity of the points of B.

6.10.5 Velocity of a Point Moving Relative to a Moving Body (see also Section 3.16.)

Consider a body B moving in a reference frame R and a point P moving relative to B as in Figure 6.10.4. Let Q be an arbitrary point fixed in B. Then the velocity of P in R is:

$$^R\mathbf{V}^P = {}^R\mathbf{V}^Q + {}^R\mathbf{V}^{P/Q} = {}^R\mathbf{V}^Q + {}^R d\mathbf{r}/dt \tag{6.10.5}$$

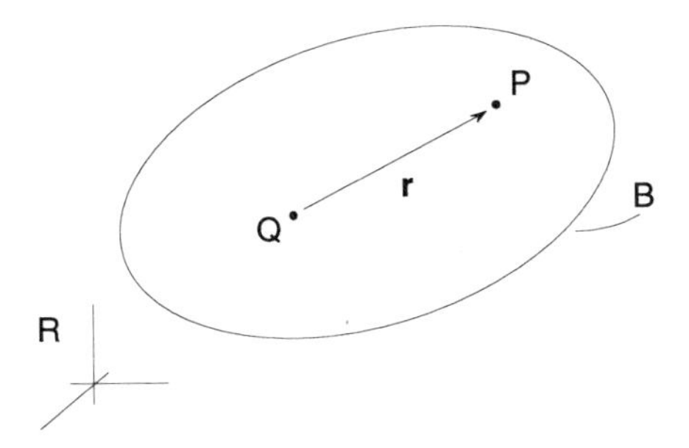

Fig. 6.10.4 A Body B and a Point P Moving Relative to B

From Equation (6.6.7) ${}^{R}d\mathbf{r}/dt$ is:

$$ {}^{R}d\mathbf{r}/dt = {}^{B}d\mathbf{r}/dt + {}^{R}\boldsymbol{\omega}^{B} \times \mathbf{r} = {}^{B}\mathbf{V}^{P} + {}^{R}\boldsymbol{\omega}^{B} \times \mathbf{r} \tag{6.10.6} $$

Hence, ${}^{R}\mathbf{V}^{P}$ is

$$ {}^{R}\mathbf{V}^{P} = {}^{R}\mathbf{V}^{Q} + {}^{B}\mathbf{V}^{P} + {}^{R}\boldsymbol{\omega}^{B} \times \mathbf{r} \tag{6.10.7} $$

Finally, if the instant of interest Q is chosen as that point P^* of B which coincides with P, then $\mathbf{r}$ is zero and ${}^{R}\mathbf{V}^{P}$ takes the reduced form [see also Equation (3.16.4)]:

$$ {}^{R}\mathbf{V}^{P} = {}^{R}\mathbf{V}^{P^*} + {}^{B}\mathbf{V}^{P} \tag{6.10.8} $$

6.11 Acceleration of Particles, or Points, of a Body

6.11.1 Relative Acceleration of Two Points of a Body (see also Sections 3.15 and 6.10)

Consider a body B moving in a reference frame R as in Figure 6.11.1. Let

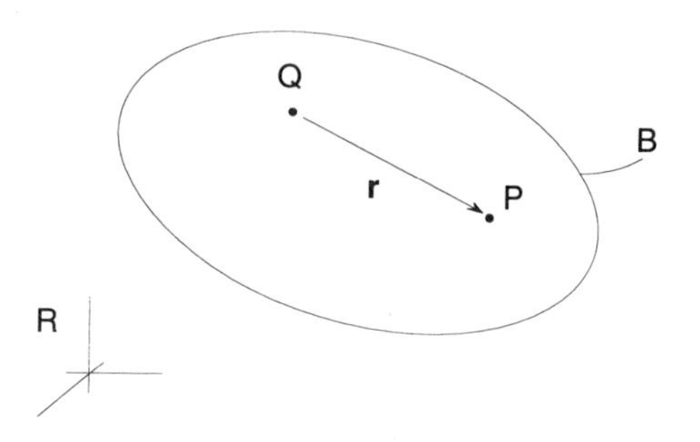

Fig. 6.11.1 A Body B, with Particles P and Q, Moving in a Reference Frame R

P and Q be typical particles, or points, of B and let **r** locate P relative to Q. Then **r** is fixed in B, and the derivative of **r** may be obtained by using Equation (6.5.9). Hence, the relative acceleration of P and Q in R is:

$$\begin{aligned} {}^R\mathbf{a}^{P/Q} &= {}^R\mathbf{a}^P - {}^R\mathbf{a}^Q = {}^Rd({}^R\mathbf{V}^{P/Q})/dt = {}^Rd^2\mathbf{r}/dt^2 \\ &= {}^Rd({}^Rd\mathbf{r}/dt)/dt = {}^Rd({}^R\boldsymbol{\omega}^B \times \mathbf{r})/dt \\ &= {}^R\boldsymbol{\alpha}^B \times \mathbf{r} + {}^R\boldsymbol{\omega}^B \times ({}^R\boldsymbol{\omega}^B \times \mathbf{r}) \end{aligned} \tag{6.11.1}$$

Then the acceleration of P is

$$ {}^R\mathbf{a}^P = {}^R\mathbf{a}^Q + {}^R\boldsymbol{\alpha}^B \times \mathbf{r} + {}^R\boldsymbol{\omega}^B \times ({}^R\boldsymbol{\omega}^B \times \mathbf{r}) \tag{6.11.2}$$

6.11.2 Acceleration of a Point Moving Relative to a Moving Body (see also Section 3.17)

Consider a body B moving in a reference frame R and a point P moving relative to B as in Figure 6.11.2. Let Q be an arbitrary point fixed in B. Then the acceleration of P in R may be expressed as:

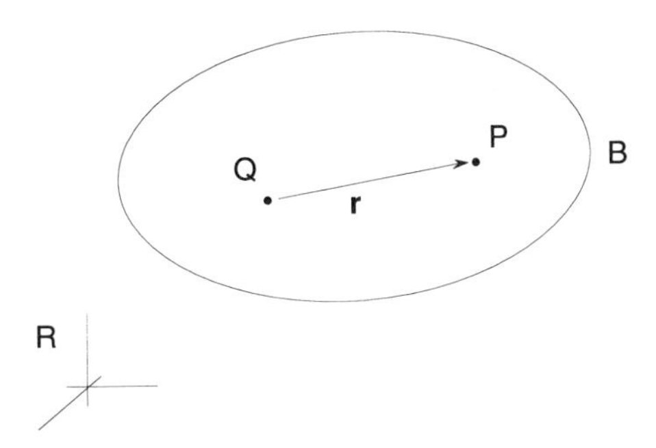

Fig. 6.11.2 A Body B and a Point Moving Relative to B.

$$ {}^{R}\mathbf{a}^{P} = {}^{R}\mathbf{a}^{Q} + {}^{R}\mathbf{a}^{P/Q} = {}^{R}\mathbf{a}^{Q} + {}^{R}d^{2}\mathbf{r}/dt^{2} \tag{6.11.3} $$

where ${}^{R}d^{2}\mathbf{r}/dt^{2}$ may be evaluated by repeated application of Equation (6.6.7). That is,

$$ \begin{aligned} {}^{R}d^{2}\mathbf{r}/dt^{2} &= {}^{R}d({}^{B}d\mathbf{r}/dt + {}^{R}\boldsymbol{\omega}^{B} \times \mathbf{r})/dt \\ &= {}^{R}d({}^{B}d\mathbf{r}/dt)/dt + {}^{R}d({}^{R}\boldsymbol{\omega}^{B} \times \mathbf{r})/dt \\ &= {}^{B}d({}^{B}d\mathbf{r}/dt)/dt + {}^{R}\boldsymbol{\omega}^{B} \times {}^{B}d\mathbf{r}/dt \\ &\quad + {}^{R}\boldsymbol{\alpha}^{B} \times \mathbf{r} + {}^{R}\boldsymbol{\omega}^{B} \times ({}^{B}d\mathbf{r}/dt + {}^{R}\boldsymbol{\omega}^{B} \times \mathbf{r}) \\ &= {}^{B}\mathbf{a}^{P} + 2{}^{R}\boldsymbol{\omega}^{B} \times {}^{B}\mathbf{V}^{P} + {}^{R}\boldsymbol{\alpha}^{B} \times \mathbf{r} \\ &\quad + {}^{R}\boldsymbol{\omega}^{B} \times ({}^{R}\boldsymbol{\omega}^{B} \times \mathbf{r}) \end{aligned} \tag{6.11.4} $$

Then the acceleration of P in R becomes

$$ {}^{R}\mathbf{a}^{P} = {}^{R}\mathbf{a}^{Q} + {}^{B}\mathbf{a}^{P} + 2\,{}^{R}\boldsymbol{\omega}^{B} \times {}^{B}\mathbf{V}^{P} + {}^{R}\boldsymbol{\alpha}^{B} \times \mathbf{r} + {}^{R}\boldsymbol{\omega}^{B} \times ({}^{R}\boldsymbol{\omega}^{B} \times \mathbf{r}) \tag{6.11.5} $$

Finally, if at an instant of interest Q is that point P^{*} of B which coincides with P then $\mathbf{r}$ is zero and ${}^{R}\mathbf{a}^{P}$ has the form [see also Equation (3.17.4)]:

$$ {}^{R}\mathbf{a}^{P} = {}^{B}\mathbf{a}^{P} + {}^{R}\mathbf{a}^{P^*} + 2\,{}^{R}\boldsymbol{\omega}^{B} \times {}^{B}\mathbf{V}^{P} \tag{6.11.6} $$

6.12 Rolling

Rolling is a special case in the kinematics of bodies, but it is a commonly occurring phenomena — particularly in machines and in multibody systems.

Rolling involves the relative moment of two bodies in contact with one another, typically at a single point, or line of points, where there is no relative velocity between the contacting points. Rolling can also involve the movement of a body in contact with a surface, where again there is no relative velocity between the contacting points.

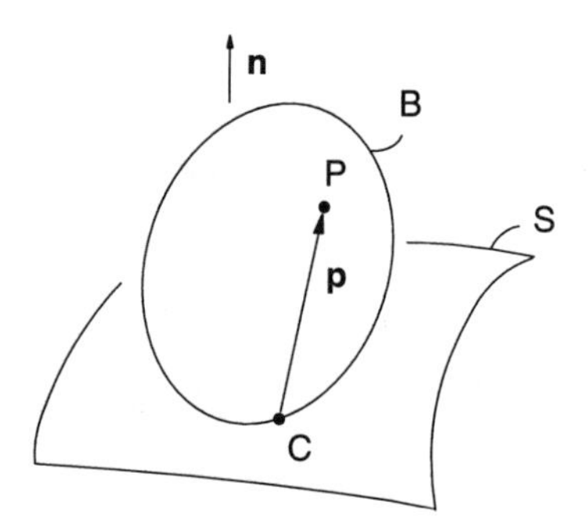

Fig. 6.12.1 A Body B Rolling on a Surface S with a Contact Point C

To develop this consider a body B and a surface S as in Figure 6.12.1. (S could be the surface of a body upon which B rolls.) Let S and B be counterformal so that they are in contact at a single point C. Then if C is regarded as being a point of B, "rolling" is said to occur if [6.4, 6.5, 6.6]:

$$^{S}\mathbf{V}^{C} = 0 \tag{6.12.1}$$

Let P be a particle (or "point") of B, and let **p** locate P relative to C. Then since P and C are both fixed in B, **p** is also fixed in B. The velocity of P in S is then [see Equation (6.5.9)]

$$^{S}\mathbf{V}^{P} = {}^{S}\mathbf{V}^{C} + {}^{S}\mathbf{V}^{P/C} = 0 + \boldsymbol{\omega} \times \mathbf{p} \tag{6.12.2}$$

where $\boldsymbol{\omega}$ is the angular velocity of B in S.

The acceleration of P in S may be obtained by differentiation in Equation (6.12.2).

Finally, let **n** be a unit vector perpendicular to S at C. Then if $\boldsymbol{\omega}$ is parallel to **n**, B is said to be "pivoting" on S. If $\boldsymbol{\omega}$ is perpendicular to **n**, B is said to have "pure rolling" on S.

Perhaps the most common occurence of rolling is with a wheel on a flat surface. This is illustrated in the following example:

Example 6.12.1 Wheel or Disk Rolling on a Flat Surface

Let D be a sharp-edged circular disk with radius r, rolling on the horizontal X-Y plane as in Figure 6.12.2 [6.4, 6.5, 6.6]. Let C be the contact point between D and the X-Y plane, and let G be the center of D. Let L be a line in the X-Y plane passing through C and tangent to D. Let ϕ be the angle between L and the X-axis (the "turning angle" of D). Let $\hat{L}$ be a radial line fixed in D, and let ψ be the angle between $\hat{L}$ and a diametral line passing through C and G (the "rolling angle" of D). Let θ be the angle between the diametral line through C and G and a vertical line through G (the "leaning angle" of D). Let $\mathbf{N}_1$, $\mathbf{N}_2$, and $\mathbf{N}_3$ be unit vectors parallel to the X, Y, and Z axes; let $\hat{\mathbf{n}}_1$, $\hat{\mathbf{n}}_2$ and $\hat{\mathbf{n}}_3$ be mutually perpendicular dextral unit vectors with $\mathbf{n}_1$ parallel to L and $\mathbf{n}_2$ parallel to the axis of D, and $\mathbf{n}_3$ parallel to the diametral line through C and G. Finally, let $\mathbf{d}_1$, $\mathbf{d}_2$, and $\mathbf{d}_3$ be mutually perpendicular unit vectors (not shown in Figure 6.12.2) fixed in D, with

$\mathbf{d}_2$ parallel to $\mathbf{n}_2$ and $\mathbf{d}_3$ parallel to $\hat{L}$.

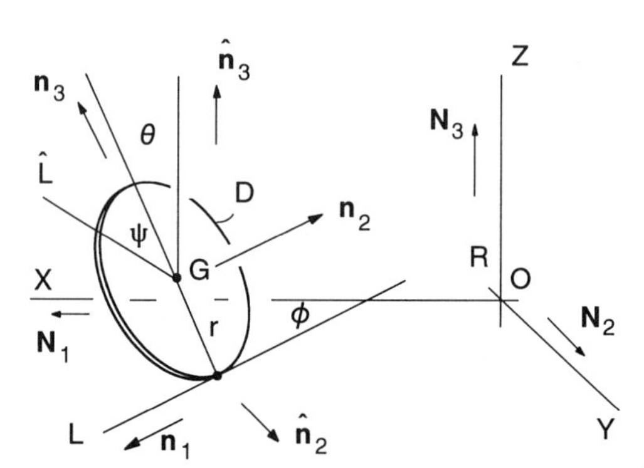

Fig. 6.12.2 Circular Disk Rolling on a Flat Horizontal Surface

Let $R(\mathbf{N}_1,\mathbf{N}_2,\mathbf{N}_3)$, $\hat{N}(\hat{\mathbf{n}}_1,\hat{\mathbf{n}}_2,\hat{\mathbf{n}}_3)$, $N(\mathbf{n}_1,\mathbf{n}_2,\mathbf{n}_3)$ and $D(\mathbf{d}_1,\mathbf{d}_2,\mathbf{d}_3)$ be reference frames containing the unit vector sets as indicated. A configuration graph relating these unit vectors is then shown in Figure 6.12.3.

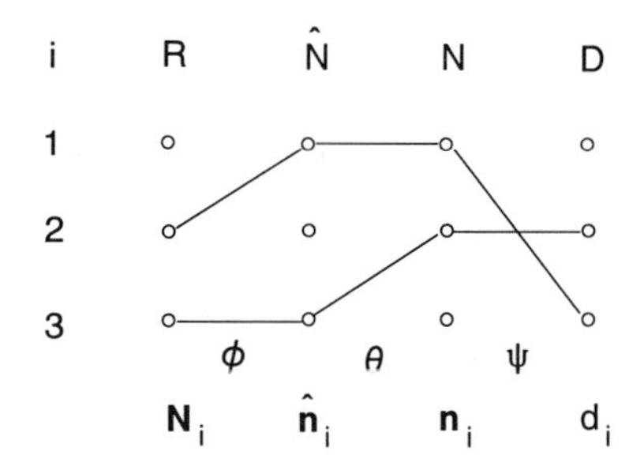

Fig. 6.12.3 Configuration Graph for the Orientation of D in R (see Figure 6.12.2)

From the addition theorem for angular velocity [Equation (6.7.5)], the angular velocity of D in R is:

$$^R\boldsymbol{\omega}^D = \dot{\phi}\mathbf{N}_3 + \dot{\theta}\hat{\mathbf{n}}_1 + \dot{\psi}\mathbf{n}_2 = \dot{\phi}\hat{\mathbf{n}}_3 + \dot{\theta}\hat{\mathbf{n}}_1 + \dot{\psi}\mathbf{d}_2 \tag{6.12.3}$$

Using the configuration graph of Figure 6.12.3 we can express the angular velocity in any one or all of the unit vector sets. The results are:

$$^{R}\boldsymbol{\omega}^{D} = (\dot{\theta}\, c_{\phi} - \dot{\psi}\, c_{\theta} s_{\phi})\mathbf{N}_1 + (\dot{\theta}\, s_{\phi} + \dot{\psi}\, c_{\theta} c_{\phi})\mathbf{N}_2 + (\dot{\phi} + \dot{\psi}\, s_{\theta})\mathbf{N}_3 \tag{6.12.4}$$

$$^{R}\boldsymbol{\omega}^{D} = \dot{\theta}\,\hat{\mathbf{n}}_1 + \dot{\psi}\, c_{\theta}\hat{\mathbf{n}}_2 + (\dot{\phi} + \dot{\psi}\, s_{\theta})\hat{\mathbf{n}}_3 \tag{6.12.5}$$

$$^{R}\boldsymbol{\omega}^{D} = \dot{\theta}\,\mathbf{n}_1 + (\dot{\phi}\, s_{\theta} + \dot{\psi})\mathbf{n}_2 + \dot{\phi}\, c_{\theta}\mathbf{n}_3 \tag{6.12.6}$$

and

$$^{R}\boldsymbol{\omega}^{D} = (\dot{\theta}\, c_{\psi} - \dot{\phi}\, s_{\psi} c_{\theta})\mathbf{d}_1 + (\dot{\phi}\, s_{\theta} + \dot{\psi})\mathbf{d}_2 + (\dot{\theta}\, s_{\psi} + \dot{\phi}\, c_{\theta} c_{\psi})\mathbf{d}_3 \tag{6.12.7}$$

where, as before, s and c are abbreviations for sine and cosine.

Using Equations (6.12.2) and (6.12.6), the velocity of the center of the disk is

$$^{R}\mathbf{V}^{G} = {}^{R}\boldsymbol{\omega}^{D} \times r\mathbf{n}_3 = r(\dot{\phi}\, s_{\theta} + \dot{\psi})\mathbf{n}_1 - r\dot{\theta}\,\mathbf{n}_2 \tag{6.12.8}$$

By differentiating in Equations (6.12.6) and (6.12.8) the angular acceleration and center acceleration of the disk relative to $\mathbf{n}_1$, $\mathbf{n}_2$, and $\mathbf{n}_3$ are:

$$^{R}\boldsymbol{\alpha}^{D} = (\ddot{\theta} - \dot{\phi}\,\dot{\psi}\, c_{\theta})\mathbf{n}_1 + (\ddot{\psi} + \dot{\theta}\,\dot{\phi}\, c_{\theta} + \ddot{\phi}\, s_{\theta})\mathbf{n}_2 + (\dot{\theta}\,\dot{\psi} - \dot{\theta}\,\dot{\phi}\, s_{\theta} + \ddot{\phi}\, c_{\theta})\mathbf{n}_3 \tag{6.12.9}$$

$$
\begin{aligned}
{}^{R}\mathbf{a}^{D} = {} & r(2\dot{\theta}\dot{\phi}c_{\theta} + \ddot{\phi}s_{\theta} + \ddot{\psi})\mathbf{n}_1 + r(-\ddot{\theta} + \dot{\phi}^2 s_{\theta}c_{\theta} + \dot{\phi}\dot{\psi}c_{\theta})\mathbf{n}_2 \\
& + r(-\dot{\theta}^2 - \dot{\phi}^2 s_{\theta}^2 - \dot{\phi}\dot{\psi}s_{\theta})\mathbf{n}_3
\end{aligned}
\tag{6.12.10}
$$

As in Equations (6.12.4) through (6.12.7) the disk center velocity and acceleration and the angular acceleration of the disk may be expressed in terms of the various unit vector sets, using the configuration graph of Figure 6.12.3. The results are listed in Tables 6.12.1 to 6.12.4.

Table 6.12.1 Components of ${}^{R}\boldsymbol{\omega}^{D}$, the Angular Velocity of the Disk in an Inertial Reference Frame R

i	$\mathbf{N}_i$	$\hat{\mathbf{n}}_i$	$\mathbf{n}_i$	$\mathbf{d}_i$
1	$\dot{\theta}c_{\phi} - \dot{\psi}c_{\theta}s_{\phi}$	$\dot{\theta}$	$\dot{\theta}$	$\dot{\theta}c_{\psi} - \dot{\phi}s_{\psi}c_{\theta}$
2	$\dot{\theta}s_{\phi} + \dot{\psi}c_{\theta}c_{\phi}$	$\dot{\psi}c_{\theta}$	$\dot{\phi}s_{\theta} + \dot{\psi}$	$\dot{\phi}s_{\theta} + \dot{\psi}$
3	$\dot{\phi} + \dot{\psi}s_{\theta}$	$\dot{\phi} + \dot{\psi}s_{\theta}$	$\dot{\phi}c_{\theta}$	$\dot{\theta}s_{\psi} + \dot{\phi}c_{\theta}c_{\psi}$

Table 6.12.2 Components of ${}^{R}\mathbf{V}^{G}$, the Disk Center Velocity in an Inertial Reference Frame R

i	$\mathbf{N}_i$	$\hat{\mathbf{n}}_i$	$\mathbf{n}_i$	$\mathbf{d}_i$
1	$r(\dot{\theta}c_{\theta}s_{\phi} + \dot{\phi}c_{\phi}s_{\theta} + \dot{\psi}c_{\phi})$	$r(\dot{\phi}s_{\theta} + \dot{\psi})$	$r(\dot{\phi}s_{\theta} + \dot{\psi})$	$r(\dot{\phi}s_{\theta}c_{\psi} + \dot{\psi}c_{\psi})$
2	$r(-\dot{\theta}c_{\theta}c_{\phi} + \dot{\phi}s_{\theta}s_{\phi} + \dot{\psi}s_{\phi})$	$-r\dot{\theta}c_{\theta}$	$-r\dot{\theta}$	$-r\dot{\theta}$
3	$r(-\dot{\theta}s_{\theta})$	$-r\dot{\theta}s_{\theta}$	0	$r(\dot{\phi}s_{\theta}s_{\psi} + \dot{\psi}s_{\psi})$

Table 6.12.3 Components of ${}^{R}\boldsymbol{\alpha}^{D}$, the Angular Acceleration of the Disk in an Inertial Reference Frame R

i	$\mathbf{N}_i$	$\hat{\mathbf{n}}_i$	$\mathbf{n}_i$	$\mathbf{d}_i$
1	$\ddot{\theta} c_\phi - \dot{\theta}\dot{\phi} s_\phi + \dot{\theta}\dot{\phi} s_\theta s_\phi - \dot{\phi}\dot{\psi} c_\theta c_\phi - \ddot{\psi} c_\theta s_\phi$	$\ddot{\theta} - \dot{\phi}\dot{\psi} c_\theta$	$\ddot{\theta} - \dot{\phi}\dot{\psi} c_\theta$	$\ddot{\theta} c_\psi - \dot{\theta}\dot{\psi} s_\psi - \ddot{\phi} s_\psi c_\theta - \dot{\phi}\dot{\psi} c_\psi c_\theta + \dot{\phi}\dot{\theta} s_\psi s_\theta$
2	$\ddot{\theta} s_\phi + \dot{\theta}\dot{\phi} c_\psi + \ddot{\psi} c_\theta c_\phi - \dot{\psi}\dot{\theta} s_\theta c_\phi - \dot{\psi}\dot{\phi} c_\theta s_\psi$	$\dot{\theta}\dot{\phi} - \dot{\theta}\dot{\psi} s_\theta + \ddot{\psi} c_\theta$	$\ddot{\psi} + \dot{\theta}\dot{\phi} c_\theta + \ddot{\phi} c_\theta$	$\ddot{\phi} s_\theta + \dot{\phi}\dot{\theta} c_\theta + \ddot{\psi}$
3	$\ddot{\phi} + \dot{\psi} s_\theta + \dot{\psi}\dot{\theta} c_\theta$	$\ddot{\phi} + \ddot{\psi} s_\theta + \dot{\psi}\dot{\theta} c_\theta$	$\dot{\theta}\dot{\psi} - \dot{\theta}\dot{\phi} s_\theta + \dot{\phi} c_\theta$	$\ddot{\theta} s_\psi + \dot{\theta}\dot{\psi} c_\psi + \ddot{\phi} c_\theta c_\psi - \dot{\phi}\dot{\theta} s_\theta c_\psi - \dot{\phi}\dot{\psi} c_\theta s_\psi$

Table 6.12.4 Components of ${}^{R}\mathbf{a}^{G}$, the Disk Center Acceleration in an Inertial Reference Frame R

i	$\mathbf{N}_i$	$\hat{\mathbf{n}}_i$	$\mathbf{n}_i$	$\mathbf{d}_i$
1	$r[\ddot{\theta} c_\theta s_\phi - \dot{\theta}^2 s_\theta s_\phi + \dot{\theta}\dot{\phi} c_\theta c_\phi - \dot{\phi}^2 s_\theta s_\phi - \dot{\phi}\dot{\psi} s_\phi + \ddot{\phi} c_\phi s_\theta + \dot{\phi}\dot{\theta} c_\theta c_\phi + \ddot{\psi} c_\phi]$	$r[2\dot{\theta}\dot{\phi} c_\theta + \ddot{\phi} s_\theta + \ddot{\psi}]$	$r[2\dot{\theta}\dot{\phi} c_\theta + \ddot{\phi} s_\theta + \ddot{\psi}]$	$r[2\dot{\theta}\dot{\phi} c_\theta c_\psi + \ddot{\phi} s_\theta c_\psi + \ddot{\psi} c_\psi + \dot{\theta}^2 s_\psi + \dot{\phi}^2 s_\theta^2 s_\psi + \dot{\phi}\dot{\psi} s_\theta s_\psi]$
2	$r[-\ddot{\theta} c_\theta c_\phi + \dot{\theta}^2 s_\theta c_\phi + \dot{\theta}\dot{\phi} c_\theta s_\phi + \ddot{\phi} s_\theta + \dot{\phi}\dot{\theta} c_\theta + \ddot{\psi} + \dot{\phi}^2 s_\theta c_\phi + \dot{\psi}\dot{\phi} c_\phi]$	$r[-\ddot{\theta} c_\theta + \dot{\theta}^2 s_\theta + \dot{\phi}^2 s_\theta + \dot{\phi}\dot{\psi}]$	$r[-\ddot{\theta} + \dot{\phi}^2 s_\theta c_\theta + \dot{\phi}\dot{\psi} c_\theta]$	$r[-\ddot{\theta} + \dot{\phi}^2 s_\theta c_\theta + \dot{\phi}\dot{\psi} c_\theta]$
3	$r[-\ddot{\theta} s_\theta - \dot{\theta}^2 c_\theta]$	$r[-\ddot{\theta} s_\theta - \dot{\theta}^2 c_\theta]$	$r[-\dot{\theta}^2 - \dot{\phi}^2 s_\theta^2 - \dot{\phi}\dot{\psi} s_\theta]$	$r[2\dot{\theta}\dot{\phi} s_\psi c_\theta + \ddot{\phi} s_\theta s_\psi + \ddot{\psi} s_\psi - \dot{\theta}^2 c_\psi - \dot{\phi}^2 s_\theta^2 c_\psi - \dot{\phi}\dot{\psi} s_\theta c_\psi]$

Observe in Tables 6.12.1 to 6.12.4 that the unit vector set $\mathbf{n}_i$ $(i = 1,2,3)$, although neither fixed in D nor R, has simpler components than the other unit vector sets. The $\mathbf{n}_i$ are parallel to principal inertia axes of D (see Chapter 7).

Example 6.12.2 <u>Vertical Disk Rolling in a Straight Line</u>

Consider the simple case of a vertical disk D with radius r rolling in a straight line. From Figure 6.12.2 the lean angle θ is then zero and the turning angle ϕ is a constant, say ϕ_0. The kinematics of D in reference frame R are then readily obtained from Tables 6.12.1 through 6.12.4. In terms of the $\mathbf{n}_i$ (i = 1,2,3) unit vectors, the angular velocity, the angular acceleration, the center velocity, and the center acceleration of D in R are:

$$^R\boldsymbol{\omega}^D = \dot{\psi}\,\mathbf{n}_2 \tag{6.12.11}$$

$$^R\boldsymbol{\alpha}^D = \ddot{\psi}\,\mathbf{n}_2 \tag{6.12.12}$$

$$^R\mathbf{V}^G = r\dot{\psi}\,\mathbf{n}_1 \tag{6.12.13}$$

and

$$^R\mathbf{a}^G = r\ddot{\psi}\,\mathbf{n}_1 \tag{6.12.14}$$

Example 6.12.3 Disk Rolling in a Circle

Consider next a disk D with radius r rolling in a circle while leaning toward the interior of the circle with a constant lean angle θ_0. Let D roll at a uniform rate so that the roll angle derivative $\dot{\psi}$ and the turning angle derivative $\dot{\phi}$ are constants, say $\dot{\psi}_0$ and $\dot{\phi}_0$. Then the kinematics of D in R are obtained from Tables 6.12.1 through 6.12.4 as:

$$^R\boldsymbol{\omega}^D = (\dot{\phi}_0 s_{\theta_0} + \dot{\psi}_0)\mathbf{n}_2 + \dot{\phi}_0 c_{\theta_0}\mathbf{n}_3 \tag{6.12.15}$$

$$^R\boldsymbol{\alpha}^D = -\dot{\phi}_0\dot{\psi}_0 c_{\theta_0}\mathbf{n}_1 \tag{6.12.16}$$

$$^R\mathbf{V}^G = r(\dot{\phi}_0 s_{\theta_0} + \dot{\psi}_0)\mathbf{n}_1 \tag{6.12.17}$$

and

$$^R\mathbf{a}^G = r(\dot{\phi}_0^2 s_{\theta_0} c_{\theta_0} + \dot{\phi}_0\dot{\psi}_0 c_{\theta_0})\mathbf{n}_2 + r(-\dot{\phi}_0^2 s_{\theta_0}^2 - \dot{\phi}_0\dot{\psi}_0 s_{\theta_0})\mathbf{n}_3 \tag{6.12.18}$$

Example 6.12.4 Pivoting (Spinning) Disk

Finally, consider a disk D pivoting about a diametral line. Let the lean angle θ be zero, and let the rolling angle ψ be a constant, say ψ_0. Then, as in the previous examples, the kinematics of D may be obtained from Tables 6.12.1 through 6.12.4 as:

$$^{R}\boldsymbol{\omega}^{D} = \dot{\phi}\mathbf{n}_3 \tag{6.12.19}$$

$$^{R}\boldsymbol{\alpha}^{D} = \ddot{\phi}\mathbf{n}_3 \tag{6.12.20}$$

$$^{R}\mathbf{V}^{G} = 0 \tag{6.12.21}$$

and

$$^{R}\mathbf{a}^{G} = 0 \tag{6.12.22}$$

6.13 Partial Angular Velocity

Consider a mechanical system S with n degrees of freedom represented by n coordinates q_r $(r = 1,\ldots,n)$. Let B be a body of S and let **c** be a vector fixed in B as in Figure 6.13.1, where as before, R is an inertial reference frame. Then the derivative of **c** in R may be expressed as:

$$\begin{aligned} {}^{R}d\mathbf{c}/dt &= {}^{R}\boldsymbol{\omega}^{B} \times \mathbf{c} \\ &= \frac{\partial \mathbf{c}}{\partial t} + \sum_{r=1}^{n} \frac{\partial \mathbf{c}}{\partial q_r}\dot{q}_r \end{aligned} \tag{6.13.1}$$

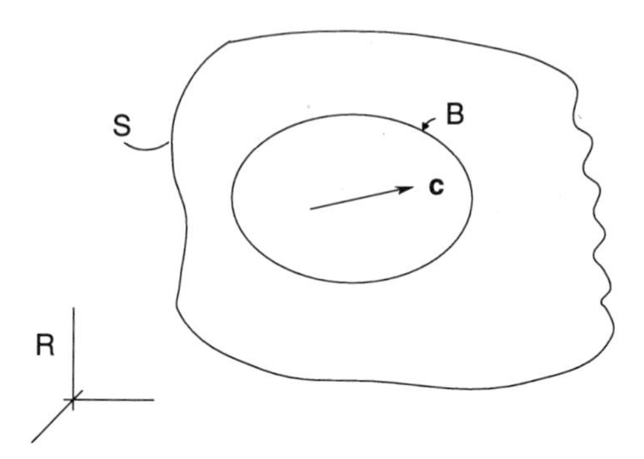

Fig. 6.13.1 A Mechanical System S with n Degrees of Freedom and Containing a Body B with Vector **c** Fixed in B

Since **c** is fixed in B the partial derivative of **c** may be developed in the same way that the time derivative of **c** was developed in Section 6.5.4. Specifically, let **a** and **b** be vectors fixed in B, as in Figure 6.13.2. Let vectors $\boldsymbol{\omega}_t$ and $\boldsymbol{\omega}_{\dot{q}_r}$ be defined as:

$$\boldsymbol{\omega}_t = \frac{\dfrac{\partial \mathbf{a}}{\partial t} \times \dfrac{\partial \mathbf{b}}{\partial t}}{\dfrac{\partial \mathbf{a}}{\partial t} \cdot \mathbf{b}} \quad \text{and} \quad \boldsymbol{\omega}_{\dot{q}_r} = \frac{\dfrac{\partial \mathbf{a}}{\partial q_r} \times \dfrac{\partial \mathbf{b}}{\partial q_r}}{\dfrac{\partial \mathbf{a}}{\partial q_r} \cdot \mathbf{b}} \tag{6.13.2}$$

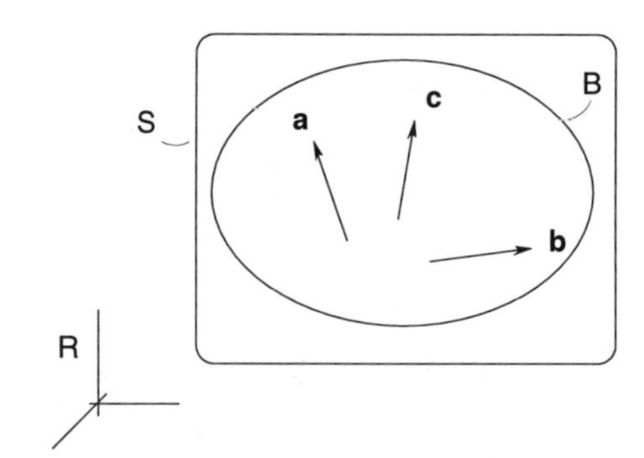

Fig. 6.13.2 A Body B of a Mechanical System S, with Fixed Vectors **a**, **b**, and **c**, and Moving in a Reference Frame R.

Then, by following the procedures outlined by Equations (6.5.13) to (6.5.22), it is readily seen that:

$$\frac{\partial \mathbf{c}}{\partial t} = \boldsymbol{\omega}_t \times \mathbf{c} \quad \text{and} \quad \frac{\partial \mathbf{c}}{\partial q_r} = \boldsymbol{\omega}_{\dot{q}_r} \times \mathbf{c} \quad (r = 1, \ldots, n) \tag{6.13.3}$$

By substituting into Equation (6.13.1) ${}^R d\mathbf{c}/dt$ may be expressed as:

$$\begin{aligned} {}^R d\mathbf{c}/dt &= {}^R\boldsymbol{\omega}^B \times \mathbf{c} = \boldsymbol{\omega}_t \times \mathbf{c} + \sum_{r=1}^{n} (\boldsymbol{\omega}_{\dot{q}_r} \times \mathbf{c})\dot{q}_r \\ &= (\boldsymbol{\omega}_t + \sum_{r=1}^{n} \boldsymbol{\omega}_{\dot{q}_r}\dot{q}_r) \times \mathbf{c} \end{aligned} \tag{6.13.4}$$

Then since $\mathbf{c}$ is arbitrary, so that Equation (6.13.4) must hold for all vectors fixed in B, ${}^R\boldsymbol{\omega}^B$ is:

$${}^R\boldsymbol{\omega}^B = \boldsymbol{\omega}_t + \sum_{r=1}^{n} \boldsymbol{\omega}_{\dot{q}_r}\dot{q}_r \tag{6.13.5}$$

The vectors $\boldsymbol{\omega}_t$ and $\boldsymbol{\omega}_{\dot{q}_r}$ $(r = 1, \ldots, n)$, which could be more formally written as ${}^R\boldsymbol{\omega}_t^B$ and ${}^R\boldsymbol{\omega}_{\dot{q}_r}^B$, are called "partial angular velocity vectors of B relative to t and $\dot{q}_r$" [6.3].

6.14 Summary

Table 6.14.1 provides a summary description of fundamental equations and formulas associated with body kinematics.

Table 6.14.1 Basic Equations and Formulas for Body Kinematics

Name	Equation	Reference
1. Angular velocity of a body B in a reference frame R	${}^{R}\boldsymbol{\omega}^{B} = \dfrac{\dfrac{d\mathbf{a}}{dt} \times \dfrac{d\mathbf{b}}{dt}}{\dfrac{d\mathbf{a}}{dt} \cdot \mathbf{b}}$ (**a**, **b** fixed in B)	Equation (6.5.13)
2. Angular velocity of a body B in a reference frame R	${}^{R}\boldsymbol{\omega}^{B} = \left(\dfrac{d\mathbf{n}_2}{dt} \cdot \mathbf{n}_3\right)\mathbf{n}_1 + \left(\dfrac{d\mathbf{n}_3}{dt} \cdot \mathbf{n}_1\right)\mathbf{n}_2 + \left(\dfrac{d\mathbf{n}_1}{dt} \cdot \mathbf{n}_2\right)\mathbf{n}_3$ ($\mathbf{n}_1$, $\mathbf{n}_2$, $\mathbf{n}_3$ are mutually perpendicular unit vectors fixed in B.)	Equation (6.5.4)
3. Differentiation Algorithm	$\dfrac{{}^{R}d(\ )}{dt} = \dfrac{{}^{\hat{R}}d(\ )}{dt} + {}^{R}\boldsymbol{\omega}^{\hat{R}} \times (\ )$ [( ) is any vector quantity and R, $\hat{R}$ are reference frames.]	Equation (6.6.7)
4. Addition Formula for Angular Velocity	${}^{R}\boldsymbol{\omega}^{B} = {}^{R}\boldsymbol{\omega}^{R_1} + {}^{R}\boldsymbol{\omega}^{R_2} + \ldots + {}^{R_{n-1}}\boldsymbol{\omega}^{R_n} + {}^{R_n}\boldsymbol{\omega}^{B}$ (R, R_1, ... , R_n are reference frames)	Equation (6.7.5)
5. Simple Angular Velocity	${}^{R}\boldsymbol{\omega}^{B} = \dot{\theta}\mathbf{k}$ (**k** is a unit vector fixed in both B and R and θ is the rotation angle.)	Equation (6.5.23)
6. Angular Acceleration	${}^{R}\boldsymbol{\alpha}^{B} = {}^{R}d\,{}^{R}\boldsymbol{\omega}^{B}/dt$	Equation (6.9.1)
7. Addition Formula for Angular Acceleration	${}^{R}\boldsymbol{\alpha}^{B} \neq {}^{R}\boldsymbol{\alpha}^{R_1} + {}^{R_1}\boldsymbol{\alpha}^{R_2} + \ldots + {}^{R_{n-1}}\boldsymbol{\alpha}^{R_n} + {}^{R_n}\boldsymbol{\alpha}^{B}$ (Equality holds only if the vectors are parallel.)	Equation (6.9.5)
8. Angular Acceleration Algorithm	If ${}^{R}\boldsymbol{\omega}^{B} = \omega_i\mathbf{n}_i$ and ${}^{R}\boldsymbol{\alpha}^{B} = \alpha_i\mathbf{n}_i$ then $\alpha_i = \dot{\omega}_i$ if $\mathbf{n}_i$ are fixed in B or R.	Equation (6.9.8)

Name	Equation	Reference
9. Relative Velocity of Points fixed on a Body	$\mathbf{V}^{P/Q} = {}^R\boldsymbol{\omega}^B \times \mathbf{r}$ ($\mathbf{r}$ locates P relative to Q)	Equation (6.10.3)
10. Velocity of a Point P Moving on a Body	${}^R\mathbf{V}^P = {}^B\mathbf{V}^P + {}^R\mathbf{V}^{P^*}$ (P^* is that point of B that coincides with P at an instant.)	Equation (6.10.8)
11. Relative Acceleration of Points fixed on a Body	${}^R\mathbf{a}^{P/Q} = {}^R\boldsymbol{\alpha}^B \times \mathbf{r} + {}^R\boldsymbol{\omega}^B \times ({}^R\boldsymbol{\omega}^B \times \mathbf{r})$ ($\mathbf{r}$ locates P relative to Q.)	Equation (6.11.2)
12. Acceleration of a Point P Moving on a Body	${}^R\mathbf{a}^P = {}^B\mathbf{a}^P + {}^R\mathbf{a}^{P^*} + 2\,{}^R\boldsymbol{\omega}^B \times {}^B\mathbf{V}^P$ (P^* is that point of B that coincides with P at an instant.)	Equation (6.11.6)
13. Rolling	$\mathbf{V}^C = 0$ (contact point of a rolling body has zero velocity)	Equation (6.12.1)
	$\mathbf{V}^P = \boldsymbol{\omega}^B \times \mathbf{p}$ (Velocity of point P of a rolling body B. $\mathbf{p}$ locates P relative to contact point C)	Equation (6.12.2)
14. Partial Angular Velocity	${}^R\boldsymbol{\omega}^B = \boldsymbol{\omega}_t + \sum_{r=1}^{n} \boldsymbol{\omega}_{\dot{q}_r}\dot{q}_r$ [System with n degrees of freedom with coordinates $\dot{q}_r$ $(r = 1,\ldots,n)$.]	Equation (6.13.5)

References

6.1 R. L. Huston, *Multibody Dynamics*, Butterworth-Heinemann, Stoneham, MA, 1990.

6.2 T. R. Kane, P. W. Likins, and D. A. Levinson, *Spacecraft Dynamics*, McGraw Hill, New York, 1983, pp. 422-431.

6.3 T. R. Kane and D. A. Levinson, *Dynamics, Theory and Applications*, McGraw Hill, New York, 1985, p. 45.

6.4 T. R. Kane, *Dynamics*, Holt, Rinehart, and Winston, New York, 1968, p. 45.

6.5 T. R. Kane, *Analytical Elements of Mechanics, Volume 2: Dynamics*, Academic Press, New York, 1961, p. 97.

6.6 H. Josephs and R. L. Huston, *Dynamics of Mechanical Systems*, CRC Press, Boca Raton, FL, 1999.

Chapter 7

ADDITIONAL TOPICS/FORMULAS IN KINEMATICS OF BODIES

7.1 Introduction

There are a number of additional topics and formulas associated with the kinematics of bodies which are important for advanced analyses—particularly computationally based analysis and multibody dynamics analysis [7.1, 7.2]. These topics and the corresponding equations are discussed and listed in this chapter.

7.2 Rotation Dyadics

A fundamental principle in the rotation of bodies is that any change in orientation of a body B relative to a reference frame can be duplicated by a simple rotation of B about a line L fixed in both B and R [7.2]. To develop this, consider the simple rotation of a body B through an angle θ about a fixed line L as in Figure 7.2.1. Let P be a typical particle of B not on L. Let Q be that point of L at the intersection of the line through P and perpendicular to L. Then during the rotation of B P moves on a circle about L with center at Q and with radius $|\mathbf{QP}|$. Finally, let O be a fixed point on L distinct from Q, let $\mathbf{p}$ locate P relative to O, and let $\boldsymbol{\lambda}$ be a unit vector parallel to L as shown.

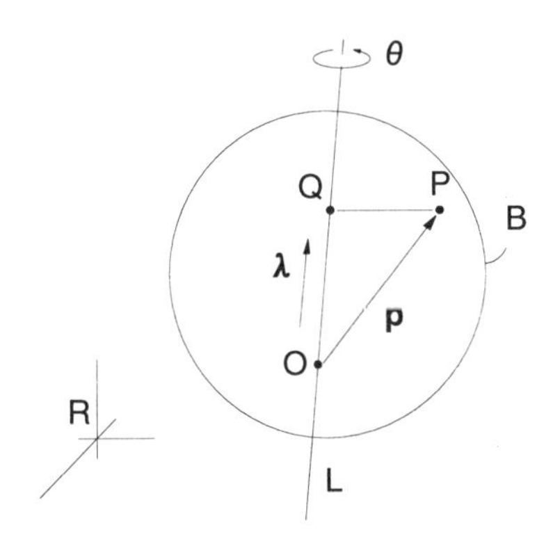

Fig. 7.2.1 Simple Rotation of a Body B in a Reference Frame R

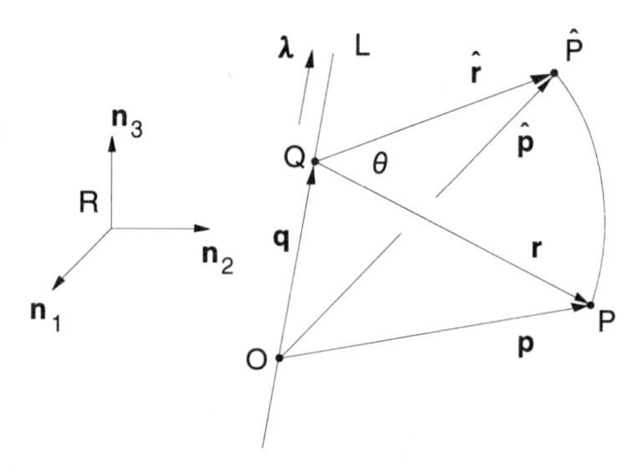

Fig. 7.2.2 Rotation of a Particle P about L

During the rotation of B through θ let P move to point $\hat{P}$ as represented in Figure 7.2.2 where $\mathbf{r}$, $\hat{\mathbf{r}}$, and $\mathbf{q}$ are the vectors $\mathbf{QP}$, $\mathbf{Q\hat{P}}$, and $\mathbf{OQ}$. Then we have the relation

$$\mathbf{p} = \mathbf{q} + \mathbf{r} \quad \text{and} \quad \hat{\mathbf{p}} = \mathbf{q} + \hat{\mathbf{r}} \tag{7.2.1}$$

Observe that the magnitude of $\mathbf{q}$ is equal to the projection of $\mathbf{p}$ (and hence, also $\hat{\mathbf{p}}$) onto L. Thus we have

$$\mathbf{q} = (\mathbf{p} \cdot \boldsymbol{\lambda})\,\boldsymbol{\lambda} \tag{7.2.2}$$

and then from Equation (7.2.1) $\mathbf{r}$ is

$$\mathbf{r} = \mathbf{p} - \mathbf{q} = \mathbf{p} - (\mathbf{p} \cdot \boldsymbol{\lambda})\,\boldsymbol{\lambda} \tag{7.2.3}$$

Also, from the perspective of a point view of L in Figure 7.2.2, we can visualize $\mathbf{r}$ and $\hat{\mathbf{r}}$ as in Figure 7.2.3. Let A be the base of the perpendicular to $\hat{P}$ along $\mathbf{r}$, and let $\boldsymbol{\mu}$ and $\mathbf{v}$ be unit vectors parallel and perpendicular to $\mathbf{r}$ as shown. Then $\boldsymbol{\mu}$, $\mathbf{v}$ and $\boldsymbol{\lambda}$ form a mutually perpendicular set of unit vectors. $\hat{\mathbf{r}}$ may then be expressed in terms of $\boldsymbol{\mu}$ and $\mathbf{v}$ as:

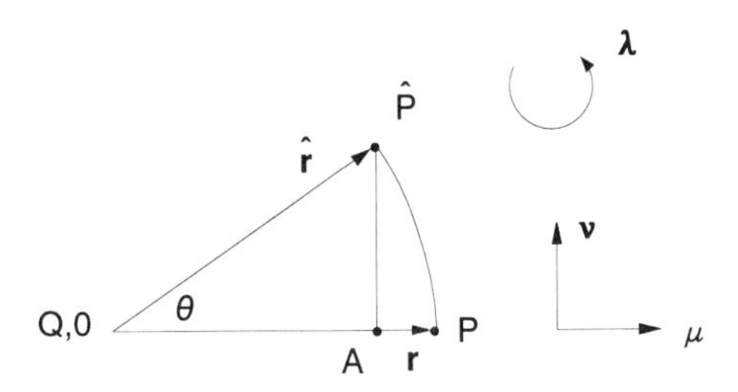

Fig. 7.2.3 Top View of Axis L and of Plane $QP\hat{P}$ (See Figure 7.2.2).

$$\hat{\mathbf{r}} = \mathbf{QA} + \mathbf{A\hat{P}} = |\mathbf{r}|\cos\theta\,\boldsymbol{\omega} + |\mathbf{r}|\sin\theta\,\mathbf{v} \tag{7.2.4}$$

(Note that $\hat{\mathbf{r}}$ and $\mathbf{r}$ have the same magnitude.) Observe in Equation (7.2.4) that $|\mathbf{r}|\boldsymbol{\mu}$ is $\mathbf{r}$ and that $|\mathbf{r}|\mathbf{v}$ is $\boldsymbol{\lambda}\times\mathbf{r}$. Then using Equation (7.2.3), Equation (7.2.4) may be rewritten in the form:

$$\begin{aligned}\hat{\mathbf{r}} &= \cos\theta\,\mathbf{r} = \sin\theta\,\boldsymbol{\lambda}\times\mathbf{r} \\ &= \cos\theta\,[\mathbf{p} - (\mathbf{p}\bullet\boldsymbol{\lambda})\boldsymbol{\lambda}] + \sin\theta\,\boldsymbol{\lambda}\times\mathbf{p}\end{aligned} \tag{7.2.5}$$

By substituting from Equations (7.2.2) and (7.2.4) into (7.2.1) we have

$$\hat{\mathbf{p}} = (\mathbf{p}\bullet\boldsymbol{\lambda})\boldsymbol{\lambda} + \cos\theta\,[\mathbf{p} - (\mathbf{p}\bullet\boldsymbol{\lambda})\boldsymbol{\lambda}] + \sin\theta\,\boldsymbol{\lambda}\times\mathbf{p} = \mathbf{R}\bullet\mathbf{p} \tag{7.2.6}$$

where $\mathbf{R}$ is the dyadic defined as:

$$\boldsymbol{R} = (1-\cos\theta)\,\boldsymbol{\lambda}\boldsymbol{\lambda} + \cos\theta\,\boldsymbol{I} + \sin\theta\,\boldsymbol{\lambda}\times\mathbf{I} \tag{7.2.7}$$

where $\boldsymbol{I}$ is the identity dyadic.

If $\boldsymbol{R}$ is expressed in component form as $R_{ij}\mathbf{n}_i\mathbf{n}_j$ where the $\mathbf{n}_i$ $(i = 1,2,3)$ are mutually perpendicular unit vectors fixed in R, then the R_{ij} are

$$R_{ij} = (1 - \cos\theta)\lambda_i\lambda_j + \cos\theta\,\delta_{ij} - \sin\theta\, e_{ijk}\lambda_k \tag{7.2.8}$$

where the λ_i and the δ_{ij} are the $\mathbf{n}_i$ components of $\boldsymbol{\lambda}$ and $\boldsymbol{I}$ and where the e_{ijk} are the components of the permutation symbol.

$\boldsymbol{R}$ is a "rotation dyadic." $\boldsymbol{R}$ may be viewed as an operator which when multiplied with a vector will rotate the vector about L through an angle θ. To see this consider that P is an arbitrary particle of B not on L. Then the rotation of B through θ rotates P about L through θ. Consequently vector $\mathbf{p}$ is rotated about L through θ. In like manner if A is another particle of B not on L the rotation of B through θ will rotate A about L through θ. Consequently if $\mathbf{a}$ locates A relative to O as in Figure 7.2.4 then the rotation of B will also rotate $\mathbf{a}$ about L through θ. Thus as $\mathbf{p}$ is rotated into $\hat{\mathbf{p}}$, $\mathbf{a}$ is rotated into $\hat{\mathbf{a}}$ (see Figure 7.2.4) and therefore $\mathbf{AP}$ is rotated about L into $\hat{\mathbf{A}}\hat{\mathbf{P}}$. We then have the relations

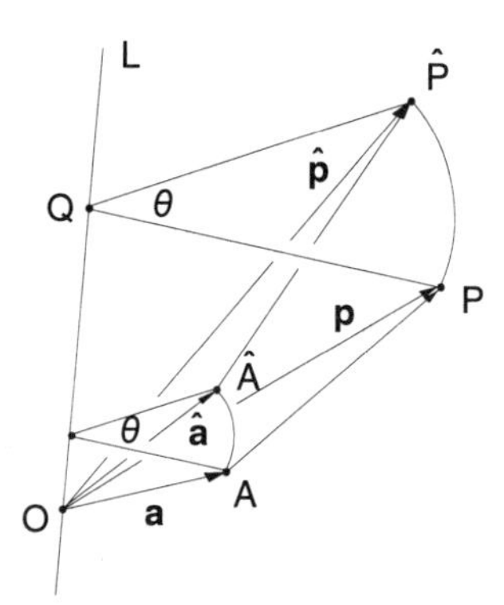

Fig. 7.2.4 Rotation of Particles A and P About L.

$$\hat{\mathbf{p}} = \boldsymbol{R}\cdot\mathbf{p} \qquad \hat{\mathbf{a}} = \boldsymbol{R}\cdot\mathbf{a} \qquad \hat{\mathbf{A}}\hat{\mathbf{P}} = \boldsymbol{R}\cdot\mathbf{AP} \tag{7.2.9}$$

Finally, if $\mathbf{V}$ is an arbitrary vector fixed in B, we could choose A and P so that $\mathbf{AP}$ coincides with $\mathbf{V}$. Then the rotation of B about L through θ will also rotate $\mathbf{V}$ about L through θ into a vector $\hat{\mathbf{V}}$ such that

$$\hat{\mathbf{V}} = \boldsymbol{R}\cdot\mathbf{V} \tag{7.2.10}$$

7.3 Properties of Rotation Dyadics

Consider again the rotation of a body B in a reference frame R about a line L fixed in both B and R. Let θ be the rotation angle, as in Figure 7.3.1. Let $\mathbf{n}_i$ $(i = 1,2,3)$ be a mutually perpendicular unit vector set fixed in R. Let $\mathbf{b}_i$ $(i = 1,2,3)$ be mutually perpendicular unit vectors fixed in B. At the beginning of

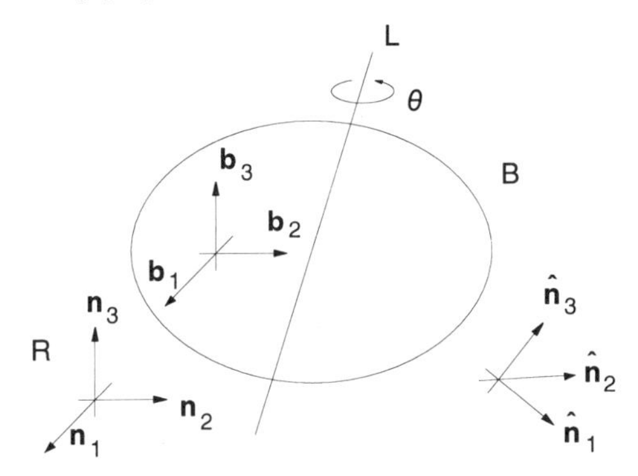

Fig. 7.3.1 Rotation of a Body B in a Reference Frame R and Unit Vector Sets Fixed in B and R.

the rotation let the $\mathbf{b}_i$ be respectively parallel to the $\mathbf{n}_i$. After the rotation let the $\mathbf{b}_i$ be respectively parallel to $\hat{\mathbf{n}}_i$, a second set of mutually perpendicular unit vectors fixed in R. Then from Equation (7.2.10) we have

$$\hat{\mathbf{n}}_i = \boldsymbol{R} \cdot \mathbf{b}_i = \boldsymbol{R} \cdot \mathbf{n}_i \tag{7.3.1}$$

If $\boldsymbol{R}$ is expressed in the form $R_{mn}\mathbf{n}_m\mathbf{n}_n$, then Equation (7.3.1) becomes

$$\hat{\mathbf{n}}_i = R_{mn}\mathbf{n}_m\mathbf{n}_n \cdot \mathbf{n}_i = R_{mi}\mathbf{n}_m \tag{7.3.2}$$

This expression is identical to the transformation matrix expression [see Equation (6.2.5)].

$$\hat{\mathbf{n}}_i = S_{mi}\mathbf{n}_m \tag{7.3.3}$$

Therefore, we have the identification of the rotation dyadic components with the transformation matrix elements. That is,

$$R_{ij} = S_{ij} \tag{7.3.4}$$

This does not mean however that the rotation dyadic and the transformation matrix are the same. On the contrary, they are quite different. The R_{ij} in Equation (7.3.4) are referred to the unit vectors $\mathbf{n}_i$ fixed in R, whereas the δ_{ij} are referred to both sets of unit vectors: $\mathbf{n}_i$ and $\hat{\mathbf{n}}_i$. The δ_{ij} are simply elements of the identity dyadic referred to two different unit vector sets. Viewed as operators on vectors, the transformation matrix does not change the vector, but instead changes its unit vector basis. The rotation dyadic however changes the vector by rotating it into a vector with the same magnitude but having a different direction.

It is readily seen from Equations (7.2.7) and (7.3.4) that the matrix of rotation dyadic components is orthogonal. That is

$$R_{ik}R_{jk} = \delta_{ij} \quad \text{and} \quad R_{ki}R_{kj} = \delta_{ij} \tag{7.3.5}$$

Finally, from Equation (7.3.2) we see that $\boldsymbol{R}$ may also be expressed in the form

$$\boldsymbol{R} = \hat{\mathbf{n}}_i\mathbf{n}_i \tag{7.3.6}$$

Table 7.3.1 lists a summary of properties of the rotation dyadic.

Table 7.3.1 Rotation Dyadic Properties

1.	The rotation dyadic $\boldsymbol{R}$ as an operator rotates a vector $\mathbf{V}$ about a line L through an angle θ, where L is fixed in a reference frame R.
2.	If $\mathbf{n}_i$ $(i = 1,2,3)$ are mutually perpendicular unit vectors, the $\boldsymbol{R}$ rotates the $\mathbf{n}_i$ into mutually perpendicular unit vectors $\hat{\mathbf{n}}_i$.
3.	$\boldsymbol{R}$ may be expressed in the following forms (1) $\boldsymbol{R} = R_{ij}\mathbf{n}_i\mathbf{n}_j$ (2) $\boldsymbol{R} = (1 - \cos\theta)\boldsymbol{\lambda}\boldsymbol{\lambda} + \cos\theta\boldsymbol{I} + \sin\theta\,\boldsymbol{\lambda} \times \boldsymbol{I}$ [$\boldsymbol{\lambda}$ is a unit vector parallel to L and $\boldsymbol{I}$ is the identity dyadic. See Equation (7.2.7).] (3) $R_{ij} = (1 - \cos\theta)\lambda_i\lambda_j + \cos\theta\,\delta_{ij} - \sin\theta\, e_{ijk}\lambda_k$ [The λ_i are the $\mathbf{n}_i$ components of $\boldsymbol{\lambda}$. See Equation (7.2.8).] (4) $R_{ij} = S_{ij}$ [The S_{ij} are $\mathbf{n}_i \bullet \hat{\mathbf{n}}_j$. See Equation (7.3.4).] (5) $\boldsymbol{R} = \hat{\mathbf{n}}_i\mathbf{n}_i$ [See Equation (7.3.6).]

7.4 Body Rotation and Rotation Dyadics

We can now establish by construction the fundamental principle stated in Section 7.2. That is, any change in orientation of a body B in a reference frame R may be duplicated by a simple rotation of B about a line L fixed in both B and R. To see this consider a body B having an arbitrary orientation change in a reference frame R as depicted in Figure 7.4.1. Let t_1 and t_2 be two times during the orientation change of B in R. Let $\mathbf{b}_1$, $\mathbf{b}_2$, and $\mathbf{b}_3$ be mutually perpendicular unit vectors fixed in B. Let $\mathbf{n}_1$, $\mathbf{n}_2$, and $\mathbf{n}_3$ be mutually perpendicular unit vectors of R which coincide with the $\mathbf{b}_i$ $(i = 1,2,3)$ at time t_2. We can then construct a rotation dyadic $\boldsymbol{R}$ between the $\mathbf{n}_i$ and $\hat{\mathbf{n}}_i$ as:

$$\boldsymbol{R} = \hat{\mathbf{n}}_i\mathbf{n}_i \tag{7.4.1}$$

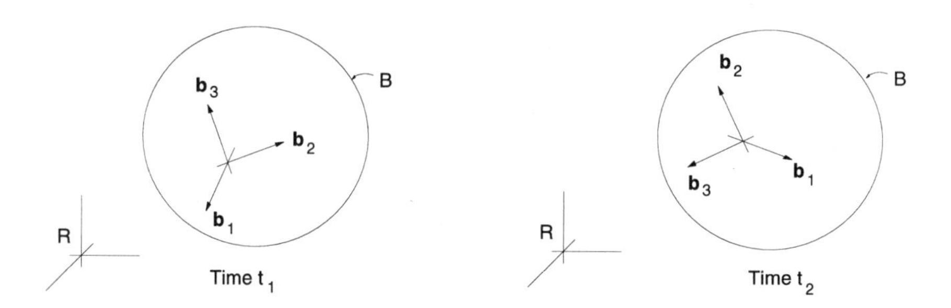

Fig. 7.4.1 Body Orientation in a Reference Frame R at Times t_1 and t_2

Then from Equation (7.2.7) we can find a unit vector $\boldsymbol{\lambda}$ parallel to a line L, fixed in B and R, and a rotation angle θ, such that if B is rotated about L through θ, then the $\mathbf{b}_i$ of B initially parallel to $\mathbf{n}_i$ $(i = 1,2,3)$ respectively, became parallel to the $\hat{\mathbf{n}}_i$ respectively after the rotation. Specifically, if the $\mathbf{n}_i$ and $\hat{\mathbf{n}}_i$ are known then the components R_{ij} of $\boldsymbol{R}$ will be known. In matrix form the R_{ij} may be listed as:

$$R_{ij} = \begin{bmatrix} [\lambda_1^2(1-\cos\theta)+\cos\theta] & [\lambda_1\lambda_2(1-\cos\theta)-\lambda_3\sin\theta] & [\lambda_1\lambda_3(1-\cos\theta)+\lambda_2\sin\theta] \\ [\lambda_1\lambda_2(1-\cos\theta)+\lambda_3\sin\theta] & [\lambda_2^2(1-\cos\theta)+\cos\theta] & [\lambda_2\lambda_3(1-\cos\theta)-\lambda_1\sin\theta] \\ [\lambda_3\lambda_1(1-\cos\theta)-\lambda_2\sin\theta] & [\lambda_2\lambda_3(1-\cos\theta)+\lambda_1\sin\theta] & [\lambda_3^2(1-\cos\theta)+\cos\theta] \end{bmatrix} \tag{7.4.2}$$

By adding the diagonal terms (the "trace") we have

$$R_{ii} = 1 + 2\cos\theta \tag{7.4.3}$$

Then

$$\cos\theta = (R_{ii} - 1)/2 \quad \text{or} \quad \theta = \cos^{-1}[(R_{ii} - 1)/2] \tag{7.4.4}$$

Similarly, by subtracting off-diagonal entries we have

$$\lambda_1 = (R_{32} - R_{23})/2\sin\theta \qquad \lambda_2 = (R_{13} - R_{31})/2\sin\theta \qquad \lambda_3 = (R_{21} - R_{12})/2\sin\theta \qquad (7.4.5)$$

Equations (7.4.4) and (7.4.5) thus determine $\boldsymbol{\lambda}$ and θ, unless θ is either zero or π. If θ is zero there is no rotation and the rotation dyadic $\boldsymbol{R}$ is the identity dyadic $\boldsymbol{I}$ ($R_{ij} = \delta_{ij}$). If θ is π, the R_{ij} become

$$R_{ij} = \begin{bmatrix} 2\lambda_1^2 - 1 & 2\lambda_1\lambda_2 & 2\lambda_1\lambda_3 \\ 2\lambda_1\lambda_2 & 2\lambda_2^2 - 1 & 2\lambda_2\lambda_3 \\ 2\lambda_1\lambda_3 & 2\lambda_2\lambda_3 & 2\lambda_3^2 - 1 \end{bmatrix} \qquad (7.4.6)$$

Then λ_1, λ_2, and λ_3 may be determined from the expressions:

$$\lambda_1^2 = (1 + R_{11})/2 \qquad \lambda_2^2 = (1 + R_{22})/2 \qquad \lambda_3^2 = (1 + R_{33})/2 \qquad (7.4.7)$$

or

$$\lambda_1 = \pm\sqrt{(1+R_{11})/2} \qquad \lambda_2 = \pm\sqrt{(1+R_{22})/2} \qquad \lambda_3 = \pm\sqrt{(1+R_{33})/2} \qquad (7.4.8)$$

where the signs are chosen so that the values are consistent with the signs of the off-diagonal terms in Equation (7.4.6). Observe that when θ is π, the sense of $\boldsymbol{\lambda}$ is not unique.

7.5 Singularities Occurring with Orientation Angles

If orientation angles are used to describe the orientation of a body B in a reference frame R, singularities can occur when angle derivatives are expressed in terms of angular velocity components. To illustrate this, consider again the orientation of a body B described by dextral orientation angles (body-fixed, 1-2-3, rotation sequence) as described in Sections 6.2, 6.3, and 6.7. From Equations (6.7.13) and (6.7.14) the angular velocity of B in a reference frame R may be expressed as

$$^R\boldsymbol{\omega}^B = \omega_1\mathbf{n}_1 + \omega_2\mathbf{n}_2 + \omega_3\mathbf{n}_3 \tag{7.5.1}$$

where ω_1, ω_2, and ω_3 are

$$\begin{aligned}\omega_1 &= \dot{\alpha}c_\beta c_\gamma + \dot{\beta}s_\gamma \\ \omega_2 &= -\dot{\alpha}c_\beta s_\gamma + \dot{\beta}c_\gamma \\ \omega_3 &= \dot{\alpha}s_\beta + \dot{\gamma}\end{aligned} \tag{7.5.2}$$

where $\mathbf{n}_1$, $\mathbf{n}_2$, and $\mathbf{n}_3$ are mutually perpendicular unit vectors fixed in B and where α, β, and γ are the dextral orientation angles. If Equations (7.5.2) are solved for $\dot{\alpha}$, $\dot{\beta}$, and $\dot{\gamma}$, we have

$$\begin{aligned}\dot{\alpha} &= \omega_1 c_\gamma - \omega_2 s_\gamma)/c_\beta \\ \dot{\beta} &= \omega_1 s_\gamma + \omega_2 c_\gamma \\ \dot{\gamma} &= (-\omega_1 c_\gamma + \omega_2 s_\gamma)s_\beta/c_\beta + \omega_3\end{aligned} \tag{7.5.3}$$

Observe that when β is $\pi/2$ or $3\pi/2$ radians there is a division by zero (or a singularity) in the expression for $\dot{\alpha}$ and $\dot{\gamma}$.

These singularities are important in that they can arise in the numerical solution of the governing differential equations. Indeed, most differential equation solvers require expression for dependent variable derivatives as in Equations (7.5.3). Moreover, it happens that singularities such as in Equations (7.5.3) occur regardless of which set of orientation angles are selected to describe the orientation of a body B (see Reference [7.1]). Tables 7.5.1 and 7.5.2 provide a listing of equations analogous to Equations (7.5.3) for the various choices of orientation angles showing where the singularities occur.

There are two ways of overcoming this singularity problem: 1) Selecting the unit vectors and unit vector orientation so that in the anticipated movement of a body B the singularity does not occur, and 2) the use of Euler parameters. The first approach, although simpler, requires prior knowledge of the expected movement of the body. In general, however, the body motion will not be known beforehand, and thus one cannot always be certain that a singularity will not occur. The second approach with Euler parameters is discussed in the following section.

Table 7.5.1 Angular Velocity Components and Orientation Angle Derivatives [7.1]. (Body Axes)

Rotation Sequence	Angular Velocity Components $({}^R\boldsymbol{\omega}^B = \Omega_i\mathbf{N}_i = \omega_i\mathbf{n}_i)$ ($\mathbf{N}_i$ fixed in R, $\mathbf{n}_i$ fixed in B)	Orientation Angle Derivatives
1. 1-2-3 or x-y-z (Dextral or Bryan Angles) (Body Axes)	$\Omega_1 = \dot{\theta}_1 + \dot{\theta}_3 s_2$ $\Omega_2 = \dot{\theta}_2 c_1 - \dot{\theta}_3 c_2 s_1$ $\Omega_3 = \dot{\theta}_2 s_1 + \dot{\theta}_3 c_2 c_1$	$\dot{\theta}_1 = (-c_1\Omega_3 + s_1\Omega_2)s_2/c_2 + \Omega_1$ $\dot{\theta}_2 = c_1\Omega_2 + s_1\Omega_3$ $\dot{\theta}_3 = (c_1\Omega_3 - s_1\Omega_2)/c_2$
	$\omega_1 = \dot{\theta}_1 c_2 c_3 + \dot{\theta}_2 s_3$ $\omega_2 = -\dot{\theta}_1 c_2 s_3 + \dot{\theta}_2 c_3$ $\omega_3 = \dot{\theta}_1 s_2 + \dot{\theta}_3$	$\dot{\theta}_1 = (\omega_1 c_3 - \omega_2 s_3)/c_2$ $\dot{\theta}_2 = \omega_1 s_1 + \omega_2 c_3$ $\dot{\theta}_3 = \omega_3 + (\omega_2 s_3 - \omega_1 c_3)s_2/c_2$
2. 2-3-1 or y-z-x (Body Axes)	$\Omega_1 = \dot{\theta}_2 s_1 + \dot{\theta}_3 c_1 c_2$ $\Omega_2 = \dot{\theta}_1 + \dot{\theta}_3 s_2$ $\Omega_3 = \dot{\theta}_2 c_1 - \dot{\theta}_3 s_2$	$\dot{\theta}_1 = \Omega_2 + (s_1\Omega_3 - c_1\Omega_1)s_2/c_2$ $\dot{\theta}_2 = s_1\Omega_1 + c_1\Omega_3$ $\dot{\theta}_3 = (c_1\Omega_1 - s_1\Omega_3)/c_2$
	$\omega_1 = \dot{\theta}_1 s_3 + \dot{\theta}_3$ $\omega_2 = \dot{\theta}_1 c_2 c_3 + \dot{\theta}_2 s_3$ $\omega_3 = -\dot{\theta}_1 c_2 s_3 + \dot{\theta}_2 c_3$	$\dot{\theta}_1 = (\omega_2 c_3 - \omega_3 s_3)/c_2$ $\dot{\theta}_2 = \omega_2 s_3 + \omega_3 c_3$ $\dot{\theta}_3 = \omega_1 + (\omega_3 s_3 - \omega_2 c_3)s_2/c_2$

Continued

Table 7.5.1 Continued

Rotation Sequence	Angular Velocity Components	Orientation Angle Derivatives
3. 3-1-2 or z-y-x (Body Axes)	$\Omega_1 = \dot{\theta}_2 c_1 - \dot{\theta}_3 s_1 c_2$ $\Omega_2 = \dot{\theta}_2 s_1 + \dot{\theta}_3 c_1 c_2$ $\Omega_3 = \dot{\theta}_1 + \dot{\theta}_3 s_2$ $\omega_1 = -\dot{\theta}_1 c_2 s_3 + \dot{\theta}_2 c_3$ $\omega_2 = \dot{\theta}_1 s_2 + \dot{\theta}_3$ $\omega_3 = \dot{\theta}_1 c_2 c_3 + \dot{\theta}_2 s_3$	$\dot{\theta}_1 = \Omega_3 + (s_1\Omega_1 - c_1\Omega_2)s_2/c_2$ $\dot{\theta}_2 = c_1\Omega_1 + s_1\Omega_2$ $\dot{\theta}_3 = (c_1\Omega_2 - s_1\Omega_1)/c_2$ $\dot{\theta}_1 = (c_3\omega_3 - s_3\omega_1)/c_2$ $\dot{\theta}_2 = c_3\omega_1 + s_3\omega_3$ $\dot{\theta}_3 = \omega_2 + (s_3\omega_1 - c_3\omega_3)s_2/c_2$
4. 1-3-2 or x-z-y (Body Axes)	$\Omega_1 = \dot{\theta}_1 - \dot{\theta}_3 s_2$ $\Omega_2 = -\dot{\theta}_2 s_1 + \dot{\theta}_3 c_1 c_2$ $\Omega_3 = \dot{\theta}_2 c_1 + \dot{\theta}_3 s_1 c_2$ $\omega_1 = \dot{\theta}_1 c_2 c_3 - \dot{\theta}_2 s_3$ $\omega_2 = -\dot{\theta}_1 s_2 + \dot{\theta}_3$ $\omega_3 = \dot{\theta}_1 c_2 s_3 + \dot{\theta}_2 c_3$	$\dot{\theta}_1 = \Omega_1 + (c_1\Omega_2 + s_1\Omega_3)s_2/c_2$ $\dot{\theta}_2 = c_1\Omega_3 - s_1\Omega_2$ $\dot{\theta}_3 = (c_1\Omega_2 + s_1\Omega_3)/c_2$ $\dot{\theta}_1 = (c_3\omega_1 + s_3\omega_3)/c_2$ $\dot{\theta}_2 = c_3\omega_3 - s_3\omega_1$ $\dot{\theta}_3 = \omega_2 + (c_3\omega_1 + s_3\omega_3)s_2/c_2$

5. 2-3-1 or y-z-x (Body Axes)	$\Omega_1 = \dot{\theta}_2 c_1 + \dot{\theta}_3 s_1 c_2$ $\Omega_2 = \dot{\theta}_1 - \dot{\theta}_3 s_2$ $\Omega_3 = -\dot{\theta}_2 s_1 + \dot{\theta}_3 c_1 c_2$ $\omega_1 = \dot{\theta}_1 c_2 s_3 + \dot{\theta}_2 c_3$ $\omega_2 = \dot{\theta}_1 c_2 c_3 - \dot{\theta}_2 s_3$ $\omega_3 = -\dot{\theta}_1 s_2 + \dot{\theta}_3$	$\dot{\theta}_1 = \Omega_2 + (s_1\Omega_1 + c_1\Omega_3) s_2/c_2$ $\dot{\theta}_2 = c_1\Omega_1 - s_1\Omega_3$ $\dot{\theta}_3 = (s_1\Omega_1 + c_1\Omega_3)/c_2$ $\dot{\theta}_1 = (s_3\omega_1 + c_3\omega_2)/c_2$ $\dot{\theta}_2 = c_3\omega_1 - s_3\omega_2$ $\dot{\theta}_3 = \omega_3 + (s_3\omega_1 + c_3\omega_2) s_2/c_2$
6. 3-2-1 or z-y-x (Body Axes)	$\Omega_1 = -\dot{\theta}_2 s_1 + \dot{\theta}_3 c_1 c_2$ $\Omega_2 = \dot{\theta}_2 c_1 + \dot{\theta}_3 s_1 c_2$ $\Omega_3 = \dot{\theta}_1 - \dot{\theta}_3 s_2$ $\omega_1 = -\dot{\theta}_1 s_2 + \dot{\theta}_3$ $\omega_2 = \dot{\theta}_1 c_2 s_3 + \dot{\theta}_2 c_3$ $\omega_3 = \dot{\theta}_1 c_2 c_3 - \dot{\theta}_2 s_3$	$\dot{\theta}_1 = \Omega_3 + (c_1\Omega_1 + s_1\Omega_2) s_2/c_2$ $\dot{\theta}_2 = c_1\Omega_2 - s_1\Omega_1$ $\dot{\theta}_3 = (c_1\Omega_1 + s_1\Omega_2)/c_2$ $\dot{\theta}_1 = (s_3\omega_2 + c_3\omega_3)/c_2$ $\dot{\theta}_2 = c_3\omega_2 - s_3\omega_3$ $\dot{\theta}_3 = \omega_1 + (s_3\omega_2 + c_3\omega_3) s_2/c_2$
7. 1-2-1 or x-y-x (Body Axes)	$\Omega_1 = \dot{\theta}_1 + \dot{\theta}_3 c_2$ $\Omega_2 = \dot{\theta}_2 c_1 + \dot{\theta}_3 s_1 s_2$ $\Omega_3 = \dot{\theta}_2 s_1 - \dot{\theta}_3 c_1 s_2$ $\omega_1 = \dot{\theta}_1 c_2 + \dot{\theta}_3$ $\omega_2 = \dot{\theta}_1 s_2 s_3 + \dot{\theta}_2 c_3$ $\omega_3 = \dot{\theta}_1 s_2 c_3 - \dot{\theta}_2 s_3$	$\dot{\theta}_1 = \Omega_1 + (c_1\Omega_3 - s_1\Omega_2) c_2/s_2$ $\dot{\theta}_2 = c_1\Omega_2 + s_1\Omega_3$ $\dot{\theta}_3 = (s_1\Omega_2 - c_1\Omega_3)/s_2$ $\dot{\theta}_1 = (s_3\omega_2 + c_3\omega_3)/s_2$ $\dot{\theta}_2 = c_3\omega_2 - s_3\omega_3$ $\dot{\theta}_3 = \omega_1 + (-s_3\omega_2 - c_3\omega_3) c_2/s_2$

Continued

Table 7.5.1 Continued

Rotation Sequence	Angular Velocity Components	Orientation Angle Derivatives
8. 1-3-1 or x-z-x (Body Axes)	$\Omega_1 = \dot{\theta}_1 + \dot{\theta}_3 c_2$ $\Omega_2 = -\dot{\theta}_2 s_1 + \dot{\theta}_3 c_1 s_2$ $\Omega_3 = \dot{\theta}_2 c_1 + \dot{\theta}_3 s_1 s_2$	$\dot{\theta}_1 = \Omega_1 + (-c_1\Omega_2 - s_1\Omega_3)c_2/s_2$ $\dot{\theta}_2 = c_1\Omega_3 - s_1\Omega_2$ $\dot{\theta}_3 = (c_1\Omega_2 + s_1\Omega_3)/s_2$
	$\omega_1 = \dot{\theta}_1 c_2 + \dot{\theta}_3$ $\omega_2 = -\dot{\theta}_1 s_2 c_3 + \dot{\theta}_2 s_3$ $\omega_3 = \dot{\theta}_1 s_2 s_3 + \dot{\theta}_2 c_3$	$\dot{\theta}_1 = (s_3\omega_3 - c_3\omega_2)/s_2$ $\dot{\theta}_2 = s_3\omega_2 + c_3\omega_3$ $\dot{\theta}_3 = \omega_1 + (c_3\omega_2 - s_3\omega_3)c_2/s_2$
9. 2-1-2 or y-x-y (Body Axes)	$\Omega_1 = \dot{\theta}_2 c_1 + \dot{\theta}_3 s_1 s_2$ $\Omega_2 = \dot{\theta}_1 + \dot{\theta}_3 c_2$ $\Omega_3 = -\dot{\theta}_2 s_1 + \dot{\theta}_3 c_1 s_2$	$\dot{\theta}_1 = \Omega_2 + (-s_1\Omega_1 - c_1\Omega_3)c_2/s_2$ $\dot{\theta}_2 = c_1\Omega_1 - s_1\Omega_3$ $\dot{\theta}_3 = (s_1\Omega_1 + c_1\Omega_3)/s_2$
	$\omega_1 = \dot{\theta}_1 s_2 s_3 + \dot{\theta}_2 c_3$ $\omega_2 = \dot{\theta}_1 c_2 + \dot{\theta}_3$ $\omega_3 = -\dot{\theta}_1 s_2 c_3 + \dot{\theta}_2 s_3$	$\dot{\theta}_1 = (s_3\omega_1 - c_3\omega_3)/s_2$ $\dot{\theta}_2 = c_2\omega_1 + s_3\omega_3$ $\dot{\theta}_3 = \omega_2 + (c_3\omega_3 - s_3\omega_1)c_2/s_2$

10. 2-3-2 or y-z-y (Body Axes)	$\Omega_1 = \dot{\theta}_2 s_1 - \dot{\theta}_3 c_1 s_2$ $\Omega_2 = \dot{\theta}_1 + \dot{\theta}_3 c_2$ $\Omega_3 = \dot{\theta}_2 c_1 + \dot{\theta}_3 s_1 s_2$ $\omega_1 = \dot{\theta}_1 s_2 c_3 - \dot{\theta}_2 s_3$ $\omega_2 = \dot{\theta}_1 c_2 + \dot{\theta}_3$ $\omega_3 = \dot{\theta}_1 s_2 s_3 + \dot{\theta}_2 c_3$	$\dot{\theta}_1 = \Omega_2 + (c_1\Omega_1 - s_1\Omega_3)c_2/s_2$ $\dot{\theta}_2 = s_1\Omega_1 + c_1\Omega_3$ $\dot{\theta}_3 = (s_1\Omega_3 - c_1\Omega_1)/s_2$ $\dot{\theta}_1 = (c_3\omega_1 + s_3\omega_3)/s_2$ $\dot{\theta}_2 = c_3\omega_3 - s_3\omega_1$ $\dot{\theta}_3 = \omega_2 + (-c_3\omega_1 - s_3\omega_3)c_2 s_2$
11. 3-1-3 or z-x-z (Euler Angles) (Body Axes)	$\Omega_1 = \dot{\theta}_2 c_1 + \dot{\theta}_3 s_1 s_2$ $\Omega_2 = \dot{\theta}_2 s_1 - \dot{\theta}_3 c_1 s_2$ $\Omega_3 = \dot{\theta}_1 + \dot{\theta}_3 c_2$ $\omega_1 = \dot{\theta}_1 - \dot{\theta}_3 s_2$ $\omega_2 = \dot{\theta}_1 s_2 c_3 - \dot{\theta}_2 s_3$ $\omega_3 = \dot{\theta}_3 + \dot{\theta}_1 c_2$	$\dot{\theta}_1 = \Omega_3 + (c_1\Omega_2 - s_1\Omega_1)c_2/s_2$ $\dot{\theta}_2 = c_1\Omega_1 + s_1\Omega_2$ $\dot{\theta}_3 = (s_1\Omega_1 - c_1\Omega_2)/s_2$ $\dot{\theta}_1 = (s_3\omega_1 + c_3\omega_2)/s_2$ $\dot{\theta}_2 = c_3\omega_1 - s_3\omega_2$ $\dot{\theta}_3 = \omega_3 + (-s_3\omega_1 - c_3\omega_2)c_2/s_2$
12. 3-2-3 or z-y-z (Body Axes)	$\Omega_1 = -\dot{\theta}_2 s_1 + \dot{\theta}_3 c_1 s_2$ $\Omega_2 = \dot{\theta}_2 c_1 + \dot{\theta}_3 s_1 s_2$ $\Omega_3 = \dot{\theta}_1 + \dot{\theta}_3 c_2$ $\omega_1 = -\dot{\theta}_1 s_2 c_3 + \dot{\theta}_2 c_3$ $\omega_2 = \dot{\theta}_1 s_2 s_3 + \dot{\theta}_2 c_3$ $\omega_3 = \dot{\theta}_3 + \dot{\theta}_1 c_2$	$\dot{\theta}_1 = \Omega_3 + (-c_1\Omega_1 - s_1\Omega_2)c_2/s_2$ $\dot{\theta}_2 = c_1\Omega_2 - s_1\Omega_1$ $\dot{\theta}_3 = (c_1\Omega_1 + s_1\Omega_2)/s_2$ $\dot{\theta}_1 = (s_3\omega_2 - c_3\omega_1)/s_2$ $\dot{\theta}_2 = s_3\omega_1 + c_3\omega_2$ $\dot{\theta}_3 = \omega_3 + (c_3\omega_1 - s_3\omega_2)c_2/s_2$

Table 7.5.2 Angular Velocity Components and Orientation Angles Derivatives [7.1]. (Space Axes)

Rotation Sequence	Angular Velocity Components $({}^R\boldsymbol{\omega}^B = \Omega_i \mathbf{N}_i = \omega_i \mathbf{n}_i)$ ($\mathbf{N}_i$ fixed in R, $\mathbf{n}_i$ fixed in B)	Orientation Angle Derivatives
1. 1-2-3 or x-y-z (Dextral or Bryan Angles) (Space Axes)	$\Omega_1 = \dot{\theta}_1 c_2 c_3 - \dot{\theta}_2 s_3$ $\Omega_2 = \dot{\theta}_1 c_2 s_3 + \dot{\theta}_2 c_3$ $\Omega_3 = -\dot{\theta}_1 s_2 + \dot{\theta}_3$	$\dot{\theta}_1 = (c_3\Omega_1 + s_3\Omega_2)/c_2$ $\dot{\theta}_2 = c_3\Omega_2 - s_3\Omega_1$ $\dot{\theta}_3 = \Omega_3 + (c_3\Omega_1 + s_3\Omega_2)s_2/c_2$
	$\omega_1 = \dot{\theta}_1 - \dot{\theta}_3 s_2$ $\omega_2 = \dot{\theta}_2 c_1 + \dot{\theta}_3 s_1 c_2$ $\omega_3 = -\dot{\theta}_2 s_1 + \dot{\theta}_3 c_1 c_2$	$\dot{\theta}_1 = \omega_1 + (s_1\omega_2 + c_1\omega_3)s_2/c_2$ $\dot{\theta}_2 = c_1\omega_2 - s_1\omega_3$ $\dot{\theta}_3 = (s_1\omega_2 + c_1\omega_3)/c_2$
2. 2-3-1 or y-z-x (Space Axes)	$\Omega_1 = -\dot{\theta}_1 s_2 + \dot{\theta}_3$ $\Omega_2 = \dot{\theta}_1 c_2 c_3 - \dot{\theta}_2 s_3$ $\Omega_3 = \dot{\theta}_1 c_2 s_3 + \dot{\theta}_2 c_3$	$\dot{\theta}_1 = (c_3\Omega_2 + s_3\Omega_3)/c_2$ $\dot{\theta}_2 = c_3\Omega_3 - s_3\Omega_2$ $\dot{\theta}_3 = \Omega_1 + (c_3\Omega_2 + s_3\Omega_3)s_2/c_2$
	$\omega_1 = -\dot{\theta}_2 s_1 + \dot{\theta}_3 c_1 c_2$ $\omega_2 = \dot{\theta}_1 - \dot{\theta}_3 s_2$ $\omega_3 = \dot{\theta}_1 c_1 + \dot{\theta}_3 s_1 c_2$	$\dot{\theta}_1 = \omega_2 + (c_1\omega_1 + s_1\omega_3)s_2/c_2$ $\dot{\theta}_2 = c_1\omega_3 - s_1\omega_1$ $\dot{\theta}_3 = (c_1\omega_1 + s_1\omega_3)/c_2$

3. 3-1-2 or z-y-x (Space Axes)	$\Omega_1 = \dot{\theta}_1 c_2 s_3 + \dot{\theta}_2 c_3$ $\Omega_2 = -\dot{\theta}_1 s_2 + \dot{\theta}_3$ $\Omega_3 = \dot{\theta}_1 c_2 c_3 - \dot{\theta}_2 s_3$ $\omega_1 = \dot{\theta}_2 c_1 + \dot{\theta}_3 s_1 c_2$ $\omega_2 = -\dot{\theta}_2 s_1 + \dot{\theta}_3 c_1 c_2$ $\omega_3 = \dot{\theta}_1 - \dot{\theta}_3 s_2$	$\dot{\theta}_1 = (s_3\Omega_1 + c_3\Omega_3)/c_2$ $\dot{\theta}_2 = c_3\Omega_1 - s_3\Omega_3$ $\dot{\theta}_3 = \Omega_2 + (s_3\Omega_1 + c_3\Omega_3)s_2/c_2$ $\dot{\theta}_1 = \omega_3 + (s_1\omega_1 + c_1\omega_2)s_2/c_2$ $\dot{\theta}_2 = c_1\omega_1 - s_1\omega_2$ $\dot{\theta}_3 = (s_1\omega_1 + c_1\omega_2)/c_2$
4. 1-3-2 or x-z-y (Space Axes)	$\Omega_1 = \dot{\theta}_1 c_2 c_3 + \dot{\theta}_2 s_3$ $\Omega_2 = \dot{\theta}_1 s_2 + \dot{\theta}_3$ $\Omega_3 = -\dot{\theta}_1 c_2 s_3 + \dot{\theta}_2 c_3$ $\omega_1 = \dot{\theta}_1 + \dot{\theta}_3 s_2$ $\omega_2 = \dot{\theta}_2 s_1 + \dot{\theta}_3 c_1 c_2$ $\omega_3 = \dot{\theta}_2 c_1 - \dot{\theta}_3 s_1 c_2$	$\dot{\theta}_1 = (c_3\Omega_1 - s_3\Omega_3)/c_2$ $\dot{\theta}_2 = s_3\Omega_1 + c_3\Omega_3$ $\dot{\theta}_3 = \Omega_2 + (s_3\Omega_3 - c_3\Omega_1)s_2/c_2$ $\dot{\theta}_1 = \omega_1 + (s_1\omega_3 - c_1\omega_2)s_2/c_2$ $\dot{\theta}_2 = s_1\omega_2 + c_1\omega_3$ $\dot{\theta}_3 = (c_1\omega_2 - s_1\omega_3)/c_2$
5. 2-1-3 or y-x-z (Space Axes)	$\Omega_1 = -\dot{\theta}_1 c_2 s_3 + \dot{\theta}_2 c_3$ $\Omega_2 = \dot{\theta}_1 c_2 c_3 + \dot{\theta}_2 s_3$ $\Omega_3 = \dot{\theta}_1 s_2 + \dot{\theta}_3$ $\omega_1 = \dot{\theta}_2 c_1 - \dot{\theta}_3 s_1 c_2$ $\omega_2 = \dot{\theta}_1 + \dot{\theta}_3 s_2$ $\omega_3 = \dot{\theta}_2 s_1 + \dot{\theta}_3 c_1 c_2$	$\dot{\theta}_1 = (-s_3\Omega_1 + c_3\Omega_2)/c_2$ $\dot{\theta}_2 = c_3\Omega_1 + s_3\Omega_2$ $\dot{\theta}_3 = \Omega_3 + (s_3\Omega_1 - c_3\Omega_2)s_2/c_2$ $\dot{\theta}_1 = \omega_2 + (s_1\omega_1 - c_1\omega_3)s_2/c_2$ $\dot{\theta}_2 = c_1\omega_1 + s_1\omega_3$ $\dot{\theta}_3 = (-s_1\omega_1 + c_1\omega_3)/c_2$

Continued

Table 7.5.2 Continued

Rotation Sequence	Angular Velocity Components	Orientation Angle Derivatives
6. 3-2-1 or z-y-x (Space Axes)	$\Omega_1 = \dot{\theta}_1 s_2 + \dot{\theta}_3$ $\Omega_2 = -\dot{\theta}_1 c_2 s_3 + \dot{\theta}_2 c_3$ $\Omega_3 = \dot{\theta}_1 c_2 c_3 + \dot{\theta}_2 s_3$ $\omega_1 = \dot{\theta}_2 s_1 + \dot{\theta}_3 c_1 c_2$ $\omega_2 = \dot{\theta}_2 c_1 - \dot{\theta}_3 s_1 c_2$ $\omega_3 = \dot{\theta}_1 + \dot{\theta}_3 s_2$	$\dot{\theta}_1 = (-s_3\Omega_2 + c_3\Omega_3)/c_2$ $\dot{\theta}_2 = c_3\Omega_2 + s_3\Omega_3$ $\dot{\theta}_3 = \Omega_1 + (s_3\Omega_2 - c_3\Omega_3)s_2/c_2$ $\dot{\theta}_1 = \omega_3 + (s_1\omega_2 - c_1\omega_1)s_2/c_2$ $\dot{\theta}_2 = s_1\omega_1 + c_1\omega_2$ $\dot{\theta}_3 = (c_1\omega_1 - s_1\omega_2)/c_2$
7. 1-2-1 or x-y-x (Space Axes)	$\Omega_1 = \dot{\theta}_1 c_2 + \dot{\theta}_3$ $\Omega_2 = \dot{\theta}_1 s_2 s_3 + \dot{\theta}_2 c_3$ $\Omega_3 = -\dot{\theta}_1 s_2 c_3 + \dot{\theta}_2 s_3$ $\omega_1 = \dot{\theta}_1 + \dot{\theta}_3 c_2$ $\omega_2 = \dot{\theta}_2 c_1 + \dot{\theta}_3 s_1 s_2$ $\omega_3 = -\dot{\theta}_2 s_1 + \dot{\theta}_3 c_1 s_2$	$\dot{\theta}_1 = (s_3\Omega_2 - c_3\Omega_3)/s_2$ $\dot{\theta}_2 = c_3\Omega_2 + s_3\Omega_3$ $\dot{\theta}_3 = \Omega_1 + (c_3\Omega_3 - s_3\Omega_2)c_2/s_2$ $\dot{\theta}_1 = \omega_1 + (-s_1\omega_2 - c_1\omega_3)c_2/s_2$ $\dot{\theta}_2 = c_1\omega_2 - s_1\omega_3$ $\dot{\theta}_3 = (s_1\omega_2 + c_1\omega_3)/s_2$

8. 1-3-1 or X-Z-X (Euler Angles) (Space Axes)	$\Omega_1 = \dot{\theta}_1 c_2 + \dot{\theta}_3$ $\Omega_2 = \dot{\theta}_1 s_2 c_3 - \dot{\theta}_2 s_3$ $\Omega_3 = \dot{\theta}_1 s_2 s_3 - \dot{\theta}_2 c_3$ $\omega_1 = \dot{\theta}_1 + \dot{\theta}_3 c_2$ $\omega_2 = \dot{\theta}_2 s_1 - \dot{\theta}_3 c_1 s_2$ $\omega_3 = \dot{\theta}_2 c_1 + \dot{\theta}_3 s_1 s_2$	$\dot{\theta}_1 = (c_3\Omega_2 + s_3\Omega_3)/s_2$ $\dot{\theta}_2 = -s_3\Omega_2 + c_3\Omega_3$ $\dot{\theta}_3 = \Omega_1 + (-c_3\Omega_2 - s_3\Omega_3)c_2/s_2$ $\dot{\theta}_1 = \omega_1 + (c_1\omega_2 - s_1\omega_3)c_2/s_2$ $\dot{\theta}_2 = s_1\omega_2 + c_1\omega_3$ $\dot{\theta}_3 = (-c_1\omega_2 + s_1\omega_3)/s_2$
9. 2-1-2 or Y-X-Y (Space Axes)	$\Omega_1 = \dot{\theta}_1 s_2 s_3 + \dot{\theta}_2 c_3$ $\Omega_2 = \dot{\theta}_1 c_2 + \dot{\theta}_3$ $\Omega_3 = \dot{\theta}_1 s_2 c_3 - \dot{\theta}_2 s_3$ $\omega_1 = \dot{\theta}_2 c_1 + \dot{\theta}_3 s_1 s_2$ $\omega_2 = \dot{\theta}_1 + \dot{\theta}_3 c_2$ $\omega_3 = \dot{\theta}_2 s_1 - \dot{\theta}_3 s_2 c_1$	$\dot{\theta}_1 = (s_3\Omega_1 + c_3\Omega_3)/s_2$ $\dot{\theta}_2 = c_3\Omega_1 - s_3\Omega_3$ $\dot{\theta}_3 = \Omega_2 + (-s_3\Omega_1 - c_3\Omega_3)c_2/s_2$ $\dot{\theta}_1 = \omega_2 + (c_1\omega_3 - s_1\omega_1)c_2/s_2$ $\dot{\theta}_2 = c_1\omega_1 + s_1\omega_3$ $\dot{\theta}_3 = (s_1\omega_1 - c_1\omega_3)/s_2$
10. 2-3-2 or Y-Z-Y (Space Axes)	$\Omega_1 = -\dot{\theta}_1 s_2 c_3 + \dot{\theta}_2 s_3$ $\Omega_2 = \dot{\theta}_1 c_2 + \dot{\theta}_3$ $\Omega_3 = \dot{\theta}_1 s_2 s_3 + \dot{\theta}_2 c_3$ $\omega_1 = -\dot{\theta}_2 s_1 + \dot{\theta}_3 c_1 s_2$ $\omega_2 = \dot{\theta}_1 + \dot{\theta}_3 c_2$ $\omega_3 = \dot{\theta}_2 c_1 + \dot{\theta}_3 s_1 s_2$	$\dot{\theta}_1 = (-c_3\Omega_1 + s_3\Omega_3)/s_2$ $\dot{\theta}_2 = s_3\Omega_1 + c_3\Omega_3$ $\dot{\theta}_3 = \Omega_2 + (c_3\Omega_1 - s_3\Omega_3)c_2/s_2$ $\dot{\theta}_1 = \omega_2 + (-c_1\omega_1 - s_1\omega_3)c_2/s_2$ $\dot{\theta}_2 = -s_1\omega_1 + c_1\omega_3$ $\dot{\theta}_3 = (c_1\omega_1 + s_1\omega_3)/s_2$

Continued

Table 7.5.2 Continued

Rotation Sequence	Angular Velocity Components	Orientation Angle Derivatives
11. 3-1-3 or Z-X-Z (Space Axes)	$\Omega_1 = \dot{\theta}_1 s_2 s_3 + \dot{\theta}_2 c_3$ $\Omega_2 = -\dot{\theta}_1 s_2 c_3 + \dot{\theta}_2 s_3$ $\Omega_3 = \dot{\theta}_1 c_2 + \dot{\theta}_3$	$\dot{\theta}_1 = (s_3\Omega_1 - c_3\Omega_2)/s_2$ $\dot{\theta}_2 = c_3\Omega_1 + s_3\Omega_2$ $\dot{\theta}_3 = \Omega_3 + (c_3\Omega_2 - s_3\Omega_1)c_2/s_2$
	$\omega_1 = \dot{\theta}_2 c_1 + \dot{\theta}_3 s_1 s_2$ $\omega_2 = -\dot{\theta}_2 s_1 + \dot{\theta}_3 c_1 s_2$ $\omega_3 = \dot{\theta}_1 + \dot{\theta}_3 c_2$	$\dot{\theta}_1 = \omega_3 + (-s_1\omega_1 - c_1\omega_2)c_2/s_2$ $\dot{\theta}_2 = c_1\omega_1 - s_1\omega_2$ $\dot{\theta}_3 = (s_1\omega_1 + c_1\omega_2)/s_2$
12. 3-2-3 or Z-Y-Z (Space Axes)	$\Omega_1 = \dot{\theta}_1 s_2 c_3 - \dot{\theta}_2 s_3$ $\Omega_2 = \dot{\theta}_1 s_2 s_3 + \dot{\theta}_2 c_3$ $\Omega_3 = \dot{\theta}_1 c_2 + \dot{\theta}_3$	$\dot{\theta}_1 = (c_3\Omega_1 + s_3\Omega_2)/s_2$ $\dot{\theta}_2 = -s_3\Omega_1 + c_3\Omega_2$ $\dot{\theta}_3 = \Omega_3 + (-c_3\Omega_1 - s_3\Omega_2)c_2/s_2$
	$\omega_1 = \dot{\theta}_2 s_1 - \dot{\theta}_3 c_1 s_2$ $\omega_2 = \dot{\theta}_2 c_1 + \dot{\theta}_3 s_1 s_2$ $\omega_3 = \dot{\theta}_1 + \dot{\theta}_3 c_2$	$\dot{\theta}_1 = \omega_3 + (c_1\omega_1 - s_1\omega_2)c_2/s_2$ $\dot{\theta}_2 = s_1\omega_1 + c_1\omega_2$ $\dot{\theta}_3 = (-c_1\omega_1 + s_1\omega_2)/s_2$

7.6 Euler Parameters

Consider again a body B moving in a reference frame R as in Figure 7.6.1. The analysis and discussions of Sections 7.2, 7.3, and 7.4 show that any change of

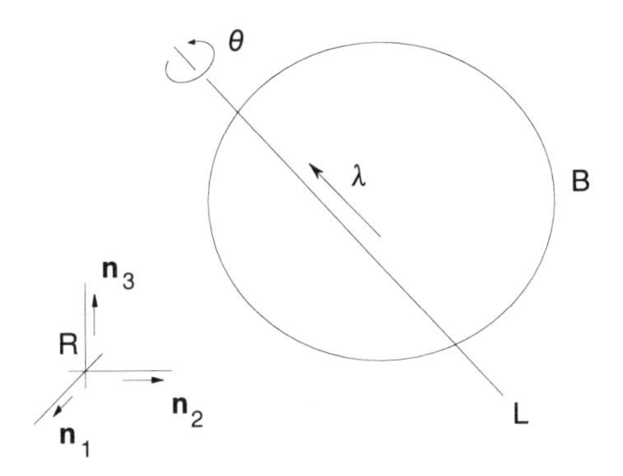

Fig. 7.6.1 A Body B Moving in a Reference Frame R

orientation of B in R may be attained by a simple rotation of B about a line L fixed in both B and R. The orientation of L, characterized by a unit vector $\boldsymbol{\lambda}$ parallel to L, and the rotation angle θ are determined once the orientation change of B in R is known. Specifically, if R_{ij} are the elements of the rotation dyadic (see Section 7.2) of B in R then θ and $\boldsymbol{\lambda}$ are obtained from Equations (7.4.4) and (7.4.5) as:

$$\theta = \cos^{-1}[(R_{ij} - 1)/2] \tag{7.6.1}$$

and

$$\lambda_1 = (R_{32} - R_{23})/2\sin\theta \quad \lambda_2 = (R_{13} - R_{31})/2\sin\theta \quad \lambda_3 = (R_{21} - R_{12})/2\sin\theta \tag{7.6.2}$$

where the λ_i are the components of $\boldsymbol{\lambda}$ relative to unit vectors $\mathbf{n}_i$ $(i = 1,2,3)$ fixed in R (see Figure 7.6.1). Furthermore from Equation (7.3.4) the values of the R_{ij} may be found from the values of the transformation matrix elements S_{ij}.

With the foregoing terminology, the orientation of B in R may also be defined in terms of four parameters ε_i $(i = 1,\ldots,4)$, called "Euler parameters," defined as:

$$\begin{aligned} \varepsilon_1 &= \lambda_1 \sin(\theta/2) \\ \varepsilon_2 &= \lambda_2 \sin(\theta/2) \\ \varepsilon_3 &= \lambda_3 \sin(\theta/2) \\ \varepsilon_4 &= \cos(\theta/2) \end{aligned} \tag{7.6.3}$$

To explore this, observe first that the Euler parameters are not independent. They are related by the expression:

$$\varepsilon_1^2 + \varepsilon_2^2 + \varepsilon_3^2 + \varepsilon_4^2 = 1 \tag{7.6.4}$$

(Note that since the orientation of a body can be defined in terms of three orientation angles, we would expect that only three Euler parameters would be independent.)

Next, recall from Equation (7.4.2) that the rotation matrix elements R_{ij} (and hence also the transformation matrix elements S_{ij}) may be expressed in terms of θ and the components of $\boldsymbol{\lambda}$ as:

$$S_{ij} = \begin{bmatrix} [\lambda_1^2(1-\cos\theta)+\cos\theta] & [\lambda_1\lambda_2(1-\cos\theta)-\lambda_3\sin\theta] & [\lambda_1\lambda_3(1-\cos\theta)+\lambda_2\sin\theta] \\ [\lambda_1\lambda_2(1-\cos\theta)+\lambda_3\sin\theta] & [\lambda_2^2(1-\cos\theta)+\cos\theta] & [\lambda_2\lambda_3(1-\cos\theta)-\lambda_1\sin\theta] \\ [\lambda_3\lambda_1(-\cos\theta)-\lambda_2\sin\theta] & [\lambda_2\lambda_3(1-\cos\theta)+\lambda_1\sin\theta] & [\lambda_3^2(1-\cos\theta)+\cos\theta] \end{bmatrix} \tag{7.6.5}$$

Then by using the definitions of the Euler parameters in Equations (7.6.3) we can express the elements of the transformation matrix entirely in terms of the Euler parameters as (see example analysis following):

$$S_{ij} = \begin{bmatrix} (\varepsilon_1^2-\varepsilon_2^2-\varepsilon_3^2+\varepsilon_4^2) & 2(\varepsilon_1\varepsilon_2-\varepsilon_3\varepsilon_4) & 2(\varepsilon_1\varepsilon_3+\varepsilon_2\varepsilon_4) \\ 2(\varepsilon_1\varepsilon_2+\varepsilon_3\varepsilon_4) & (-\varepsilon_1^2+\varepsilon_2^2-\varepsilon_3^2+\varepsilon_4^2) & 2(\varepsilon_2\varepsilon_3-\varepsilon_1\varepsilon_4) \\ 2(\varepsilon_1\varepsilon_3-\varepsilon_2\varepsilon_4) & 2(\varepsilon_2\varepsilon_3+\varepsilon_1\varepsilon_4) & (-\varepsilon_1^2-\varepsilon_2^2+\varepsilon_3^2+\varepsilon_4^2) \end{bmatrix} \tag{7.6.6}$$

Example 7.6.1: Use Equations (7.6.3) and (7.6.5) to develop the elements of Equation (7.6.6).

Solution: From Equation (7.6.5) S_{11} is:

$$S_{11} = \lambda_1^2(1 - \cos\theta) + \cos\theta$$

From Equation (7.6.3) ε_1^2 and ε_4^2 are

$$\varepsilon_1^2 = \lambda_1^2 \sin^2\theta/2 = \lambda_1^2(1 - \cos\theta)/2$$

$$\varepsilon_4^2 = \cos^2\theta/2 = (1 + \cos\theta)/2$$

Then by substitution S_{11} may be expressed as:

$$S_{11} = 2\varepsilon_1^2 + 2\varepsilon_4^2 - 1$$

Finally by substituting from Equation (7.6.4), S_{11} becomes:

$$S_{11} = \varepsilon_1^2 - \varepsilon_2^2 - \varepsilon_3^2 + \varepsilon_4^2 \tag{7.6.7}$$

Next, from Equation (7.6.5) S_{21} is:

$$S_{21} = \lambda_1\lambda_2(1 - \cos\theta) + \lambda_3 \sin\theta$$

From Equation (7.6.3) $\varepsilon_1\varepsilon_2$ and $\varepsilon_3\varepsilon_4$ are

$$\varepsilon_1\varepsilon_2 = \lambda_1\lambda_2 \sin^2\theta/2 = \lambda_1\lambda_2(1 - \cos\theta)/2$$

$$\varepsilon_3\varepsilon_4 = \lambda_3 \sin\theta/2 \cos\theta/2 = \lambda_3 (\sin\theta)/2$$

Hence, S_{21} becomes

$$S_{21} = 2\varepsilon_1\varepsilon_2 + 2\varepsilon_3\varepsilon_4 \tag{7.6.8}$$

Each of the remaining S_{ij} can be similarly expressed in terms of the Euler parameters, leading to Equation (7.6.6).

7.7 Differentiation of Transformation Matrices

Let B be a body moving in a reference frame R and let $\mathbf{N}_i$ and $\mathbf{n}_i$ $(i = 1,2,3)$ be unit vectors fixed in R and B as in Figure 7.7.1. Then the transformation matrix S_{ij} between R and B is defined as:

$$S_{ij} = \mathbf{N}_i \cdot \mathbf{n}_j \tag{7.7.1}$$

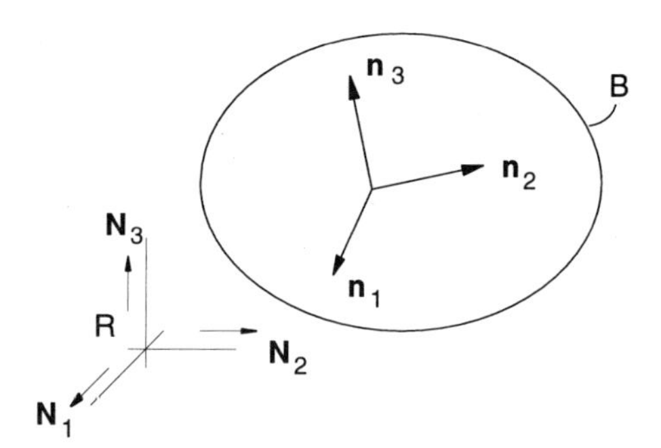

Fig. 7.7.1 A Body B Moving in a Reference Frame R

The derivative of S_{ij} is then:

$$\begin{aligned} dS_{ij}/dt &= d(\mathbf{N}_i \bullet \mathbf{n}_j)/dt \\ &= \mathbf{N}_i \bullet \boldsymbol{\omega} \times \mathbf{n}_j \end{aligned} \tag{7.7.2}$$

where $\boldsymbol{\omega}$ is the angular velocity of B in R. By expressing $\boldsymbol{\omega}$ and $\mathbf{n}_j$ in terms of the unit vectors $\mathbf{N}_i$ of R (that is, $\boldsymbol{\omega} = \Omega_k \mathbf{N}_k$ and $\mathbf{n}_j = S_{\ell j} \mathbf{N}_\ell$) we can express the derivative of S_{ij} in the form:

$$\begin{aligned} dS_{ij}/dt &= \mathbf{N}_i \bullet \Omega_k \mathbf{N}_k \times S_{\ell j} \mathbf{N}_\ell \\ &= e_{ik\ell} \Omega_k S_{\ell j} \\ &= -e_{i\ell k} \Omega_k S_{\ell j} \\ &= W_{i\ell} S_{\ell j} \end{aligned} \tag{7.7.3}$$

where the $W_{i\ell}$ are elements of the "dual matrix" W of $\boldsymbol{\omega}$ defined as

$$W_{i\ell} = -e_{i\ell k} \Omega_k = \begin{bmatrix} 0 & -\Omega_3 & \Omega_2 \\ \Omega_3 & 0 & -\Omega_1 \\ -\Omega_2 & \Omega_1 & 0 \end{bmatrix} \tag{7.7.4}$$

Equation (7.7.3) may be written in matrix form as

$$dS/dt - WS \tag{7.7.5}$$

where W is sometimes also called the "angular velocity matrix."

Equations (7.7.3) and (7.7.4) may be used to relate the derivatives of the transformation matrix elements dS_{ij}/dt and the space fixed angular velocity components Ω_k. Specifically, by expanding Equation (7.7.3) we obtain

$$\begin{aligned} dS_{11}/dt &= -\Omega_3 S_{21} + \Omega_2 S_{31} \\ dS_{12}/dt &= -\Omega_3 S_{22} + \Omega_2 S_{32} \\ dS_{13}/dt &= -\Omega_3 S_{23} + \Omega_2 S_{33} \end{aligned} \tag{7.7.6}$$

$$\begin{aligned} dS_{21}/dt &= -\Omega_1 S_{31} + \Omega_3 S_{11} \\ dS_{22}/dt &= -\Omega_1 S_{32} + \Omega_3 S_{12} \\ dS_{23}/dt &= -\Omega_1 S_{33} + \Omega_3 S_{13} \end{aligned} \tag{7.7.7}$$

$$\begin{aligned} dS_{31}/dt &= -\Omega_2 S_{11} + \Omega_1 S_{21} \\ dS_{32}/dt &= -\Omega_2 S_{12} + \Omega_1 S_{22} \\ dS_{33}/dt &= -\Omega_2 S_{13} + \Omega_1 S_{23} \end{aligned} \tag{7.7.8}$$

Recalling that the transformation matrix is orthogonal, we can solve Equations (7.7.6), (7.7.7), and (7.7.8) for Ω_1, Ω_2, and Ω_3. Specifically, the orthogonality property of S gives the expressions:

$$SS^T = I \quad \text{and} \quad S^TS = I \tag{7.7.9}$$

or

$$S_{ik}S_{jk} = \delta_{ij} \quad \text{and} \quad S_{ki}S_{kj} = \delta_{ij} \tag{7.7.10}$$

Then by multiplying Equations (7.7.8) by S_{21}, S_{22} and S_{23} respectively and adding, we have

$$\Omega_1 = S_{21}dS_{31}/dt + S_{22}dS_{32}/dt + S_{23}dS_{33}/dt \tag{7.7.11}$$

Similarly, by multiplying Equations (7.7.6) by S_{31}, S_{32}, and S_{33} respectively and adding, we have

$$\Omega_2 = S_{31}dS_{11}/dt + S_{32}dS_{12}/dt + S_{33}dS_{13}/dt \tag{7.7.12}$$

Finally, by multiplying Equations (7.7.7) by S_{11}, S_{12}, and S_{13} respectively and adding, we have

$$\Omega_3 = S_{11}dS_{21}/dt + S_{12}dS_{22}/dt + S_{13}dS_{23}/dt \tag{7.7.13}$$

7.8 Euler Parameters and Angular Velocity

Equations (7.7.11), (7.7.12), and (7.7.13) together with Equation (7.6.6) may be used to obtain relations between the angular velocity components and the Euler parameter derivatives. Specifically, by substituting from Equation (7.6.6) into Equations (7.7.11), (7.7.12), and (7.7.13), we have

$$\begin{aligned}
\Omega_1 &= 2(\varepsilon_4\dot{\varepsilon}_1 - \varepsilon_3\dot{\varepsilon}_2 + \varepsilon_2\dot{\varepsilon}_3 - \varepsilon_1\dot{\varepsilon}_4) \\
\Omega_2 &= 2(\varepsilon_3\dot{\varepsilon}_1 + \varepsilon_4\dot{\varepsilon}_2 - \varepsilon_1\dot{\varepsilon}_3 - \varepsilon_2\dot{\varepsilon}_4) \\
\Omega_3 &= 2(-\varepsilon_2\dot{\varepsilon}_1 + \varepsilon_1\dot{\varepsilon}_2 + \varepsilon_4\dot{\varepsilon}_3 - \varepsilon_3\dot{\varepsilon}_4)
\end{aligned} \tag{7.8.1}$$

Equations (7.8.1) may in turn be used to obtain relations between the Euler parameter derivatives and the angular velocity components. To do this, however, an additional equation is needed, since there are four Euler parameter derivatives. The fourth equation may be obtained by differentiating Equation (7.6.4) giving the expression:

$$0 = 2(\varepsilon_1\dot{\varepsilon}_1 + \varepsilon_2\dot{\varepsilon}_2 + \varepsilon_3\dot{\varepsilon}_3 + \varepsilon_4\dot{\varepsilon}_4) \tag{7.8.2}$$

Equations (7.8.1) and (7.8.2) then constitute four simultaneous linear algebraic equations for the four Euler parameter derivatives $\dot{\varepsilon}_i$ $(i = 1,...,4)$. In matrix form these equations may be written as

$$\begin{bmatrix} \Omega_1 \\ \Omega_2 \\ \Omega_3 \\ 0 \end{bmatrix} = 2 \begin{bmatrix} \varepsilon_4 & -\varepsilon_3 & \varepsilon_2 & -\varepsilon_1 \\ \varepsilon_3 & \varepsilon_4 & -\varepsilon_1 & -\varepsilon_2 \\ -\varepsilon_2 & \varepsilon_1 & \varepsilon_4 & -\varepsilon_3 \\ \varepsilon_1 & \varepsilon_2 & \varepsilon_3 & \varepsilon_4 \end{bmatrix} \begin{bmatrix} \dot{\varepsilon}_1 \\ \dot{\varepsilon}_2 \\ \dot{\varepsilon}_3 \\ \dot{\varepsilon}_4 \end{bmatrix} \tag{7.8.3}$$

or as

$$\Omega = 2E\dot{\varepsilon} \tag{7.8.4}$$

where Ω, E, and $\dot{\varepsilon}$ are arrays defined by inspection. Then solving for $\dot{\varepsilon}$ we have

$$\dot{\varepsilon} = (1/2)E^{-1}\Omega \tag{7.8.5}$$

Remarkably, the inverse of E is readily obtained since E happens to be orthogonal. That is,

$$E^{-1} = E^T = \begin{bmatrix} \varepsilon_4 & \varepsilon_3 & -\varepsilon_2 & \varepsilon_1 \\ -\varepsilon_3 & \varepsilon_4 & \varepsilon_1 & \varepsilon_2 \\ \varepsilon_2 & -\varepsilon_1 & \varepsilon_4 & \varepsilon_3 \\ -\varepsilon_1 & -\varepsilon_2 & -\varepsilon_3 & \varepsilon_4 \end{bmatrix} \tag{7.8.6}$$

Then by substituting into Equation (7.8.5), the Euler parameter derivatives become:

$$\begin{aligned}
\dot{\varepsilon}_1 &= 1/2\,(\varepsilon_4\Omega_1 + \varepsilon_3\Omega_2 - \varepsilon_2\Omega_3) \\
\dot{\varepsilon}_2 &= 1/2\,(-\varepsilon_3\Omega_1 + \varepsilon_4\Omega_2 + \varepsilon_1\Omega_3) \\
\dot{\varepsilon}_3 &= 1/2\,(\varepsilon_2\Omega_1 - \varepsilon_1\Omega_2 + \varepsilon_4\Omega_3) \\
\dot{\varepsilon}_4 &= 1/2\,(-\varepsilon_1\Omega_1 - \varepsilon_2\Omega_2 - \varepsilon_3\Omega_3)
\end{aligned} \tag{7.8.7}$$

Observe the linearity in Equations (7.8.1) and (7.8.7). This means that the singularities which occur with orientation angles are avoided with Euler parameters. The cost is the introduction of a fourth variable.

References

7.1 T. R. Kane, P. W. Likins, and D. A. Levinson, *Spacecraft Dynamics*, McGraw Hill, New York, 1983, pp. 422-431.

7.2 R. L. Huston, *Multibody Dynamics*, Butterworth-Heinemann, Stoneham, Massachusetts, 1990.

Chapter 8

MASS DISTRIBUTION AND INERTIA

8.1 Introduction

In Newton's second law, often considered symbolically as: F = ma, we may regard the force (F), mass (m), and acceleration (a) as representing branches of dynamics: kinetics, inertia, and kinematics respectively. In this chapter we review and summarize formulas and procedures related to mass distribution and inertia. Our sources for the majority of these formulas are References 8.1 to 8.11.

Since mass distribution properties and inertia are well documented in many texts and handbooks, we will not repeat the details herein, but in keeping with our objectives we will primarily summarize results.

8.2 First Moment Vectors

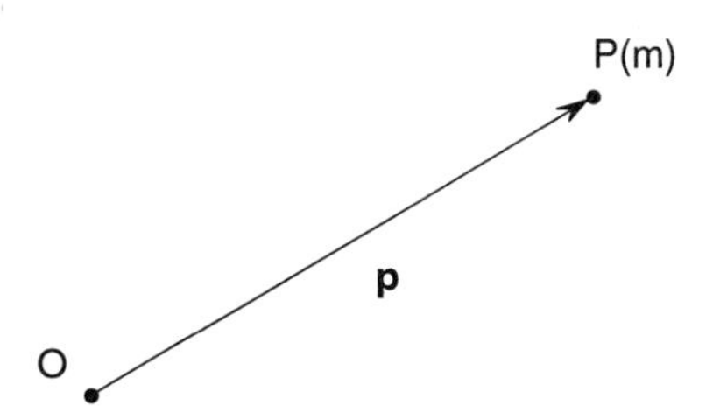

Fig. 8.2.1 A Particle P and Reference Point O

Consider a particle P having as mass m and a point O as in Figure 8.2.1. Let **p** locate P relative to O as shown. The *first moment* $\boldsymbol{\phi}^{P/O}$ of P relative to O is defined as:

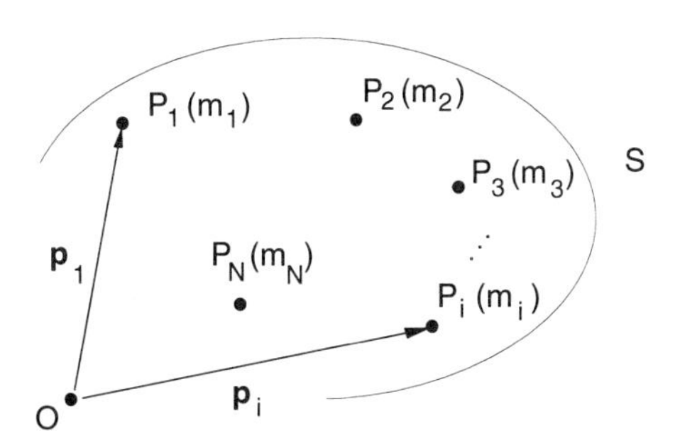

Fig. 8.2.2 A Set S of Particles and a Reference Point O

$$\phi^{P/O} = m\mathbf{p} \tag{8.2.1}$$

Consider next a set S of particles P_i $(i = 1, \ldots, N)$ and a reference point O as in Figure 8.2.2. Let m_i be the mass of typical particle P_i and let $\mathbf{p}_i$ locate P_i relative to O. Then the first moment $\phi^{S/O}$ of S relative to O is defined as:

$$\phi^{S/O} = \sum_{i=1}^{N} \phi^{P_i/O} = \sum_{i=1}^{N} m_i \mathbf{p}_i \tag{8.2.2}$$

Finally consider a body B and a reference point O as in Figure 8.2.3. Let B be modeled as a set of particles P_i $(i = 1, \ldots, N)$ connected and fixed relative to one another with masses m_i. As before let $\mathbf{p}_i$ locate P_i relative to O. Then the first moment $\phi^{B/O}$ of B for O is defined as:

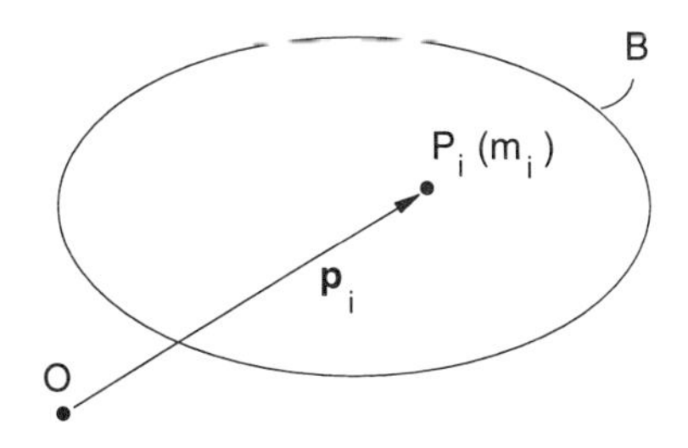

Fig. 8.2.3 A Body B and a Reference Point O and a Typical Particle P_i of B

$$\phi^{B/O} = \sum_{i=1}^{N} \phi^{P_i/O} = \sum_{i=1}^{N} m_i \mathbf{p}_i \tag{8.2.3}$$

If B can be regarded as a smooth body as in a continuum then the summation in Equation (8.2.3) can be replaced by an integral:

$$\phi^{B/O} = \int_B \rho \mathbf{p} \, dV \tag{8.2.4}$$

where ρ is the mass density of B at a typical point P of B, $\mathbf{p}$ locates P relative to O and dV is a differential volume element of B containing P.

8.3 Mass Center/Center of Gravity

The first moment vector ϕ is dependent upon the location of the reference point O. For example, the magnitude of ϕ could be zero. Indeed, for a set of particles the point G for which the first moment is zero, is called the "mass center." Specifically, the mass center is defined as that reference point for which the first moment is zero.

A procedure for locating the mass center may readily be determined from this definition. Specifically, let S be a set of N particles P_i with masses m_i $(i = 1,\ldots,N)$ and let O be an arbitrary reference point as in Figure 8.3.1. Let G be the mass center of S and let $\mathbf{p}_G$ be the position vector locating G relative to O. (Hence, G is located once $\mathbf{p}_G$ is known.) Let

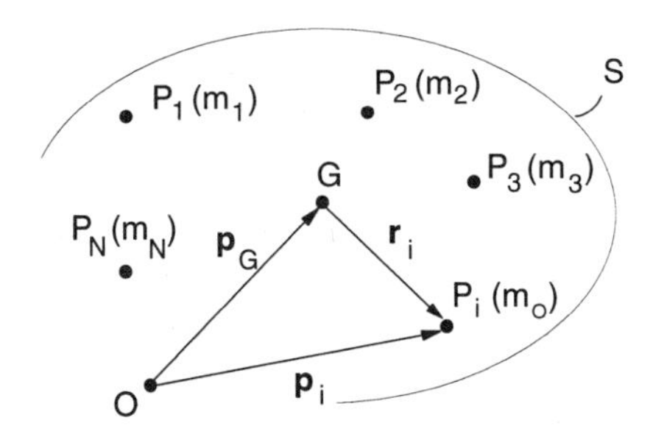

Fig. 8.3.1 A Set S of Particles with Mass Center G

$\mathbf{p}_i$ and $\mathbf{r}_i$ locate P_i relative to O and G. Then from Figure 8.3.1 we have

$$\mathbf{p}_i = \mathbf{p}_G + \mathbf{r}_i \tag{8.3.1}$$

Since G is the mass center, the first moment of S relative to G is zero. That is,

$$\sum_{i=1}^{N} m_i \mathbf{r}_i = 0 \tag{8.3.2}$$

Substituting from Equation (8.3.1) into (8.3.2) we have

$$\sum_{i=1}^{N} m_i \mathbf{p}_i - \sum_{i=1}^{N} m_i \mathbf{p}_G = 0$$

or

$$\sum_{i=1}^{N} m_i \mathbf{p}_i = \sum_{i=1}^{N} m_i \mathbf{p}_G = M\mathbf{p}_G$$

or

$$\mathbf{p}_G = (1/M)\sum_{i=1}^{N} m_i \mathbf{p}_i \tag{8.3.3}$$

where M is the total mass of the particles of S.

It is readily seen that the mass center always exists and is unique.

For a body B Equation (8.3.3) may be expressed in the form

$$\mathbf{p}_G = (1/M)\int_B \rho \mathbf{p} \, dV \tag{8.3.4}$$

where, as before, ρ is the mass density of B and **p** locates a typical point P of B relative to the reference point O as in Figure 8.3.2, and M is the mass of B. If the mass of B is uniformly distributed p is a constant and then M is simply ρV where V is the volume of B. Then P_G has the form

$$P_G = (1/V) \int_B \mathbf{p}\, dV \tag{8.3.5}$$

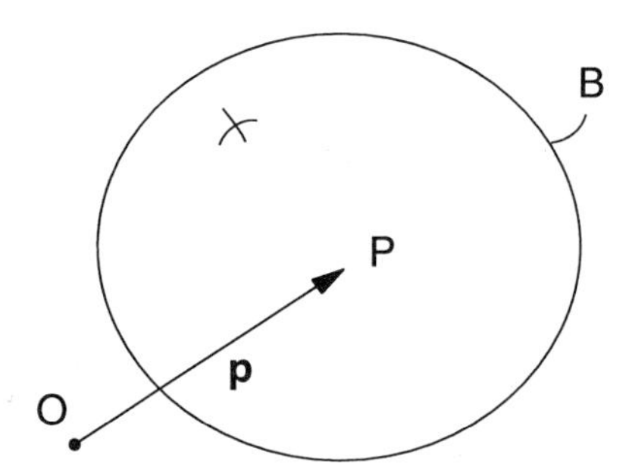

Fig. 8.3.2 A Body B and a Typical Point P

For a body with uniform mass distribution, the mass center is located at the centroid of the geometrical figure occupied by the body. Equation (8.3.5) may be used to locate the centroid and such analyses have been made for common geometrical figures [8.1 to 8.11]. Table 8.3.1 summarizes results of these analyses.

The term "center of gravity" is often used to designate the mass center. Generally there is no important difference between the center of gravity and the mass center, but technically they are different. The center of gravity of a body is defined as that point $\hat{G}$ through which all the gravitational forces could be replaced by a single force. To see that $\hat{G}$ is distinct from G consider a body consisting of a light rod and two identical particles P_1 and P_2 each with mass m as in Figure 8.3.3. then if **k** is a unit vector in the vertical direction, P_1 will be further away from the

Table 8.3.1 Centroid and Mass Center Location for commonly shaped Bodies with Uniform Mass Distribution

I. **Curves, Wires, Thin Rods**:

1. *Straight Line, Rod*:

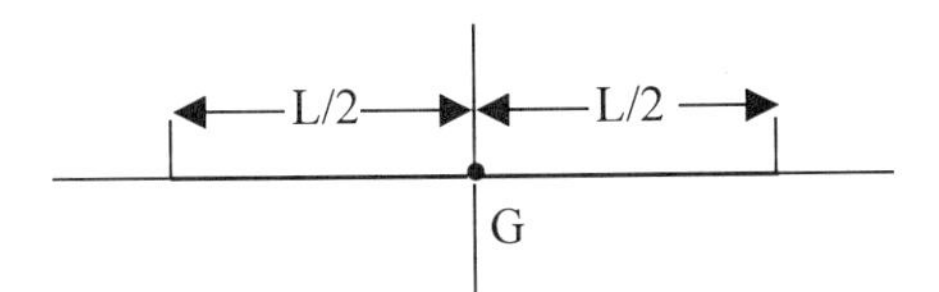

2. *Circular Arc, Circular Rod*:

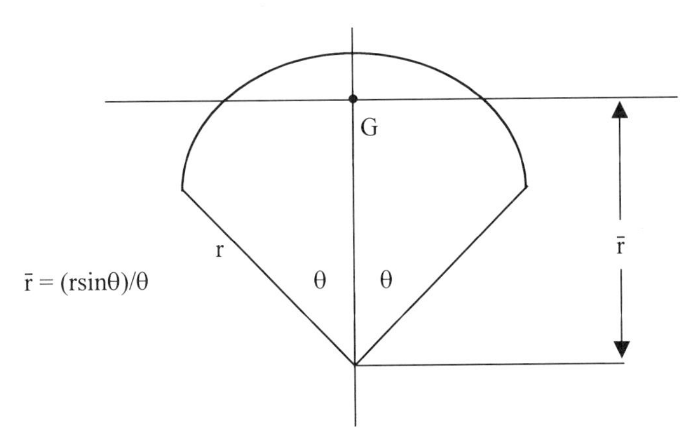

3. *Semi Circular Arc, Semi Circular Rod*:

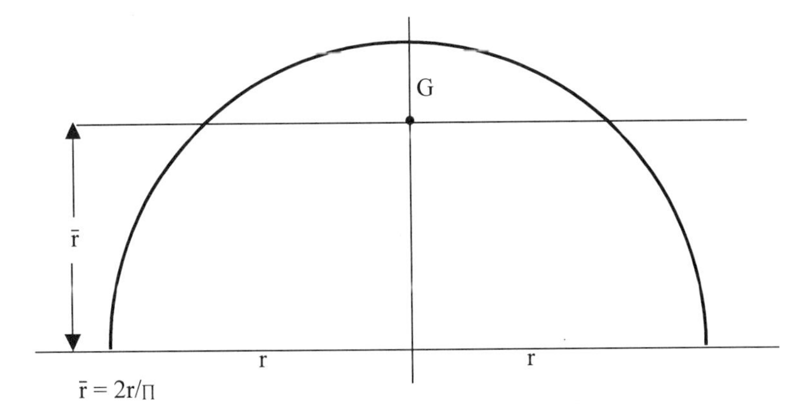

$\bar{r} = 2r/\pi$

4. *Circle, Hoop*:

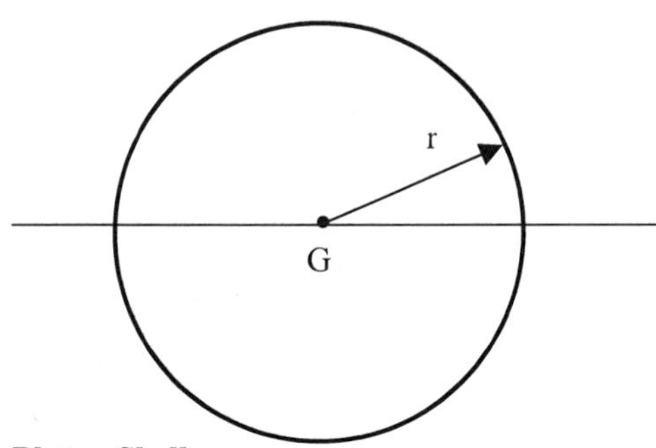

II. **Surfaces, Thin Plates, Shells**:

1. *Triangle, Triangular Plate*:

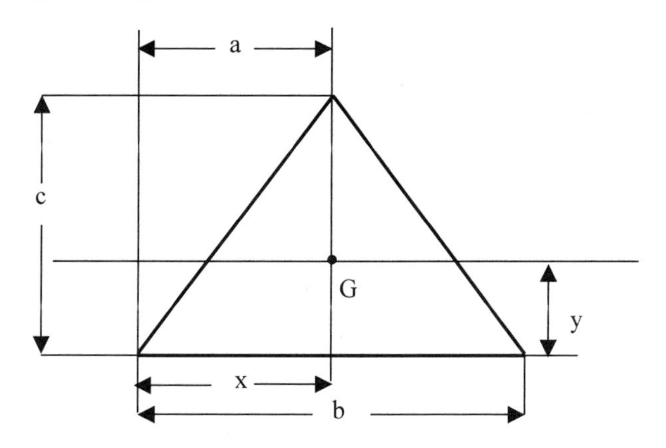

x = (a+b)/3
y = c/3

2. *Rectangle, Rectangular Plate*:

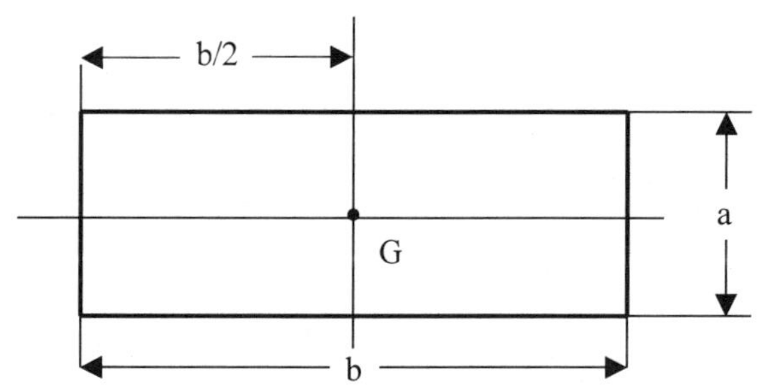

3. *Circular Sector, Circular Section Plate*:

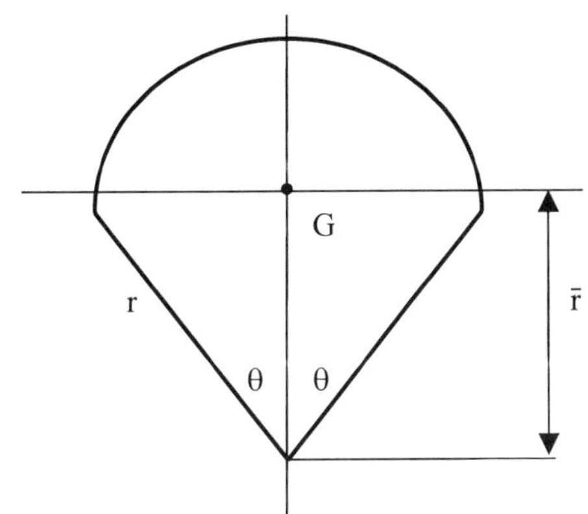

$\bar{r} = (2r/3)(\sin\theta)/\theta$

4. *Semicircle, Semicircular Plate*:

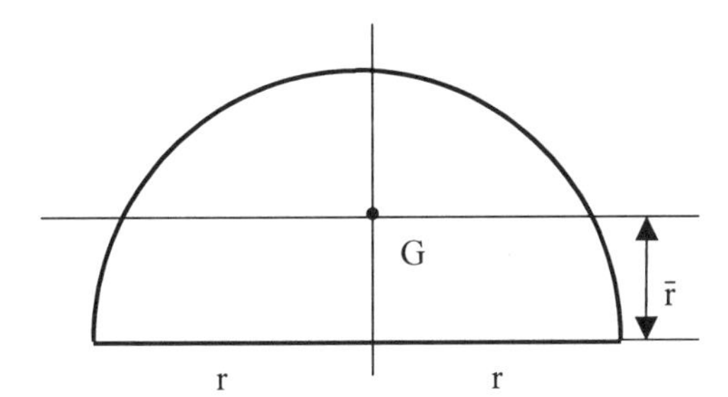

$\bar{r} = 4r/3\pi$

5. *Circle, Circular Plate*:

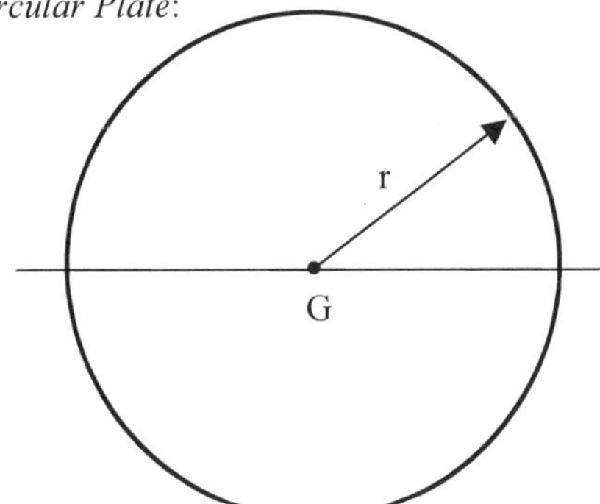

6. *Circular Segment, Circular Segment Plate*:

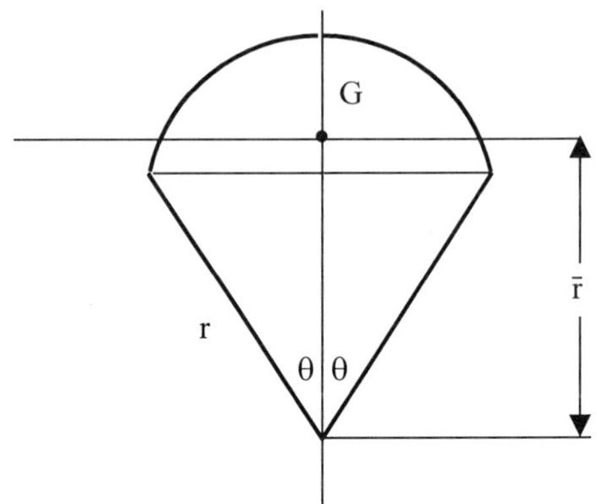

$\bar{r} = (2r/3)(\sin^3\theta)/(\theta - \sin\theta\cos\theta)$

7. *Cylinder, Cylindrical Shell*:

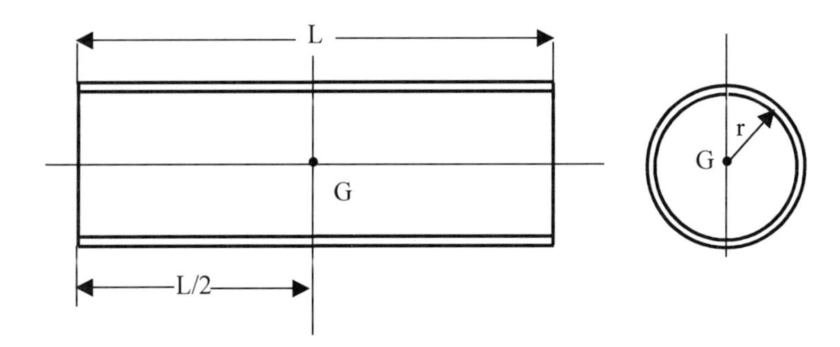

8. *Semicylinder, Semicylindrical Shell*:

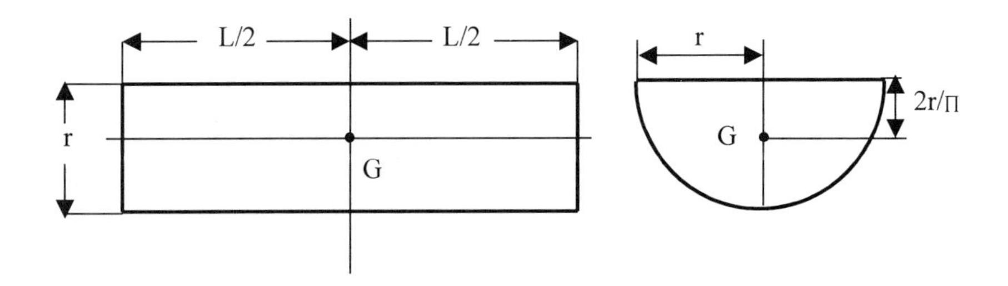

9. *Sphere, Spherical Shell*:

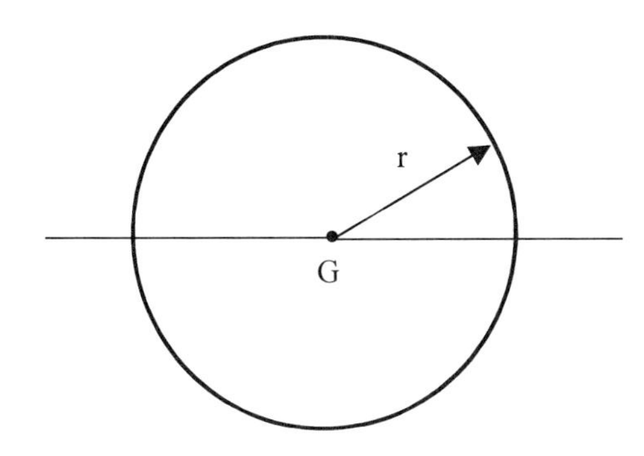

10. *Hemisphere, Hemispherical Shell*:

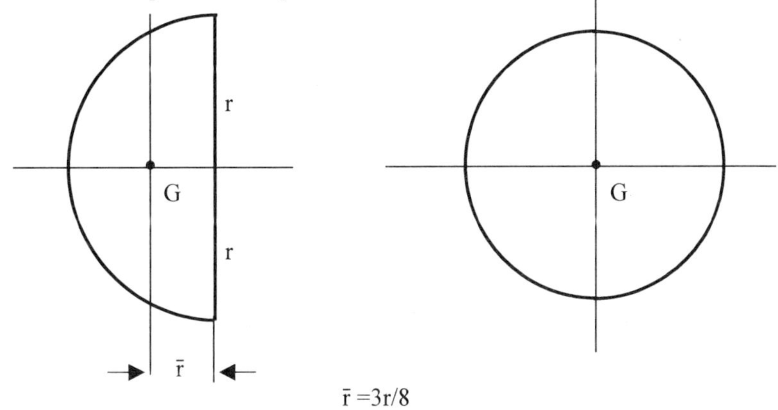

11. *Cone, Conical Shell*:

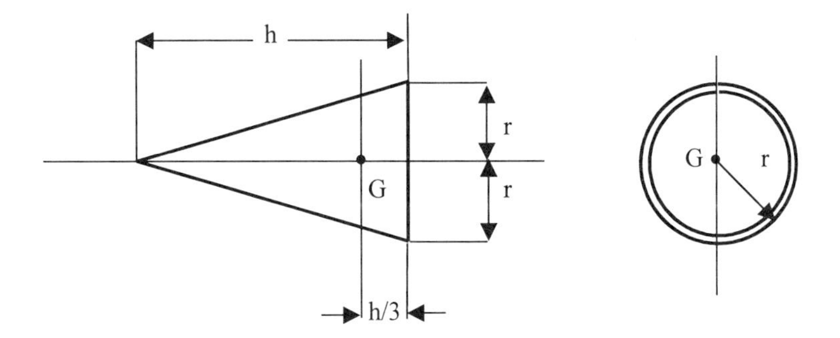

12. *Half-cone, Half Conical Shell*:

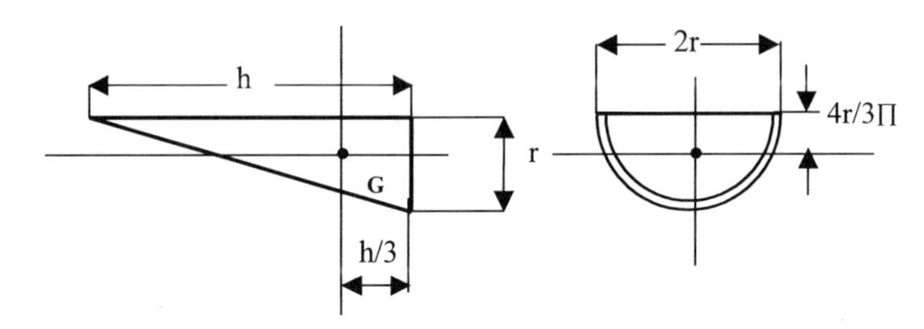

III **Solids, Bodies**:

1. *Parallelopiped, Block*:

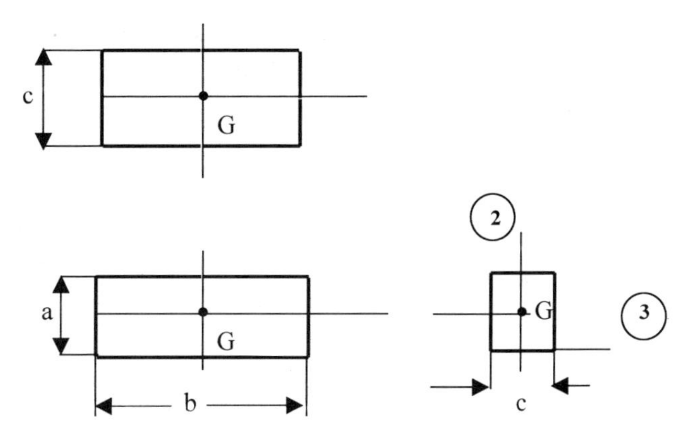

2. *Cylinder*:

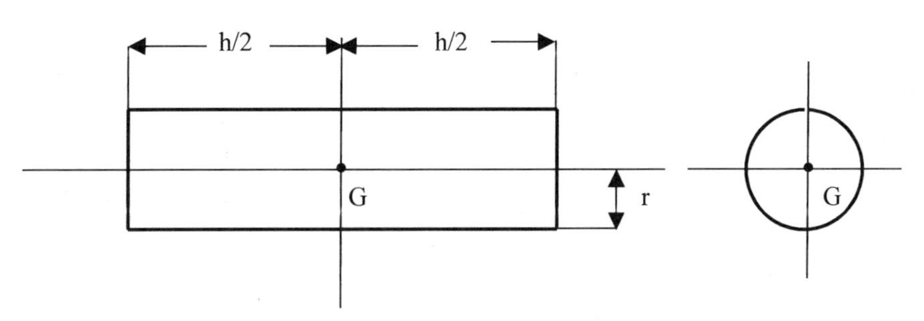

3. *Half Cylinder*:

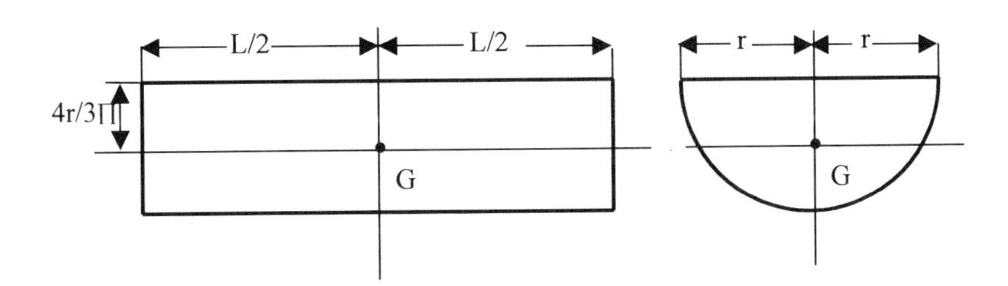

4. *Cone*:

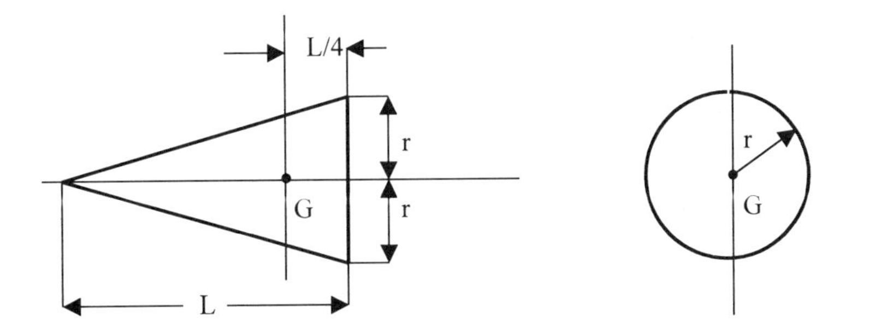

5. *Half cone*:

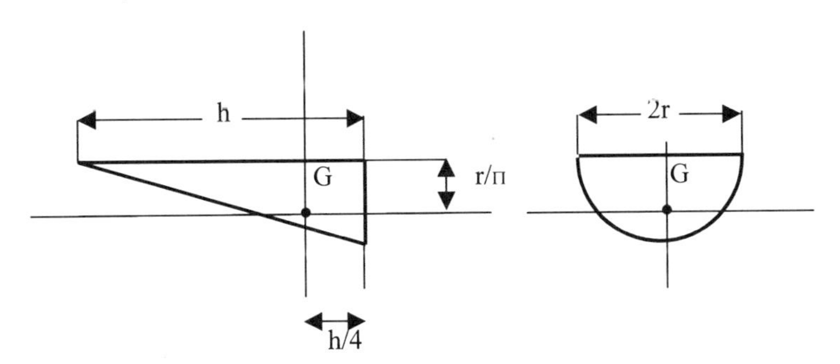

6. *Sphere*:

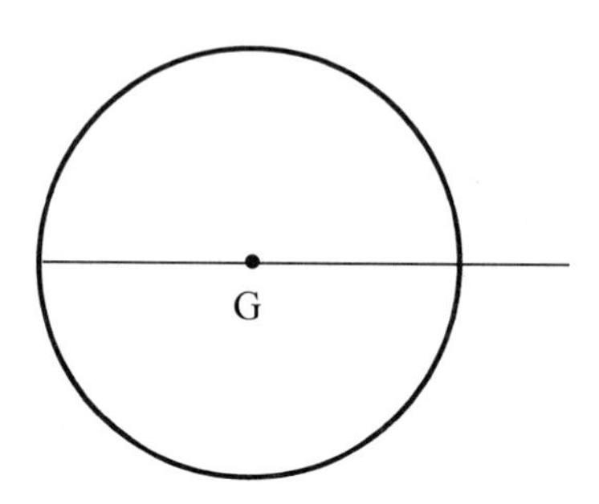

7. *Hemisphere*:

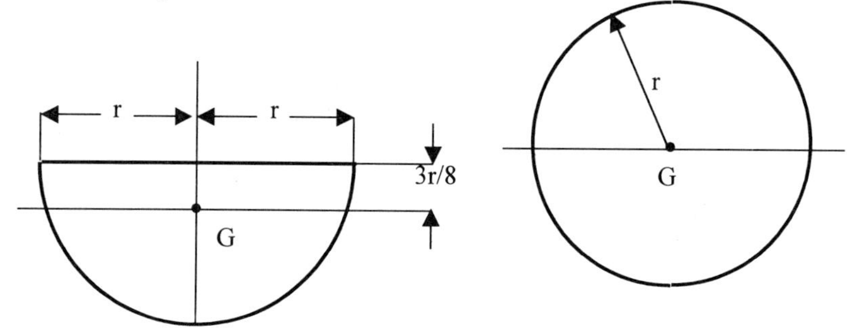

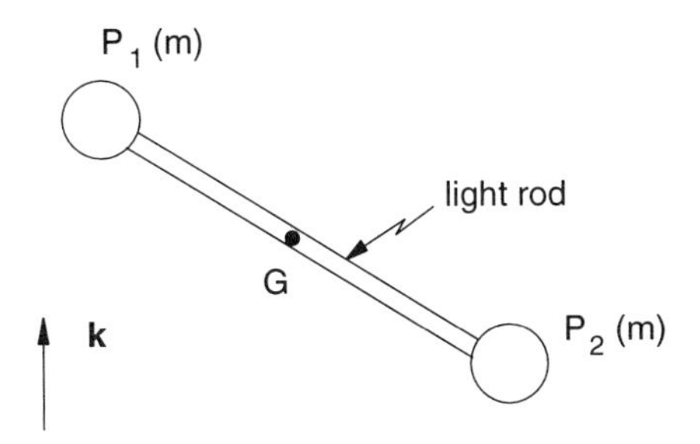

Fig. 8.3.3 Identical Particles at the Ends of a Light Rod

earth's center* than P_2. Thus the weight (gravitational attraction) of P_1 will be less than that of P_2. Therefore, to replace the gravitational forces by a single force (without a couple) would require the line of action of the force to pass through a point $\hat{G}$ which is slightly higher than the center point G (see Section 9.6.1).

For small bodies near the earth's surface this distinction is generally unimportant. For larger unsymmetrical bodies such as a spacecraft in orbit there may be a small gravitational moment if the weight force is placed through the mass center. More precisely, if the gravitational forces acting on a body are represented by an equivalent force system consisting of a single force passing through the mass center together with a couple, the torque of the couple will generally not be zero, but instead it will be a small gravitational moment (see Section 9.6.2).

8.4 Second Moment Vectors

Consider a particle P with mass m, a reference point O, and a unit vector $\mathbf{n}_a$ as in Figure 8.4.1. As before, let $\mathbf{p}$ locate P relative to O. Then the *second moment* $\mathbf{I}_a^{P/O}$ of P relative to O for the direction of $\mathbf{n}_a$ is defined as

*A sphere, such as the earth, has a gravitational attraction as though it were a single particle located at the center, with all the mass concentrated in the particle.

$$\mathbf{I}_a^{P/O} = m\mathbf{p} \times (\mathbf{n}_a \times \mathbf{p}) \tag{8.4.1}$$

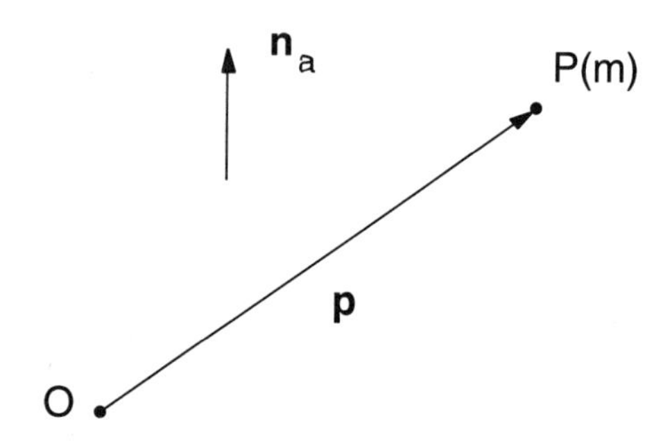

Fig. 8.4.1 A Point P, a Reference Point O, and a Unit Vector $\mathbf{n}_a$

Next, consider a set S of particles P_i $(i = 1, \ldots, N)$ with masses m_i, a reference point O, and a unit vector $\mathbf{n}_a$ as in Figure 8.4.2.

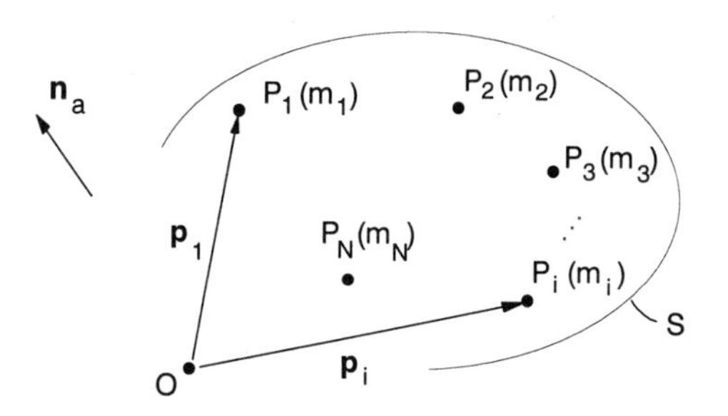

Fig. 8.4.2 A Set S of Particles, a Reference Point O, and a Unit Vector $\mathbf{n}_a$

The second moment $\mathbf{I}_a^{S/O}$ of S relative to O for the direction $\mathbf{n}_a$, is then defined as:

$$\mathbf{I}_a^{S/O} = \sum_{i=1}^{N} m_i \mathbf{p}_i \times (\mathbf{n}_a \times \mathbf{p}_i) \tag{8.4.2}$$

Finally, consider a body B, a reference point O, and a unit vector $\mathbf{n}_a$ as in Figure 8.4.3. As before, let B be modeled as a set of particles P_i $(i = 1,\ldots,N)$ connected and fixed relative to one another, with masses m_i. Let $\mathbf{p}_i$ locate P_i relative to O. Then the second moment $\mathbf{I}_a^{B/O}$ of B relative to O for the direction $\mathbf{n}_a$ is

$$\mathbf{I}_a^{B/O} = \sum_{i=1}^{N} m_i \mathbf{p}_i \times (\mathbf{n}_a \times \mathbf{p}_i) \tag{8.4.3}$$

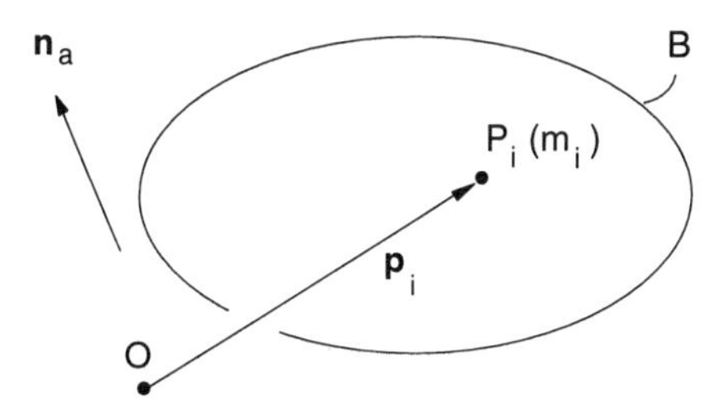

Fig. 8.4.3 A Body B, a Reference Point O, and a Unit Vector $\mathbf{n}_a$

If B can be regarded as a smooth body as in a continuum, then the summation of Equation (8.4.3) may be replaced by an integral as:

$$\mathbf{I}_a^{B/O} = \int_B \rho\, \mathbf{p} \times (\mathbf{n}_a \times \mathbf{p})\, dV \tag{8.4.4}$$

where ρ is the mass density of B at a typical point P, $\mathbf{p}$ locates P relative to O, and dV is a differential volume element of B containing P.

Several comments might be informative: First, the positioning of the parentheses in Equations (8.4.1) through (8.4.4) is not critical. That is, due to the properties of the triple vector product, we have

$$\mathbf{p}_i \times (\mathbf{n}_a \times \mathbf{p}_i) = (\mathbf{p}_i \times \mathbf{n}_a) \times \mathbf{p}_i \tag{8.4.5}$$

Next, in general $\mathbf{I}_a$ is not parallel to $\mathbf{n}_a$. If it happens, however, that $\mathbf{I}_a$ is parallel to $\mathbf{n}_a$ then $\mathbf{n}_a$ is said to be a "principal unit vector" or "eigen unit vector."

Finally, the second moment vector $\mathbf{I}_a$ is sometimes called an "inertia vector" (analogous to the stress vector of continuum mechanics). Second moment vectors are useful in defining and obtaining moments and products of inertia, as in the following section, and in obtaining inertia dyadics as in Section 8.8.

8.5 Moments and Products of Inertia

Consider a particle P with mass m, a reference point O, and unit vectors $\mathbf{n}_a$ and $\mathbf{n}_b$ as in Figure 8.5.1. Let $\mathbf{I}_a^{P/O}$ be the second moment vector (inertia vector) of P for O for the direction $\mathbf{n}_a$ (see previous section). Then the *product of inertia* $I_{ab}^{P/O}$ of P for O for the directions of $\mathbf{n}_a$ and $\mathbf{n}_b$ is defined as:

$$I_{ab}^{P/O} = \mathbf{I}_a^{P/O} \bullet \mathbf{n}_b \tag{8.5.1}$$

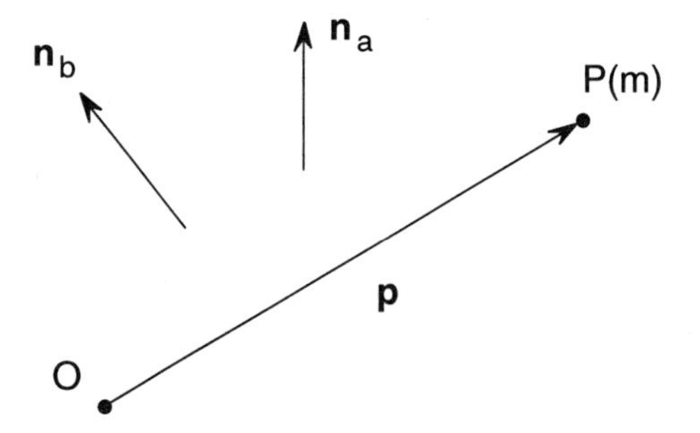

Fig. 8.5.1 A Particle P, a Reference Point O, and Unit Vectors $\mathbf{n}_a$ and $\mathbf{n}_b$

By substituting from Equation (8.4.1) we have

$$\begin{aligned} I_{ab}^{P/O} &= m\mathbf{p} \times (\mathbf{n}_a \times \mathbf{p}) \cdot \mathbf{n}_b \\ &= m(\mathbf{p} \times \mathbf{n}_a) \cdot (\mathbf{p} \times \mathbf{n}_b) \end{aligned} \quad (8.5.2)$$

Then by inspection, we have

$$I_{ab}^{P/O} = I_{ba}^{P/O} \quad (8.5.3)$$

In an analogous manner the *moment of inertia* of P for O for direction of $\mathbf{n}_a$ is defined as:

$$I_{aa}^{P/O} = \mathbf{I}_a^{P/O} \cdot \mathbf{n}_a \quad (8.5.4)$$

Then also from Equation (8.4.1) we have

$$\begin{aligned} I_{aa}^{P/O} &= m\mathbf{p} \times (\mathbf{n}_a \times \mathbf{p}) \cdot \mathbf{n}_a \\ &= m(\mathbf{p} \times \mathbf{n}_a) \cdot (\mathbf{p} \times \mathbf{n}_a) \\ &= m(\mathbf{p} \times \mathbf{n}_a)^2 \end{aligned} \quad (8.5.5)$$

Observe from Equations (8.5.2) and (8.5.5) that the moment of inertia is always positive or zero, whereas the product of inertia may be positive, negative, and/or zero. Observe further that the unit vectors $\mathbf{n}_a$ and $\mathbf{n}_b$ need not be perpendicular although generally products of inertia are referred to as perpendicular unit vectors (that is, perpendicular directions).

Example 8.5.1 A Particle in Cartesian Reference Frame

Let P be a particle with mass m located at a point P in a Cartesian reference frame R as in Figure 8.5.2. Let the coordinates of P be (x,y,z). Find the second moment vectors for the X, Y, and Z directions and the corresponding moments and products of inertia.

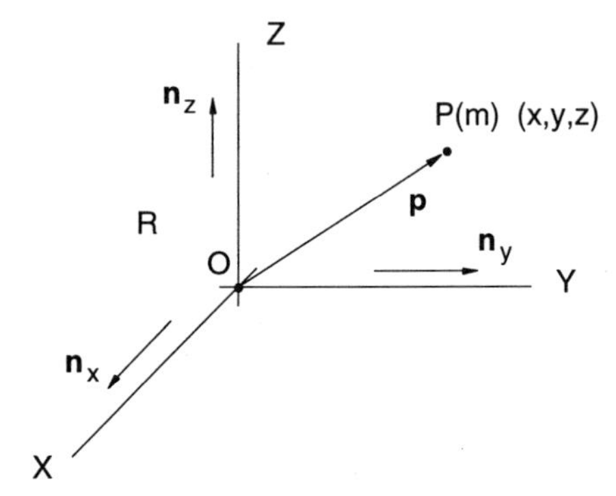

Fig. 8.5.2 A Particle P with Mass m in a Cartesian Reference Frame R

Solution: Let $\mathbf{n}_x$, $\mathbf{n}_y$, and $\mathbf{n}_z$ be unit vectors parallel to X, Y, and Z as shown. Then from Equation (8.4.1), the second moment vectors (inertia vectors) $\mathbf{I}_x^{P/O}$, $\mathbf{I}_y^{P/O}$, and $\mathbf{I}_z^{P/O}$ are

$$\begin{aligned}\mathbf{I}_x &= m\mathbf{p} \times (\mathbf{n}_x \times \mathbf{p}) \\ &= m(x\mathbf{n}_x + y\mathbf{n}_y + z\mathbf{n}_z) \times [\mathbf{n}_x \times (x\mathbf{n}_x + y\mathbf{n}_y + z\mathbf{n}_z)] \\ &= m(y^2 + z^2)\mathbf{n}_x - mxy\mathbf{n}_y - mxz\mathbf{n}_z \end{aligned} \quad (8.5.6)$$

Similarly,

$$\mathbf{I}_y = -myx\mathbf{n}_x + m(x^2 + y^2)\mathbf{n}_y - myz\mathbf{n}_z \quad (8.5.7)$$

and

$$\mathbf{I}_z = -mzx\mathbf{n}_x - mzy\mathbf{n}_y + m(y^2 + z^2)\mathbf{n}_z \quad (8.5.8)$$

where the superscript P/O has been deleted for simplicity in writing and reading.

Then from Equations (8.5.1) and (8.5.4) the moments and products of inertia are simply the components of the inertia vectors. That is,

$$\begin{aligned}
I_{xx} &= m(y^2+z^2) & I_{xy} &= -mxy & I_{xz} &= -mxz \\
I_{yx} &= -myx & I_{yy} &= m(z^2+x^2) & I_{yz} &= -myz \\
I_{zx} &= -mzx & I_{zy} &= -mzy & I_{zz} &= m(x^2+y^2)
\end{aligned} \tag{8.5.9}$$

Example 8.5.2 A Particle in a Cartesian Reference Frame — Use of Index Notation

See Example 8.5.1. Use index notation in the Example of 8.5.1. That is, express the second moment vectors and the moments and products of inertia of a particle in a reference frame in terms of coordinates (x_1, x_2, x_3) as in Figure 8.5.3.

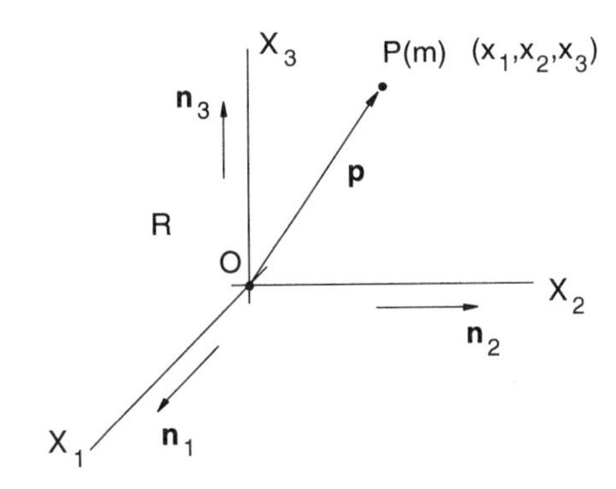

Fig. 8.5.3 A Particle P with Mass m in a Cartesian Reference Frame — Index Notation

In this notation, the inertia vectors and moments and products of inertia may be obtained by the substitutions: $x \rightarrow x_1$, $y \rightarrow x_2$, $z \rightarrow x_3$.

Hence, we have

$$\begin{aligned}
\mathbf{I}_1 &= m(x_2^2+x_3^2)\mathbf{n}_1 - mx_1x_2\mathbf{n}_2 - mx_1x_3\mathbf{n}_3 \\
\mathbf{I}_2 &= -mx_2x_1\mathbf{n}_1 + m(x_1^2+x_2^2)\mathbf{n}_2 - mx_2x_3\mathbf{n}_3 \\
\mathbf{I}_3 &= -mx_3x_1\mathbf{n}_1 - mx_3x_2\mathbf{n}_2 + m(x_2^2+x_3^2)\mathbf{n}_3
\end{aligned} \tag{8.5.10}$$

and

$$
\begin{array}{lll}
I_{11} = m(x_2^2 + x_3^2) & I_{12} = -mx_1x_2 & I_{13} = -mx_1x_3 \\
I_{21} = -mx_2x_1 & I_{22} = m(x_3^2 + x_1^2) & I_{23} = -mx_2x_3 \\
I_{31} = -mx_3x_1 & I_{32} = -mx_3x_2 & I_{33} = m(x_1^2 + x_2^2)
\end{array}
\tag{8.5.11}
$$

By inspection of the indices of Equations (8.5.10) and (8.5.11) we see that the moments and products of inertia may be represented by the single expression

$$I_{ij} = -mx_ix_j + m(x_1^2 + x_2^2 + x_3^2)\delta_{ij} \tag{8.5.12}$$

where as before δ_{ij} is Kronecker's delta function. Then the inertia vectors of Equation (8.5.10) may be expressed as:

$$\mathbf{I}_i = I_{ij}\mathbf{n}_j \tag{8.5.13}$$

Equation (8.5.12) gives rise to the "inertia matrix" I (not to be confused with the identity matrix) defined with elements I_{ij} as:

$$
I = m\begin{bmatrix}
(x_2^2+x_3^2) & -x_1x_2 & -x_1x_3 \\
-x_2x_1 & (x_1^2+x_3^2) & -x_2x_3 \\
-x_3x_1 & -x_3x_2 & (x_1^2+x_2^2)
\end{bmatrix}
\tag{8.5.14}
$$

Observe that I is symmetric and that the diagonal elements are non-negative.

Next, consider a set S of particles P_i $(i = 1,\ldots,N)$, a reference point O, and unit vectors $\mathbf{n}_a$ and $\mathbf{n}_b$ as in Figure 8.5.4. Then the moments and products of inertia of S for O for the directions of $\mathbf{n}_a$ and $\mathbf{n}_b$ are defined as the sums of the moments and products of inertia of the individual particles. That is,

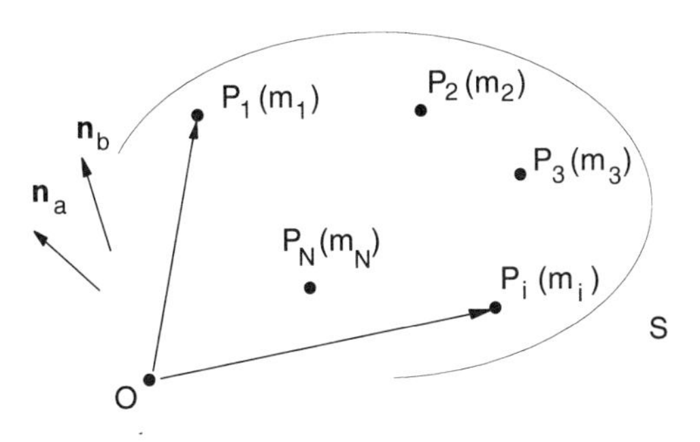

Fig. 8.5.4 A Set S of Particles, a Reference Point O and Unit Vectors $\mathbf{n}_a$ and $\mathbf{n}_b$

$$I_{aa}^{S/O} = \sum_{i=1}^{N} I_{aa}^{P_i/O} = \sum_{i=1}^{N} m_i(\mathbf{p}_i \times \mathbf{n}_a)^2 \tag{8.5.15}$$

and

$$\mathbf{I}_{ab}^{S/O} = \sum_{i=1}^{N} \mathbf{I}_{ab}^{P_i/O} = \sum_{i=1}^{N} m_i (\mathbf{p}_i \times \mathbf{n}_a) \bullet (\mathbf{p}_i \times \mathbf{n}_b) \tag{8.5.16}$$

Finally, consider a body B, a reference point O, and unit vectors $\mathbf{n}_a$ and $\mathbf{n}_b$ as in Figure 8.5.5. As before, let B be modeled as a set of particles P_i $(i = 1,\ldots,N)$ connected and fixed relative to one another, with masses m_i. Then the moments and products of inertia of B for O for the directions of $\mathbf{n}_a$ and $\mathbf{n}_b$ are given by Equations (8.5.15) and (8.5.16).

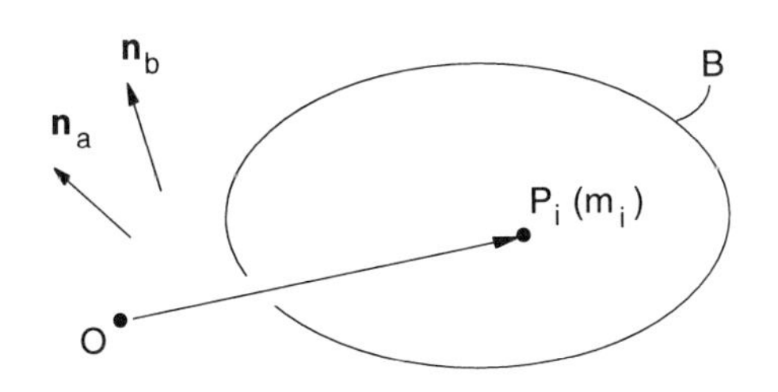

Fig. 8.5.5 A Body B, a Reference Point O, and Unit Vectors $\mathbf{n}_a$ and $\mathbf{n}_b$

If B can be regarded as a smooth body as in a continuum, then the summations of Equations (8.5.15) and (8.5.16) may be replaced by integrals as

$$I_{aa}^{B/O} = \int_B \rho\,(\mathbf{p} \times \mathbf{n}_a)^2 dV \tag{8.5.17}$$

and

$$I_{ab}^{B/O} = \int_B \rho\,(\mathbf{p} \times \mathbf{n}_a) \cdot (\mathbf{p} \times \mathbf{n}_b) dV \tag{8.5.18}$$

where ρ is the mass density of B at a typical point P, $\mathbf{p}$ locates P relative to O, and dV is a differential volume element of B containing P.

8.6 Geometric Interpretation of Moments and Products of Inertia, Axes of Inertia

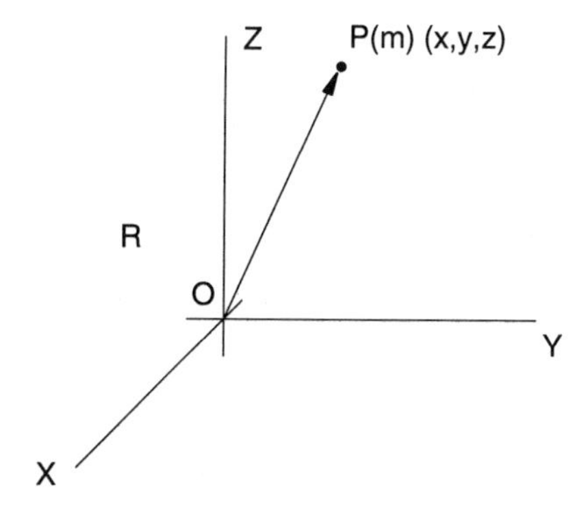

Fig. 8.6.1 A Particle P with Mass m in a Cartesian Reference Frame

Consider again a particle P with mass m located at a point P in a Cartesian reference frame R as in Figure 8.5.2 and shown again in Figure 8.6.1. From Equations (8.5.9) and (8.5.14) the moments and products of inertia may be represented in matrix form as:

$$I = m\begin{bmatrix} (y^2+z^2) & -xy & -xz \\ -yx & (x^2+z^2) & -yz \\ -zx & -zy & (x^2+y^2) \end{bmatrix} \tag{8.6.1}$$

The diagonal elements, containing the moments of inertia, may be regarded as the product of the mass and the square of the distance of P from the coordinate axes. That is,

$$I_{xx} = md_x^2 \qquad I_{yy} = md_y^2 \qquad I_{zz} = md_z^2 \tag{8.6.2}$$

where d_x, d_y, and d_z are the distances of P from the X, Y, and Z axes, respectively.

Observe that I_{xx}, being equal to $m(y^2+z^2)$, is independent of x. This means that the moment of inertia of P for O in the X direction is the same for all points on the X-axis. That is, if Q is any point on the X-axis other than O, we have

$$I_{xx}^{P/Q} = I_{xx}^{P/O} \tag{8.6.3}$$

Therefore, since the reference point could be any point on the axis, $I_{xx}^{P/O}$ may be called "the moment of inertia of P about the X-axis." Alternatively, the X-axis may be regarded as an "axis of inertia."

These observations also hold for a set of particles or a body. For example, consider the body B represented as a set of particles P_i $(i = 1,\ldots,N)$ as in Figure 8.6.2. Let O be a reference point and let L be a line passing through O. Let $\mathbf{n}_L$ be a unit vector parallel to L and let Q be any point on L other

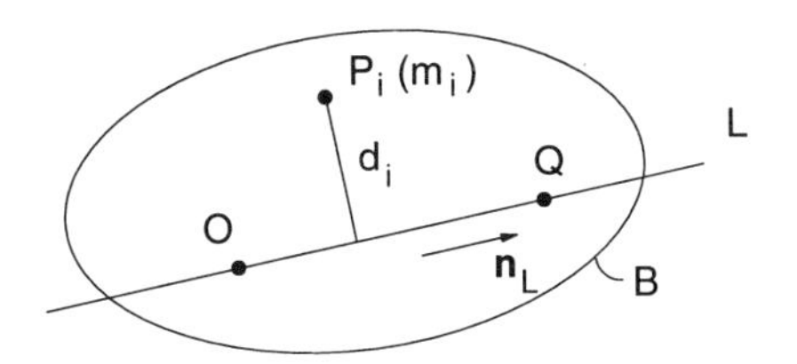

Fig. 8.6.2 A Body B, Line L and Reference Points O and Q

than O. Then the moment of inertia of B relative to O for the direction of $\mathbf{n}_L$ is equal to the moment of inertia of B relative to Q for the direction of $\mathbf{n}_L$. Specifically,

$$I_{LL}^{B/O} = I_{LL}^{B/Q} = \sum_{i=1}^{N} m_i d_i^2 \qquad (8.6.4)$$

where m_i is the mass of P_i and d_i is the distance of P_i from L. The terms $I_{XX}^{B/O}$, $I_{XX}^{B/Q}$, etc. are also referred to as the "moment of inertia of B about L."

The off-diagonal elements of the matrix of Equation (8.6.1), representing the products of inertia, may be regarded as the negative products of the particle mass and coordinate distances of P from the coordinate planes. For example, I_{xy} may be regarded as the negative of the particle mass multiplied by the distances from P to the Y-Z and the X-Z planes.

8.7 Radius of Gyration

Observe again that moment of inertia is inherently positive — or at least, non-negative. Therefore, on occasion it is convenient to normalize a moment of inertia by dividing by the mass — be it the mass of a particle, or the total mass of a set of particles, or of a rigid body. With mass also being positive, the normalized moment of inertia is positive and called the square of the "radius of gyration." Specifically if I is the moment of inertia and m is the mass, the radius of gyration k is defined as:

$$K^2 = I/m \qquad (8.7.1)$$

The radius of gyration is a purely geometric quantity and often used in tabular representations of moments of inertia.

8.8 Inertia Dyadic

Consider again the inertia matrix of Section 8.5, as represented in Equation (8.5.14). Although this matrix was developed for a single particle, a corresponding matrix would exist for a set of particles or a rigid body. Such matrices all have the form:

$$I = [I_{ij}] = \begin{bmatrix} I_{11} & I_{12} & I_{13} \\ I_{21} & I_{22} & I_{23} \\ I_{31} & I_{32} & I_{33} \end{bmatrix} \tag{8.8.1}$$

The matrix is symmetrical and the diagonal elements are all positive. The elements of the rows (and hence also of the columns) are components of the inertia vectors (that is, the second moment vectors). These relations may be conveniently summarized by introduction and use of the inertia dyadic. Specifically, the inertia dyadic $\boldsymbol{I}$ is defined as:

$$\boldsymbol{I} = I_{ij}\mathbf{n}_i\mathbf{n}_j \tag{8.8.2}$$

where the $\mathbf{n}_i$ ($i = 1,2,3$) are mutually perpendicular unit vectors, and as before, there is a summation over repeated indices. Then for any direction represented by a unit vector $\mathbf{n}_k$, the second moment vector $\mathbf{I}_k$ is

$$\mathbf{I}_k = \boldsymbol{I} \cdot \mathbf{n}_k \tag{8.8.3}$$

Equation (8.8.3) is valid for a particle, a set of particles, or a rigid body, all relative to a reference point O.

In view of Equations (8.8.2) and (8.8.3) we can express the inertia dyadic in the forms:

$$\boldsymbol{I} = \mathbf{I}_k\mathbf{n}_k = \mathbf{n}_k\mathbf{I}_k \tag{8.8.4}$$

Next, suppose that $\mathbf{n}_a$ and $\mathbf{n}_b$ are any two unit vectors with representations relative to the $\mathbf{n}_i$ as:

$$\mathbf{n}_a = a_i \mathbf{n}_i \quad \text{and} \quad \mathbf{n}_b = b_i \mathbf{n}_i \tag{8.8.5}$$

Then from Equations (8.5.1), (8.5.4), and (8.8.3) we have

$$I_{ab} = a_i b_j I_{ij} \tag{8.8.6}$$

Finally, suppose $\hat{\mathbf{n}}_i$ $(i = 1,2,3)$ form a set of mutually perpendicular unit vectors with different inclinations than the $\mathbf{n}_i$. As before, let the $\mathbf{n}_i$ and the $\hat{\mathbf{n}}_i$ be related by the expressions:

$$\mathbf{n}_i = S_{ij} \hat{\mathbf{n}}_j \quad \text{and} \quad \hat{\mathbf{n}}_i = S_{ji} \mathbf{n}_j \tag{8.8.7}$$

where the S_{ij} are elements of the orthogonal transformation matrix defined as:

$$S_{ij} = \mathbf{n}_i \bullet \hat{\mathbf{n}}_j \tag{8.8.8}$$

Then the inertia dyadic $\boldsymbol{I}$ may be expressed in terms of either the $\mathbf{n}_i$ or the $\hat{\mathbf{n}}_i$ as

$$\boldsymbol{I} = I_{ij} \mathbf{n}_i \mathbf{n}_j = \hat{I}_{k\ell} \hat{\mathbf{n}}_k \hat{\mathbf{n}}_\ell \tag{8.8.9}$$

The elements of the inertia matrices are then related by the expressions:

$$\hat{I}_{k\ell} = S_{ik} S_{j\ell} I_{ij} \quad \text{and} \quad I_{ij} = S_{ik} S_{j\ell} \hat{I}_{k\ell} \tag{8.8.10}$$

8.9 Parallel Axis Theorem

The inertia quantities (the inertia vector, the moments and products of inertia, the inertia dyadic, and the inertia matrix) are all defined relative to an arbitrary reference point O. The parallel axes theorem provides a means for conveniently obtaining the inertia quantities relative to different reference points.

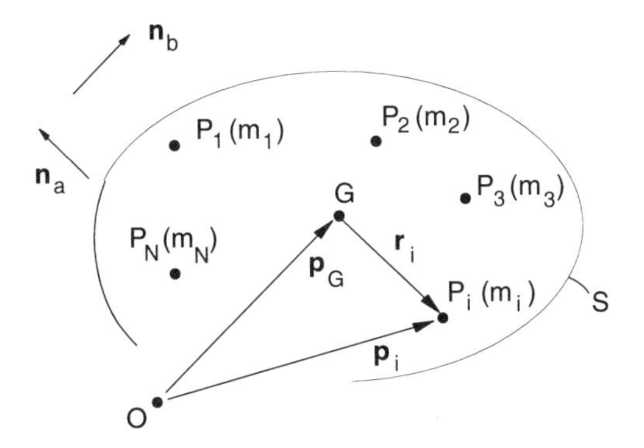

Fig. 8.9.1 A Set of Particles with Mass Center G and Reference Point O

To develop this, consider a set S of particles P_i $(i = 1, ..., N)$ having masses m_i as in Figure 8.9.1. Let G be the mass center of S, let O be an arbitrary reference point, and let $\mathbf{n}_a$ and $\mathbf{n}_b$ be arbitrarily directed unit vectors. Then from Equation (8.4.2) the second moment of S for O $\mathbf{I}_a^{S/O}$ for the direction of $\mathbf{n}_a$ is

$$\mathbf{I}_a^{S/O} = \sum_{i=1}^{N} m_i \mathbf{p}_i \times (\mathbf{n}_a \times \mathbf{p}_i) \tag{8.9.1}$$

From Figure 8.9.1 $\mathbf{p}_i$ is seen to be

$$\mathbf{p}_i = \mathbf{p}_G + \mathbf{r}_i \tag{8.9.2}$$

By substituting into Equation (8.9.1) $\mathbf{I}_a^{S/O}$ is seen to be:

$$\mathbf{I}_a^{S/O} = \sum_{i=1}^{N} m_i (\mathbf{p}_G + \mathbf{r}_i) \times [\mathbf{n}_a \times (\mathbf{p}_G \times \mathbf{r}_i)]$$

$$= \sum_{i=1}^{N} m_i \mathbf{p}_G \times (\mathbf{n}_a \times \mathbf{p}_G) + \sum_{i=1}^{N} m_i \mathbf{p}_G \times (\mathbf{n}_a \times \mathbf{r}_i)$$

$$+ \sum_{i=1}^{N} m_i \mathbf{r}_i \times (\mathbf{n}_a \times \mathbf{p}_G) + \sum_{i=1}^{N} m_i \mathbf{r}_i \times (\mathbf{n}_a \times \mathbf{r}_i)$$

$$= \overbrace{\left(\sum_{i=1}^{N} m_i\right)}^{M} \mathbf{p}_G \times (\mathbf{n}_a \times \mathbf{p}_G) + \mathbf{p}_G \times \left[\mathbf{n}_a \times \overbrace{\left(\sum_{i=1}^{N} m_i \mathbf{r}_i\right)}^{O}\right]$$

$$+ \overbrace{\left(\sum_{i=1}^{N} m_i \mathbf{r}_i\right)}^{O} \times (\mathbf{n}_a \times \mathbf{p}_G) + \sum_{i=1}^{N} m_i \mathbf{r}_i \times (\mathbf{n}_a \times \mathbf{r}_i)$$

or

$$\mathbf{I}_a^{S/O} = \mathbf{I}_a^{G/O} + \mathbf{I}_a^{S/G} \tag{8.9.3}$$

where $\mathbf{I}_a^{G/O}$ is defined as:

$$\mathbf{I}_a^{G/O} = M \mathbf{p}_G \times (\mathbf{n}_a \times \mathbf{p}_G) \tag{8.9.4}$$

where M is the mass center of S.

Equation (8.9.3) and various forms derived from it are called the "parallel axis theorem." The other forms are readily obtained from Equation (8.9.3) by taking the scalar products with $\mathbf{n}_a$ and $\mathbf{n}_b$, and by forming a dyadic with inertia vectors obtained from Equation (8.9.3). These computations lead to the expressions

$$I_{aa}^{S/O} = I_{aa}^{S/G} + I_{aa}^{G/O} \tag{8.9.5}$$

$$I_{ab}^{S/O} = I_{ab}^{S/G} + I_{ab}^{G/O} \tag{8.9.6}$$

$$\mathbf{I}^{S/O} = \mathbf{I}^{S/G} + \mathbf{I}^{G/O} \tag{8.9.7}$$

Equations (8.9.5), (8.9.6), and (8.9.7) may also be used for a body by regarding the body as a set of particles.

Equation (8.9.5) is the best known of the parallel axis theorems. In view of Equations (8.9.4) and (8.9.5) $I_{aa}^{G/O}$ may be written as

$$I_{aa}^{G/O} = \mathbf{I}_a^{G/O} \bullet \mathbf{n}_a = M(\mathbf{p}_G \times \mathbf{n}_a)^2 = Md^2 \tag{8.9.8}$$

where d is the distance between lines through O and G and parallel to $\mathbf{n}_a$ as in Figure 8.9.2.

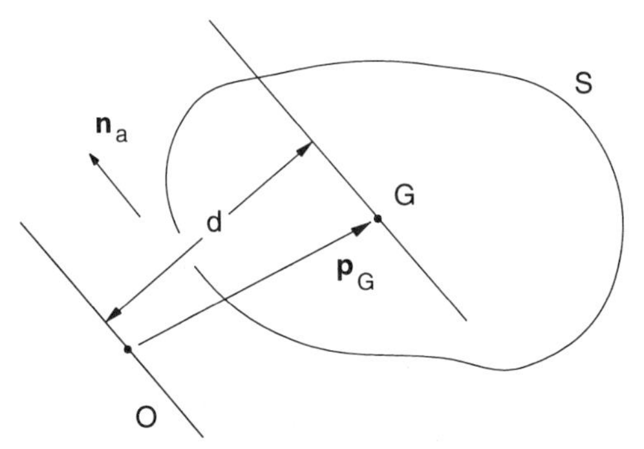

Fig. 8.9.2 Parallel Lines (Axes) Through Reference Point O and Mass Center G

Equations (8.9.5), (8.9.6), and (8.9.7) may readily be used to change the reference point for an inertia quantity as illustrated in the following example.

Example 8.9.1 Change in Reference Point for the Inertia Dyadic

Consider a body B and Reference Points O and Q, as in Figure 8.9.3. Suppose $\mathbf{I}^{B/O}$ is given and $\mathbf{I}^{B/G}$ is to be determined.

Solution: From Equation (8.9.7) we have:

$$\mathbf{I}^{B/O} = \mathbf{I}^{B/G} + \mathbf{I}^{G/O} \quad \text{and} \quad \mathbf{I}^{B/Q} = \mathbf{I}^{B/G} + \mathbf{I}^{G/Q} \tag{8.9.9}$$

By eliminating $\mathbf{I}^{B/G}$ between these expressions and by solving for $\mathbf{I}^{B/Q}$ we have

$$\mathbf{I}^{B/Q} = \mathbf{I}^{B/O} + \mathbf{I}^{G/O} - \mathbf{I}^{G/Q} \tag{8.9.10}$$

Observe that the "correction" terms $\mathbf{I}^{G/O}$ and $\mathbf{I}^{G/Q}$ are readily obtained from Equation (8.9.3).

8.10 Principal Direction, Principal Axes, and Principal Moments of Inertia

Recall again the definition of the second moment vector for a set of particles as given by Equation (8.4.2):

$$\mathbf{I}_a^{S/O} = \sum_{i=1}^{N} m_i \mathbf{p}_i \times (\mathbf{n}_a \times \mathbf{p}_i) \tag{8.10.1}$$

The direction of $\mathbf{I}_a^{S/O}$ will, of course, depend upon the positions of the particles, their masses, the direction of unit vector $\mathbf{n}_a$, and also the location of the reference point O. In general, $\mathbf{I}_a^{S/O}$ will not be parallel to $\mathbf{n}_a$. If it should happen that $\mathbf{I}_a^{S/O}$ is parallel to $\mathbf{n}_a$, then $\mathbf{n}_a$ is said to designate a "principal direction" of S for O, and $\mathbf{n}_a$ is called "principal unit vector" or "unit eigenvector." A line L through O and parallel to $\mathbf{n}_a$ is called a "principal axis of inertia."

Observe that if $\mathbf{n}_a$ is a unit eigenvector and if $\mathbf{n}_b$ is perpendicular to $\mathbf{n}_a$ then the product of inertia of S for O for the direction of $\mathbf{n}_a$ and $\mathbf{n}_b$ is zero. Observe also that the moment of inertia of S for O for $\mathbf{n}_a$ is equal to the magnitude of the second moment vector $\mathbf{I}_a^{S/O}$. That is

$$I_{aa}^{S/O} = 0 \quad \text{and} \quad I_{aa}^{S/O} = |\mathbf{I}_a^{S/O}| \tag{8.10.2}$$

If $\mathbf{n}_a$ is a unit eigenvector or principal unit vector, then $I_{aa}^{S/O}$ is called an "eigenvalue of inertia" or "principal value of inertia."

Observe from Equation (8.10.2) that if $\mathbf{n}_a$ is a unit eigenvector then the inertia vector $\mathbf{I}_a^{S/O}$ being parallel to $\mathbf{n}_a$ has the form:

$$\mathbf{I}_a^{S/O} = \lambda \mathbf{n}_a \tag{8.10.3}$$

where λ, a scalar, is the principal value, or eigenvalue, of inertia. In like manner, recall from Equation (8.8.3) that the inertia vector for $\mathbf{n}_a$ is the projection of the inertia dyadic along $\mathbf{n}_a$. That is,

$$\mathbf{I}_a^{S/O} = \boldsymbol{I}^{S/O} \bullet \mathbf{n}_a \tag{8.10.4}$$

Thus Equation (8.10.3) and (8.10.4) lead to

$$\boldsymbol{I}^{S/O} \bullet \mathbf{n}_a = \lambda \mathbf{n}_a \tag{8.10.5}$$

In view of the computational efficiency obtained with unit eigenvectors, as seen in Equations (8.10.2) (the products of inertia are zero), the central issue becomes: How do we find the unit eigenvectors, $\mathbf{n}_a$? To answer this question, let $\mathbf{n}_1$, $\mathbf{n}_2$, and $\mathbf{n}_3$ be any convenient set of mutually perpendicular unit vectors and let $\mathbf{I}^{S/O}$, $\mathbf{I}_a^{S/O}$, $I_{aa}^{S/O}$, and $\mathbf{n}_a$ be expressed as

$$\begin{aligned} \mathbf{I}^{S/O} &= I_{ij}\mathbf{n}_i\mathbf{n}_j \qquad & \mathbf{I}_a^{S/O} &= I_{aj}\mathbf{n}_j \\ I_{aa}^{S/O} &= \lambda \qquad & \mathbf{n}_a &= a_i\mathbf{n}_i \end{aligned} \tag{8.10.6}$$

Then Equation (8.10.5) becomes

$$(I_{ij}\mathbf{n}_i\mathbf{n}_j) \bullet \mathbf{n}_a = \lambda a_i \mathbf{n}_i \tag{8.10.7}$$

or

$$I_{ij}a_j\mathbf{n}_i = \lambda a_i \mathbf{n}_i \tag{8.10.8}$$

or

$$I_{ij}a_j = \lambda a_i \quad (i = 1,2,3) \tag{8.10.9}$$

Equations (8.10.9) form the basis for finding $\mathbf{n}_a$ specifically, $\mathbf{n}_a$ is known once the components a_i $(i = 1,2,3)$ of $\mathbf{n}_a$ relative to the $\mathbf{n}_i$ are known. Equations (8.10.9) form three equations for the three a_i. However, the equations while being linear and algebraic are homogeneous. Thus no solution exists unless the determinant of the coefficients is zero, in which case Equations (8.10.9) are dependent. Thus, a unique solution will be found only with an additional equation. Such an equation is readily obtained by recalling that since $\mathbf{n}_a$ is a unit vector, we have:

$$a_i a_i = 1 \tag{8.10.10}$$

The procedure for determining the a_i and for finding the principal moments of inertia is outlined in the following paragraphs.

Let Equation (8.10.9) be written explicitly as:

$$\begin{aligned} (I_{11} - \lambda)a_1 + I_{12}a_2 + I_{13}a_3 &= 0 \\ I_{21}a_1 + (I_{22} - \lambda)a_2 + I_{23}a_3 &= 0 \\ I_{31}a_1 + I_{32}a_2 + (I_{33} - \lambda)a_3 &= 0 \end{aligned} \tag{8.10.11}$$

Then since these equations are linear and homogeneous a non-trivial solution for the a_i will exist only if the determinant of the coefficients is zero. That is,

$$\begin{vmatrix} (I_{11}-\lambda) & I_{12} & I_{13} \\ I_{21} & (I_{22}-\lambda) & I_{23} \\ I_{31} & I_{32} & (I_{33}-\lambda) \end{vmatrix} = 0 \tag{8.10.12}$$

Expanding this determinant leads to the cubic equation (Hamilton-Cayley equation):

$$\lambda^3 - I_I\lambda^2 + I_{II}\lambda - I_{III} = 0 \tag{8.10.13}$$

where the coefficients I_I, I_{II}, and I_{III} are

$$I_I = I_{11} + I_{22} + I_{33} \tag{8.10.14}$$

$$I_{II} = I_{22}I_{33} - I_{32}I_{23} + I_{33}I_{11} - I_{13}I_{31} + I_{11}I_{22} - I_{21}I_{12} \tag{8.10.15}$$

$$I_{III} = I_{11}I_{22}I_{33} - I_{11}I_{32}I_{23} + I_{12}I_{31}I_{23} - I_{12}I_{21}I_{33} + I_{21}I_{32}I_{13} - I_{31}I_{13}I_{22} \tag{8.10.16}$$

These coefficients are called the first, second, and third scalar invariants of **I** and they may be described in terms of the inertia matrix $[I_{ij}]$ as:

I_I — The sum of the diagonal elements of $[I_{ij}]$
I_{II} — The sum of the diagonal elements of the cofactor matrix of $[I_{ij}]$
I_{III} — The determinant of $[I_{ij}]$

Equation (8.10.13) is a cubic equation for λ. There are thus in general three roots. Indeed, since $[I_{ij}]$ may be recognized as being symmetric and positive-definite, the roots of Equations (8.10.13) are seen to be real and positive [8.7]. From Equations (8.10.5) and (8.10.6) these roots are principal moments of inertia and associated with each, there is a corresponding principal unit vector (or unit eigenvector) and a principal direction. Due to the symmetry of $[I_{ij}]$ these unit eigenvectors are seen to be mutually perpendicular [8.7].

Let the three roots, λ_1, λ_2, and λ_3, of Equation (8.10.13) be I_a, I_b, and I_c, and let the corresponding principal unit vectors be $\mathbf{n}_a$, $\mathbf{n}_b$, and $\mathbf{n}_c$. Then from Equations (8.8.4) and (8.10.5) the inertia dyadic $I^{S/O}$ may be expressed as:

$$I^{S/O} = I_a \mathbf{n}_a \mathbf{n}_a + I_b \mathbf{n}_b \mathbf{n}_b + I_c \mathbf{n}_c \mathbf{n}_c \tag{8.10.17}$$

The procedure for obtaining the principal moments of inertia and the associated unit eigenvectors is as follows: First, given the matrix $[I_{ij}]$ use Equations (8.10.14), (8.10.15), and (8.10.16) to determine the coefficients of Equation (8.10.13). Next, solve Equations (8.10.13) for the three λ. Finally, use Equation (8.10.10) and (8.10.11) to obtain the components of the unit eigenvectors. In this last step observe that the three equations of Equations (8.10.11) are dependent and thus any two of them may be selected and combined with Equation (8.10.10) to determine the a_i $(i = 1,2,3)$. (This procedure is repeated for each of the three values of λ.)

The foregoing is illustrated in the following numerical example:

Example 8.10.1 Determination of Principal Moments of Inertia and Unit Eigenvectors

Consider a body B and a reference point O as in Figure 8.10.1, where $\mathbf{n}_1$, $\mathbf{n}_2$, and $\mathbf{n}_3$ are mutually perpendicular. Let the matrix $[I_{ij}]$ of the inertia dyadic of B for

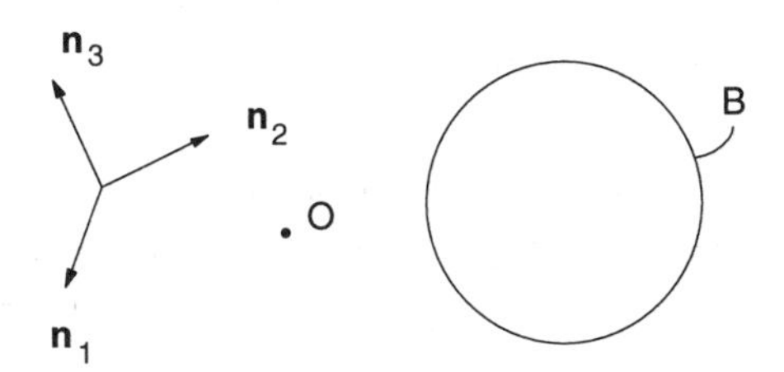

Fig. 8.10.1 A Body B, a Reference Point O, and Mutually Perpendicular Unit Vectors

O relative to the $\mathbf{n}_i$ be:

$$[I_{ij}] = \frac{1}{8}\begin{bmatrix} 30 & -\sqrt{6} & -\sqrt{6} \\ -\sqrt{6} & 41 & -15 \\ -\sqrt{6} & -15 & 41 \end{bmatrix} \text{ slug ft}^2 \tag{8.10.18}$$

Find the principal values of $\boldsymbol{I}^{B/O}$ and the corresponding principal directions.

Solution: The determinantal equation [Equation (8.10.12)] for the principal values becomes

$$\begin{vmatrix} (30-\lambda) & -\sqrt{6} & -\sqrt{6} \\ -\sqrt{6} & (41-\lambda) & -15 \\ -\sqrt{6} & -15 & (41-\lambda) \end{vmatrix} = 0 \tag{8.10.19}$$

By expansion of the determinant, or equivalently, by using Equations (8.10.4), (8.10.15), and (8.10.16), the Hamilton-Cayley equation [Equation (8.10.13)] for λ becomes:

$$\lambda^3 - 112\lambda^2 + 3904\lambda - 43008 = 0 \tag{8.10.20}$$

Solving for λ we obtain

$$\lambda = \lambda_1, \lambda_2, \lambda_3 = 3,4,7 \text{ slug ft}^2 \tag{8.10.21}$$

As expected, we obtained three principal inertia values. To keep them distinct, let them be renamed I_a, I_b, and I_c. That is,

$$\lambda_1 = I_a = 3 \qquad \lambda_2 = I_b = 4 \qquad \lambda_3 = I_c = 7 \text{ slug ft}^2 \tag{8.10.22}$$

To obtain unit vectors parallel to the principal directions, we can use Equations (8.10.10) and (8.10.11) to determine the components. That is, for $\lambda = I_a = 3$, we have

$$a_1^2 + a_2^2 + a_3^2 = 1 \tag{8.10.23}$$

and

$$\begin{aligned} [(30/8) - 3]a_1 - (\sqrt{6}/8)a_2 - (\sqrt{6}/8)a_3 &= 0 \\ -(\sqrt{6}/8)a_1 + [(41/8) - 3]a_2 - (15/8)a_3 &= 0 \\ -(\sqrt{6}/8)a_1 - (15/8)a_2 + [(41/8) - 3]a_3 &= 0 \end{aligned} \tag{8.10.24}$$

Equations (8.10.24) are seen to be dependent. By using two of them, say the first two, together with Equation (8.10.23) and solving for a_1, a_2, and a_3 we obtain

$$a_1 = 1/2 \qquad a_2 = -\sqrt{3}/2 \qquad a_3 = 0 \tag{8.10.25}$$

Hence, the corresponding unit eigenvector $\mathbf{n}_a$ is

$$\mathbf{n}_a = (1/2)\mathbf{n}_1 - (\sqrt{3}/2)\mathbf{n}_2 \tag{8.10.26}$$

Similarly, for $\lambda = I_b = 4$, Equation (8.10.10) and (8.10.11) become:

$$b_1^2 + b_2^2 + b_3^2 = 1 \tag{8.10.27}$$

and

$$\begin{aligned} [(30/8) - 4]b_1 - (\sqrt{6}/8)b_2 - (\sqrt{6}/8)b_3 &= 0 \\ -(\sqrt{6}/8)b_1 + [(41/8) - 4]b_2 - (15/8)b_3 &= 0 \\ -(\sqrt{6}/8)b_1 - (15/8)b_2 + [(41/8) - 4]b_3 &= 0 \end{aligned} \tag{8.10.28}$$

Solving for b_1, b_2, and b_3, we obtain

$$b_1 = \sqrt{6}/4 \qquad b_2 = \sqrt{2}/4 \qquad b_3 = -\sqrt{2}/2 \tag{8.10.29}$$

Therefore, the second unit eigenvector is

$$\mathbf{n}_b = (\sqrt{6}/4)\mathbf{n}_1 + (\sqrt{2}/4)\mathbf{n}_2 - (\sqrt{2}/2)\mathbf{n}_3 \tag{8.10.30}$$

Finally, for $\lambda = I_c = 7$, Equations (8.10.10) and (8.10.11) become:

$$c_1^2 + c_2^2 + c_3^2 = 1 \tag{8.10.31}$$

and

$$\begin{aligned} &[(30/8) - 7]c_1 - (\sqrt{6}/8)c_2 - (\sqrt{6}/8)c_3 = 0 \\ &-(\sqrt{6}/8)c_1 + [(41/8) - 7]c_2 - (15/8)c_3 = 0 \\ &-(\sqrt{6}/8)c_1 - (15/8)c_2 + [(41/8) - 7]c_3 = 0 \end{aligned} \tag{8.10.32}$$

Solving for c_1, c_2, and c_3, we obtain

$$c_1 = \sqrt{6}/4 \qquad c_2 = \sqrt{2}/4 \qquad c_3 = \sqrt{2}/2 \tag{8.10.33}$$

Therefore, the third unit eigenvector is:

$$\mathbf{n}_c = (\sqrt{6}/4)\mathbf{n}_1 + (\sqrt{2}/4)\mathbf{n}_2 + (\sqrt{2}/2)\mathbf{n}_3 \tag{8.10.34}$$

Observe from Equation (8.10.26), (8.10.30), and (8.10.34) that the unit eigenvectors $\mathbf{n}_a$, $\mathbf{n}_b$ and $\mathbf{n}_c$ are mutually perpendicular. Observe further that the transformation matrix S between the $\mathbf{n}_i$ and $\mathbf{n}_a$, $\mathbf{n}_b$ and $\mathbf{n}_c$ is

$$S = \begin{bmatrix} 1/2 & -\sqrt{3}/2 & 0 \\ \sqrt{6}/4 & \sqrt{2}/4 & -\sqrt{2}/2 \\ \sqrt{6}/4 & \sqrt{2}/4 & \sqrt{2}/2 \end{bmatrix} \tag{8.10.35}$$

Hence, the transpose of S is

$$S^T = \begin{bmatrix} 1/2 & \sqrt{6}/4 & \sqrt{6}/4 \\ -\sqrt{3}/2 & \sqrt{2}/4 & \sqrt{2}/4 \\ 0 & -\sqrt{2}/2 & \sqrt{2}/2 \end{bmatrix} \tag{8.10.36}$$

Then by pre-multiplying the inertia matrix of Equation (8.10.17) by S^T and post-multiplying the result by S we have

$$\begin{bmatrix} 1/2 & \sqrt{6}/4 & \sqrt{6}/4 \\ -\sqrt{3}/2 & \sqrt{2}/4 & \sqrt{2}/4 \\ 0 & -\sqrt{2}/2 & \sqrt{2}/2 \end{bmatrix} \begin{bmatrix} 15/4 & -\sqrt{6}/8 & -\sqrt{6}/8 \\ -\sqrt{6}/8 & 41/8 & -15/8 \\ -\sqrt{6}/8 & -15/8 & 41/8 \end{bmatrix} \begin{bmatrix} 1/2 & -\sqrt{3}/2 & 0 \\ \sqrt{6}/4 & \sqrt{2}/4 & -\sqrt{2}/2 \\ \sqrt{6}/4 & \sqrt{2}/4 & \sqrt{2}/2 \end{bmatrix} = \begin{bmatrix} 3 & 0 & 0 \\ 0 & 4 & 0 \\ 0 & 0 & 7 \end{bmatrix} \tag{8.10.37}$$

Therefore, in terms of $\mathbf{n}_a$, $\mathbf{n}_b$, and $\mathbf{n}_c$, the inertia dyadic of B for O becomes

$$\boldsymbol{I}^{B/O} = 3\mathbf{n}_a\mathbf{n}_a + 4\mathbf{n}_b\mathbf{n}_b + 7\mathbf{n}_c\mathbf{n}_c \text{ slug ft}^2$$

8.11 Discussion: Principal Directions, Principal Axes, and Principal Moments of Inertia — Additional Formulas and Interpretations

References 8.1 to 8.7 and 8.9 provide detailed and extensive discussion of inertia including the concepts and procedures for principal moments of inertia. These references develop and state results concerning principal moments of inertia. In this section we summarize some of these results.

8.11.1 Maximum and Minimum Moments of Inertia

A particularly useful result is that of all the directions in space, the principal directions are the ones having the maximum and minimum moments of inertia. Specifically, if an inertia dyadic has three distinct principal moments of inertia, for a given reference point, one of these moments of inertia will have a value greater than all other moments of inertia, regardless of the associated unit vector direction. Also, one of the principal moments of inertia will have a value smaller than all other moments of inertia. The third principal moment of inertia will then have a value intermediate to these maximum and minimum values.

The proof of this result is readily obtained using Lagrange multipliers as in Reference 8.7.

8.11.2 Inertia Ellipsoid

The foregoing results for maximum and minimum moments of inertia can be visualized using the concept of an "inertia ellipsoid." To develop this, consider a body B, a referent point O, the inertia dyadic $\boldsymbol{I}^{B/O}$, and an arbitrarily directed unit vector $\mathbf{n}_a$ as in Figure 8.11.1. Then from Equations (8.5.4) and (8.8.3), the moment of inertia of B for O for the direction $\mathbf{n}_a$ is

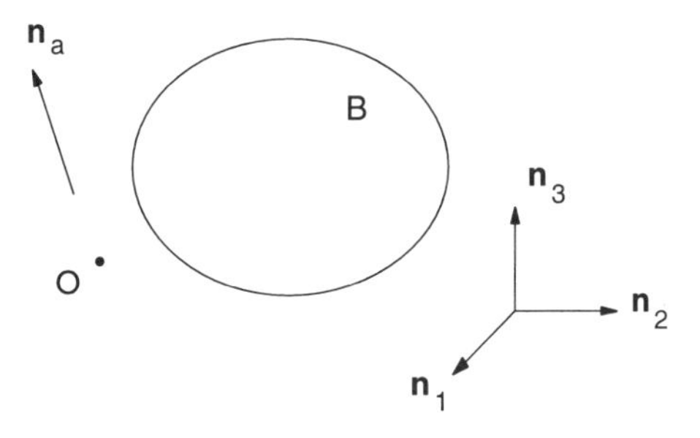

Fig. 8.11.1 A Body B, Reference Point O, Unit Vector $\mathbf{n}_a$ and Mutually Perpendicular Principal Unit Vectors $\mathbf{n}_1$, $\mathbf{n}_2$, and $\mathbf{n}_3$

$$I_{aa}^{B/O} = \mathbf{n}_a \cdot \boldsymbol{I}^{B/O} \cdot \mathbf{n}_a \tag{8.11.1}$$

That is, the moment of inertia I_{aa} may be viewed as the double projection of the inertia dyadic $\boldsymbol{I}^{B/O}$ onto the direction of $\mathbf{n}_a$.

Let $\mathbf{n}_1$, $\mathbf{n}_2$, and $\mathbf{n}_3$ be principal unit vectors of B for O with corresponding principal moments of inertia I_{11}, I_{22}, and I_{33}. Then from Equations (8.10.17) the moment of inertia I_{aa} may be expressed as:

$$I_{aa} = a_1^2 I_{11} + a_2^2 I_{22} + a_3^2 I_{33} \tag{8.11.2}$$

where the a_i $(i = 1,2,3)$ are the $\mathbf{n}_i$ components of $\mathbf{n}_a$, and where the superscripts B/O have been deleted for simplicity in writing. By dividing by I_{aa} Equation (8.11.2) may be written in the form:

$$1 = \frac{a_1^2}{(I_{aa}/I_{11})} + \frac{a_2^2}{(I_{aa}/I_{22})} + \frac{a_3^2}{(I_{aa}/I_{33})} \tag{8.11.3}$$

or as

$$1 = a_1^2/a^2 + a_2^2/b^2 + a_3^2/c^2 \tag{8.11.4}$$

where a^2, b^2, and c^2 are defined by inspection as the moment of inertia ratios (I_{aa}/I_{11}), (I_{aa}/I_{22}), (I_{aa}/I_{22}), and (I_{aa}/I_{33}).

If we consider a_1, a_2, and a_3 as Cartesian coordinate variables, Equation (8.11.4) defines an ellipsoid in the a_1, a_2, and a_3 space, with semi-major axes a, b, and c. This ellipsoid is commonly called the "inertia ellipsoid."

Observe that since $\mathbf{n}_a$ is a unit vector the sum of the squares of the a_i is 1.0. That is,

$$a_1^2 + a_2^2 + a_3^2 = 1 \tag{8.11.5}$$

Hence, if for example, a_1 is 1.0, then a_2 and a_3 are zero; and then $\mathbf{n}_a$ is parallel to $\mathbf{n}_1$; and then, in turn, from Equation (8.11.3) I_{aa} is I_{11}. Similarly, if a_2 is 1.0, I_{aa} is I_{22}; and if a_3 is 1.0, I_{aa} is I_{33}.

Observe further that if I_{aa}, I_{11}, I_{22}, and I_{33} are expressed in terms of radii of gyration as in Equation (8.7.1), as

$$I_{aa} = mk_a^2 \qquad I_{11} = mk_1^2 \qquad I_{22} = mk_2^2 \qquad I_{33} = mk_3^2 \tag{8.11.6}$$

where m is the mass of B, then Equations (8.11.3) and (8.11.4) may be expressed as:

$$1 = a_1^2/a^2 + a_2^2/b^2 + a_3^2/c^2 = \frac{a_1^2}{(k_a/k_1)^2} + \frac{a_2^2}{(k_a/k_2)^2} + \frac{a_3^2}{(k_a/k_3)^2} \tag{8.11.7}$$

where by inspection a, b, and c are

$$a = k_a/k_1 \qquad b = k_a/k_2 \qquad c = k_a/k_3 \tag{8.11.8}$$

These results show that the semi-major axes of the inertia ellipsoid are inversely proportional to the principal radii of gyration. Moreover, for any point P on the surface of the ellipsoid, the distance to the origin is proportional to the radii of gyration k_a with $\mathbf{n}_a$ being parallel to the line through the origin and passing through P.

8.11.3 Non-Distinct Roots of the Hamilton-Cayley Equation

Due to the symmetry and positive definiteness of the inertia matrix dyadic it may be shown that the roots of the Hamilton-Cayley equations are positive and real [8.7, 8.9]. However, they are not necessarily distinct. That is, there may be repeated roots. For example, the roots could be

$$\lambda = \lambda_1 = I_a \qquad \lambda = \lambda_2 = \lambda_3 = I_b \tag{8.11.9}$$

In this case the inertia ellipsoid of the foregoing subsection has a circular cross-section. Hence, there is a complete circle, or entire plane of principal directions.

If all three roots are equal, the inertia ellipsoid becomes a sphere and all directions are principal directions with all having the same principal moment of inertia.

8.11.4 Invariants of the Inertia Dyadic

Consider again the cubic Hamilton-Cayley equation for λ of Equation (8.10.13):

$$\lambda^3 - I_I \lambda^2 + I_{II} \lambda - I_{III} = 0 \tag{8.11.10}$$

where I_I, I_{II}, and I_{III} are the invariants of the inertia dyadic (see Section 8.10). Solving Equation (8.11.10) for λ is equivalent to factoring the equation. That is, if λ_a, λ_b, and λ_c are the roots (or "zeros") of Equation (8.11.10) then the equation may be written in the equivalent form:

$$(\lambda - \lambda_a)(\lambda - \lambda_b)(\lambda - \lambda_c) = 0 \tag{8.11.11}$$

By expanding Equation (8.11.11) and by comparing the coefficients of λ^3, λ^2, and λ with those of Equations (8.11.10) we have:

$$I_I = \lambda_a + \lambda_b + \lambda_c \tag{8.11.12}$$

$$I_{II} = \lambda_b \lambda_c + \lambda_c \lambda_a + \lambda_a \lambda_b \tag{8.11.13}$$

$$I_{III} = \lambda_a \lambda_b \lambda_c \tag{8.11.14}$$

Observe from Equation (8.10.17) that if the inertia dyadic is written in the form

$$\boldsymbol{I} = \lambda_a \mathbf{n}_a \mathbf{n}_a + \lambda_b \mathbf{n}_b \mathbf{n}_b + \lambda_c \mathbf{n}_c \mathbf{n}_c \tag{8.11.15}$$

where $\mathbf{n}_a$, $\mathbf{n}_b$, and $\mathbf{n}_c$ are the unit eigenvectors, then the corresponding inertia matrix is

$$I_{ij} = \begin{pmatrix} \lambda_a & 0 & 0 \\ 0 & \lambda_b & 0 \\ 0 & 0 & \lambda_c \end{pmatrix} \tag{8.11.16}$$

Then by inspection we see that I_I, I_{II}, and I_{III} are the trace (sum of the diagonal elements), the trace of the matrix of cofactors, and the determinant respectively. Indeed, I_I, I_{II}, and I_{III} have the same values independent of the unit vector basis [8.7, 8.11], hence the name "invariant."

8.11.5 Hamilton-Cayley Equation

Recall that the dot product of a dyadic $\boldsymbol{D}$ with itself is also a dyadic which may be written as $\boldsymbol{D}^2$. Specifically, if $\boldsymbol{D}$ is expressed in the form $\mathbf{n}_i d_{ij} \mathbf{n}_j$, where the $\mathbf{n}_i$ $(i = 1,2,3)$ are mutually perpendicular unit vectors, then $\boldsymbol{D}^2$ is

$$\boldsymbol{D}^2 = \boldsymbol{D} \cdot \boldsymbol{D} = \mathbf{n}_i d_{ij} \mathbf{n}_j \cdot \mathbf{n}_k d_{k\ell} \mathbf{n}_\ell = \mathbf{n}_i d_{ik} d_{k\ell} \mathbf{n}_\ell \tag{8.11.17}$$

Suppose the inertia dyadic $\boldsymbol{I}$ of a body B for a reference point O is expressed in terms of its principal values λ_1, λ_2, and λ_3 and mutually perpendicular principal unit vectors $\mathbf{n}_1$, $\mathbf{n}_2$, and $\mathbf{n}_3$ as:

$$\boldsymbol{I} = \lambda_1 \mathbf{n}_1 \mathbf{n}_1 + \lambda_2 \mathbf{n}_2 \mathbf{n}_2 + \lambda_3 \mathbf{n}_3 \mathbf{n}_3 \tag{8.11.18}$$

Then from Equation (8.11.17), $\boldsymbol{I}^2$ is

$$\boldsymbol{I}^2 = \lambda_1^2 \mathbf{n}_1 \mathbf{n}_1 + \lambda_2^2 \mathbf{n}_2 \mathbf{n}_2 + \lambda_3^2 \mathbf{n}_3 \mathbf{n}_3 \tag{8.11.19}$$

Similarly, $\boldsymbol{I}^3$ is

$$I^3 = I^2 \bullet I = \lambda_1^3 \mathbf{n}_1 \mathbf{n}_1 + \lambda_2^3 \mathbf{n}_2 \mathbf{n}_2 + \lambda_3^3 \mathbf{n}_3 \mathbf{n}_3 \qquad (8.11.20)$$

Recall the Hamilton-Cayley equation of Equation (8.10.13):

$$\lambda^3 - I_I \lambda^2 + I_{II} \lambda - I_{III} = 0 \qquad (8.11.21)$$

Noting that λ_1, λ_2, and λ_3 are roots of this equation, we see from Equations (8.11.18), (8.11.19), and (8.11.20) that

$$I^3 - I_I I^2 + I_{II} I - I_{III} U = 0 \qquad (8.11.22)$$

where U is the identity (or unit) dyadic. That is, the inertia dyadic itself satisfies the Hamilton-Cayley equations.

8.11.6 Central Inertia Dyadic and Other Geometrical Results

If the reference point of an inertia dyadic of a body B is its mass center G, as in the parallel axis theorem, then the inertia dyadic $I^{B/G}$ is called the "central inertia dyadic." Similarly, the inertia vector, and the moments and products of inertia and principal, referred to the mass center G are called "central inertia vectors" and "central moments and products of inertia."

There are several other geometrical results which are helpful in finding principal directions. One of the most useful of these is symmetry [8.9]. Specifically:

1) A plane of symmetry for a body or a set of points is perpendicular to principal directions and principal axes for all reference points in the plane. Alternatively, principal directions may be identified as directions normal to planes of symmetry for all points in the plane.

In addition to symmetry there are the following results [8.3, 8.4, 8.9]

2) A central principal axis is a principal axis for each of its points.

3) If a principal axis L for a reference point other than the mass center, passes through the mass center then L is a central principal axis.

4) If a line L is a principal axis for two reference points, then L is a central principal axis.

8.12 Planar Bodies; Polar Moments of Inertia

If a body is planar, such as a flat plate, then a plane of the planar body is a plane of symmetry. Thus, from Section 8.11, 1) above, lines perpendicular to this plane are principal axes of inertia. The corresponding principal moments of inertia about these lines are sometimes called "polar moments of inertia."

To discuss this further, consider the planar body B represented in Figure 8.12.1. Let P_i be a particle of B with mass m_i. Then from Equation (8.6.4) the moment of inertia of B about the X-axis is

$$I_{XX} = \sum_i m_i y_i^2 \quad (8.12.1)$$

where the sum is for all the particles of B and where y_i is the distance of P_i from the X-axis. Similarly, the moments of inertia of B about the Y and Z axes are:

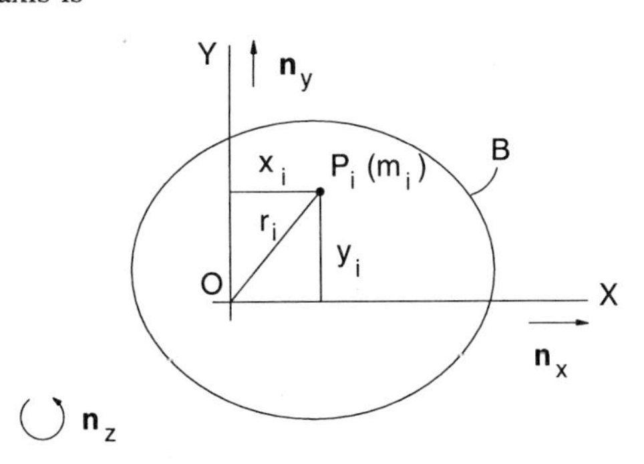

Fig. 8.12.1 A Planar Body B

$$I_{yy} = \sum_i m_i x_i^2 \quad (8.12.2)$$

and

$$I_{zz} = \sum_i m_i r_i^2 \quad (8.12.3)$$

where as usual the Z-axis is perpendicular to the X and Y-axes and is here also perpendicular to the plan of B, and where r_i is the distance from P_i to the Z-axis.

Since $r_i^2 = x_i^2 + y_i^2$, we immediately have the result:

$$I_{zz} = I_{xx} + I_{yy} \tag{8.12.4}$$

Since the Z-axis is a principal axis of inertia, there will be at least two other principal axes of inertia perpendicular to the Z-axis, and thus in the X-Y plane. These axes can be found by using the procedures of Section 8.10. Specifically, since the Z-axis is a principal axis the inertia dyadic matrix of B for O has the form

$$I_{ij} = \begin{bmatrix} I_{xx} & I_{yy} & 0 \\ I_{yx} & I_{yy} & 0 \\ 0 & 0 & I_{zz} \end{bmatrix} \tag{8.12.5}$$

Let $\mathbf{n}_a$ be a principal unit vector, as before, and let $\mathbf{n}_a$ be expressed as

$$\mathbf{n}_a = a_x \mathbf{n}_x + a_y \mathbf{n}_y + a_z \mathbf{n}_z \tag{8.12.6}$$

where $\mathbf{n}_x$, $\mathbf{n}_y$, and $\mathbf{n}_z$ are unit vectors parallel to the X, Y, and Z axes, as in Figure 8.12.1. Then from Equations (8.12.5), the equations determining a_x, a_y, and a_z are [Equations (8.10.11)]:

$$\begin{aligned} (I_{xx} - \lambda) a_x + I_{xy} a_y + O a_z &= 0 \\ I_{yx} a_x + (I_{yy} - \lambda) a)y + O a_z &= 0 \\ O a_x + O a_y + (I_{zz} - \lambda) a_z &= 0 \end{aligned} \tag{8.12.7}$$

Setting the determinant of the coefficients equal to zero lead to the Hamilton-Cayley equation:

$$(I_{zz} - \lambda)[(I_{xx} - \lambda)(I_{yy} - \lambda) - I_{xy}^2] = 0 \tag{8.12.8}$$

or

$$\lambda^3 - I_I \lambda^2 + I_{II} \lambda - I_{III} = 0 \tag{8.12.9}$$

where the invariants I_I, I_{II}, and I_{III} are

$$I_I = I_{xx} + I_{yy} + I_{zz} \tag{8.12.10}$$

$$I_{II} = I_{yy} I_{zz} + I_{xx} I_{zz} + I_{xx} I_{yy} - I_{xy}^2 \tag{8.12.11}$$

$$I_{III} = I_{zz}(I_{xx} I_{yy} - I_{xy}^2) \tag{8.12.12}$$

Solving for λ, we obtain

$$\lambda_1, \lambda_2 = \left(\frac{I_{xx} + I_{yy}}{2}\right) \pm \left[\left(\frac{I_{xx} - I_{yy}}{2}\right)^2 + I_{xy}^2\right]^{1/2} \tag{8.12.13}$$

and

$$\lambda_3 = I_{zz} \tag{8.12.14}$$

The corresponding principal directions are then determined by back substituting for λ into Equations (8.12.7). This leads to the equations:

$$\text{For } \lambda = \lambda_1, \lambda_2 \qquad a_y/a_x = -I_{xy}/(I_{yy} - \lambda) \quad \text{and} \quad a_z = 0 \tag{8.12.15}$$

$$\text{For } \lambda = \lambda_3 = I_{zz} \qquad a_x = a_y = 0 \qquad a_z = 1 \tag{8.12.16}$$

In Equation (8.12.15), if we let a_x and a_y have the forms

$$a_x = \cos\theta \quad \text{and} \quad a_y = \sin\theta \tag{8.12.17}$$

then

$$\tan\theta = -I_{xy}/(I_{yy} - \lambda) \tag{8.12.18}$$

8.13 Inertia Properties for Commonly Shaped Uniform Bodies

Table 8.13.1 provides a listing of central moments of inertia for bodies with uniform mass distribution and occupying common geometrical shapes. Except for the triangular body, the moments of inertia are also principal moments of inertia with the principal directions as indicated.

Table 8.13.1 Inertia Properties (Moments and Products of Inertia) for Commonly Shaped Bodies with Uniform Mass Distribution

I. **Curves, Wires, Thin Rods**:

1. *Straight Line, Rod*:

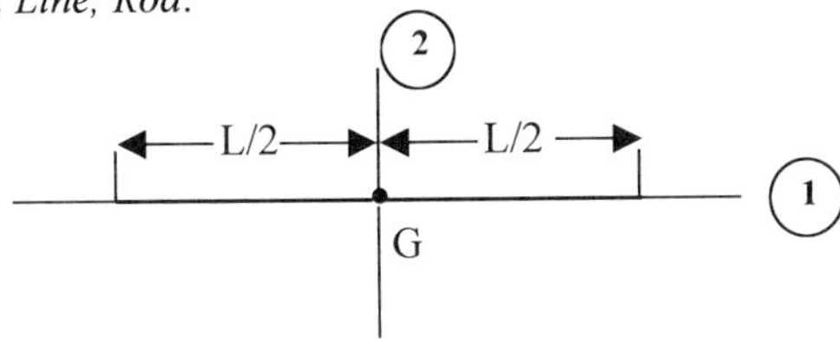

$I_{11} = 0$
$I_{22} = ML^2/12$

2. *Circular Arc, Circular Rod*:

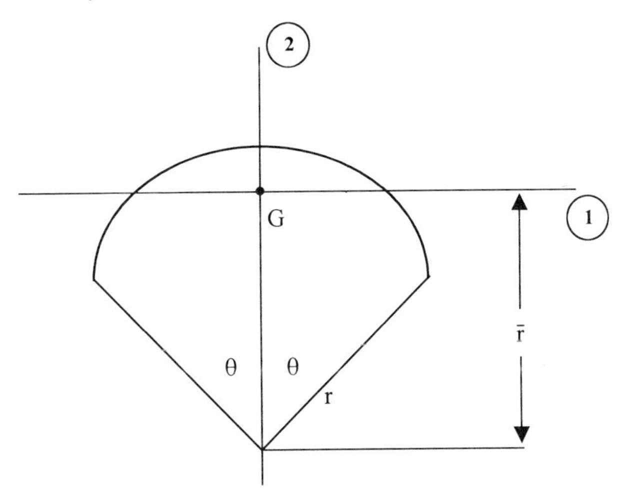

$\bar{r} = (r\sin\theta)/\theta$
$I_{11} = mr^2(1+\sin2\theta/2\theta-2\sin^2\theta/\theta^2)/2$
$I_{22} = mr^2(1-\sin2\theta/2\theta)/2$

3. *Semi Circular Arc, Semi Circular Rod*:

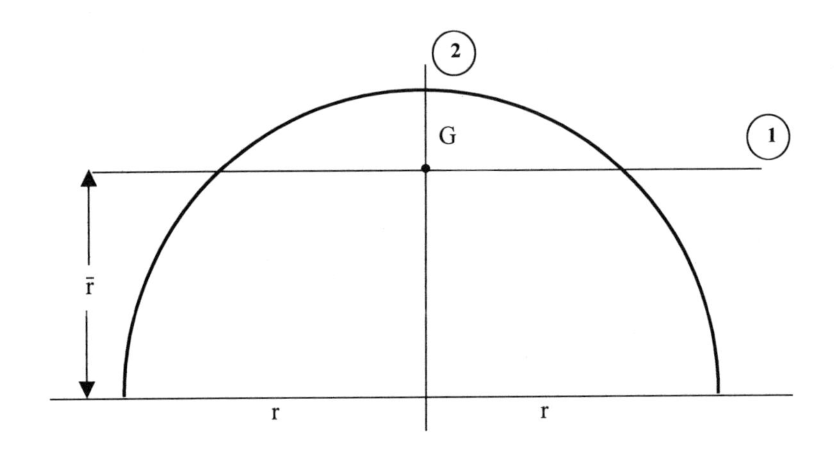

$I_{11} = mr^2(1-8/\pi^2)/2$
$I_{22} = mr^2/2$
$\bar{r} = 2r/\pi$

4. *Circle, Hoop*:

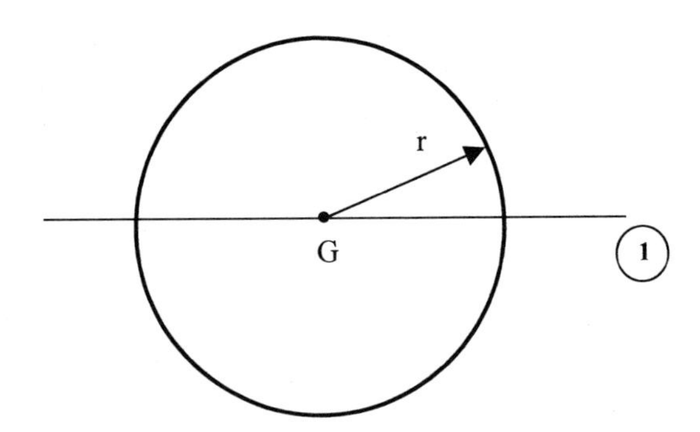

$I_{11} = mr^2/2$

II. **Surfaces, Thin Plates, Shells**:

1. *Triangle, Triangular Plate:*

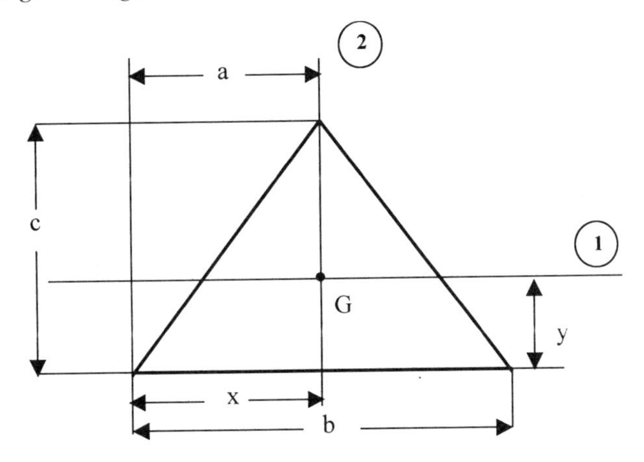

$x = (a+b)/3$
$y = c/3$
$I_{11} = mc^2/18$
$I_{22} = m(b^2-ab+a^2)/18$
$I_{12} = mc(2a-b)/18$

2. *Rectangle, Rectangular Plate*:

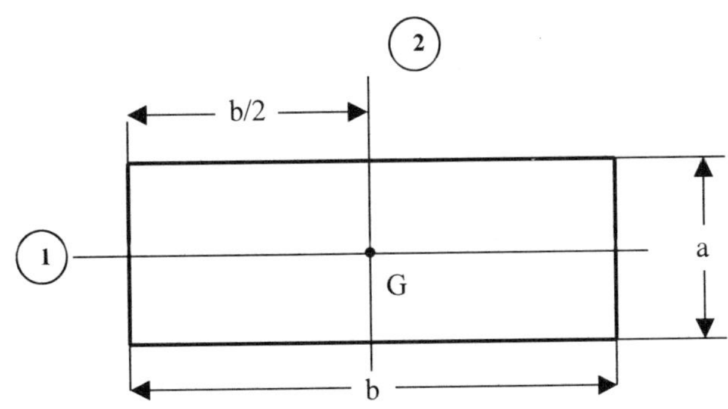

$I_{11} = ma^2/12$
$I_{22} = mb^2/12$

3. *Circular Sector, Circular Section Plate*:

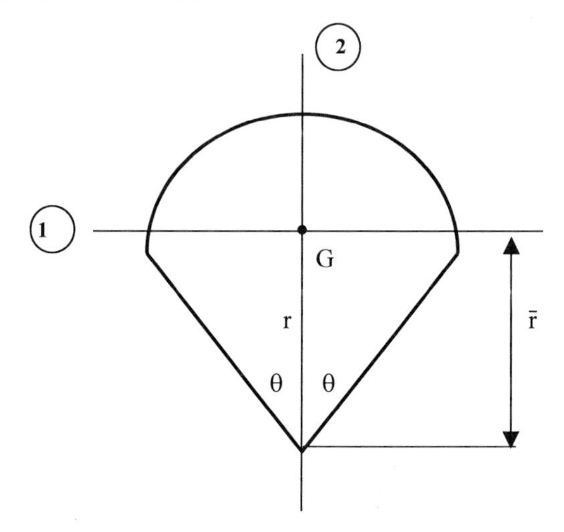

$\bar{r} = (2r/3)(\sin\theta)/\theta$
$I_{11} = (mr^2/4)(1+\sin\theta\cos\theta/\theta - 16\sin^2\theta/9\theta^2)$
$I_{22} = (mr^2/4)(1-\sin\theta\cos\theta/\theta)$

4. *Semicircle, Semicircular Plate*:

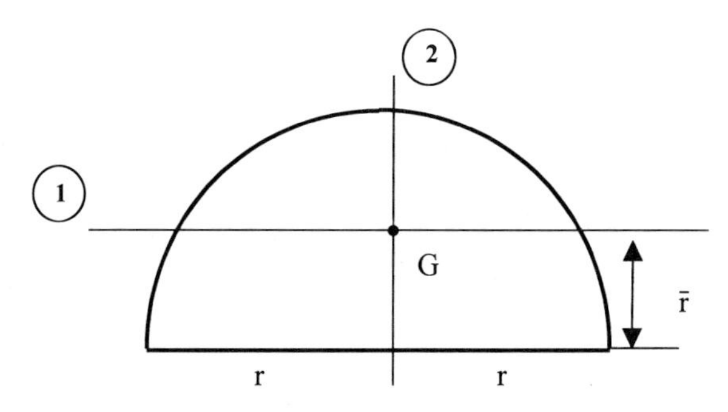

$\bar{r} = 4r/3\pi$
$I_{11} = mr^2(9\pi^2-64)/36\pi^2$
$I_{22} = mr^2/4$

5. *Circle, Circular Plate*:

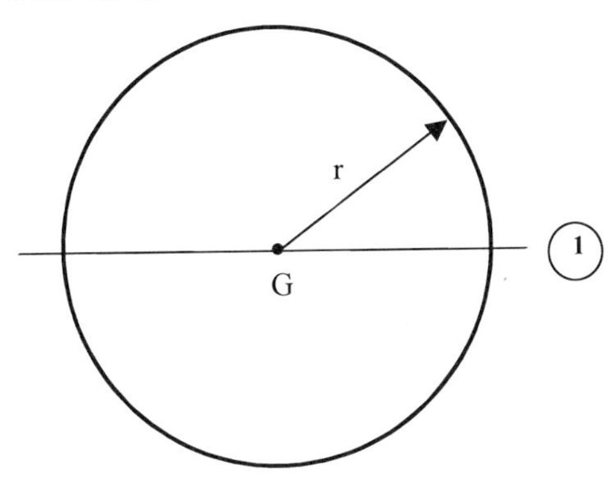

$I_{11} = mr^2/4$

6. *Circular Segment, Circular Segment Plate*:

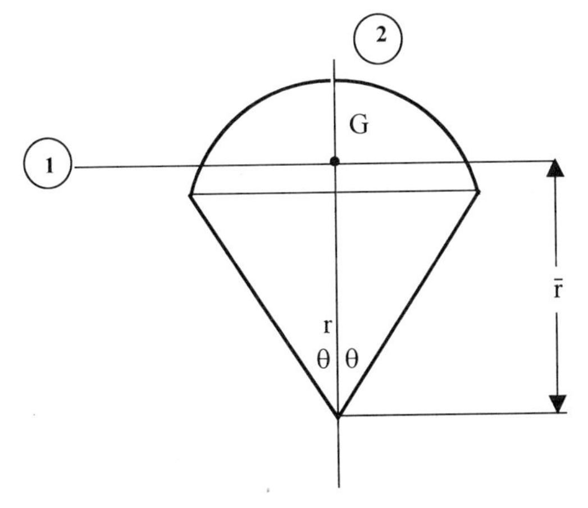

$\bar{r} = (2r/3)(\sin^3\theta)/(\theta - \sin\theta\cos\theta)$

$I_{11} = mr^2(9\theta^2 + 9\sin^2\theta\cos^2\theta - 36\sin^4\theta\cos^4\theta - 18\theta\sin\theta\cos\theta + 36\sin^3\theta\cos\theta - 8\sin^6\theta)/18(\theta - \sin\theta\cos\theta)^2$

$I_{22} = mr^2[1 - 2\sin^3\theta\cos\theta/3(\theta - \sin\theta\cos\theta)]/4$

7. *Cylinder, Cylindrical Shell*:

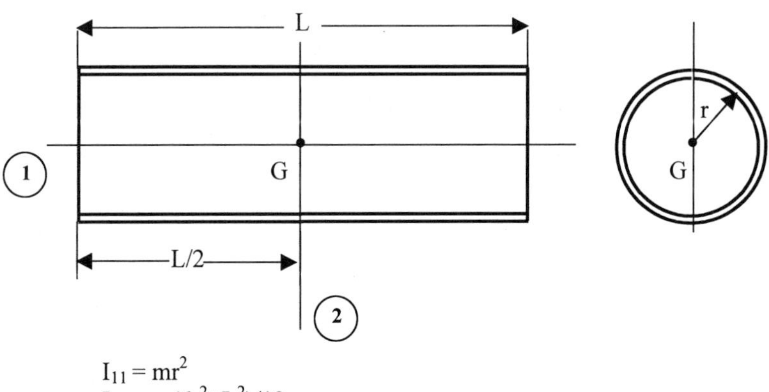

$I_{11} = mr^2$
$I_{22} = m(6r^2+L^2)/12$

8. *Semicylinder, Semicylindrical Shell*:

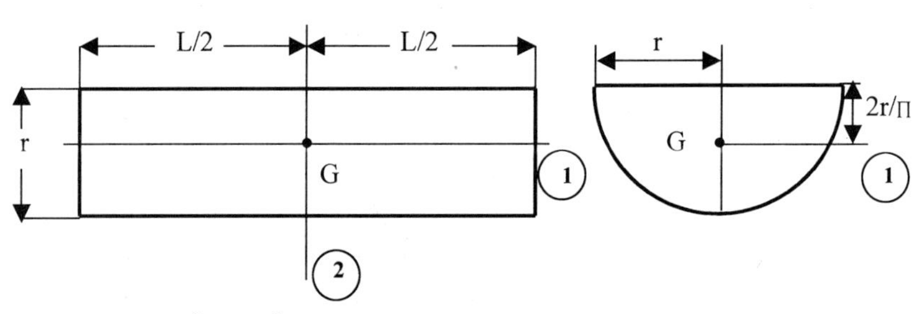

$I_{11} = mr^2(1-4/\Pi^2)$
$I_{22} = m(r^2+L^2/6)/2$
$I_{33} = mr^2(1/2-4/\Pi^2)+mL^2/12$

9. *Sphere, Spherical Shell*:

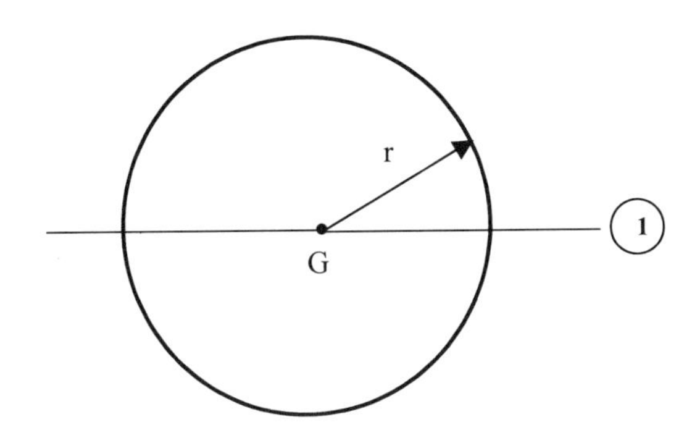

$I_{11} = 2mr^2/3$

10. *Hemisphere, Hemispherical Shell*:

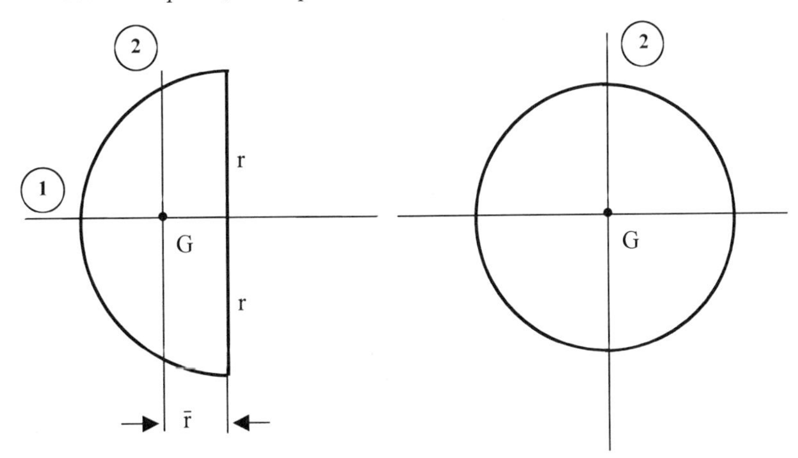

$\bar{r} = 3r/8$
$I_{11} = 2mr^2/3$
$I_{22} = 5mr^2/12$

11. *Cone, Conical Shell*:

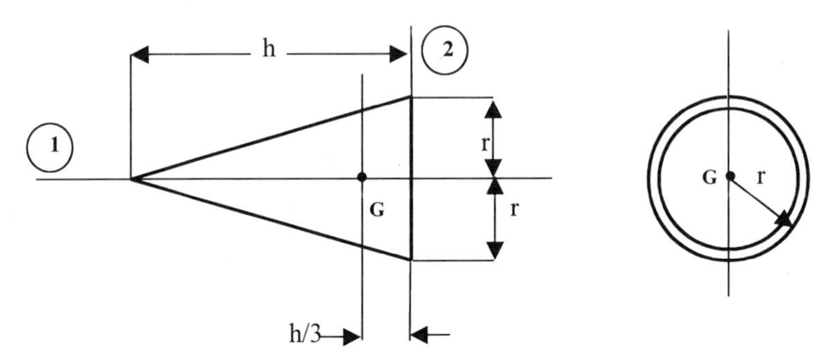

$I_{11} = mr^2/2$
$I_{22} = m(r^2/2 + h^2/9)/2$

12. *Half-cone, Half Conical Shell*:

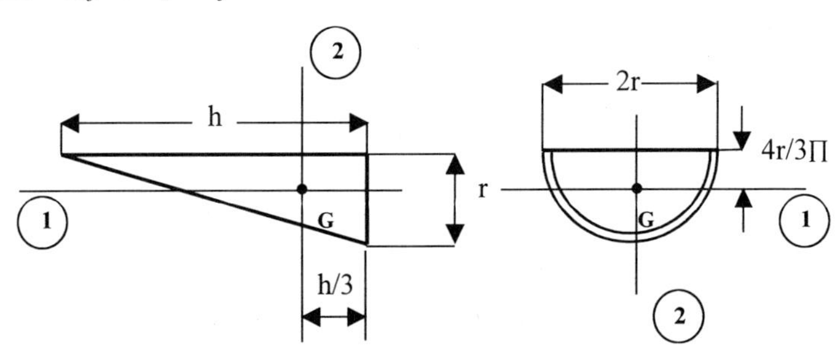

$I_{11} = mr^2(1-8/9\Pi^2)/2$
$I_{22} = (m/36)(9r^2+2h^2)$
$I_{33} = mh^2/18+mr^2(1-16/9\Pi^2)/4$
$I_{12} = -mrh/9\Pi$

III. **Solids, Bodies:**

1. *Parallelopiped, Block*:

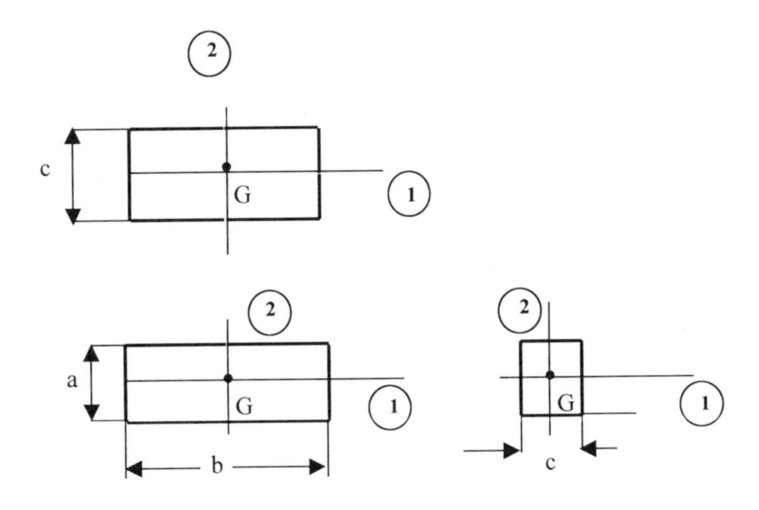

$I_{11} = m(a^2+c^2)/12$
$I_{22} = m(b^2+c^2)/12$
$I_{33} = m(a^2+b^2)/12$

2. *Cylinder*:

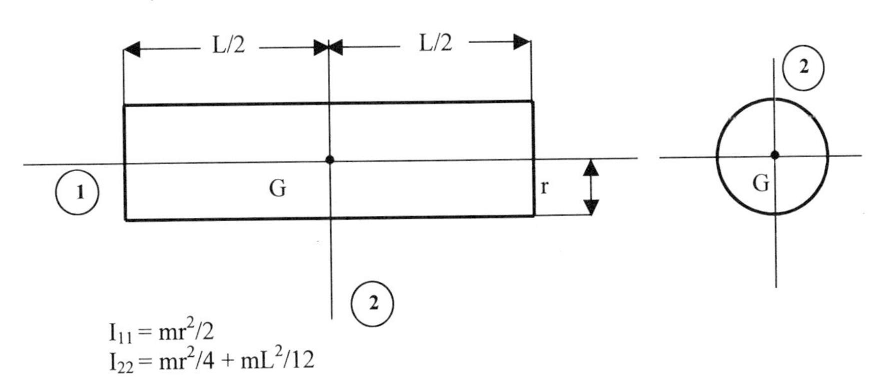

$I_{11} = mr^2/2$
$I_{22} = mr^2/4 + mL^2/12$

3. *Half Cylinder*:

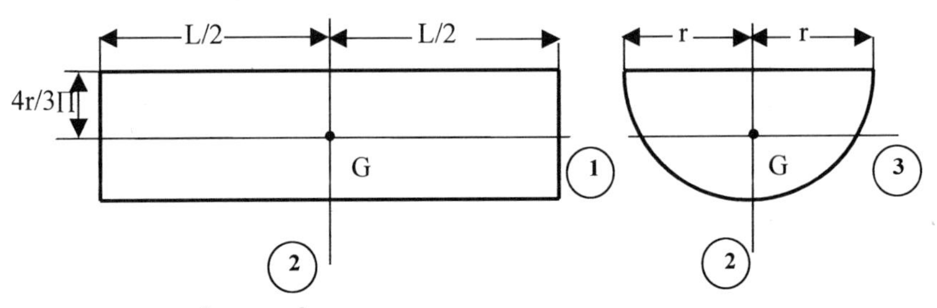

$I_{11} = mr^2(9-32/\Pi^2)/18$
$I_{22} = m(3r^2 + L^2)/12$
$I_{33} = mL^2/12+mr^2(9-64/\Pi^2)/36$

4. *Cone*:

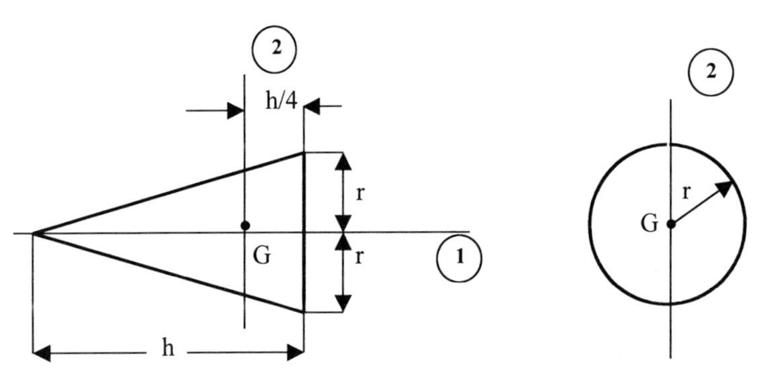

$I_{11} = 3mr^2/10$
$I_{22} = 3m(4r^2 + h^2)/80$

5. *Half cone*:

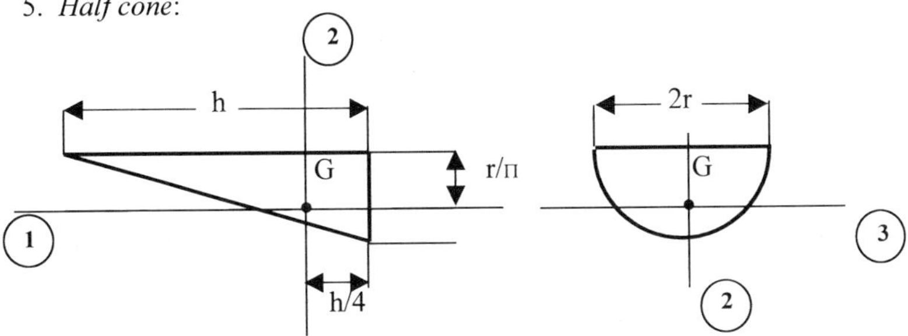

$I_{11} = mr^2(3/10-1/\Pi^2)$
$I_{22} = 3m(4r^2 + h^2)/80$
$I_{33} = mr^2(3/20-1/\Pi^2) + 3mh^2/80$
$I_{12} = -mrh/20\Pi$

6. *Sphere*:

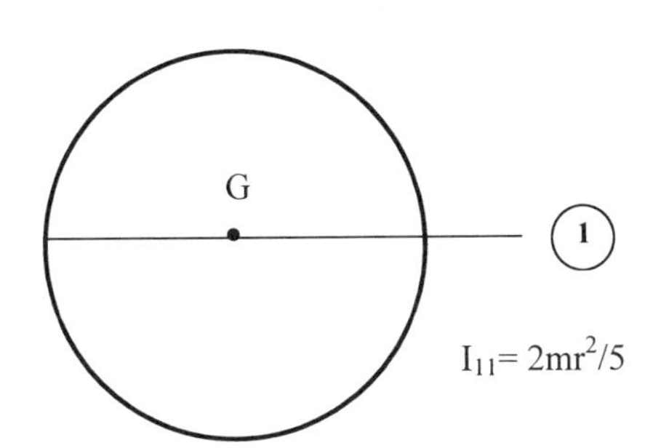

$I_{11}= 2mr^2/5$

7. *Hemisphere*:

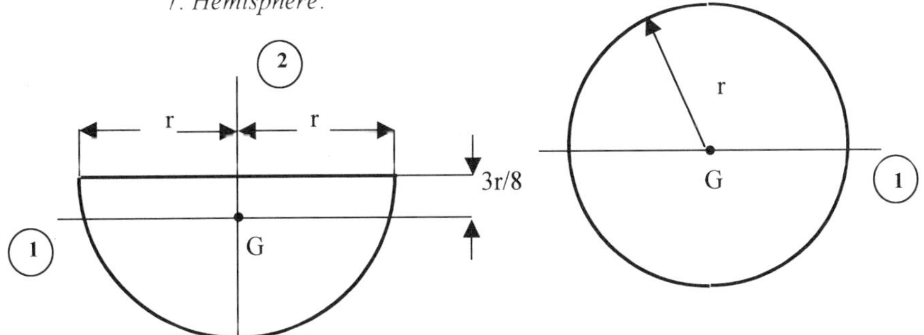

$I_{11} = 83mr^2/320$
$I_{22} = 2mr^2/5$

References

8.1 J. L. Meriam, *Engineering Mechanics: Statics and Dynamics*, Wiley, New York, 1978, pp. 526-530.

8.2 J. H. Ginsberg and J. Genin, *Dynamics* (Second Edition), Wiley, New York, 1984, pp. A1-A8.

8.3 T. R. Kane, *Analytical Elements of Mechanics, Volume 2, Dynamics*, Academic Press, New York, pp. 113-170, 304, 327-330.

8.4 T. R. Kane, *Dynamics*, Holt, Rinehart and Winston, New York, 1968, pp. 92-115, 161.

8.5 T. R. Kane and D. A. Levinson, *Dynamics: Theory and Applications*, McGraw Hill, New York, 1985, pp. 57-87, 361-371.

8.6 P. W. Likins, *Elements of Engineering Mechanics*, McGraw Hill, New York, 1973, pp. 522-526.

8.7 R. L. Huston, *Multibody Dynamics*, Butterworth-Heinemann, Boston, 1990, pp. 153-213.

8.8 B. D. Tapley and T. R. Poston (Editors), *Eshbach's Handbook of Engineering Fundamentals*, Wiley Engineering Handbook Series, John Wiley & Sons, New York, 1990, pp. 3.12-3.24.

8.9 H. Josephs and R. L. Huston, *Dynamics of Mechanical Systems*, CRC Press, Boca Raton, FL, 2001.

8.10 H. A. Rothbart (Editor-in-Chief), *Mechanical Design and Systems Handbook*, 1964, McGraw Hill, New York, pp. 1-36 to 1-41.

8.11 L. Brand, *Vector and Tensor Analysis*, John Wiley & Sons, New York, 1964, pp. 136-167.

Chapter 9

RIGID BODY KINETICS

9.1 Introduction

In this chapter we review formulas and expressions for forces on rigid bodies. We consider them in two categories: active (applied) and passive (inertia) forces. We examine systems of forces and their replacement by equivalent force systems. We also develop and summarize expressions for generalized forces and generalized force systems.

9.2 Useful Formulas from the Kinematics of Bodies

Chapters 6 and 7 summarize the formulations and resulting equations for the basic kinematic quantities of interest for rigid bodies. Table 9.2.1 summarizes selected kinematic expressions which are useful in developing the kinetics of bodies (see Table 6.14.1 for a more complete listing of equations for rigid body kinematics).

9.3 Summary of Concepts and Formulas for Force Systems on Bodies

Consider a rigid body B and let S be a force system exerted on B. Let S be represented by a set of N forces $\mathbf{F}_i$ ($i = 1,\ldots,N$) as in Figure 9.3.1. From the examples of Section 2.10 it is seen that the principal vectors representing S are the resultant $\mathbf{R}$ and the moment of S about a reference point O $\mathbf{M}_O$ as

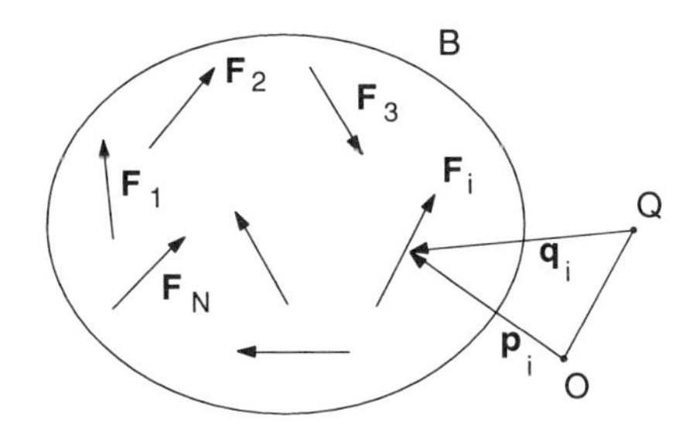

Fig. 9.3.1 A System of Forces on a Body B and Reference Points O and Q

Table 9.2.1 Summary of Useful Kinematic Expressions for Rigid Body Kinetic Formulations (see also Table 6.14.1)

Name	Equation	Reference Equation
1. Angular Velocity of a body B in a Reference Frame R	${}^R\boldsymbol{\omega}^B = \left(\frac{d\mathbf{n}_2}{dt}\bullet\mathbf{n}_3\right)\mathbf{n}_1 + \left(\frac{d\mathbf{n}_3}{dt}\bullet\mathbf{n}_1\right)\mathbf{n}_2 + \left(\frac{d\mathbf{n}_1}{dt}\bullet\mathbf{n}_2\right)\mathbf{n}_3$	(6.5.4)
2. Simple Angular Velocity	${}^R\boldsymbol{\omega}^B = \dot{\theta}\mathbf{k}$ (**k** is a unit vector fixed in both B and R and θ is the rotation angle)	(6.5.23)
3. Addition Formula for Angular Velocity	${}^R\boldsymbol{\omega}^B = {}^R\boldsymbol{\omega}^{R_1} + {}^{R_1}\boldsymbol{\omega}^{R_2} + \ldots + {}^{R_{n-1}}\boldsymbol{\omega}^{R_n} + {}^R\boldsymbol{\omega}^B$ (R, $R_1,\ldots,R_n$ are reference frames)	(6.7.5)
4. Angular Acceleration	${}^R\boldsymbol{\alpha}^B = {}^Rd\,{}^R\boldsymbol{\omega}^B/dt$	(6.9.1)
5. Relative Velocity of Points Fixed on a Body	${}^R\mathbf{V}^{P/Q} = {}^R\boldsymbol{\omega}^B \times \mathbf{r}$ (**r** locates P relative to Q)	(6.10.3)
6. Relative Acceleration of Points Fixed on a Body	${}^R\mathbf{a}^{P/Q} = {}^R\boldsymbol{\alpha}^B \times \mathbf{r} + {}^R\boldsymbol{\omega}^B \times ({}^R\boldsymbol{\omega}^B \times \mathbf{r})$ (**r** locates P relative to Q)	(6.11.2)
7. Partial Angular Velocity	${}^R\boldsymbol{\omega}^B = \boldsymbol{\omega}_t + \sum_{r=1}^{n} \boldsymbol{\omega}_{\dot{q}_r}\dot{q}_r$ [System with n degrees of freedom with coordinates $\dot{q}_r$ (r=1,...,n)]	(6.13.5)
8. Rolling	$\mathbf{V}^C = 0$ (contact point of a rolling body has zero velocity) $\mathbf{V}^P = \boldsymbol{\omega}^B \times \mathbf{P}$ (Velocity of a point P of a roll)	(6.12.1)

given by:

$$\mathbf{R} = \sum_{i=1}^{N} \mathbf{F}_i \tag{9.3.1}$$

and

$$\mathbf{M}_O = \sum_{i=1}^{N} \mathbf{p}_i \times \mathbf{F}_i \tag{9.3.2}$$

The reference point O is arbitrary. However, by knowing the moment of S about O we can obtain the moment of S about any other reference point Q through the expression (see Example 2.10.3):

$$\mathbf{M}_O = \mathbf{M}_Q + \mathbf{OQ} \times \mathbf{R} \tag{9.3.3}$$

Finally, any force system S may be shown to be mechanically equivalent to a single force **F** with a couple with torque **T**, where **F** and **T** are (see Example 2.10.7)

$$\mathbf{F} = \mathbf{R} \quad \text{and} \quad \mathbf{T} = \mathbf{M}_O \tag{9.3.4}$$

9.4 Partial Velocity and Partial Angular Velocity

Consider a mechanical system S, containing particle and rigid bodies, moving in an inertial reference frame R. Let S be free of kinematic constraints (a "holonomic system" [9.1]) and let S have n degrees of freedom in R. Let these degrees of freedom be represented by n variables q_r $(r = 1,..,n)$, called generalized coordinates.

Consider a typical particle P of S. The position vector **p** locating P relative to a fixed point O of R will then be a function of the q_r $(r = 1,...,n)$ and time t. That is,

$$\mathbf{p} = \mathbf{p}(q_r, t) \tag{9.4.1}$$

By differentiating, the velocity **V** of P in R may be expressed as:

$$\mathbf{V} = \frac{d\mathbf{p}}{dt} = \frac{\partial \mathbf{p}}{\partial t} + \sum_{r=1}^{n} \frac{\partial \mathbf{p}}{\partial q_r} \dot{q}_r \tag{9.4.2}$$

where the partial derivatives $\partial \mathbf{p}/\partial t$ and $\partial \mathbf{p}/\partial q_r$ are often designated as $\mathbf{V}_t$ and $\mathbf{V}_{\dot{q}_r}$ and called "partial velocity" vectors [9.1, 9.2, 9.3].

In this context **V** may be expressed as

$$\mathbf{V} = \mathbf{V}_t + \sum_{r=1}^{n} \mathbf{V}_{\dot{q}_r} \dot{q}_r \tag{9.4.3}$$

By differentiating with respect to $\dot{q}_r$ we have

$$\frac{\partial \mathbf{V}}{\partial \dot{q}_r} = \mathbf{V}_{\dot{q}_r} = \frac{\partial \mathbf{p}}{\partial q_r} \tag{9.4.4}$$

Observe further that by differentiating in Equation (9.4.2) with respect to q_s we have

$$\frac{\partial \mathbf{V}}{\partial q_s} = \frac{\partial}{\partial q_s}\left[\frac{\partial \mathbf{p}}{\partial t} + \sum_{r=1}^{n} \frac{\partial \mathbf{p}}{\partial q_r} \dot{q}_r\right] = \frac{\partial}{\partial q_s}\left(\frac{\partial \mathbf{p}}{\partial t}\right) + \sum_{r=1}^{n} \frac{\partial}{\partial q_s}\left(\frac{\partial \mathbf{p}}{\partial q_r}\right)\dot{q}_r$$

$$= \frac{\partial}{\partial t}\left(\frac{\partial \mathbf{p}}{\partial q_s}\right) + \sum_{r=1}^{n} \frac{\partial}{\partial q_r}\left(\frac{\partial \mathbf{p}}{\partial q_s}\right)\dot{q}_r = \frac{d}{dt}\left(\frac{\partial \mathbf{p}}{\partial q_s}\right)$$

or

$$\frac{\partial \mathbf{V}}{\partial q_s} = \frac{d}{dt}(\mathbf{V}_{\dot{q}_r}) \tag{9.4.5}$$

where the final equality follows from Equation (9.4.4).

The partial velocity vectors are useful in developing generalized forces as defined in the following section. To that end it is useful to consider the projection of a particle's acceleration along the partial velocity:

$$\mathbf{a} \bullet \mathbf{V}_{\dot{q}_r} = \frac{d\mathbf{V}}{dt} \bullet \mathbf{V}_{\dot{q}_r} = \frac{d}{dt}(\mathbf{V} \bullet \mathbf{V}_{\dot{q}_r}) - \mathbf{V} \bullet \frac{d}{dt}(\mathbf{V}_{\dot{q}_r})$$

$$= \frac{1}{2}\frac{d}{dt}\left(\frac{\partial \mathbf{V}^2}{\partial \dot{q}_r}\right) - \mathbf{V} \bullet \frac{\partial \mathbf{V}}{\partial q_r}$$

or

$$\mathbf{a} \bullet \mathbf{V}_{\dot{q}_r} = \frac{1}{2}\frac{d}{dt}\left(\frac{\partial \mathbf{V}^2}{\partial \dot{q}_r}\right) - \frac{1}{2}\frac{\partial \mathbf{V}^2}{\partial q_r} \tag{9.4.6}$$

Observe that each term on the right side of Equation (9.4.6) contains the common factor: $(1/2)\mathbf{V}^2$, which is proportional to the kinetic energy. This observation is useful for the development of Lagrange's equation, as in the following chapter.

Next, consider a rigid body B moving in a reference frame R as represented in Figure 9.4.1. Let P and Q be particles of B and let $\mathbf{r}$ locate P relative to Q. Let O be a point fixed in R and let $\mathbf{p}$ and $\mathbf{q}$ locate P and Q relative to O. Finally, let B be part of a mechanical system S having n degrees of freedom represented by the variables q_r $(r = 1, \ldots, n)$. Then from Equations (9.4.2) and (9.4.3) the velocities of P and Q in R may be expressed as:

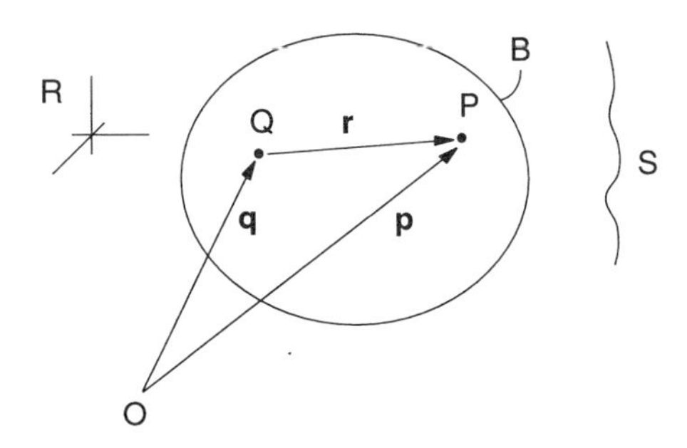

Fig. 9.4.1 A Rigid Body B with Particles P and Q Moving in a Reference Frame R

$$ {}^{R}\mathbf{V}^{P} = \frac{\partial \mathbf{p}}{\partial t} + \sum_{r=1}^{n} \frac{\partial \mathbf{p}}{\partial q_r}\dot{q}_r = \mathbf{V}_t^P + \sum_{r=1}^{n} \mathbf{V}_{\dot{q}_r}^{P}\dot{q}_r \tag{9.4.7} $$

and

$$ {}^{R}\mathbf{V}^{Q} = \frac{\partial \mathbf{q}}{\partial t} + \sum_{r=1}^{n} \frac{\partial \mathbf{q}}{\partial q_r}\dot{q}_r = \mathbf{V}_t^Q + \sum_{r=1}^{n} \mathbf{V}_{\dot{q}_r}^{Q}\dot{q}_r \tag{9.4.8} $$

The velocity of P relative to Q in R is then

$$ \begin{aligned} {}^{R}\mathbf{V}^{P/Q} &= {}^{R}\mathbf{V}^{P} - {}^{R}\mathbf{V}^{Q} = \frac{\partial \mathbf{p}}{\partial t} + \sum_{r=1}^{n} \frac{\partial \mathbf{p}}{\partial q_r}\dot{q}_r - \frac{\partial \mathbf{q}}{\partial t} - \sum_{r=1}^{n} \frac{\partial \mathbf{q}}{\partial q_r}\dot{q}_r \\ &= \frac{\partial}{\partial t}(\mathbf{p} - \mathbf{q}) + \sum_{r=1}^{n} \frac{\partial}{\partial q_r}(\mathbf{p} - \mathbf{q})\dot{q}_r \\ &= \frac{\partial \mathbf{r}}{\partial t} + \sum_{r=1}^{n} \frac{\partial \mathbf{r}}{\partial q_r}\dot{q}_r = \frac{{}^{R}d\mathbf{r}}{dt} \\ &= {}^{R}\boldsymbol{\omega}^{B} \times \mathbf{r} = (\boldsymbol{\omega}_t + \sum_{r=1}^{n} \boldsymbol{\omega}_{q_r}\dot{q}_r) \times \mathbf{r} \\ &= \boldsymbol{\omega}_t \times \mathbf{r} + \sum_{r=1}^{n} (\boldsymbol{\omega}_{q_r} \times \mathbf{r})\dot{q}_r \end{aligned} \tag{9.4.9} $$

where use has been made of Equations (6.10.3) and (6.13.5) [see Table 9.2.1].

Recall from the development of Section 6.13 that the partial angular velocities may be expressed as [see Equations (6.13.2)]:

$$ \boldsymbol{\omega}_t = \frac{\dfrac{\partial \mathbf{a}}{\partial t} \times \dfrac{\partial \mathbf{b}}{\partial t}}{\dfrac{\partial \mathbf{a}}{\partial t} \bullet \mathbf{b}} \quad \text{and} \quad \boldsymbol{\omega}_{\dot{q}_r} = \frac{\dfrac{\partial \mathbf{a}}{\partial q_r} \times \dfrac{\partial \mathbf{b}}{\partial q_r}}{\dfrac{\partial \mathbf{a}}{\partial q_r} \bullet \mathbf{b}} \tag{9.4.10} $$

where **a** and **b** are non-zero, non-parallel vectors fixed in body B. Alternatively, if $\mathbf{n}_1$, $\mathbf{n}_2$, and $\mathbf{n}_3$ are mutually perpendicular unit vectors fixed in B, it is readily seen that $\boldsymbol{\omega}_t$ or $\boldsymbol{\omega}_{\dot{q}_r}$ may also be expressed as:

$$\boldsymbol{\omega}_t = \left(\frac{\partial \mathbf{n}_2}{\partial t} \cdot \mathbf{n}_3\right)\mathbf{n}_1 + \left(\frac{\partial \mathbf{n}_3}{\partial t} \cdot \mathbf{n}_1\right)\mathbf{n}_2 + \left(\frac{\partial \mathbf{n}_1}{\partial t} \cdot \mathbf{n}_2\right)\mathbf{n}_3 \tag{9.4.11}$$

and

$$\boldsymbol{\omega}_{\dot{q}_r} = \left(\frac{\partial \mathbf{n}_2}{\partial q_r} \cdot \mathbf{n}_3\right)\mathbf{n}_1 + \left(\frac{\partial \mathbf{n}_3}{\partial q_r} \cdot \mathbf{n}_1\right)\mathbf{n}_2 + \left(\frac{\partial \mathbf{n}_1}{\partial q_r} \cdot \mathbf{n}_2\right)\mathbf{n}_3 \tag{9.4.12}$$

Table 9.4.1 provides a summary of the principal equations and formulas for partial velocity and partial angular velocity expressions.

Table 9.4.1 Partial Velocity and Partial Angular Velocity Expressions for a Holonomic Mechanical System S with n Degrees of Freedom Represented by Generalized Coordinates q_r $(r = 1, \ldots, n)$

Name	Expression	Reference Equation
Particle Velocity	$\mathbf{V} = \mathbf{V}_t + \sum_{r=1}^{n} \mathbf{V}_{\dot{q}_r} \dot{q}_r$	(9.4.3)
Partial Velocity of Particle P (**p** locates P in a reference frame R)	$\mathbf{V}_{\dot{q}_r} = \frac{\partial \mathbf{p}}{\partial \dot{q}_r} = \frac{\partial \mathbf{V}}{\partial \dot{q}_r}$	(9.4.4)
Derivative Relation	$\frac{\partial \mathbf{V}}{\partial q_s} = \frac{d}{dt} \mathbf{V}_{\dot{q}_r}$	(9.4.5)

Name	Expression	Reference Equation
Acceleration Projection	$\mathbf{a} \bullet \mathbf{V}_{\dot{q}_r} = \frac{1}{2}\frac{d}{dt}\left(\frac{\partial \mathbf{V}^2}{\partial \dot{q}_r}\right) - \frac{1}{2}\frac{\partial \mathbf{V}^2}{\partial \dot{q}_r}$	(9.4.6)
Body Angular Velocity in Reference Frame R	${}^R\boldsymbol{\omega}^B = \boldsymbol{\omega}_t + \sum_{r=1}^{n} \boldsymbol{\omega}_{q_r}\dot{q}_r$	(6.13.5)
Partial Angular Velocity	$\boldsymbol{\omega}_{q_r} = \left(\frac{\partial \mathbf{n}_2}{\partial q_r} \bullet \mathbf{n}_3\right)\mathbf{n}_1 + \left(\frac{\partial \mathbf{n}_3}{\partial q_r} \bullet \mathbf{n}_1\right)\mathbf{n}_2 + \left(\frac{\partial \mathbf{n}_1}{\partial q_r} \bullet \mathbf{n}_2\right)\mathbf{n}_3$	(9.4.12)

($\mathbf{n}_1$, $\mathbf{n}_2$, $\mathbf{n}_3$ are mutually perpendicular unit vectors fixed in body B).

9.5 Generalized Forces

Consider a mechanical system S composed of particles and rigid bodies and having n degrees of freedom represented by the generalized coordinates q_r $(r = 1,\ldots,n)$. Let P be a particle of S and let **F** be a force exerted on P as represented in Figure 9.5.1. Then the generalized force F_{q_r} on P, for the coordinate q_r, due to **F**, is defined as the projection of **F** onto the partial velocity of P for q_r. That is

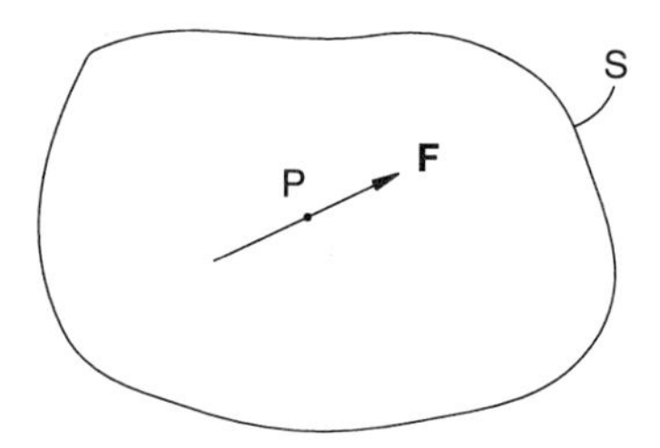

Fig. 9.5.1 A Force **F** Acting on a Particle P of a Mechanical System S

$$F_{q_r} = \mathbf{F} \bullet \mathbf{V}_{\dot{q}_r} \quad (r = 1, \ldots, n) \tag{9.5.1}$$

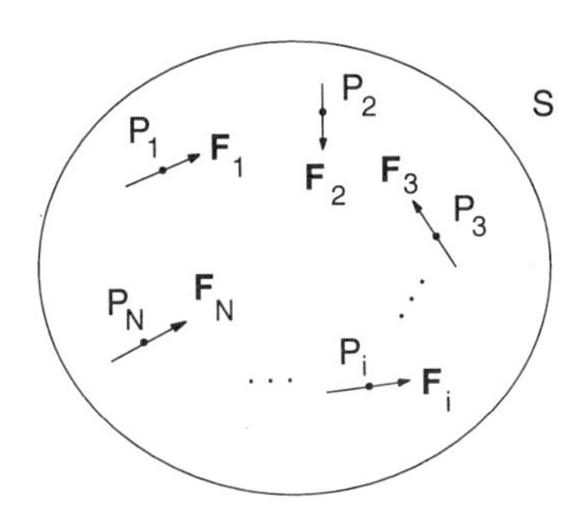

Fig. 9.5.2 Forces $\mathbf{F}_i$ Acting on Particles P_i $(i = 1, \ldots, N)$ of a Mechanical System S

Next consider a set of forces $\mathbf{F}_i$ $(i = 1, \ldots, N)$ exerted on particles P_i $(i = 1, \ldots, N)$ of S as in Figure 9.5.2. The generalized force F_r, for the coordinate q_r, due to the set of forces is then simply the addition of the projection of the forces along the respective partial velocities of the particles for q_r. That is

$$F_{q_r} = \sum_{i=1}^{N} \mathbf{F}_i \bullet \mathbf{V}_{\dot{q}_r}^{P_i} \quad (r = 1, \ldots, n) \tag{9.5.2}$$

Finally, consider a rigid body B of S. Let B be subjected to N forces $\mathbf{F}_i$ $(i = 1, \ldots, N)$ acting on particles P_i $(i = 1, \ldots, N)$ of B as in Figure 9.5.3. As in Equation (9.5.2) the generalized force F_{q_r} on B may be expressed as:

$$F_{q_r} = \sum_{i=1}^{N} \mathbf{F}_i \bullet \mathbf{V}_{\dot{q}_r}^{P_i} \quad (r = 1, \ldots, n) \tag{9.5.3}$$

If Q is an arbitrary reference point of B, then the velocity of a typical particle P_i of B may be expressed in terms of the velocity of Q as

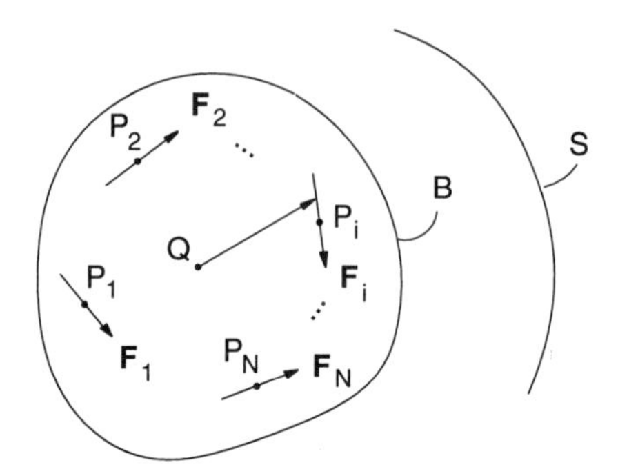

Fig. 9.5.3 Forces $\mathbf{F}_i$ Acting on Particles P_i of a Body B of a Mechanical System S

$$\mathbf{V}^{P_i} = \mathbf{V}^Q + \boldsymbol{\omega} \times \mathbf{r}_i \qquad (i = 1,\ldots,n) \tag{9.5.4}$$

where $\boldsymbol{\omega}$ is the angular velocity of B and $\mathbf{r}_i$ locates P_i relative to Q. (Note that unless otherwise stated all kinematical quantities are evaluated relative to an inertial reference frame.) Since $\mathbf{r}_i$ is not explicitly dependent upon the $\dot{q}_r$, differentiation with respect to $\dot{q}_r$ in Equation (9.5.4) leads to the expression:

$$\mathbf{V}^{P_i}_{\dot{q}_r} = \mathbf{V}^Q_{\dot{q}_r} + \boldsymbol{\omega}_{\dot{q}_r} \times \mathbf{r}_i \qquad (i = 1,\ldots,n) \tag{9.5.5}$$

Then by substituting for $\mathbf{V}^{P_i}_{\dot{q}_r}$ in Equation (9.5.3) F_r becomes:

$$\begin{aligned} F_{\dot{q}_r} &= \sum_{i=1}^{N} \mathbf{F}_i \cdot (\mathbf{V}^Q_{\dot{q}_r} + \boldsymbol{\omega}_{\dot{q}_r} \times \mathbf{r}_i) \\ &= \left(\sum_{i=1}^{N} \mathbf{F}_i\right) \cdot \mathbf{V}^Q_{\dot{q}_r} + \sum_{i=1}^{N} \mathbf{F}_i \cdot (\boldsymbol{\omega}_{\dot{q}_r} \times \mathbf{r}_i) \\ &= \left(\sum_{i=1}^{N} \mathbf{F}_i\right) \cdot \mathbf{V}^Q_{\dot{q}_r} + \boldsymbol{\omega}_{\dot{q}_r} \cdot \left(\sum_{i=1}^{N} \mathbf{r}_i \times \mathbf{F}_i\right) \qquad (r = 1,\ldots,n) \end{aligned} \tag{9.5.6}$$

where the last term follows from the cyclic permutation of terms of the triple scalar product.

Equation (9.5.6) may be written in the compact form:

$$F_{q_r} = \mathbf{R} \bullet \mathbf{V}^Q_{\dot{q}_r} + \mathbf{M}_Q \bullet \boldsymbol{\omega}_{\dot{q}_r} \quad (r = 1,\ldots,n) \tag{9.5.7}$$

where $\mathbf{R}$ is the resultant of the force system and $\mathbf{M}_Q$ is its moment about the reference point Q. [See Equations (9.3.1) and (9.3.2).]

9.6 Applied (Active) Forces

As with particles the forces on rigid bodies may be divided into two major categories: Applied (or "active") forces and Inertia (or "passive") forces. Here we consider the applied forces. In Section 9.12 we will consider the inertia forces.

The applied forces on a body arise from entities external to the body in the form of gravity, contact, and electromagnetic forces. If a body is modeled as a set of rigidly attached particles, then the system of applied forces on the body is composed of the applied forces exerted on the particles of the body. Forces exerted by the particles of the body on one another are reflexive or "self-equilibrating." They are internal to the body and cancel one another. Thus for rigid bodies these internal forces may be ignored.

By regarding the applied force system on a body as a set of forces acting on the particles of the body, we may represent the force system by an equivalent force system consisting of a single force $\mathbf{F}$ passing through an arbitrary point Q of the body (or the body extended) together with a couple with torque $\mathbf{T}$. In the following paragraphs we will consider such representations for a few examples of gravitational and contact forces.

9.6.1 Gravitational Forces Exerted by the Earth on a Body

Consider a body B on the surface of the earth E as represented in Figure 9.6.1. Let B be composed of particles P_i having masses m_i $(i = 1,\ldots,N)$.

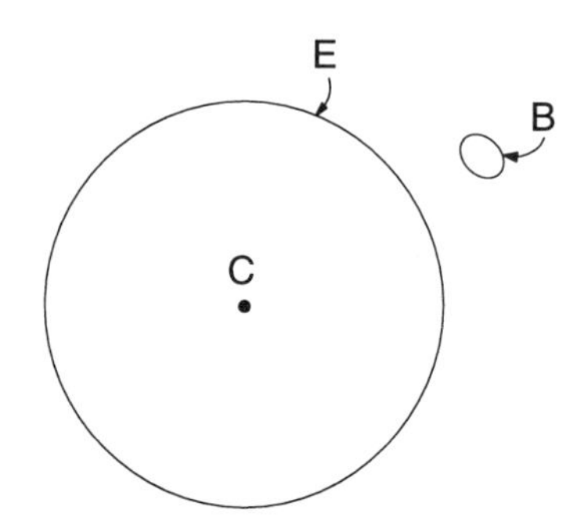

Fig. 9.6.1 A Body B Near the Surface of the Earth E

From a practical perspective it is reasonable to regard the earth as a sphere with symmetrical physical properties and to regard B as being small compared to E. Then from Example 4.3.1 [Equation (4.3.9)] we can represent the gravitational attraction of E on the particles of B as though E were itself a particle located at its center C with a mass M equal to the mass of the entire earth. That is, the earth attracts as a point at its center with concentrated mass.

Next, if B is small compared with E, the lines connecting the particles P_i of B with C (the "vertical" lines) are nearly parallel. Also, the distances of the particles from C are all nearly equal to the radius r of the earth (6371 kilometers or 3960 miles).

With these assumptions, the gravitational forces exerted by E on the particles of B may be represented by downward vertical weight forces $\mathbf{W}_i$ equal to the particle mass m_i and local gravity acceleration g. That is,

$$\mathbf{W}_i = -\frac{GMm_i}{r^2}\mathbf{k} = -m_i g \mathbf{k} \tag{9.6.1}$$

where G is the universal gravitation constant, **k** is a vertical unit vector and g is defined by inspection as

$$g = GM/r^2 \tag{9.6.2}$$

Recall from Section 4.3 that G and M have the values

$$G = 6.673 \times 10^{-11}\ \mathrm{m^3/kg\,s^2} = 3.438 \times 10^{-8}\ \mathrm{ft^2/slug\,s^2} \tag{9.6.3}$$

and

$$M = 5.976 \times 10^{24}\ \mathrm{kg} = 4.096 \times 10^{23}\ \mathrm{slug} \tag{9.6.4}$$

Then g becomes

$$g = 9.81\ \mathrm{m/s^2} = 32.2\ \mathrm{ft/sec^2} \tag{9.6.5}$$

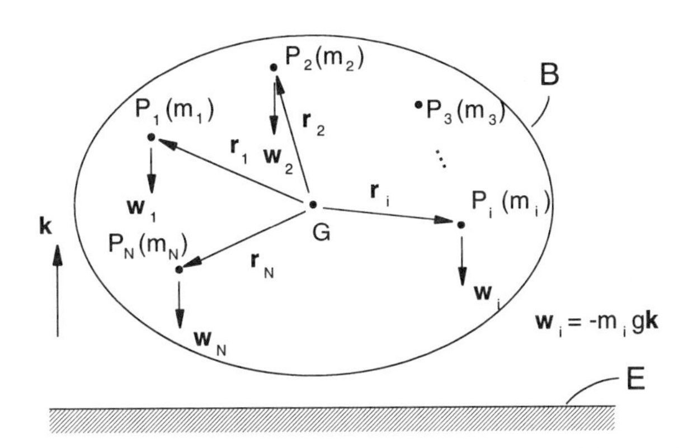

Fig. 9.6.2 A Representation of the Attraction of the Earth E on the Particles of a Body B

Figure 9.6.2 depicts the attraction on B. Since all of the $\mathbf{w}_i$ are parallel they are equivalent to a single force **W** passing through the mass center G, where **W** is

$$\mathbf{W} = -mg\mathbf{k} \quad \text{where} \quad m = \sum_{i=1}^{N} m_i \tag{9.6.6}$$

This is seen by noting the resultant of the $\mathbf{w}_i$ is

$$\sum_{i=1}^{N} \mathbf{w}_i = -\sum_{i=1}^{N} m_i g \mathbf{k} = -\left(\sum_{i=1}^{N} m_i\right) g \mathbf{k} = -mg\mathbf{k} = \mathbf{W} \tag{9.6.7}$$

Also note that the moment of the $\mathbf{w}_i$ about G is

$$\mathbf{M}_G = \sum_{i=1}^{N} \mathbf{r}_i \times \mathbf{w}_i = -\sum_{i=1}^{N} \mathbf{r}_i \times m_i g \mathbf{k} = -g\left(\sum_{i=1}^{N} m_i \mathbf{r}_i\right) \times \mathbf{k} = 0 \tag{9.6.8}$$

where the last equality is due to G being the mass center so that $\sum_{i=1}^{N} m_i \mathbf{r}_i$ is zero [see Equation (8.3.2)]. Figure 9.6.3 shows the equivalent force system, with equivalence being established by Equations (9.6.7) and (9.6.8).

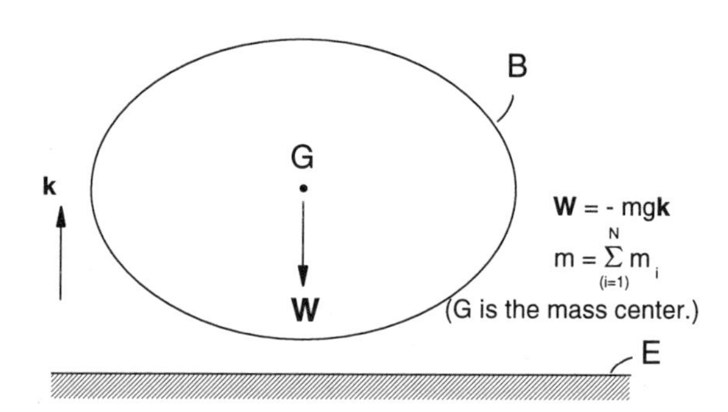

Fig. 9.6.3 Force System Equivalent to the Forces of the Earth E on a Body B (See Fig. 9.6.2)

The equivalence of the gravitational force systems of Figures 9.6.2 and 9.6.3 is based upon the assumptions of B being small compared to E, of all vertical lines being parallel, and of all particles P_i of B being the same distance from the center C of E. While these assumptions are quite reasonable and generally acceptable for most applications, a question which arises is: What is the error introduced by these assumptions? To address this question, consider first that even if the assumptions are relaxed, the gravitational forces on B by E are still equivalent to a single force **W** passing through a point Q of B called the "center of gravity." Observe, however, that Q is then not at the mass center of G of B. That is, the center of gravity and the mass center (or "center of mass") are distinct points. To see this consider a representation of the gravitational forces on a body B by the earth E with an exaggerated representation of the size of B as in Figure 9.6.4. As before let C be the center of E with the concentrated mass of E. Then the attraction of E on the particles P_i of B develop forces $\mathbf{w}_i$ $(i = 1, \ldots, N)$ each with line of action passing through C. Let $\mathbf{W}$ be the resultant of the $\mathbf{w}_i$ and let $\mathbf{W}$ pass through C. Then $\mathbf{W}$ is equivalent to the $\mathbf{w}_i$ since the moment of the $\mathbf{w}_i$ about C and the moment of $\mathbf{W}$ about C are both zero.

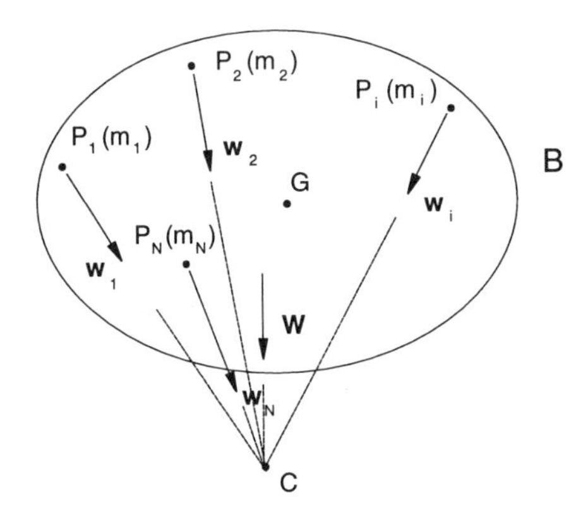

Fig. 9.6.4 Exaggerated Representation of the Size of a Body and Gravitational Attraction of the Earth on the Particles of B

To see that the line of action of $\mathbf{W}$ will not in general pass through the mass center G of B observe that the particles of B closer to C will have slightly greater attraction to E than particles of equal mass but further away from C, thus creating an unbalance in the moment about G. This is illustrated by the following example.

Example 9.6.1 <u>Gravitational Forces on a Dumbbell</u>

Consider a dumbbell D consisting of two identical particles P_1 and P_2, each having mass m at opposite ends of a light (massless) rod of length 2ℓ as in Figure

9.6.5. Let D be located on the surface of the earth E and let D be supported by a frictionless pin at a point O in the interior of D as shown. Let O be approximately in the center of D with the distance OP_1 being ℓ_1 and OP_2 being ℓ_2. Thus we have

$$\ell_1 + \ell_2 = 2\ell \qquad (9.6.9)$$

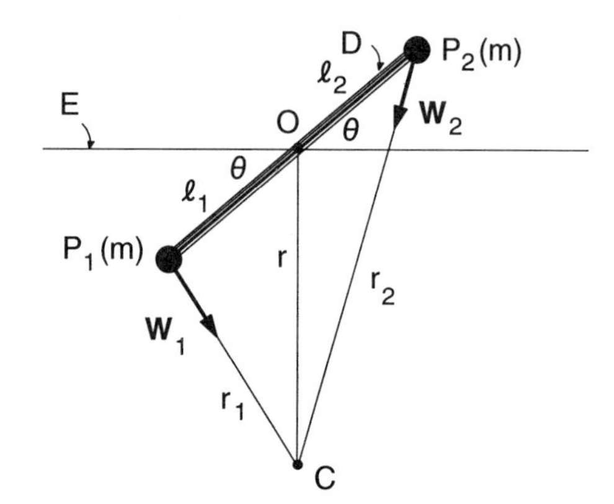

Fig. 9.6.5 A Dumbbell Pinned on the Surface of the Earth E and Inclined at an Angle θ Relative to E

Let C be the center of the earth and let the distance CE be r. Let D be inclined to E at an angle θ as shown in Figure 9.6.5. Let the distance from C to P_1 be r_1 and from C to P_2 be r_2. Then for the angle θ shown in Figure 9.6.5 r_1 is smaller than r where r_2 is larger than r. Let $\mathbf{W}_1$ and $\mathbf{W}_2$ be the gravity (or weight) forces on P_1 and P_2. From Equation (9.6.1) with r_1 being smaller than r_2 the magnitude of $\mathbf{W}_1$ will be larger than the magnitude of $\mathbf{W}_2$. Therefore, if O is in the center of D there will be an imbalance of forces—or a "gravitational moment"—about O.

To explore this further consider the geometrical parameters of Figure 9.6.6. Specifically, let ϕ_1 and ϕ_2 be the angles between the "vertical" lines at C, let β_1 and β_2 be the angles at P_1 and P_2 and let θ_2 be the complement of θ at O. Let Unit vectors $\mathbf{v}_1$ and $\mathbf{v}_2$ be parallel to P_1C and P_2C; let $\boldsymbol{\lambda}$ be a unit vector parallel to P_1P_2; let $\mathbf{k}$ be a vertical (directed downward) unit vector at O; let $\mathbf{i}$ be a horizontal unit vector parallel to the P_1P_2C plane; and let $\mathbf{j}$ be normal to the P_1P_2C plane so that $\mathbf{j}$ is $\mathbf{k} \times \mathbf{i}$.

Observe that if ℓ is much smaller than r, then r_1, r, and r_2 are nearly equal to each other and ϕ_1 and ϕ_2 are small.

From Figure 9.6.6 with ϕ_1 and ϕ_2 being small we have from the law of sines that

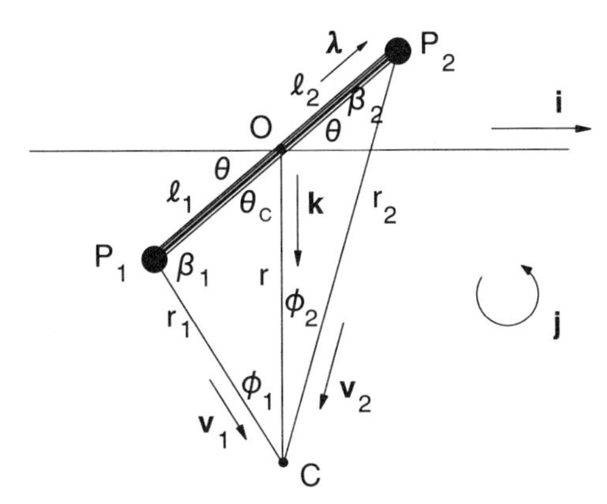

Fig. 9.6.6 Dumbbell System Geometry

$$\phi_1/\ell_1 = (1/r)\sin\beta_1 \quad \text{and} \quad \phi_2/\ell_2 = (1/r)\sin\beta_2 \tag{9.6.10}$$

Also since ϕ_1 and ϕ_2 are small $\sin\beta_1$ and $\sin\beta_2$ are each approximately equal to $\cos\theta$. [That is, $\beta_1 \approx \pi/2 + \theta$ and $\beta_2 \approx \pi/2 - \theta$.] Hence ϕ_1 and ϕ_2 are approximately

$$\phi_1 = (\ell_1/r)\cos\theta \quad \text{and} \quad \phi_2 = (\ell_2/r)\cos\theta \tag{9.6.11}$$

From Equation (9.6.1) the weight forces $\mathbf{W}_1$ and $\mathbf{W}_2$ may be expressed as

$$\mathbf{W}_1 = (\kappa/r_1^2)\mathbf{v}_1 \quad \text{and} \quad \mathbf{W}_2 = (\kappa/r_2^2)\mathbf{v}_2 \tag{9.6.12}$$

where κ is

$$\kappa = GMm = mgr^2 \tag{9.6.13}$$

where as before G is the universal gravity constant and M is the mass of E. [See Equations (9.6.2), (9.6.3), and (9.6.4).

Consider the moments $\mathbf{M}_O$ of $\mathbf{W}_1$ and $\mathbf{W}_2$ about O: From Figures 9.6.5 and 9.6.6 we have

$$\begin{aligned}\mathbf{M}_O &= -\ell_1 \boldsymbol{\lambda} \times \mathbf{W}_1 + \ell_2 \boldsymbol{\lambda} \times \mathbf{W}_2 \\ &= \kappa[-(\ell_1/r_1^2)\, \boldsymbol{\lambda} \times \mathbf{v}_1 + (\ell_2/r_2^2)\, \boldsymbol{\lambda} \times \mathbf{v}_2]\end{aligned} \tag{9.6.14}$$

From Figure 9.6.6 we see that with ϕ_1 and ϕ_2 being small $\boldsymbol{\lambda}$, $\mathbf{v}_1$ and $\mathbf{v}_2$ may be expressed as:

$$\boldsymbol{\lambda} = \cos\theta\,\mathbf{i} - \sin\theta\,\mathbf{k} \qquad \mathbf{v}_1 = \mathbf{k} + \phi_1\mathbf{i} \qquad \mathbf{v}_2 = \mathbf{k} - \phi_2\mathbf{i} \tag{9.6.15}$$

Then by substituting into Equation (9.6.14) $\mathbf{M}_O$ becomes:

$$\mathbf{M}_O = \kappa[(\ell_1/r_1^2)(\cos\theta + \phi_1 \sin\theta) - (\ell_2/r_2^2)(\cos\theta - \phi_2 \sin\theta)]\mathbf{j} \tag{9.6.16}$$

Next, by referring again to Figure 9.6.6, from the law of sines we see that r_1 and r_2 may be expressed in terms of r as

$$(1/r_1)\sin\theta_C = (1/r)\sin\beta_1 \quad \text{and} \quad (1/r_2)\sin(\pi/2 + \theta) = (1/r)\sin\beta_2 \tag{9.6.17}$$

Since θ_C is the complement of θ (that is, $\pi/2 - \theta$), β_1 is $\pi/2 + \theta - \phi_1$. Therefore, from Equation (9.6.17) $1/r_1^2$ is approximately

$$\begin{aligned}1/r_1^2 &= (1/r^2)\frac{\sin^2\beta_1}{\sin^2\theta_C} = (1/r^2)\frac{\sin^2(\pi/2+\theta-\phi_1)}{\sin^2(\pi/2-\theta)} = (1/r^2)\frac{(\cos\theta+\phi_1\sin\theta)^2}{\cos^2\theta} \\ &= (1/r^2)(1+\phi_1\tan\theta)^2 = (1/r^2)(1+2\phi_1\tan\theta)\end{aligned} \tag{9.6.18}$$

Similarly, β_2 is $\pi/2 - \theta - \phi_2$ and thus from Equation (9.6.17) $1/r_2^2$ is

$$1/r_2^2 = (1/r^2)\frac{\sin^2\beta_2}{\sin^2(\pi/2+\theta)} = (1/r^2)\frac{\sin^2(\pi/2-\theta-\phi_2)}{\sin^2(\pi/2+\theta)} = (1/r^2)\frac{(\cos\theta-\phi_2\sin\theta)^2}{\cos^2\theta}$$

$$= (1/r^2)(1-\phi_2\tan\theta)^2 = (1/r^2)(1-2\phi_2\tan\theta) \tag{9.6.19}$$

By substituting from Equations (9.6.18) and (9.6.19) into Equation (9.6.16) $\mathbf{M}_O$ becomes after neglecting small terms

$$\begin{aligned}\mathbf{M}_O &= \kappa[(\ell_1/r^2)(\cos\theta + 3\phi_1\sin\theta) - (\ell_2/r^2)(\cos\theta - 3\phi_2\sin\theta)]\mathbf{j} \\ &= mg[(\ell_1-\ell_2)\cos\theta + 3(\ell_1\phi_1 + \ell_2\phi_2)\sin\theta]\mathbf{j}\end{aligned} \tag{9.6.20}$$

where from Equation (9.6.13) κ is mgr^2.

To interpret the result of Equation (9.6.20) let ℓ_1 and ℓ_2 be expressed as

$$\ell_1 = \ell - \epsilon \quad \text{and} \quad \ell_2 = \ell + \epsilon \tag{9.6.21}$$

where ε is small. Then by substituting into Equation (9.6.20) $\mathbf{M}_O$ becomes:

$$\mathbf{M}_O = mg[-2\varepsilon\cos\theta + (6\ell^2/r)\sin\theta\cos\theta]\mathbf{j} \tag{9.6.22}$$

If ε is zero, that is, if the dumbbell is supported at its mass center the imbalance (or "gravity moment") is

$$\mathbf{M}_O = (6mg\ell^2/r)\sin\theta\cos\theta\,\mathbf{j} \tag{9.6.23}$$

The maximum imbalance magnitude M_{max} thus occurs when θ is 45 degrees. That is

$$M_{max} = 3mg\ell^2/r \tag{9.6.24}$$

If, for example, the dumbbell length is 3 feet (0.91 m) with 100 pound (45.36 kg) masses, M_{max} is

$$M_{max} = 1.3 \times 10^{-4} \text{ ft lb} \quad (9.59 \times 10^{-1} \text{ Nm}) \tag{9.6.25}$$

Alternatively, if $\mathbf{M}_O$ is to be zero, that is, if the support is to be at the "center of gravity," ε is

$$\varepsilon = (3\ell^2/r)\sin\theta \tag{9.6.26}$$

For θ equal to 45 degrees, with a 3 ft dumbbell, ε is

$$\varepsilon = 9.1 \times 10^{-7} \text{ ft} \quad (2.8 \times 10^{-7} \text{ m}) \tag{9.6.27}$$

Thus there is little difference between the center of gravity and the mass center.

9.7 Gravitational Moment of Orthogonal, Nonintersecting Rods

Consider the mutual gravitational attraction of two rods R_1 and R_2 aligned along non-intersecting orthogonal lines as in Figure 9.7.1 (see Reference [9.4]), where h is the distance between the lines along the common perpendicular. Let the rods have lengths L_1 and L_2 and let each have uniform density ρ per unit length.

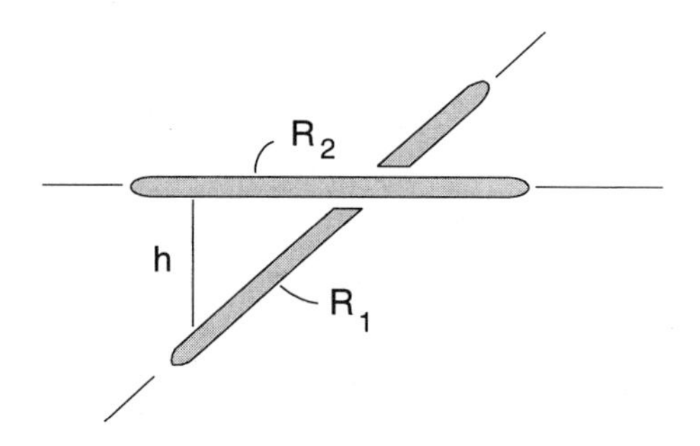

Fig. 9.7.1 Non-Intersecting Orthogonal Rods

Let the common perpendicular intersect the rods at points O_1 and O_2 and let the rod lengths on either side of O_1 and O_2 be ℓ_1, ℓ_2, ℓ_3, and ℓ_4 as in

Figure 9.7.2. Let $\mathbf{n}_1$ and $\mathbf{n}_2$ be unit vectors along R_1 and R_2 and let $\mathbf{n}_3$ be a unit vector along the common perpendicular. Let ξ and η be distance coordinates measured from O_1 and O_2 along R_1 and R_2 as indicated in Figure 9.7.2.

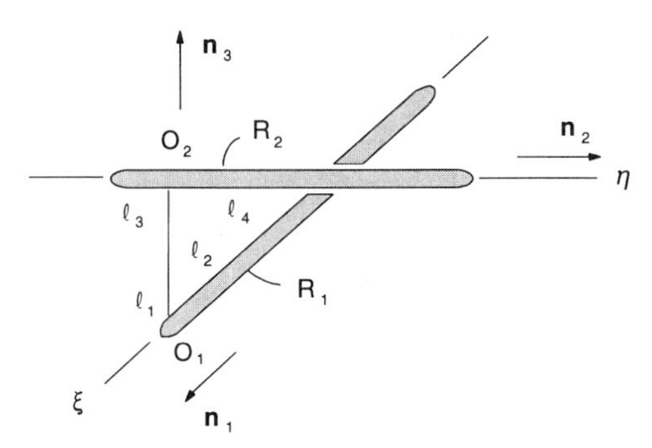

Fig. 9.7.2 Rod Geometry

Next, let the gravitational attraction between the rods be modeled by the superposition of the attraction between differential elements of the rods as represented in Figure 9.7.3. Specifically, the force $\mathbf{dF}$ of an element of R_2 exerted on an element of R_1 is:

$$\mathbf{dF} = G\frac{(\rho d\xi)(\rho d\eta)}{r^2}\boldsymbol{\lambda} \tag{9.7.1}$$

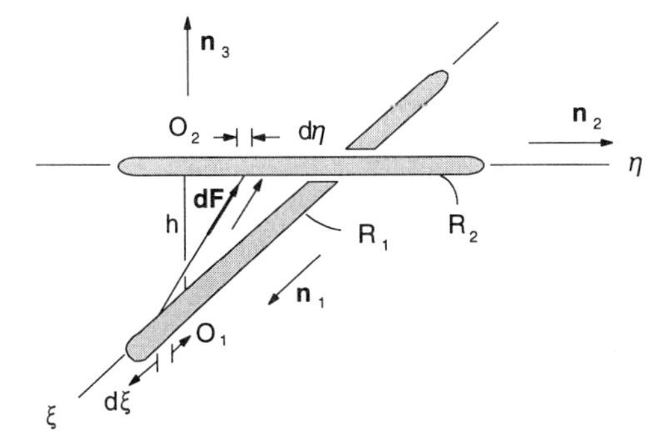

Fig. 9.7.3 Differential Gravitational Forces **dF** Between the Rods

where, as before, G is the universal gravitational constant (6.673×10^{-11} $m^3/kg\,s^2$), r is the distance between the elements, and $\boldsymbol{\lambda}$ is a unit vector along the segment connecting the elements, as shown in Figure 9.7.3. From Figure 9.7.3 we see that $\boldsymbol{\lambda}$ and r are:

$$\boldsymbol{\lambda} = (-\xi\,\mathbf{n}_1 + \eta\,\mathbf{n}_2 + h\,\mathbf{n}_3)/(\xi^2 + \eta^2 + h^2)^{1/2} \tag{9.7.2}$$

and

$$r = (\xi^2 + \eta^2 + h^2)^{1/2} \tag{9.7.3}$$

Therefore, $\mathbf{dF}$ becomes

$$\mathbf{dF} = G\rho^2 \left[\frac{-\xi\,\mathbf{n}_1 + \eta\,\mathbf{n}_2 + h\,\mathbf{n}_3}{(\xi^2 + \eta^2 + h^2)^{1/2}} \right] d\eta\, d\xi \tag{9.7.4}$$

By superposing the forces from all the elements of R_2 on all the elements of R_1 we obtain the resultant gravitational force $\mathbf{F}$ exerted by R_2 on R_1 as

$$\mathbf{F} = G\rho^2 \int_{-\ell_2}^{\ell_1} \left\{ \int_{-\ell_4}^{\ell_3} \left[\frac{-\xi\mathbf{n}_1 + \eta\mathbf{n}_2 + h\mathbf{n}_3}{\xi^2 + \eta^2 + h^2)^{1/2}} \right] d\eta \right\} d\xi \tag{9.7.5}$$

By carrying out the indicated integration, $\mathbf{F}$ becomes:

$$\mathbf{F} = F_1\mathbf{n}_1 + F_2\mathbf{n}_2 + F_3\mathbf{n}_3 \tag{9.7.6}$$

where the components F_1, F_2, and F_3 are:

$$F_1 = (1/2)G\rho^2 \ell n \left| \frac{(d_{14} - \ell_4)(d_{13} - \ell_3)(d_{24} + \ell_4)(d_{23} + \ell_3)}{(d_{14} + \ell_4)(d_{13} + \ell_3)(d_{24} - \ell_4)(d_{23} - \ell_3)} \right| \tag{9.7.7}$$

$$F_2 = G\rho^2 \ell n \left| \frac{(d_{13} + \ell_1)(d_{24} - \ell_2)}{(d_{14} + \ell_1)(d_{23} - \ell_2)} \right| \tag{9.7.8}$$

and

$$F_3 = G\rho^2 \left[\tan^{-1}\left(\frac{\ell_1 \ell_4}{hd_{14}} \right) + \tan^{-1}\left(\frac{\ell_1 \ell_3}{hd_{13}} \right) + \tan^{-1}\left(\frac{\ell_2 \ell_4}{hd_{24}} \right) + \tan^{-1}\left(\frac{\ell_2 \ell_3}{hd_{23}} \right) \right] \tag{9.7.9}$$

where d_{13}, d_{14}, d_{23}, and d_{24} are defined as

$$d_{13} \stackrel{D}{=} (\ell_1^2 + h^2 + \ell_3^2)^{1/2} \qquad d_{14} \stackrel{D}{=} (\ell_1^2 + h^2 + \ell_4^2)^{1/2} \tag{9.7.10}$$

$$d_{23} \stackrel{D}{=} (\ell_2^2 + h^2 + \ell_3^2)^{1/2} \qquad d_{24} \stackrel{D}{=} (\ell_2^2 + h^2 + \ell_n^2)^{1/2} \tag{9.7.11}$$

Let the gravitational forces exerted by R_2 and R_1 be equivalent to a single force **F** passing through O_2 together with a couple with torque **T**. Then **F** is given by Equations (9.7.6) to (9.7.11) and **T** is the sum of the moments of the differential forces **dF** about O_2. Specifically, the force **dF** has a moment **dT** about O_2 given by

$$\mathbf{dT} = G\rho^2 \frac{h\eta \mathbf{n}_4 + \xi\eta \mathbf{n}_3}{(\xi^2 + h^2 + \eta^2)^{3/2}} d\eta \, d\xi \tag{9.7.12}$$

Then the torque **T** becomes:

$$\mathbf{T} = G\rho^2 \int_{-\ell_2}^{\ell_1} \left\{ \int_{-\ell_4}^{\ell_3} \left[\frac{h\eta \mathbf{n}_1 + \xi\eta \mathbf{n}_3}{(\xi^2 + h^2 \eta^2)^{3/2}} \right] d\eta \right\} d\xi \tag{9.7.13}$$

By carrying out the indicated integrations, **T** becomes:

$$\mathbf{T} = T_1\mathbf{n}_1 + T_2\mathbf{n}_2 + T_3\mathbf{n}_3 \tag{9.7.14}$$

where the components T_1, T_2, and T_3 are

$$T_1 = h\,\ell n\left|\frac{(d_{24} - \ell_2)(d_{13} + \ell_1)}{d_{14} + \ell_1)(d_{23} - \ell_2)}\right| \tag{9.7.15}$$

$$T_2 = 0 \tag{9.7.16}$$

$$T_3 = -d_{14} + d_{24} + d_{13} - d_{23} \tag{9.7.17}$$

9.8 Gravitational Forces on a Satellite

Consider the gravitational forces exerted by the earth on a satellite. Let a satellite S be in orbit about the earth E as in Figure 9.8.1 where the scale of the satellite is greatly exaggerated for analysis convenience. Let E, the earth, be modeled as a sphere with mass M concentrated at its center O. Let the mass center of S, the satellite, be G and let $\mathbf{p}_G$ locate G relative to O. Let **k** be a unit vector parallel to $\mathbf{p}_G$, and thus a local vertical of E. Let d be the magnitude of $\mathbf{p}_G$ so that

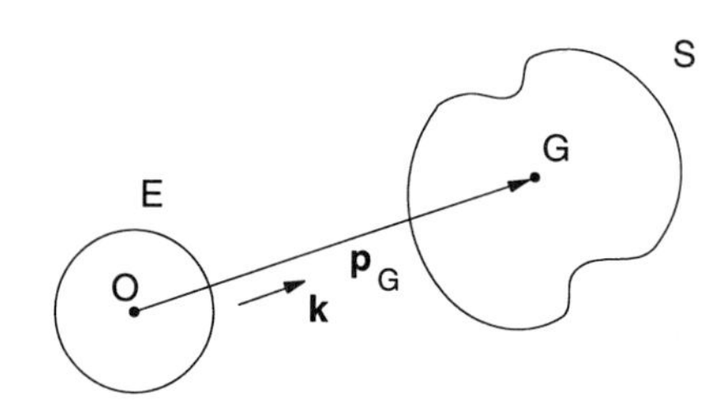

Fig. 9.8.1 A Satellite in Orbit about the Earth

$$\mathbf{p}_G = d\mathbf{k} \tag{9.8.1}$$

To examine the gravitational attraction of E on S, let S be modeled as a body composed of particles as in Figure 9.8.2 where P_i is a typical particle of S. Let $\mathbf{r}_i$ locate P_i relative to G, and let $\mathbf{p}_i$ locate P_i relative to the center O of E. Then $\mathbf{p}_i$, $\mathbf{p}_G$, and $\mathbf{r}_i$ are related by the expression

$$\mathbf{p}_i = \mathbf{p}_G + \mathbf{r}_i = d\mathbf{k} + \mathbf{r}_i \tag{9.8.2}$$

Observe that the magnitude of $\mathbf{r}_i$ is much less than d (that is, $|\mathbf{r}_i| \ll d$).

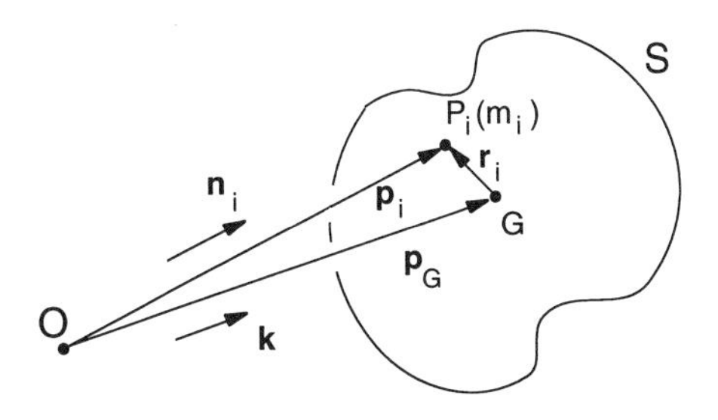

Fig. 9.8.2 A Satellite Represented by a System of Particles

The gravitational attraction of E on S may be represented as the superposition of the gravitational attraction of E on the individual particles of S. The gravitational attraction force $\mathbf{F}_i$ by E on a typical particle P_i of S is

$$\mathbf{F}_i = -GMm_i/|\mathbf{p}_i|^2\,\mathbf{n}_i \tag{9.8.3}$$

where as before G is the universal gravitational constant, m_i is the mass of P_i, and $\mathbf{n}_i$ is a unit vector parallel to $\mathbf{p}_i$ as in Figure 9.8.2. Specifically, $\mathbf{n}_i$ may be expressed as

$$\mathbf{n}_i = \mathbf{p}_i / |\mathbf{p}_i| \tag{9.8.4}$$

Hence, $\mathbf{F}_i$ may be expressed as:

$$\mathbf{F}_i = -GMm_i / |\mathbf{p}_i|^3 \, \mathbf{p}_i = -GMm_i (\mathbf{p}_i \cdot \mathbf{p}_i)^{-3/2} \mathbf{p}_i \tag{9.8.5}$$

Let the superposition of all the gravitational forces on the particles of S be represented by an equivalent force system consisting of a single force **F** passing through G together with a couple with torque **T**. Then **F** is simply the resultant of the forces $\mathbf{F}_i$ of Equation (9.8.5) and **T** is the resultant of the moment of the forces about G. That is

$$\mathbf{F} = \sum_{i=1}^{N} \mathbf{F}_i \quad \text{and} \quad \mathbf{T} = \sum_{i=1}^{N} \mathbf{r}_i \times \mathbf{F}_i \tag{9.8.6}$$

where N is the total number of particles comprising S.

We may develop the expression for **F** and **T** by expressing $\mathbf{p}_i$ in terms of $\mathbf{p}_G$ and $\mathbf{r}_i$ and by taking advantage of the large difference in magnitude between $\mathbf{p}_G$ and $\mathbf{r}_i$. Specifically, from Equations (9.8.2) and (9.8.5) $\mathbf{F}_i$ may be expressed as:

$$\mathbf{F}_i = -GMm_i [(\mathbf{p}_G + \mathbf{r}_i) \cdot (\mathbf{p}_G + \mathbf{r}_i)]^{-3/2} (\mathbf{p}_G + \mathbf{r}_i) \tag{9.8.7}$$

Then by expressing $\mathbf{p}_G$ as $d\mathbf{k}$, by expanding the radical using the binomial series, and by neglecting higher order small terms we obtain:

$$\begin{aligned}\mathbf{F}_i &= -GMm_i[d^2(1 + 2\mathbf{k}\bullet\mathbf{r}_i/d + \mathbf{r}_i^2/d^2)]^{-3/2}(d\mathbf{k} + \mathbf{r}_i)\\ &= -\frac{GMm_i}{d^3}(1 - 3\mathbf{k}\bullet\mathbf{r}_i/d)(d\mathbf{k} + \mathbf{r}_i)\\ &= -\frac{GMm_i}{d^3}[d\mathbf{k} - 3(\mathbf{k}\bullet\mathbf{r}_i)\mathbf{k} + \mathbf{r}_i]\end{aligned} \tag{9.8.8}$$

where terms of the type $(|\mathbf{r}_i|/d)^2$ and smaller have been neglected. Similarly, the term $\mathbf{r}_i \times \mathbf{F}_i$ becomes

$$\mathbf{r}_i \times \mathbf{F}_i = -\frac{GMm_i}{d^3}[d\mathbf{r}_i \times \mathbf{k} - 3(\mathbf{k}\bullet\mathbf{r}_i)\mathbf{r}_i \times \mathbf{k}] \tag{9.8.9}$$

Finally, by substituting from Equations (9.8.8) and (9.8.9) into Equation (9.8.6) **F** and **T** become:

$$\mathbf{F} = -\frac{GM}{d^3}\left\{\left(\sum_{i=1}^{N} m_i\right)d\mathbf{k} - 3\left[\mathbf{k}\bullet\left(\sum_{i=1}^{N} m_i\mathbf{r}_i\right)\right]\mathbf{k} + \sum_{i=1}^{N} m_i\mathbf{r}_i\right\} \tag{9.8.10}$$

or

$$\mathbf{F} = -\frac{GMm}{d^2}\mathbf{k} \tag{9.8.11}$$

where m is the total mass of S $\left(\sum_{i=1}^{N} m_i\right)$ and where the last two terms of Equation (9.8.10) are zero since G is the mass center with $\sum_{i=1}^{N} m_i\mathbf{r}_i$ being zero [see Equation (8.3.2)]; and

$$\mathbf{T} = -\frac{GM}{d^3}\left\{d\left(\sum_{i=1}^{N} m_i\mathbf{r}_i\right) \times \mathbf{k} - 3\sum_{i=1}^{N} m_i(\mathbf{k}\bullet\mathbf{r}_i)\mathbf{r}_i \times \mathbf{k}\right\} \tag{9.8.12}$$

or

$$\mathbf{T} = 3\frac{GM}{d^3}\sum_{i=1}^{N} m_i(\mathbf{k} \cdot \mathbf{r}_i)\mathbf{r}_i \times \mathbf{k} \tag{9.8.13}$$

where again the first term is zero. (See also [9.5].)

We can express $\mathbf{T}$ in terms of a second moment vector [see Equation (8.5.4)] by observing that

$$\mathbf{k} \times [\mathbf{r}_i \times (\mathbf{k} \times \mathbf{r}_i)] \equiv \mathbf{k} \cdot \mathbf{r}_i(\mathbf{r}_i \times \mathbf{k}) \tag{9.8.14}$$

Therefore, $\mathbf{T}$ takes the simple form

$$\mathbf{T} = 3\frac{GM}{d^3}\mathbf{k} \times \mathbf{I}_{\mathbf{k}}^{S/G} \tag{9.8.15}$$

9.9 Generalized Forces on Rigid Bodies

If we model a rigid body B as a collection of particles fixed relative to one another we can determine the generalized forces on B by superposing the generalized forces on the individual particles of B (see Section 9.5). Specifically, let B be a rigid body which is a member of a mechanical system S which in turn is moving in an inertial reference frame R, as depicted in Figure 9.9.1. Let S have n degrees of freedom in R, represented by generalized coordinates q_r $(r = 1, \ldots, n)$. Let B be composed of N particles P_i $(i = 1, \ldots, N)$ as in Figure 9.9.2 and let $\mathbf{F}_i$ $(i = 1, \ldots, N)$ be forces passing through the particles. Then from Equation (4.6.2) the generalized force F_{q_r} on B for the generalized coordinate q_r is:

$$F_{q_r} = \sum_{i=1}^{N} \mathbf{F}_i \cdot \mathbf{V}_{q_r}^{P_i} \tag{9.9.1}$$

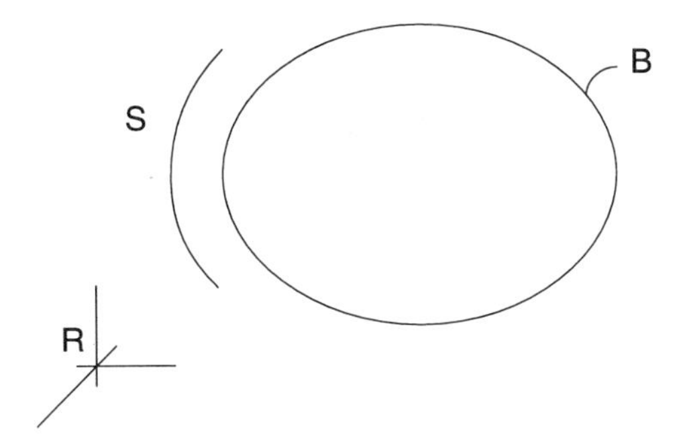

Fig. 9.9.1 A Body B as a Member of a Mechanical System S in a Reference Frame R

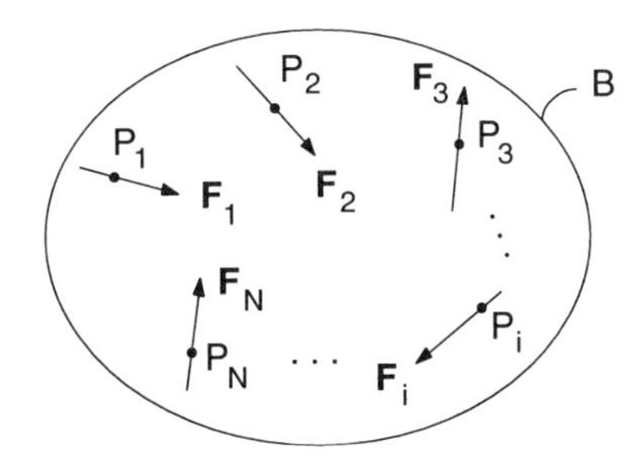

Fig. 9.9.2 A Body B Represented as a Set of Particles with Forces Passing through the Particles

where as before $\mathbf{V}_{\dot{q}_r}^{P_i}$ is the partial velocity in R of a typical particle P_i relative to the generalized coordinate q_r (see Section 4.5.4).

Let G be the mass center of B and let $\mathbf{r}_i$ locate typical particle P_i relative to G as in Figure 9.9.3. Then the partial velocity of P_i may be expressed as [see Equations (6.10.3) and (6.13.5)].

$$\mathbf{V}_{\dot{q}_r}^{P_i} = \mathbf{V}_{\dot{q}_r}^{G} + \boldsymbol{\omega}_{\dot{q}_r} \times \mathbf{r}_i \qquad (r = 1,\ldots,n;\ i = 1,\ldots,N) \tag{9.9.2}$$

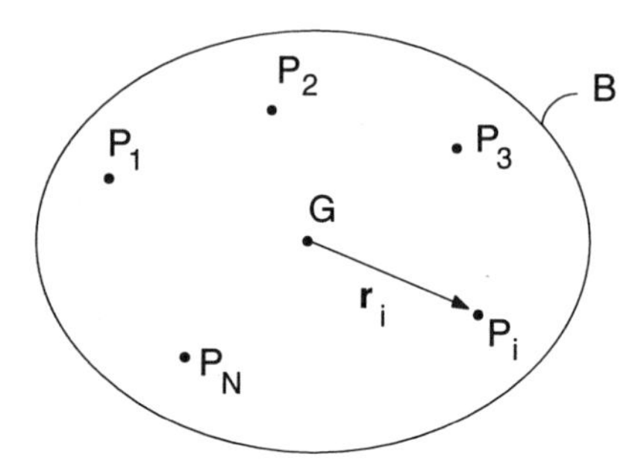

Fig. 9.9.3 A Body B with Mass Center G and Particle Position Vectors $\mathbf{r}_i$

where $\boldsymbol{\omega}_{\dot{q}_r}$ is the partial angular velocity of B in R with respect to the generalized coordinate q_r [see Equation (6.13.3)]. Then by substituting into Equation (9.9.1) F_{q_r} becomes

$$\begin{aligned} F_{q_r} &= \sum_{i=1}^{N} \mathbf{F}_i \bullet (\mathbf{V}^G_{\dot{q}_r} + \boldsymbol{\omega}_{\dot{q}_r} \times \mathbf{r}_i) \\ &= \left(\sum_{i=1}^{N} \mathbf{F}_i \right) \bullet \mathbf{V}^G_{\dot{q}_r} + \sum_{i=1}^{N} \mathbf{F}_i \bullet \boldsymbol{\omega}_{\dot{q}_r} \times \mathbf{r}_i \\ &= \left(\sum_{i=1}^{N} \mathbf{F}_i \right) \bullet \mathbf{V}^G_{\dot{q}_r} + \left(\sum_{i=1}^{N} \mathbf{r}_i \times \mathbf{F}_i \right) \bullet \boldsymbol{\omega}_{\dot{q}_r} \end{aligned} \tag{9.9.3}$$

or

$$F_{q_r} = \mathbf{F} \bullet \mathbf{V}^G_{\dot{q}_r} + \mathbf{T} \bullet \boldsymbol{\omega}_{\dot{q}_r} \tag{9.9.4}$$

where **F** is the resultant of the forces on B and where **T** is the resultant of the moments of the forces on B about G. [See also Equation (9.5.7).]

Equation (9.9.4) shows that if a force system on a body B is replaced by an equivalent force system consisting of a single force **F**, passing through the mass center G of B, together with a couple with torque **T**, then the generalized force F_{q_r} on B, for the generalized coordinate q_r, is simply the projection of **F** on the partial velocity of G added to the projection of **T** onto the partial angular velocity of B.

9.10 Applied and Inertia Forces

As noted in Chapter 4, forces on particles may be divided into two categories: applied (or "active") forces and inertia (or "passive") forces. Since rigid bodies may be modeled as collections of particles at fixed distances relative to one another, the forces on rigid bodies may also be considered as being either applied (active) or inertia (passive).

Inertia forces on a body occur when the particles of the body have accelerations in an inertial (or "Newtonian") reference frame (see Section 4.4). Applied forces on a body are all other forces. That is, applied forces are all forces which arise aside from those due to particle acceleration. Applied forces thus include gravity forces, contact forces, electromagnetic forces, and nuclear forces.

In the following sections we will review and summarize expressions for both applied and inertia forces exerted on rigid bodies.

9.11 Generalized Applied (Active) Forces

Consider a body B moving in an inertial reference frame R and subjected to an applied force field consisting of forces $\mathbf{F}_1,\ldots,\mathbf{F}_N$ exerted on line of action passing through particles $P_1,\ldots,P_N$ of B as represented in Figure 9.11.1. Let B be part of a mechanical system S having n degrees of freedom in R measured by generalized coordinates q_r $(r = 1,\ldots,n)$. Then from

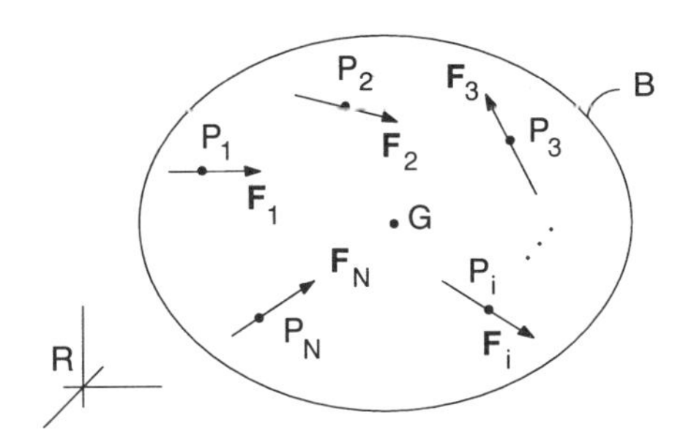

Fig. 9.11.1 A Body B Subjected to Applied Forces

Equation (9.9.1) the generalized active force F_{q_r} on B for the generalized coordinate q_r is simply the sum of the generalized forces for q_r on the individual particles of B. That is

$$F_{q_r} = \sum_{i=1}^{N} F_{q_r}^{P_i} = \sum_{i=1}^{N} \mathbf{V}_{\dot{q}_r}^{P_i} \bullet \mathbf{F}_i \tag{9.11.1}$$

The summation of Equation (9.11.1) may be avoided by using Equation (9.9.4) as follows: Let the force system on B be replaced by an equivalent force system consisting of a single force **F** passing through an arbitrary point of B, say the mass center G, together with a couple with torque **T**. Then **F** is simply the resultant of the force system on B and **T** is the sum of the moments of the forces about G. Equation (9.9.4) then gives F_{q_r} as:

$$F_{q_r} = \mathbf{F} \bullet \mathbf{V}_{\dot{q}_r}^{G} + \mathbf{T} \bullet \boldsymbol{\omega}_{\dot{q}_r} \tag{9.11.2}$$

where, as before, $\mathbf{V}_{\dot{q}_r}^{G}$ is the partial velocity of G with respect to q_r and $\boldsymbol{\omega}_{\dot{q}_r}$ is the partial angular velocity of B with respect to q_r [see Equation (6.13.5)].

9.11.1 Contribution of Gravity (or Weight) Forces to the Generalized Active Forces

Equation (9.11.2) is especially useful with gravity (or weight) forces: Suppose a rigid body B, near the surface of the earth, is modeled, as before, as a set of particles fixed relative to one another. Suppose further that the corresponding set of gravitational forces on the particles of B are represented by a single vertical force **W** passing through the mass center G of B, as represented in Figure 9.11.2 where **W** may be expressed as:

$$\mathbf{W} = -mg\,\mathbf{n}_z \tag{9.11.3}$$

where m is the mass of B, g is the gravity acceleration, and $\mathbf{n}_z$, together with $\mathbf{n}_x$ and $\mathbf{n}_y$, are mutually perpendicular unit vectors fixed in the earth (regarded as an inertial reference frame) with $\mathbf{n}_z$ being vertical.

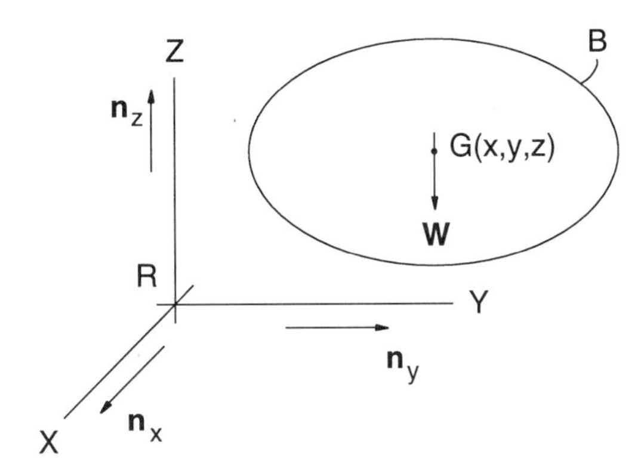

Fig. 9.11.2 A Body B in a Gravitational Force Field Whose Effect on B is Represented by a Single Force **W**

Suppose further that B is a member of a mechanical system S having n degrees of freedom represented by generalized coordinates q_r $(r = 1, \ldots, n)$. Then from Equation (9.11.2) the contribution F_{q_r} to the generalized force on B for the coordinate q_r is:

$$F_{q_r} = \mathbf{V}^G_{q_r} \bullet \mathbf{W} = -mg\mathbf{V}^G_{q_r} \bullet \mathbf{n}_z \tag{9.11.4}$$

Finally, suppose that the velocity of G in R may be expressed as

$$\mathbf{V}^G = \dot{x}\mathbf{n}_x + \dot{y}\mathbf{n}_y + \dot{z}\mathbf{n}_z \tag{9.11.5}$$

where x, y and z are the coordinates of G in a Cartesian reference frame fixed on the earth (see Figure 9.11.2). Then x, y, and z are in general functions of the coordinates q_r and time t. That is,

$$x = x(q_r, t) \qquad y = y(q_r, t) \qquad z = z(q_r, t) \tag{9.11.6}$$

Then $\dot{z}$ may be expressed as

$$\dot{z} = \frac{dz}{dt} = \sum_{r=1}^{n} \frac{\partial z}{\partial q_r} \dot{q}_r + \frac{\partial z}{\partial t} \tag{9.11.7}$$

Hence, from Equation (9.11.5) $\mathbf{V}_{\dot{q}_r}^{G}$ is

$$\mathbf{V}_{\dot{q}_r}^{G} = \frac{\partial z}{\partial q_r} \mathbf{n}_z \tag{9.11.8}$$

and then F_{q_r} becomes

$$F_{q_r} = -mg \frac{\partial z}{\partial q_r} \tag{9.11.9}$$

[Compare Equation (9.11.9) with Equation (4.6.9) for a single particle.]

9.11.2 Contribution of Internal Forces Between the Particles of a Rigid Body to the Generalized Active Forces

Consider again the modeling of a body B by a set of particles fixed relative to one another. Let the particles be small (or of negligible size) so that they may be represented by points.

Let P and Q be a typical pair of adjoining particles of B as represented in Figure 9.11.3. With the particle size being negligible, the positions of the adjacent particles are essentially the same, and thus as B moves in a reference frame R, the velocities and accelerations of the adjacent particles are the same. Therefore, the inertia forces on P and Q are the same.

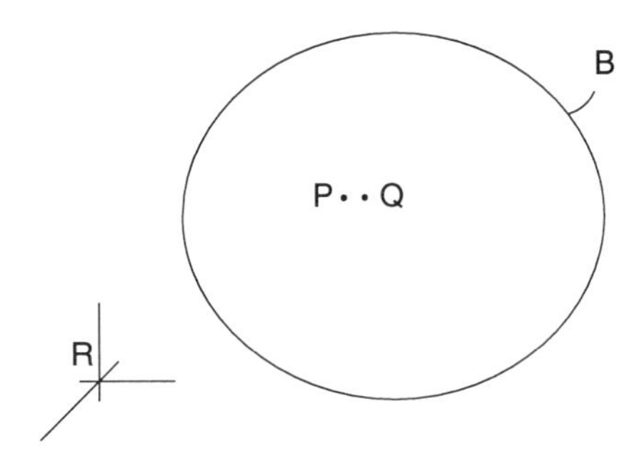

Fig. 9.11.3 A Rigid Body B and Two Typical Adjoining Particles of B

With particles P and Q being adjacent to and adjoining each other, there will in general be contact forces between them. With the particles being small, the contact forces exerted by say P on Q can be represented by a single force **C** passing through Q (with no moment, since P and Q are small). And similarly, by the law of "action-reaction" [9.4], the forces exerted by Q on P may be represented by a single force $-\mathbf{C}$ passing through P as depicted in Figure 9.11.4.

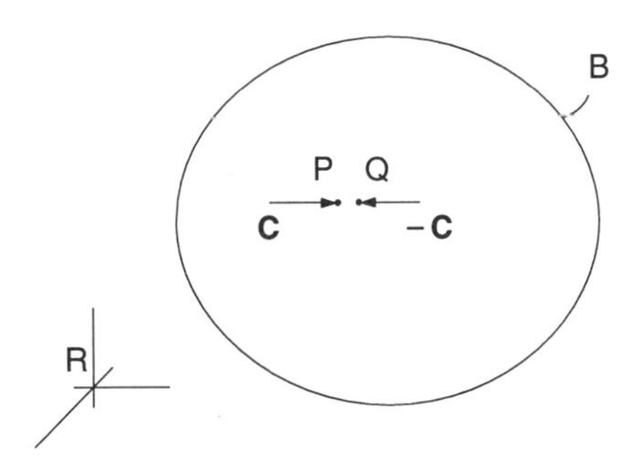

Fig. 9.11.4 Equal Magnitudc but Oppositcly Directed Contact Forces Between Adjacent Particles of a Rigid Body B

As before, let B be a member of a mechanical system S having n degrees of freedom in a reference frame R represented by generalized coordinates q_r $(r = 1, \ldots, n)$. Then the contribution of **C** and $-\mathbf{C}$ to the generalized active force F_{q_r} is zero since the partial velocities of P and Q with respect to $\dot{q}_r$ are the same. That is, $\mathbf{V}^Q_{\dot{q}_r} = \mathbf{V}^P_{\dot{q}_r}$ and then

$$F_{q_r} = \mathbf{C} \bullet \mathbf{V}^Q_{\dot{q}_r} + (-\mathbf{C}) \bullet \mathbf{V}^P_{\dot{q}_r} = \mathbf{V}^P_{\dot{q}_r}(\mathbf{C} - \mathbf{C}) = 0 \tag{9.11.10}$$

Equation (9.11.10) shows that internal contact forces between adjacent particles do not contribute to the generalized active forces and therefore they may be neglected in computing generalized active forces for a rigid body.

9.11.3 Contribution to Generalized Forces by Forces Exerted Across Smooth Surfaces Internal to a Mechanical System

Next, consider two rigid bodies B and $\hat{B}$ which are members of a mechanical system S with n degrees of freedom in an inertial reference frame R. As before, let these degrees of freedom be represented by generalized coordinates q_r $(r = 1, \ldots, n)$.

Let B and $\hat{B}$ be in contact with each other and let the contact occur across smooth surfaces of the bodies as represented in Figure 9.11.5. Let the contacting surfaces be convex so that the contact occurs at a single point. Let the corresponding contact points of B and $\hat{B}$ be C and $\hat{C}$. Let **n** be a unit vector normal to the plane of contact. Then the contact forces between B and $\hat{B}$ may be represented by single forces **F** and $\hat{\mathbf{F}}$ where **F** is the force on B by $\hat{B}$ and $\hat{\mathbf{F}}$ is the force on $\hat{B}$ by B.

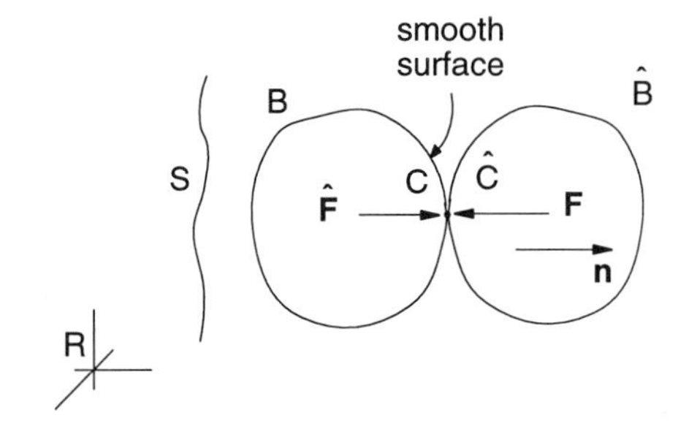

Fig. 9.11.5 Contacting Bodies with Smooth Surfaces within a Mechanical System

Since the surfaces are smooth the contact forces **F** and $\hat{\mathbf{F}}$ are parallel to **n**, and by the law of action and reaction [9.4] they have equal magnitudes but opposite direction. That is,

$$\mathbf{F} = -\hat{\mathbf{F}} = F\mathbf{n} \tag{9.11.11}$$

Since C and $\hat{C}$ are in contact, their relative velocity in the **n** direction (normal to the plane of contact) is zero. That is,

$$(\mathbf{V}^{C} - \mathbf{V}^{\hat{C}}) \cdot \mathbf{n} = 0 \tag{9.11.12}$$

Then, we also have

$$(\mathbf{V}^{C}_{\dot{q}_r} - \mathbf{V}^{\hat{C}}_{\dot{q}_r}) \cdot \mathbf{n} = 0 \tag{9.11.13}$$

From Equations (9.11.11) and (9.11.13) the contribution of **F** and $\hat{\mathbf{F}}$ to the generalized active force F_{q_r} is then:

$$\begin{aligned} F_{q_r} &= \mathbf{V}^{C}_{\dot{q}_r} \cdot \mathbf{F} + \mathbf{V}^{\hat{C}}_{\dot{q}_r} \cdot \hat{\mathbf{F}} \\ &= \mathbf{V}^{C}_{\dot{q}_r} \cdot F\mathbf{n} - \mathbf{V}^{C}_{\dot{q}_r} \cdot F\mathbf{n} \\ &= (\mathbf{V}^{C}_{\dot{q}_r} - \mathbf{V}^{\hat{C}}_{\dot{q}_r}) \cdot \mathbf{n}F \\ &= 0 \end{aligned} \tag{9.11.14}$$

Equation (9.11.14) shows that forces exerted across smooth surfaces internal to a mechanical system do not contribute to the generalized active forces.

9.11.4 Contribution to Generalized Forces by Forces Exerted at Points with Specified Motion

Consider further, a body B of a mechanical system S with n degrees of freedom in an inertial reference frame R. As before let these degrees of freedom be represented by generalized coordinates q_r $(r = 1,...,n)$. Let P be a point of B whose motion is specified in R. That is, let the velocity of P have the form:

$$^R\mathbf{V}^P = v(t)\mathbf{n} \tag{9.11.15}$$

where v(t) is a specified (or given) function of time t and where $\mathbf{n}$ is a unit vector parallel to $^R\mathbf{V}^P$. Then $^R\mathbf{V}^P$ is not dependent upon the q_r or the $\dot{q}_r$. Thus,

$$^R\mathbf{V}^P_{\dot{q}_r} = 0 \tag{9.11.16}$$

Let $\mathbf{F}$ be a force exerted on B at P. Then from Equation (9.11.16) the contribution of $\mathbf{F}$ to the generalized active force F_{q_r} for q_r is

$$F_{q_r} = \mathbf{F} \bullet {}^R\mathbf{V}^P_{\dot{q}_r} = 0 \tag{9.11.17}$$

Equation (9.11.17) shows that forces exerted on bodies at points whose motion is specified do not contribute to the generalized active forces.

9.11.5 Contribution to Generalized Forces by Forces Transmitted Across Rolling Surfaces of Bodies

Consider two bodies B_1 and B_2 of a mechanical system S which has n degrees of freedom in an inertial reference frame R. As before, let these degrees of freedom be represented by generalized coordinates q_r $(r = 1,...,n)$. Let B_1 and B_2 roll on each other as represented in Figure 9.11.6. Let C_1 and C_2 be the contact points between the bodies, and let $\mathbf{C}_1$ and $\mathbf{C}_2$ be the contact forces transmitted between the bodies at the contact points. Since B_1 and B_2 roll on each other the relative velocity of the contact points C_1 and C_2 is zero [see Equation (6.12.1)]. Then,

$$^R\mathbf{V}^{C_1} = {}^R\mathbf{V}^{C_2} \quad \text{and} \quad \mathbf{V}^{C_1}_{\dot{q}_r} = \mathbf{V}^{C_2}_{\dot{q}_r} \tag{9.11.18}$$

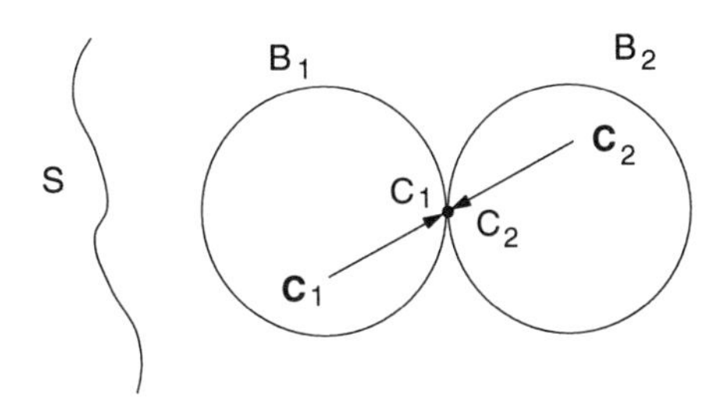

Fig. 9.11.6 Two Bodies of a Mechanical System Rolling on Each Other

Next, by the law of action-reaction [9.4] $\mathbf{C}_1$ and $\mathbf{C}_2$ are equal and opposite. That is,

$$\mathbf{C}_1 = -\mathbf{C}_2 \tag{9.11.19}$$

The contribution to the generalized force F_{q_r} by the rolling contact forces is then

$$F_{q_r} = \mathbf{C}_2 \bullet \mathbf{V}_{\dot{q}_r}^{C_1} + \mathbf{C}_1 \bullet \mathbf{V}_{\dot{q}_r}^{C_2} = (\mathbf{C}_2 + \mathbf{C}_1) \bullet \mathbf{V}_{\dot{q}_r}^{C_1} = 0 \tag{9.11.20}$$

Consider next a body B of the system S rolling on a body $\hat{B}$ whose motion is specified in R. (For example, $\hat{B}$ could be at rest in R.) Let C be the contact point of B with $\hat{B}$ and let $\mathbf{C}$ be the contact force exerted by $\hat{B}$ on B as in Figure 9.11.7. Then since the velocity of C in $\hat{B}$ is zero (B rolls on $\hat{B}$) and since the motion of $\hat{B}$ in R is specified, the velocity of C in R is independent

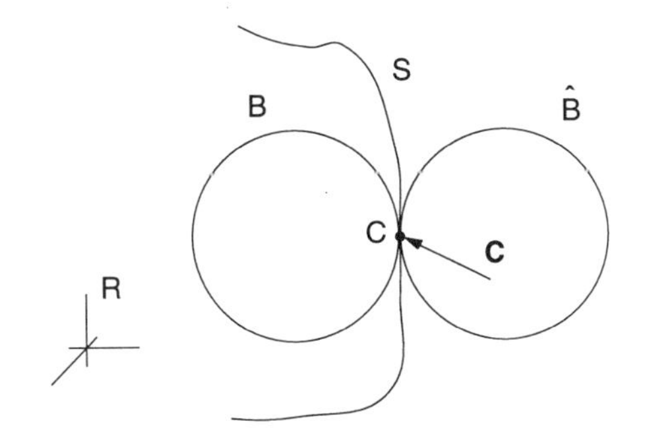

Fig. 9.11.7 A Body B of a System S Rolling on a Body $\hat{B}$ Whose Motion is Specified in an Inertial Reference Frame R

of the $\dot{q}_r$. That is,

$$ {}^{R}\mathbf{V}_{\dot{q}_r}^{C} = 0 \tag{9.11.21} $$

Therefore, the contribution to the generalized force F_{q_r} by the rolling contact force **C** is zero:

$$ F_{q_r} = {}^{R}\mathbf{V}_{\dot{q}_r}^{C} \bullet \mathbf{C} = 0 \tag{9.11.22} $$

9.11.6 Contribution to Generalized Forces by Forces Exerted by Springs Between Bodies Internal to a Mechanical System

Consider again a mechanical system S with n degrees of freedom in an inertial reference frame R. Let the degrees of freedom be represented by n generalized coordinates q_r $(r = 1, \ldots, n)$. Let B_1 and B_2 be bodies of the system and let an axial (coil) spring connect the bodies with the respective spring attachment points being P_1 and P_2 as in Figure 9.11.8. Let the spring have natural length ℓ and extension (or compression) x as represented in Figure 9.11.8. In general x will depend upon the q_r [that is, $x = x(q_r)$]. (See also Example 4.6.2.) Finally, let **n** be a unit vector along the axis of the spring.

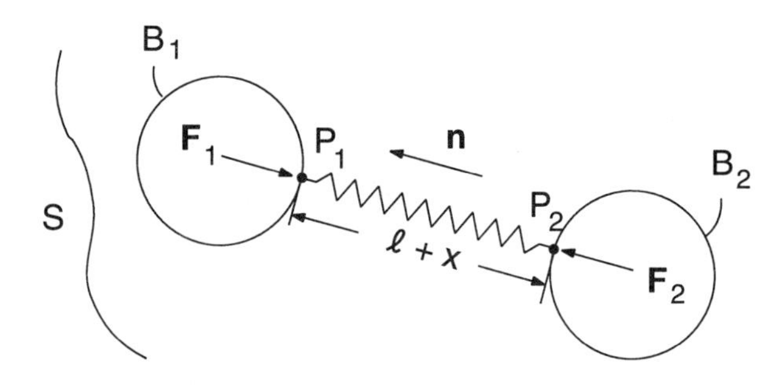

Fig. 9.11.8 A Coil Spring Connecting Bodies of a Mechanical System

When the spring is extended (or compressed) it will exert forces $\mathbf{F}_1$ and $\mathbf{F}_2$ on B_1 and B_2 at P_1 and P_2 as represented in Figure 9.11.8. $\mathbf{F}_1$ and $\mathbf{F}_2$ may be expressed as:

$$\mathbf{F}_1 = -\mathbf{F}_2 = -f(x)\mathbf{n} \qquad (9.11.23)$$

where f(x) is the spring modulus.

From Equations (4.6.2) and (9.11.1) the contribution F_{q_r} of these spring forces to the generalized active force for q_r $(r = 1,\ldots,n)$ is

$$F_{q_r} = \mathbf{V}_{\dot{q}_r}^{P_1} \bullet \mathbf{F}_1 + \mathbf{V}_{\dot{q}_r}^{P_2} \bullet \mathbf{F}_2 = -f(x)\mathbf{n} \bullet \mathbf{V}_{\dot{q}_r}^{P_1} + f(x)\mathbf{n} \bullet \mathbf{V}_{\dot{q}_r}^{P_2}$$

$$= -f(x)\mathbf{n} \bullet (\mathbf{V}_{\dot{q}_r}^{P_1} - \mathbf{V}_{\dot{q}_r}^{P_2}) \quad (r = 1,\ldots,n) \qquad (9.11.24)$$

Observe that the velocities of P_1 and P_2 in an inertia frame R may be related by the expression:

$$\mathbf{V}^{P_1} = \mathbf{V}^{P_2} + \mathbf{V}^{P_1/P_2} \qquad (9.11.25)$$

where the relative velocity $\mathbf{V}^{P_1/P_2}$ may be expressed as:

$$\mathbf{V}^{P_1/P_2} = \dot{x}\mathbf{n} + (\quad)\mathbf{n}_\perp \qquad (9.11.26)$$

where $(\quad)\mathbf{n}_\perp$ is the projection of $\mathbf{V}^{P_1/P_2}$ perpendicular to $\mathbf{n}$. By differentiating in Equation (9.11.24) with respect to $\dot{q}_r$ we obtain

$$\mathbf{V}_{\dot{q}_r}^{P_1} = \mathbf{V}_{\dot{q}_r}^{P_2} + \mathbf{V}_{\dot{q}_r}^{P_1/P_2} \qquad (9.11.27)$$

By also differentiating in Equation (9.11.20) with respect to $\dot{q}_r$ we obtain:

$$\mathbf{V}_{\dot{q}_r}^{P_1/P_2} = \frac{\partial \dot{x}}{\partial \dot{q}_r}\mathbf{n} + \frac{\partial(\ )}{\partial \dot{q}_r}\mathbf{n}_\perp \tag{9.11.28}$$

Since x is a function of the q_r, $\dot{x}$ may be expressed as

$$\dot{x} = \sum_{r=1}^{n} \frac{\partial x}{\partial q_r}\dot{q}_r \tag{9.11.29}$$

Then $\partial \dot{x}/\partial \dot{q}_r$ is $\partial x/\partial q_r$ and then from Equations (9.11.27) and (9.11.28) we have

$$\mathbf{V}_{\dot{q}_r}^{P_1} - \mathbf{V}_{\dot{q}_r}^{P_2} = \mathbf{V}_{\dot{q}_r}^{P_1/P_2} = \frac{\partial x}{\partial q_r}\mathbf{n} + \frac{\partial(\ )}{\partial \dot{q}_r}\mathbf{n}_\perp \tag{9.11.30}$$

Finally, by substituting into Equation (9.11.24) the spring contribution to F_{q_r} becomes

$$F_{q_r} = -f(x)\ \partial x/\partial q_r \tag{9.11.31}$$

If the spring modulus f(x) is linear, say kx, then F_{q_r} is

$$F_{q_r} = -kx\ \partial x/\partial q_r \tag{9.11.32}$$

Similarly, let B_1 and B_2 be connected by a torsion spring which governs the relative rotation of the bodies about the axis of the spring as represented in Figure 9.11.9. Let **n** be a unit vector parallel to the axis of the spring as shown. Let θ represent the rotation of B_1 relative to B_2 about the spring axis. Let the spring be in its natural configuration when θ is zero.

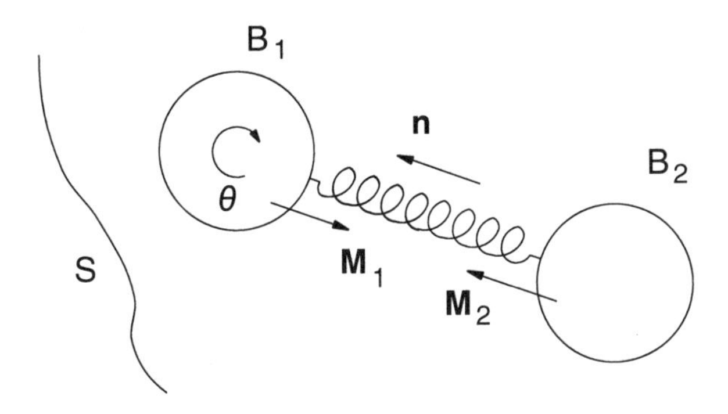

Fig. 9.11.9 A Torsion Spring Connecting Bodies of a Mechanical System

When the bodies are rotated relative to one another the spring will exert moments $\mathbf{M}_1$ and $\mathbf{M}_2$ on the bodies respectively, as represented in Figure 9.11.9. $\mathbf{M}_1$and $\mathbf{M}_2$ may be expressed as:

$$\mathbf{M}_1 = -\mathbf{M}_2 = -\phi(\theta)\mathbf{n} \tag{9.11.33}$$

where $\phi(\theta)$ is the torsion spring modulus.

From Equation (9.11.2) the contribution F_{q_r} of these spring forces to the generalized active force for q_r $(r = 1, \ldots, n)$ is:

$$\begin{aligned} F_{q_r} &= \boldsymbol{\omega}^{B_1}_{\dot{q}_r} \bullet \mathbf{M}_1 + \boldsymbol{\omega}^{B_2}_{\dot{q}_r} \bullet \mathbf{M}_2 = -\phi(\theta)\mathbf{n} \bullet \boldsymbol{\omega}^{B_1}_{\dot{q}_r} + \phi(\theta)\mathbf{n} \bullet \boldsymbol{\omega}^{B_2}_{\dot{q}_r} \\ &= -\phi(\theta)\mathbf{n} \bullet (\boldsymbol{\omega}^{B_1}_{\dot{q}_r} - \boldsymbol{\omega}^{B_2}_{\dot{q}_r}) \quad (r = 1, \ldots, n) \end{aligned} \tag{9.11.34}$$

Observe that with $\dot{\theta}\mathbf{n}$ measuring the rotation rate of B_1 relative to B_2, the angular velocities of B_1 and B_2 are related to each other by the expression

$$ {}^{R}\boldsymbol{\omega}^{B_1} = {}^{R}\boldsymbol{\omega}^{B_2} + \dot{\theta}\mathbf{n} \tag{9.11.35} $$

where, as before, R is an inertial reference frame. Then by differentiating with respect to $\dot{q}_r$ we have the partial angular velocities related as:

$$ \boldsymbol{\omega}_{\dot{q}_r}^{B_1} = \boldsymbol{\omega}_{\dot{q}_r}^{B_2} + (\partial\dot{\theta}/\partial\dot{q}_r)\mathbf{n} \tag{9.11.36} $$

Since θ is a function of the q_r, $\dot{\theta}$ may be expressed as:

$$ \dot{\theta} = \sum_{r=1}^{n} \frac{\partial\theta}{\partial q_r}\dot{q}_r \tag{9.11.37} $$

Then $\partial\dot{\theta}/\partial\dot{q}_r$ is $\partial\theta/\partial q_r$ and thus from Equation (9.11.36) we have

$$ \boldsymbol{\omega}_{\dot{q}_r}^{B_1} - \boldsymbol{\omega}_{\dot{q}_r}^{B_2} = (\partial\theta/\partial q_r)\mathbf{n} \tag{9.11.38} $$

Hence, by substituting into Equation (9.11.34) the generalized force contribution becomes:

$$ F_{q_r} = -\phi(\theta)\,\partial\theta/\partial q_r \tag{9.11.39} $$

If the spring modulus $\phi(\theta)$ is linear, say $\kappa\theta$, then F_{q_r} is

$$ F_{q_r} = -\kappa\theta\,\partial\theta/\partial q_r \tag{9.11.40} $$

9.12 Inertia Forces on a Rigid Body

If we model a rigid body B as a set of particles at fixed distances relative to one another, then as the body moves in an inertial reference frame R, the particles of B will experience accelerations in R giving rise to inertia forces on the particles

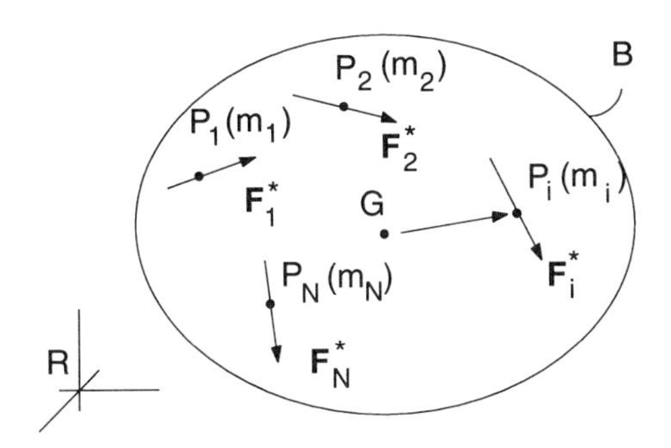

Fig. 9.12.1 Inertia Forces on Particles Making Up a Rigid Body B

as represented in Figure 9.12.1. Specifically, for a typical particle P_i with mass m_i, the inertia force $\mathbf{F}_i^*$ on P_i is:

$$\mathbf{F}_i^* = -m_i\,{}^R\mathbf{a}^{P_i} \qquad (i = 1,\ldots,N) \tag{9.12.1}$$

where as before ${}^R\mathbf{a}^{P_i}$ is the acceleration of P_i in R and N is the number of particles in B.

Since the number of particles making up a rigid body is quite large, the number of forces in the inertia force system is also quite large. Hence, it is convenient to replace these many forces with an equivalent force system consisting of a single force, say $\mathbf{F}^*$ passing through the mass center G of B together with a couple having torque $\mathbf{T}^*$. To this end, consider that since the particles are at fixed distances relative to one another, their accelerations are related to each other and specifically to G by the expression:

$${}^R\mathbf{a}^{P_i} = {}^R\mathbf{a}^G + {}^R\boldsymbol{\alpha}^B \times \mathbf{r}_i + {}^R\boldsymbol{\omega}^B \times ({}^R\boldsymbol{\omega}^B \times \mathbf{r}_i) \tag{9.12.2}$$

where as before ${}^R\mathbf{a}^G$ is the acceleration of G in R, ${}^R\boldsymbol{\alpha}^B$ and ${}^R\boldsymbol{\omega}^B$ are the angular acceleration and angular velocity of B in R, and where $\mathbf{r}_i$ locates P_i relative to G. Hence, from Equation (9.12.1) $\mathbf{F}_i^*$ becomes

$$\mathbf{F}_i^* = -m_i[{}^R\mathbf{a}^G + {}^R\boldsymbol{\alpha}^B \times \mathbf{r}_i + {}^R\boldsymbol{\omega}^B \times ({}^R\boldsymbol{\omega}^B \times \mathbf{r}_i)] \tag{9.12.3}$$

In developing the equivalent force system, $\mathbf{F}^*$ must be equal to the resultant of the $\mathbf{F}_i^*$ and $\mathbf{T}^*$ must be equal to the resultant of the moments of the $\mathbf{F}_i^*$ about G. That is,

$$\mathbf{F}^* = \sum_{i=1}^{N} \mathbf{F}_i^* \quad \text{and} \quad \mathbf{T}^* = \sum_{i=1}^{N} \mathbf{r}_i \times \mathbf{F}_i^* \tag{9.12.4}$$

(See Example 2.10.7).

By substituting from Equation (9.12.3) into the first of Equations (9.12.4), $\mathbf{F}^*$ is seen to be:

$$\begin{aligned} \mathbf{F}^* &= -\sum_{i=1}^{N} m_i\, {}^R\mathbf{a}^G - \sum_{i=1}^{N} m_i\, {}^R\boldsymbol{\alpha}^B \times \mathbf{r}_i - \sum_{i=1}^{N} m_i\, {}^R\boldsymbol{\omega}^B \times \left({}^R\boldsymbol{\omega}^B \times \mathbf{r}_i\right) \\ &= -\left(\sum_{i=1}^{N} m_i\right) {}^R\mathbf{a}^G - {}^R\boldsymbol{\alpha}^B \times \sum_{i=1}^{N} m_i\mathbf{r}_i - {}^R\boldsymbol{\omega}^B \times \left({}^R\boldsymbol{\omega}^B \times \sum_{i=1}^{N} m_i\mathbf{r}_i\right) \end{aligned} \tag{9.12.5}$$

Since G is the mass center of B, Equation (8.3.2) shows that the last two terms of Equation (9.12.5) are zero. Thus $\mathbf{F}^*$ becomes simply

$$\mathbf{F}^* = -M\, {}^R\mathbf{a}^G \tag{9.12.6}$$

where M is the mass of B (that is, the sum of the masses of the particles of B).

Similarly, by substituting from Equation (9.12.3) into the second of Equations (9.12.4), $\mathbf{T}^*$ is seen to be:

$$\mathbf{T}^* = -\sum_{i=1}^{N} m_i \mathbf{r}_i \times {}^R\mathbf{a}^G - \sum_{i=1}^{N} m_i \mathbf{r}_i \times \left({}^R\boldsymbol{\alpha}^B \times \mathbf{r}_i\right) - \sum_{i=1}^{N} m_i \mathbf{r}_i \times \left[{}^R\boldsymbol{\omega}^B \times \left({}^R\boldsymbol{\omega}^B \times \mathbf{r}_i\right)\right]$$

$$= -\left(\sum_{i=1}^{N} m_i \mathbf{r}_i\right) \times {}^R\mathbf{a}^G - \sum_{i=1}^{N} m_i \mathbf{r}_i \times \left({}^R\boldsymbol{\alpha}^B \times \mathbf{r}_i\right) - {}^R\boldsymbol{\omega}^B \times \left[\sum_{i=1}^{N} m_i \mathbf{r}_i \times \left({}^R\boldsymbol{\omega}^B \times \mathbf{r}_i\right)\right] \quad (9.12.7)$$

where the last equality is verified by expanding the vector triple products (cross products) in terms of scalar products (dot products). [See Equations (2.13.2) and (2.13.3).] Since G is the mass center of B, the first term of Equation (9.12.7) is seen to be zero. The remaining two terms of Equation (9.12.7) have similar forms and may be expressed in terms of the inertia dyadic of B for G. (See Section 8.8.) Specifically, in the second term ${}^R\boldsymbol{\alpha}^B$ may be expressed in the form

$${}^R\boldsymbol{\alpha}^B = \alpha \mathbf{n}_\alpha \quad (9.12.8)$$

where $\mathbf{n}_\alpha$ is a unit vector parallel to ${}^R\boldsymbol{\alpha}^B$. Then the second term may be expressed as [see Equations (8.4.3) and (8.8.3)]:

$$\sum_{i=1}^{N} m_i \mathbf{r}_i \times ({}^R\boldsymbol{\alpha}^B \times \mathbf{r}_i) = \sum_{i=1}^{N} m_i \mathbf{r}_i \times (\alpha \mathbf{n}_\alpha \times \mathbf{r}_i) = \alpha \sum_{i=1}^{N} m_i \mathbf{r}_i \times (\mathbf{n}_\alpha \times \mathbf{r}_i)$$

$$= \alpha \mathbf{I}_\alpha^{B/G} = \alpha \boldsymbol{I}^{B/G} \bullet \mathbf{n}_\alpha = \mathbf{I}^{B/G} \bullet {}^R\boldsymbol{\alpha}^B \quad (9.12.9)$$

By similar analysis, the third term of Equation (9.12.7) may be expressed as:

$${}^R\boldsymbol{\omega}^B \times \left[\sum_{i=1}^{N} m_i \mathbf{r}_i \times ({}^R\boldsymbol{\omega}^B \times \mathbf{r}_i)\right] = {}^R\boldsymbol{\omega}^B \times (\mathbf{I}^{B/G} \bullet {}^R\boldsymbol{\omega}^B) \quad (9.12.10)$$

Therefore, the torque $\mathbf{T}^*$ is seen to be simply:

$$\mathbf{T}^* = -\mathbf{I}^{B/G} \bullet {}^R\boldsymbol{\alpha}^B - {}^R\boldsymbol{\omega}^B \times (\mathbf{I}^{B/G} \bullet {}^R\boldsymbol{\omega}^B) \quad (9.12.11)$$

Equation (9.12.11) has a relatively simple form if the vector terms are expressed in terms of principal unit vectors of inertia (see Section 8.10). Specifically let $\mathbf{n}_1$, $\mathbf{n}_2$, and $\mathbf{n}_3$ be central principal inertia vectors. Let $\mathbf{T}^*$, $\mathbf{I}^{B/G}$, ${}^R\boldsymbol{\alpha}^B$, and ${}^R\boldsymbol{\omega}^B$ be expressed in the forms:

$$\mathbf{T}^* = T_1\mathbf{n}_1 + T_2\mathbf{n}_2 + T_3\mathbf{n}_3 \tag{9.12.12}$$

$$\mathbf{I}^{B/G} = I_{11}\mathbf{n}_1\mathbf{n}_1 + I_{22}\mathbf{n}_2\mathbf{n}_2 + I_{33}\mathbf{n}_3\mathbf{n}_3 \tag{9.12.13}$$

$${}^R\boldsymbol{\alpha}^B = \alpha_1\mathbf{n}_1 + \alpha_2\mathbf{n}_2 + \alpha_3\mathbf{n}_3 \tag{9.12.14}$$

$${}^R\boldsymbol{\omega}^B = \omega_1\mathbf{n}_1 + \omega_2\mathbf{n}_2 + \omega_3\mathbf{n}_3 \tag{9.12.15}$$

Then T_1, T_2, and T_3 have the forms

$$\begin{aligned} T_1 &= -\alpha_1 I_{11} + \omega_2\omega_3(I_{22} - I_{33}) \\ T_2 &= -\alpha_2 I_{22} + \omega_3\omega_1(I_{33} - I_{11}) \\ T_3 &= -\alpha_3 I_{33} + \omega_1\omega_2(I_{11} - I_{22}) \end{aligned} \tag{9.12.16}$$

Finally, if a body B has planar motion such that ${}^R\boldsymbol{\alpha}^B$ and ${}^R\omega^B$ have the simplified forms

$${}^R\boldsymbol{\omega}^B = \omega\mathbf{n}_3 \qquad {}^R\boldsymbol{\alpha}^B = \alpha\mathbf{n}_3 = \dot{\omega}\mathbf{n}_3 \tag{9.12.17}$$

then the components of $\mathbf{T}^*$ are

$$T_1 = T_2 = 0 \quad \text{and} \quad T_3 = -\alpha I_{33} \tag{9.12.18}$$

9.13 Generalized Inertia Forces

Consider a body B as part of a mechanical system S having n degrees of freedom in an inertial reference frame R, measured by generalized coordinates q_r $(r = 1,\ldots,n)$. Let the inertia forces on B be represented by an equivalent force system consisting of a single force $\mathbf{F}^*$ passing through the mass center G of B, together with a couple with torque $\mathbf{T}^*$. Then from Equation (9.9.4) the generalized inertia forces $F^*_{q_r}$ on B for q_r are:

$$F^*_{q_r} = \mathbf{F}^* \bullet \mathbf{V}^G_{\dot{q}_r} + \mathbf{T}^* \bullet \boldsymbol{\omega}^B_{\dot{q}_r} \tag{9.13.1}$$

where, as before, $\mathbf{V}^G_{\dot{q}_r}$ is the partial velocity of G for q_r and $\boldsymbol{\omega}^B_{\dot{q}_r}$ is the partial angular velocity of B for q_r.

9.14 Summary

Of all the equations developed in this chapter those concerned with generalized forces are expected to be the most useful in analysis of mechanical systems. Table 9.14.1 provides a summary list of these equations.

Table 9.14.1 Formulas and Equations for Generalized Forces for Rigid Bodies*

Quantity	Formula	Note	Reference Equation
1. Generalized Forces	$F_{q_r} = \mathbf{F} \cdot \mathbf{V}^{P}_{\dot{q}_r}$	**F** is a force acting at a point P of a body B	(9.5.1)
	$F_{q_r} = \sum_{i=1}^{N} \mathbf{F}_i \cdot \mathbf{V}^{P_i}_{\dot{q}_r}$	System of forces $\mathbf{F}_i$ acting at points P_i	(9.5.2)
	$F_{q_r} = \mathbf{V}^{G}_{\dot{q}_r} \cdot \mathbf{F} + \boldsymbol{\omega}^{B}_{\dot{q}_r} \cdot \mathbf{T}$	Body B acted on by a force **F** passing through a point G together with a couple with torque **T**	(9.9.4)
2. Contribution from gravity forces	$F_{q_r} = -mg\,\partial z/\partial q_r$	Body B with mass m with a mass center G at an elevation z above a reference plane	(9.11.9)
3. Contribution from forces exerted across smooth surfaces inside a mechanical system	$F_{q_r} = 0$		(9.11.14)
4. Contribution from forces exerted at a point of a body whose motion is specified	$F_{q_r} = 0$		(9.11.17)

Quantity	Formula	Note	Reference Equation
5. Contribution from forces exerted at contact points between rolling bodies	$F_{q_r} = 0$		(9.11.20) (9.11.22)
6. Contribution from axial (coil) springs	$F_{q_r} = -f(x)\frac{\partial x}{\partial q_r}$	f(x) is the spring modulus x is spring elongation	(9.11.31)
	$F_{q_r} = -kx\frac{\partial x}{\partial q_r}$	linear spring modulus	(9.11.32)
7. Contribution from torsion springs	$F_{q_r} = -\phi(\theta)\frac{\partial \theta}{\partial q_r}$	$\phi(\theta)$ is the spring modulus θ is spring rotation	(9.11.39)
	$F_{q_r} = -\kappa\theta\frac{\partial \theta}{\partial q_r}$	linear spring modulus	(9.11.40)
8. Generalized inertia force	$F^*_{q_r} = \mathbf{V}^G_{\dot{q}_r} \bullet \mathbf{F}^* + \boldsymbol{\omega}^B_{\dot{q}_r} \bullet \mathbf{T}^*$	See 8a and 8b	(9.13.1)

Quantity	Formula	Note	Reference Equation
8a. Equivalent inertia force system on a rigid body B	$\mathbf{F}^* = -m\,{}^R\mathbf{a}^G$	$\mathbf{F}^*$ passes through mass center G. m is the mass of B	(9.12.6)
	$\mathbf{T}^* = -\mathbf{I}\cdot\alpha^B - \boldsymbol{\omega}^B \times (\mathbf{I}\cdot\omega^B)$	$\mathbf{T}^*$ is the torque of the equivalent couple. $\mathbf{I}$ is the control inertia dyadic of B and $\boldsymbol{\omega}^B$ and $\boldsymbol{\alpha}^B$ are the angular velocity and angular accelerate of B in R	(9.12.11)
8b. Use of principal inertia direction for $\mathbf{T}^*$	$\mathbf{T}^* = T_1\mathbf{n}_1 + T_2\mathbf{n}_2 + T_3\mathbf{n}_3$	$\mathbf{n}_i$ $(i = 1,2,3)$ are principal unit vectors	(9.12.12)
	$T_1 = -\alpha_1 I_{11} + \omega_2\omega_3(I_{22} - I_{33})$ $T_2 = -\alpha_2 I_{22} + \omega_3\omega_1(I_{33} - I_{11})$ $T_3 = -\alpha_3 I_{33} + \omega_1\omega_2(I_{11} - I_{22})$	α_i and ω_i are $\mathbf{n}_i$ components of $\boldsymbol{\alpha}^B$ and $\boldsymbol{\omega}^B$. I_{11}, I_{22}, and I_{33} are principal moments of inertia	(9.12.16)

---

*As in the text of the chapter the bodies are members of a mechanical system S having n degrees of freedom in an inertial reference frame R. These degrees of freedom are represented by generalized coordinates q_r $(r = 1,\ldots,n)$.

References

9.1 T. R. Kane and D. A. Levinson, *Dynamics, Theory and Applications*, McGraw Hill, New York, NY, 1985, pp. 35-46.

9.2 R. L. Huston, *Multibody Dynamics*, Butterworth, Stoneham, MA, 1990, pp. 38, 285.

9.3 H. Josephs and R. L. Huston, *Dynamics of Mechanical Systems*, CRC Press, Boca Raton, FL, 2001.

9.4 T. R. Kane, *Analytical Elements of Mechanics*, Vol. 1, Academic Press, New York, NY, pp. 128, 150.

9.5 R. A. Nidey, "Gravitational Torque on a Satellite of Arbitrary Shape," *ARS Journal*, Vol. 30, 1960, AIAA, Reston, VA, pp. 203-204.

Chapter 10

RIGID BODY DYNAMICS

10.1 Introduction

In this chapter we list commonly used formulas for dynamics analyses of rigid bodies. These formulas arise from basic principles of physics (mechanics) and the corresponding laws of motion. In subsequent chapters we will extend these concepts for the analysis of systems of rigid bodies — that is, multibody systems.

Initially we review the principles themselves. We then explore the various terms in the formulas and equations expressing the principles. We illustrate the principles with a series of sample classical examples in the following chapter.

10.2 Principles of Dynamics/Laws of Motion

As noted with the dynamics of particles (see Chapter 5) most principles of dynamics have their roots in Newton's laws — particularly, Newton's second law. In this sense they are equivalent to one another. Table 10.2.1 provides a listing of commonly used principles/laws/formulas as they are applied in rigid body dynamics. References [10.2-10.32] provide sources for these and less commonly used principles.

Table 10.2.1 Principles of Dynamics/Laws of Motion and Formulas for Rigid Bodies

Name	Principle/Statement	Formulas/Notes	Reference Equations
1. Newton's Second Law	Acceleration $\mathbf{a}_i$ of each particle P_i of a body B is proportional to the resultant $\mathbf{F}_i$ of the forces on the particle and inversely proportional to the mass m_i of the particle.	$\mathbf{F}_i = m_i\mathbf{a}_i$, $i = 1,\ldots,N$ (N is the number of particles in B.) Alternatively, $\sum_{i=1}^{N} \mathbf{F}_i = m\mathbf{a}^G$ and $\mathbf{M}_G = \mathbf{I}_G\boldsymbol{\alpha} - \boldsymbol{\omega} \times (\mathbf{I}_G \cdot \boldsymbol{\omega})$. G is the mass center of B, m is the mass of B, $\mathbf{I}_G$ is the central inertia dyadic, $\mathbf{M}_G$ is the resultant of moments of the applied forces about G, $\boldsymbol{\omega}$ is the angular velocity of B in an inertial reference frame R, and $\boldsymbol{\alpha}$ is the angular acceleration of B in R.	(5.2.2) (9.12.11) (10.2.1)

Name	Principle/Statement	Formulas/Notes	Reference Equations
2. d'Alembert's Principle	The combined system of applied (active) forces and inertia (passive) forces on a body B constitute a zero system.	$\sum_{i=1}^{N} \mathbf{F}_i + F^* = 0$ $\mathbf{M}_G + \mathbf{T}^* = 0$ G is the mass center of B. $\mathbf{F}^* = -m\mathbf{a}^G$ and $\mathbf{T}^* = -\mathbf{I}_G\boldsymbol{\alpha} + \boldsymbol{\omega} \times (\mathbf{I}_G \cdot \boldsymbol{\omega})$ (See Notes under Newton's Second Law for explanation of notation.)	(5.2.5) (9.12.11) (10.2.2)
3. Work-Energy	The work done on a body B in moving it from one position and orientation to another is equal to the change in kinetic energies of B between the two positions/orientations.	${}_1W_2 = K_2 - K_1$ W is work. K is kinetic energy. [See Equations (10.7.3) and (10.3.4).]	(5.2.6) (10.2.3)

Name	Principle/Statement	Formulas/Notes	Reference Equations
4. Impulse-Momentum	The resultant of the impulses on a body B between times t_1 and t_2 is equal to the change in linear momentum of B between t_1 and t_2; and the resultant of the angular impulses on B (moment of impulses) about a fixed point O or about the mass center G of B between t_1 and t_2 is equal to the change in angular momentum (moment of momentum) of B about O or G between t_1 and t_2.	${}_1\mathbf{I}_2 = \mathbf{L}_2 - \mathbf{L}_1 = m(\mathbf{v}_2^G - \mathbf{v}_1^G)$ ${}_1\mathbf{J}_{O2} = \mathbf{A}_{O2} - \mathbf{A}_{O1}$ ${}_1\mathbf{J}_{G2} = \mathbf{A}_{G2} - \mathbf{A}_{G1}$ **I** is impulse $\mathbf{J}_O$ is angular impulse (moment of impulse about O) $\mathbf{A}_O$ is angular momentum (moment of momentum) about O.	(5.2.7) (10.2.4)
5. Kane's Equations	Let body B belong to a mechanical system S having n degrees of freedom represented by coordinates q_r $(r = 1, \ldots, n)$. Then the sum of the generalized applied (active) and inertia (passive) forces is zero.	$F_{q_r} + F_{q_r}^* = 0$ $(r = 1, \ldots, n)$	(5.2.8) (10.2.5)

Name	Principle/Statement	Formulas/Notes	Reference Equations
6. Lagrange's Equations	(See 5.) Let S be holonomic. Then the generalized inertia forces may be expressed in terms of the kinetic energy K as $F^*_{a_r} = -\frac{d}{dt}\left(\frac{\partial K}{\partial \dot{q}_r}\right) + \frac{\partial K}{\partial q_r}$, and then Kane's equations give Lagrange's equations in the form of Equations (10.2.6).	$\frac{d}{dt}\left(\frac{\partial K}{\partial \dot{q}_r}\right) - \frac{\partial K}{\partial q_r} = F_{q_r}$ $(r = 1,\ldots,n)$	(5.2.9) (10.2.6)
	Alternatively, the time integral of the Lagrangian L (kinetic minus potential energy) is a minimum (Hamilton's Principle). The calculus of variations then lead to Equation (10.2.7).	$\frac{d}{dt}\left(\frac{\partial L}{\partial \dot{q}_r}\right) - \frac{\partial L}{\partial q_r} = 0$ $(r = 1,\ldots,n)$	(5.2.10) (10.2.7)

10.3 Kinetic Energy

Let B be a rigid body moving in a reference frame R. Envision B as being composed of particles P_i $(i = 1,\ldots,N)$ having masses m_i $(i = 1,\ldots,N)$. Then the kinetic energy K of B in R is the sum of the kinetic energies of the particles P_i. That is,

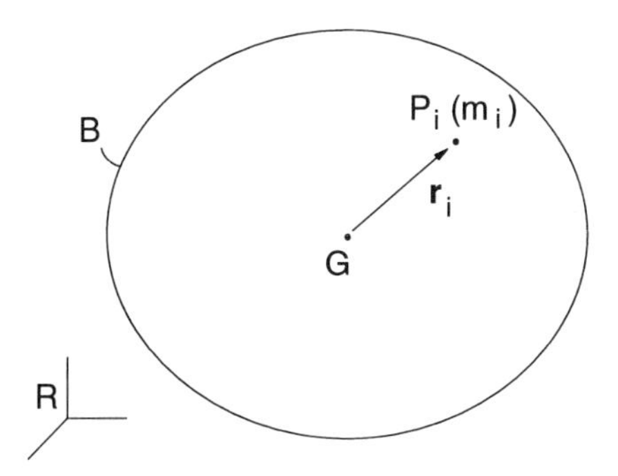

Fig. 10.3.1 A Body B Moving in a Reference Frame R

$$K = \sum_{i=1}^{N} K^{P_i} = \frac{1}{2}\sum_{i=1}^{N} m_i \left(\mathbf{v}^{P_i}\right)^2 \tag{10.3.1}$$

The sum of Equation (10.3.1) may be conveniently expressed in terms of the inertia dyadic and moments of inertia of B as follows: Let B be represented as in Figure 10.3.1 where G is the mass center of B and $\mathbf{r}_i$ locates a typical particle P_i of B relative to G. Then the velocity of P_i in a reference frame R in which B moves may be expressed as

$$^R\mathbf{V}^{P_i} = {}^R\mathbf{V}^G + {}^R\boldsymbol{\omega}^B \times \mathbf{r}_i \tag{10.3.2}$$

By substituting into Equation (10.3.1) (and by deleting the superscript R for simplicity), the kinetic energy of B has the form

$$
\begin{aligned}
K^B &= \frac{1}{2}\sum_{i=1}^{N} m_i\left(\mathbf{V}^G + \boldsymbol{\omega}^B \times \mathbf{r}_i\right)^2 \\
&= \frac{1}{2}\sum_{i=1}^{N} M_i(\mathbf{V}^G)^2 + \frac{1}{2}\sum_{i=1}^{N} m_i 2\mathbf{V}^G \bullet \boldsymbol{\omega}^B \times \mathbf{r}_i \\
&\quad + \frac{1}{2}\sum_{i=1}^{N} m_i\left(\boldsymbol{\omega}^B \times \mathbf{r}_i\right)^2 \\
&= \frac{1}{2}M(\mathbf{V}^G)^2 + \mathbf{V}^G \bullet \boldsymbol{\omega}^B \times \sum_{i=1}^{N} m_i \mathbf{r}_i \\
&\quad + \frac{1}{2}\boldsymbol{\omega}^B \bullet \sum_{i=1}^{N} m_i \mathbf{r}_i \times \left(\boldsymbol{\omega}^B \times \mathbf{r}_i\right)
\end{aligned}
$$

or

$$
K^B = \frac{1}{2}M(\mathbf{V}^G)^2 + \frac{1}{2}\boldsymbol{\omega}^B \bullet \mathbf{I}^{B/G} \bullet \boldsymbol{\omega}^B \tag{10.3.3}
$$

where the second term is zero since G is the mass center of B [see Equation (8.3.2)] and where the last term is obtained using Equation (9.12.10).

If $\mathbf{n}_1$, $\mathbf{n}_2$, and $\mathbf{n}_3$ are parallel to principal inertia directions of B for G, then Equation (10.3.3) may be expanded to the form:

$$
K^B = \frac{1}{2}M(\mathbf{V}^G)^2 + \frac{1}{2}\left(\omega_1^2 I_{11} + \omega_2^2 I_{22} + \omega_3^2 I_{33}\right) \tag{10.3.4}
$$

where ω_1, ω_2, and ω_3 are the $\mathbf{n}_1$, $\mathbf{n}_2$, and $\mathbf{n}_3$ components of $\boldsymbol{\omega}^B$, and where I_{11}, I_{22}, and I_{33} are the central principal moments of inertia of B.

10.4 Potential Energy

In elementary textbooks potential energy is often defined as "the ability to do work." That is, if a mechanical system has potential energy, it then has the ability (or "potential") to do work. "Work" in turn is defined as a force causing a particle or body to be displaced, with the value of the work being the product of the force magnitude and the displacement in the direction of the force.

In mechanical systems with several degrees of freedom it is convenient to define potential energy in terms of generalized forces. Specifically, if a mechanical system S has n degrees of freedom represented by generalized coordinates q_r $(r = 1,\ldots,n)$, potential energy P is defined as the function such that (see Section 4.8.2):

$$F_{q_r} = -\partial P/\partial q_r \quad (r = 1,\ldots,n) \tag{10.4.1}$$

Observe that with this definition potential energy is not unique: If C is a constant then P + C is also a potential energy. Also, potential energy may not exist for any given mechanical system.

If the generalized forces F_{q_r} are known, Equation (10.4.1) may be used, via integration, to obtain an expression for potential energy. However, if the utility of potential energy is to obtain generalized forces, and if generalized forces are needed to obtain potential energy, little is gained by the computation. Also, Equation (10.4.1) may not always be integrable in terms of elementary functions. When Equations (10.4.1) are integrable, and the mechanical system possesses a potential energy, the system is said to be "conservative."

There are frequently occurring forces on rigid bodies for which Equations (10.4.1) may be integrated a priori. These are gravitational and spring forces. That is, for gravitational and spring forces, Equations (10.4.1) may be formally integrated to obtain their contributions to the potential energy.

Consider first gravitational forces: Let B be a rigid body and a member of a mechanical system S having n degrees of freedom represented by generalized coordinates q_r $(r = 1,...,n)$. Let the gravitational forces on B be represented by a single vertical (downward) force **W** passing through the mass center G of B as in Figure 9.11.2 and shown again in Figure 10.4.1. Then **W** may be expressed as:

$$\mathbf{W} = -mg\mathbf{n}_z \tag{10.4.2}$$

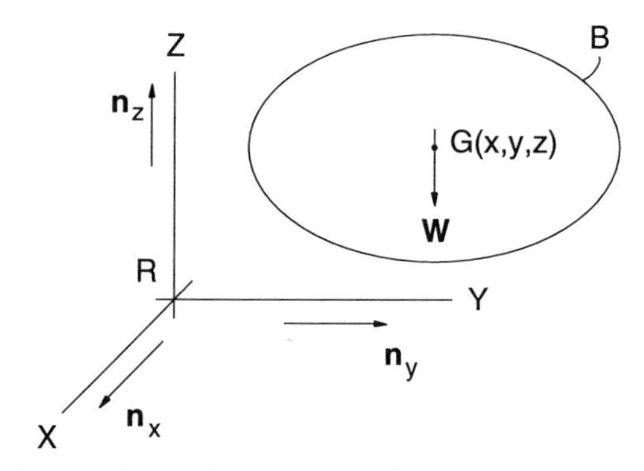

Fig. 10.4.1 A Body B in a Gravitational Force Field Whose Effect on B is Represented by a Single Force **W**

where m is the mass of B, g is the gravity constant, and $\mathbf{n}_z$ is a vertical (directed up) unit vector.

Let the elevation of G above a horizontal reference plane, say the X-Y plane, of Figure 10.4.1 be z. Then from Equation (9.11.9), the contribution of **W** to the generalized force F_{q_r} is:

$$F_{q_r} = -mg\frac{\partial z}{\partial q_r} \tag{10.4.3}$$

By comparing Equations (10.4.1) and (10.4.3) we see immediately that the contribution of the gravitational forces to the potential energy P is

$$P = mgz \tag{10.4.4}$$

Consider next spring forces: Let B_1 and B_2 be bodies of a mechanical system S and let an axial (coil) spring connect B_1 and B_2 as in Figure 9.11.8, and as shown again in Figure 10.4.2, where ℓ is the natural length of the spring and x is its extension. $\mathbf{F}_1$ and $\mathbf{F}_2$ are the forces exerted on B_1 and B_2 by the spring. From Equation (9.11.23) $\mathbf{F}_1$ and $\mathbf{F}_2$ are:

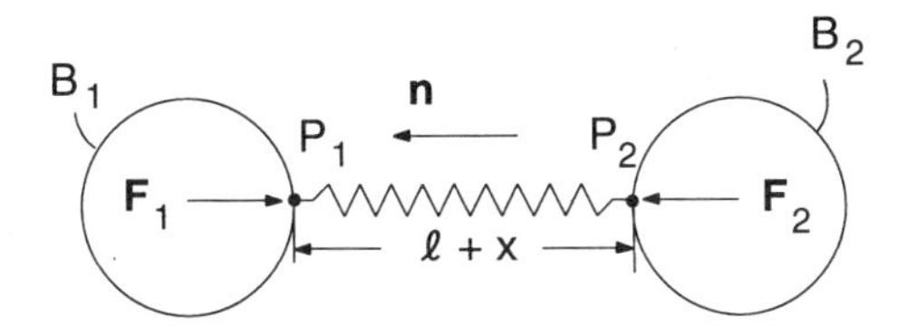

Fig. 10.4.2 A Coil Spring Connecting Bodies of a Mechanical System

$$\mathbf{F}_1 = -\mathbf{F}_2 = -f(x)\mathbf{n} \tag{10.4.5}$$

where f(x) is the spring modulus.

From Equation (9.11.31) the contribution of the spring force to the generalized active force F_{q_r} is:

$$F_{q_r} = -f(x)\ \partial x/\partial q_r \tag{10.4.6}$$

Then by comparing Equations (10.4.1) and (10.4.6) we see that the contribution to the potential energy by the spring force is

$$P = \int_0^x f(\xi)d\xi \tag{10.4.7}$$

Finally, if the spring is linear such that $f(x) = kx$, the contribution to the potential energy is

$$P = 1/2kx^2 \tag{10.4.8}$$

where as before k is the spring modulus.

Similarly, let B_1 and B_2 be connected by a torsion spring affecting the relative rotation of the bodies as in Figure 9.11.9 and as shown again in Figure 10.4.3 where

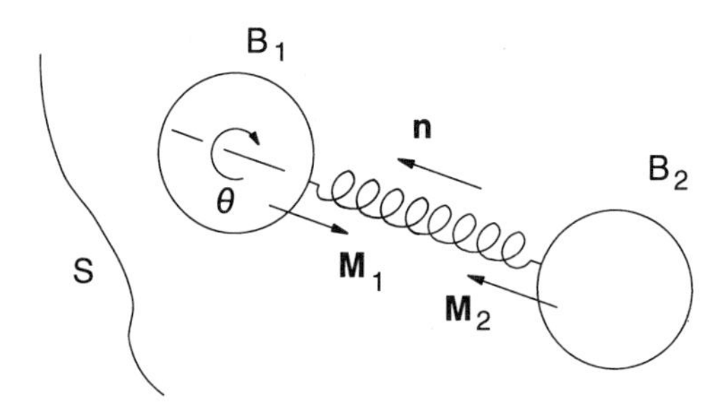

Fig. 10.4.3 A Coil Spring Connecting Bodies of a Mechanical System

n is a unit vector parallel to the axis of the spring and where θ represents the rotation of B_1 relative to B_2. As before, let the spring be in its natural configuration when θ is zero. Then, when the bodies are rotated relative to one another, the spring will exert moments $\mathbf{M}_1$ and $\mathbf{M}_2$ on the bodies as represented in Figure 10.4.3. From Equation (9.11.33) $\mathbf{M}_1$ and $\mathbf{M}_2$ are:

$$\mathbf{M}_1 = -\mathbf{M}_2 = -\phi(\theta)\mathbf{n} \tag{10.4.9}$$

where $\phi(\theta)$ is the torsion spring modulus.

From Equation (9.11.39) the contribution of the spring moment to the generalized active force F_{q_r} is:

$$F_{q_r} = -\phi(\theta)\ \partial\theta/\partial q_r \tag{10.4.10}$$

Then by comparing Equations (10.4.1) and (10.4.6) we see that the contribution to the potential energy by the spring moment is

$$P = \int_0^\theta \phi(\eta)\,d\eta \tag{10.4.11}$$

Finally, if the spring is linear such that $\phi(\theta) = \kappa\theta$, the contribution to the potential energy is:

$$P = \frac{1}{2}\kappa\theta^2 \tag{10.4.12}$$

Observe the similarities of Equations (10.4.8) and (10.4.12). Observe further that for linear springs the potential energy has the same form independent of whether the spring is in tension or compression, or of the direction of relative rotation of the bodies.

10.5 Linear Momentum

The linear momentum of a body B in a reference frame R is the sum of the linear momenta of the particles making up the body. Let B be represented as in Figure 10.5.1 where G is the mass center and P_i (i = 1,...,N) are particles comprising B. Let $\mathbf{r}_i$ locate a typical particle P_i relative to G. Then the linear momentum $\mathbf{L}^B$ of B in R is

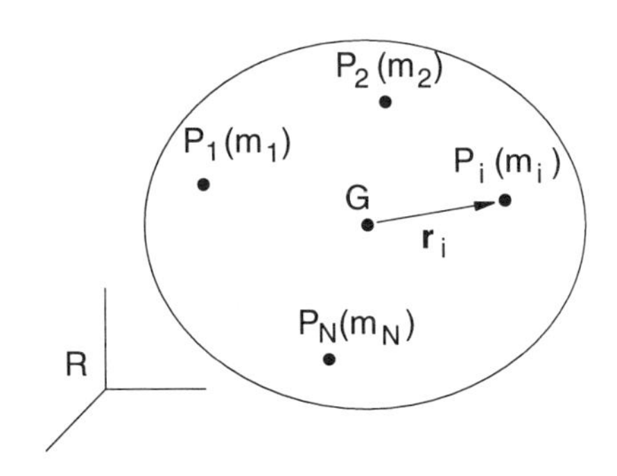

Fig. 10.5.1 A Body B Moving in a Reference Frame R

$$\mathbf{L}^B = \sum_{i=1}^{N} m_i \mathbf{V}^{P_i} \tag{10.5.1}$$

where $\mathbf{V}^{P_i}$ is the velocity of P_i in R.

Since P_i and G are both fixed in B, $\mathbf{V}^{P_i}$ may be expressed as [see Equation (3.14.1)]:

$$\mathbf{V}^{P_i} = \mathbf{V}^G + \boldsymbol{\omega} \times \mathbf{r}_i \tag{10.5.2}$$

where $\mathbf{V}^G$ is the velocity of G in R and, as before, $\boldsymbol{\omega}$ is the angular velocity of B in R. Then by substitution into Equation (10.5.1) $\mathbf{L}^B$ takes the form

$$\begin{aligned} \mathbf{L}^B &= \sum_{i=1}^{N} m_i \left(\mathbf{V}^G + \boldsymbol{\omega} \times \mathbf{r}_i \right) \\ &= \left(\sum_{i=1}^{N} m_i \right) \mathbf{V}^G + \boldsymbol{\omega} \times \left(\sum_{i=1}^{N} m_i \mathbf{r}_i \right) \\ &= M\mathbf{V}^G + 0 \end{aligned} \tag{10.5.3}$$

where M is the total mass of B and the last term is zero since G is the mass center of B [see Equation (8.3.2)]. Hence $\mathbf{L}^B$ has the simple form:

$$\mathbf{L}^B = M\mathbf{V}^G \tag{10.5.4}$$

10.6 Angular Momentum

The angular momentum of a body B relative to a reference point O in an inertial frame R is the sum of the angular momenta relative to O of the particles making up B. Observe that unlike linear momentum, angular momentum requires a specific reference point. Also, angular momentum is sometimes regarded as the "moment of momentum" about a point. Thus, it is often convenient to refer to the angular momentum "about" a point instead of "relative" to a point.

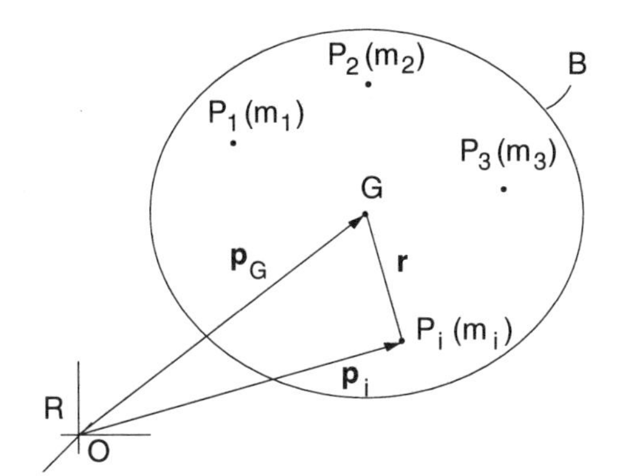

Fig. 10.6.1 A body B Moving in a Reference Frame R

As before, let B be represented by a set of particles P_i $(i = 1,\ldots,N)$ as in Figure 10.6.1. Let R be an inertial reference frame and let O be a point fixed in R. Let P_i be a typical particle of B and let m_i be the mass of P_i. Then the angular momentum of P_i relative to O is

$$\mathbf{A}^{P_i/O} = \mathbf{p}_i \times m_i \mathbf{V}^{P_i} \tag{10.6.1}$$

where $\mathbf{p}_i$ locates P_i relative to O and $\mathbf{V}^{P_i}$ is the velocity of P_i in R. Then the angular momentum of B relative to O is simply the sum:

$$\mathbf{A}^{B/O} = \sum_{i=1}^{N} \mathbf{A}^{P_i/O} = \sum_{i=1}^{N} m_i \mathbf{p}_i \times \mathbf{V}^{P_i} \tag{10.6.2}$$

The sum in Equation (10.6.2) may be conveniently expressed in terms of global inertia properties of B: Let G be the mass center of B and let $\mathbf{r}_i$ locate P_i relative to G. Then $\mathbf{p}_i$ may be expressed as:

$$\mathbf{p}_i = \mathbf{p}_G + \mathbf{r}_i \tag{10.6.3}$$

where $\mathbf{p}_G$ locates G relative to O. Also $\mathbf{V}^{P_i}$ may be expressed as:

$$\mathbf{V}^{P_i} = \mathbf{V}^G + \mathbf{V}^{P_i/G} = \mathbf{V}^G + \boldsymbol{\omega} \times \mathbf{r}_i \tag{10.6.4}$$

where as before $\boldsymbol{\omega}$ is the angular velocity of B in R. Then by substituting from Equations (10.6.3) and (10.6.4) into (10.6.2) we have:

$$\begin{aligned} \mathbf{A}^{B/O} &= \sum_{i=1}^{N} [m_i(\mathbf{p}_G + \mathbf{r}_i) \times (\mathbf{V}^G + \boldsymbol{\omega} \times \mathbf{r}_i)] \\ &= \mathbf{p}_G \times M\mathbf{V}_G + \mathbf{p}_G \times \left[\boldsymbol{\omega} \times \left(\sum_{i=1}^{N} m_i\mathbf{r}_i\right)\right]^{\nearrow 0} \\ &\quad + \left(\sum_{i=1}^{N} m_i\mathbf{r}_i\right)^{\nearrow 0} \times \mathbf{V}_G + \sum_{i=1}^{N} m_i\mathbf{r}_i \times (\boldsymbol{\omega} \times \mathbf{r}_i) \end{aligned} \tag{10.6.5}$$

where M is the mass of B and where, as before, the sum $\sum_{i=1}^{N} m_i\mathbf{r}_i$ is zero since G is the mass center [see Equation (8.3.2)].

The first term of Equation (10.6.5) may be regarded as the angular momentum of G (with mass M) about O, $\mathbf{A}^{G/O}$. In view of Equations (10.6.2) and (10.6.4) the last term of Equation (10.6.5) may be written as:

$$\sum_{i=1}^{N} m_i\mathbf{r}_i \times \mathbf{V}^{P_i/G} = \mathbf{A}^{B/G} \tag{10.6.6}$$

By using the analysis of Equation (9.12.9), the last term of Equation (10.6.5) may also be written as:

$$\sum_{i=1}^{N} m_i \mathbf{r}_i \times (\boldsymbol{\omega} \times \mathbf{r}_i) = \omega \sum_{i=1}^{N} m_i \mathbf{r}_i \times (\mathbf{n}_\omega \times \mathbf{r}_i)$$

$$= \omega \mathbf{I}_{\mathbf{n}_\omega} = \omega \mathbf{I}^{B/G} \bullet \mathbf{n}_\omega$$

$$= \mathbf{I}^{B/G} \bullet \boldsymbol{\omega} = \mathbf{A}^{B/G} \tag{10.6.7}$$

In summary, $\mathbf{A}^{B/O}$ may be written as:

$$\mathbf{A}^{B/O} = \mathbf{A}^{G/O} + \mathbf{A}^{B/G} = \mathbf{A}^{G/O} + \mathbf{I}^{B/G} \bullet \boldsymbol{\omega} \tag{10.6.8}$$

Finally, if O is also fixed in B, that is, if B is rotating about a fixed point O in R, Equation (10.6.2) may be further simplified since then the position vectors $\mathbf{p}_G$, $\mathbf{r}_i$, and $\mathbf{p}_i$ are all fixed in B. Thus, in this case $\mathbf{V}^{P_i}$ is simply $\boldsymbol{\omega} \times \mathbf{p}_i$ and then Equation (10.6.2) may be written in the form:

$$\mathbf{A}^{B/O} = \sum_{i=1}^{N} m_i \mathbf{p}_i \times (\boldsymbol{\omega} \times P_i) \tag{10.6.9}$$

By again using the analysis of Equation (9.12.9), the right side of Equation (10.6.9) is seen to be: $\mathbf{I}^{B/O} \bullet \boldsymbol{\omega}$. Therefore, if O is fixed in B, we have

$$\mathbf{A}^{B/O} = \mathbf{I}^{B/O} \bullet \boldsymbol{\omega} \tag{10.6.10}$$

A more explicit expression for the dyadic/angular velocity products of Equations (10.6.7), (10.6.8), and (10.6.10) may be obtained by using principal direction unit vectors (see Section 8.10). For example, in Equation (10.6.7), suppose that $\mathbf{n}_1$, $\mathbf{n}_2$, and $\mathbf{n}_3$ are principal unit vectors of B for G. Then the product $\mathbf{I}^{B/G} \bullet \boldsymbol{\omega}$ may be expressed as:

$$\mathbf{A}^{B/G} = \mathbf{I}^{B/G} \bullet \boldsymbol{\omega} = I_{11}^{B/G} \omega_1 \mathbf{n}_1 + I_{22}^{B/G} \omega_2 \mathbf{n}_2 + I_{33}^{B/G} \omega_3 \mathbf{n}_3 \tag{10.6.11}$$

where $I_{11}^{B/G}$, $I_{22}^{B/G}$, and $I_{33}^{B/G}$ are principal moments of inertia of B for G corresponding to the direction of $\mathbf{n}_1$, $\mathbf{n}_2$, and $\mathbf{n}_3$ and where ω_1, ω_2, and ω_3 are the $\mathbf{n}_1$, $\mathbf{n}_2$, and $\mathbf{n}_3$ components of $\boldsymbol{\omega}$.

Table 10.6.1 provides a summary of the angular momentum formulas.

Table 10.6.1 Angular Momentum Expressions

Description	Expression	Reference Equations	Comments
1. Angular Momentum of a Particle P about a Point O.	$\mathbf{A}^{P/O} = \mathbf{p} \times m\mathbf{v}$	(10.6.1)	$\mathbf{p}$ locates P relative to O, m is the mass of P, and $\mathbf{v}$ is the velocity of P relative to O.
2. Angular Momentum of a Body B about its Mass Center G.	$\mathbf{A}^{B/G} = \mathbf{I}^{B/G} \cdot \boldsymbol{\omega}$	(10.6.7)	$\mathbf{I}^{B/G}$ is the central inertia dyadic of B and $\boldsymbol{\omega}$ is the angular velocity of B.
3. Angular Momentum of a Body B about a Point O.	$\mathbf{A}^{B/O} = \mathbf{A}^{B/G} + \mathbf{A}^{G/O}$	(10.6.8)	$\mathbf{A}^{G/O}$ is the angular momentum of a particle at G, with the mass of B, about O.
4. Angular Momentum of a Body B about a point O fixed in B.	$\mathbf{A}^{B/O} = \mathbf{I}^{B/O} \cdot \boldsymbol{\omega}$	(10.6.10)	O is fixed in both B and an inertial reference frame R, $\boldsymbol{\omega}$ is the angular velocity of B in R, and $\mathbf{I}^{B/O}$ is the inertia dyadic of B relative to O.

10.7 Newton's Laws/d'Alembert's Principle

Newton's second and third law form the basis for dynamic analyses of rigid bodies: Recall that for a particle P with mass m, moving in a Newtonian (inertial) reference frame R, the force $\mathbf{F}$ on P is proportional to the acceleration $\mathbf{a}$ of P in R. That is:

$$\mathbf{F} = m\mathbf{a} \qquad (10.7.1)$$

Newton's third law states that for every action (positive force) exerted on a particle there is an equal magnitude and oppositely directed reaction (negative force) exerted by the particle. This means that if a rigid body is modeled as a set of particles, fixed relative to one another, then the forces (e.g. contact forces) exerted between adjacent particles on each other, will be self-equilibrating — that is, equal in magnitude but oppositely directed and thus adding to zero. Therefore, forces internal to a body, exerted by the particles making up the body on one another, will not contribute to the totality of forces exerted on the body.

For computational purposes it is often convenient to rewrite Equation (10.7.1) in the form:

$$\mathbf{F} + \mathbf{F}^* = 0 \qquad (10.7.2)$$

where $\mathbf{F}^*$, called an "inertia force," is defined as:

$$\mathbf{F}^* = -m\mathbf{a} \qquad (10.7.3)$$

Verbally, Equation (10.7.2) states that the sum of the applied ("active") forces and the inertia ("passive") forces on P is a zero system (d'Alembert's principle).

Consider now a body B moving in an inertial reference frame R as in Figure 10.7.1. As before, let B be composed of particles P_i $(i = 1,\ldots,N)$ having masses m_i. Let $\mathbf{F}_j$ $(j = 1,\ldots,\hat{N})$ be a force system S applied to B. Then with the internal forces of B being self-equilibrating, thc totality of all the applied forces on B is

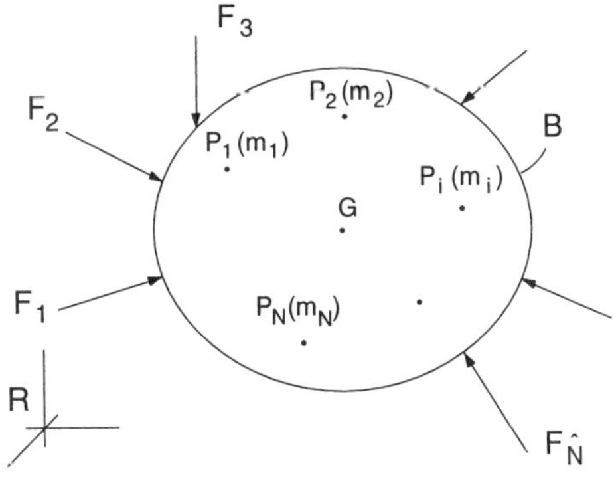

Fig. 10.7.1 A Rigid Body B Moving in an Inertial Reference Frame R

equivalent to S.

Let the resultant of S be **F** and let the moment of S about G be $\mathbf{M}_G$. Then for computational purposes it is convenient to replace S by an equivalent force system $\hat{S}$ (see Example 2.7.10) consisting of a single force **F** passing through the mass center G together with a couple with torque $\mathbf{M}_G$, as represented in Figure 10.7.2.

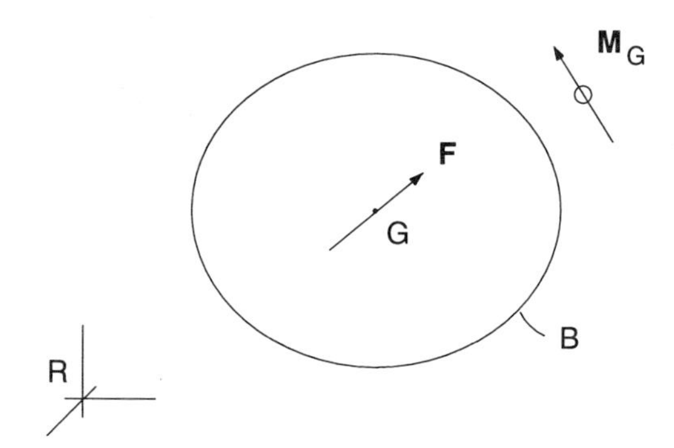

Fig. 10.7.2 Equivalent Applied Force Systems on Rigid Body B

Similarly, from Section 9.12 the inertia forces on the particles of B are seen to be equivalent to a single force $\mathbf{F}^*$ passing through mass center G together with a couple with torque $\mathbf{T}^*$ as represented in Figure 10.7.3. From Equations (9.12.6) and (9.12.11) $\mathbf{F}^*$ and $\mathbf{T}^*$ are seen to be:

$$\mathbf{F}^* = -M\,{}^R\mathbf{a}^G \tag{10.7.4}$$

and

$$\mathbf{T}^* = -\mathbf{I}^{B/G} \bullet {}^R\boldsymbol{\alpha}^B - {}^R\boldsymbol{\omega}^B \times \left(\mathbf{I}^{B/G} \bullet {}^R\boldsymbol{\omega}^B\right) \tag{10.7.5}$$

where, as before, M is the mass of B, $\mathbf{I}^{B/G}$ is the central inertia dyadic of B, and ${}^R\boldsymbol{\omega}^B$ and ${}^R\boldsymbol{\alpha}^B$ are the angular velocity and angular acceleration of B in R.

Taken together, Figures 10.7.2 and 10.7.3 constitute a free-body diagram for B. Specifically, by applying Newton's second law (or equivalently, d'Alembert's

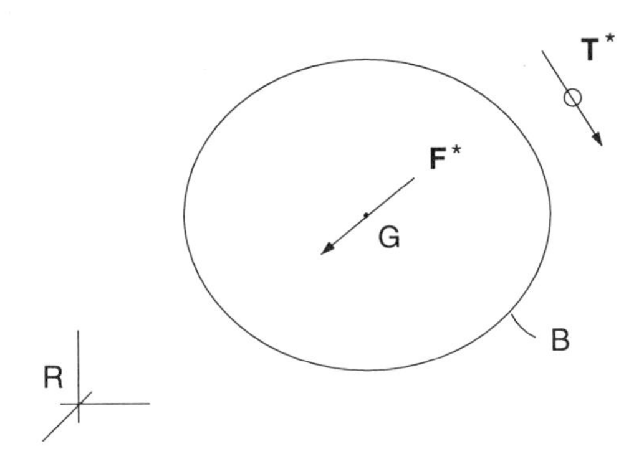

Fig. 10.7.3 Equivalent Inertia Force System on Rigid Body B

principle) for each particle P_i of B, by adding the resulting equations, and by using Newton's third law, we have:

$$\mathbf{F} + \mathbf{F}^* = 0 \quad \text{and} \quad \mathbf{M}_G + \mathbf{T}^* = 0 \tag{10.7.6}$$

or

$$\mathbf{F} = M\,{}^R\mathbf{a}^G \quad \text{and} \quad \mathbf{M}_G = \mathbf{I}^{B/G} \bullet {}^R\boldsymbol{\alpha}^B + {}^R\boldsymbol{\omega}^B \times (\mathbf{I}^{B/G} \bullet {}^R\boldsymbol{\omega}^B) \tag{10.7.7}$$

Finally, if $\mathbf{n}_1$, $\mathbf{n}_2$, and $\mathbf{n}_3$ are mutually perpendicular unit vectors parallel to principal directions of B for G, the second equation of Equation (10.7.7) may be expressed in component form as:

$$\begin{aligned} M_{G1} &= \alpha_1 I_{11} - \omega_2\omega_3(I_{22} - I_{33}) \\ M_{G2} &= \alpha_2 I_{22} - \omega_3\omega_1(I_{33} - I_{11}) \\ M_{G3} &= \alpha_3 I_{33} - \omega_1\omega_2(I_{11} - I_{22}) \end{aligned} \tag{10.7.8}$$

where, as before, the subscript indices refer to the indices of the $\mathbf{n}_i$ $(i = 1,2,3)$.

Example 10.7.1 Rod Pendulum Using Newton's Laws/d'Alembert's Principle

For an elementary illustration using Newton's laws/d'Alembert's principle consider the motion of a rod pendulum: Specifically, a thin rod or bar B of length ℓ is supported by a frictionless pin at one end and allowed to oscillate in a vertical plane as depicted in Figure 10.7.4. Let θ define the orientation of B. The objective is to determine the differential equation for θ and the reaction of the pin O.

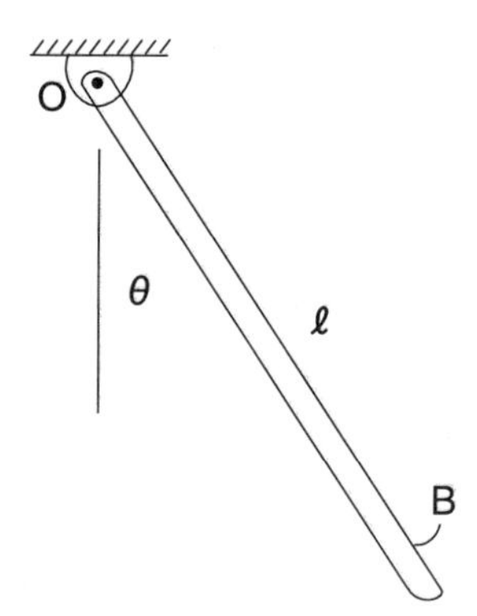

Fig. 10.7.4 A Rod Pendulum

Solution: Using d'Alembert's principle, construct a free-body diagram of B: First, observe that since B moves in a plane the kinematics is relatively simple with the angular velocity, angular acceleration, and mass center acceleration given by:

$$\boldsymbol{\omega} = \dot{\theta}\,\mathbf{n}_z \qquad \boldsymbol{\alpha} = \ddot{\theta}\,\mathbf{n}_z \qquad \mathbf{a} = (\ell/2)\ddot{\theta}\,\mathbf{n}_\theta - (\ell/2)\dot{\theta}^2\,\mathbf{n}_r \tag{10.7.9}$$

where $\mathbf{n}_r$, $\mathbf{n}_\theta$, and $\mathbf{n}_z$ are unit vectors as shown in Figure 10.7.5.

Next, the applied forces on B are equivalent to a single vertical (downward) force $-mg\,\mathbf{n}_y$, passing through the mass center G, representing the weight of B, and a pin reaction force $O_x\,\mathbf{n}_x + O_y\,\mathbf{n}_y$, passing through O, where m is the mass of B, g is the gravity constant, and $\mathbf{n}_x$ and $\mathbf{n}_y$ are horizontal and vertical unit vectors shown in Figure 10.7.5.

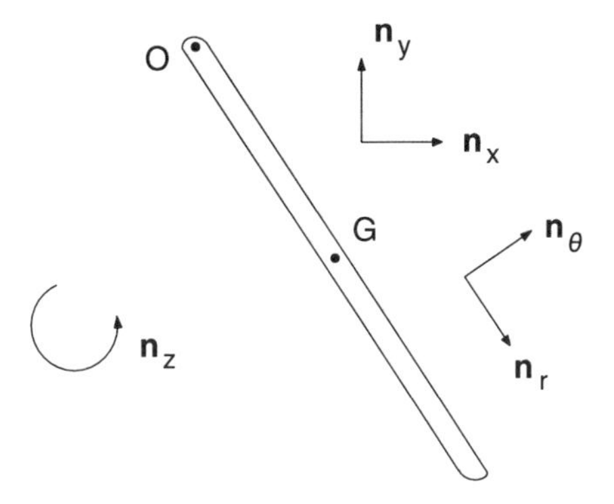

Fig. 10.7.5 Unit Vectors for the Rod Pendulum

The inertia forces on B are equivalent to a single force $-m\mathbf{a}$ passing through G [See Equation (10.7.4)] together with an inertia torque $-\mathbf{I}\boldsymbol{\alpha}$ where I is the central moment of inertia of B for the $\mathbf{n}_z$ direction $(m\ell^2/12)$ [see Equation (10.7.5)].

Figure 10.7.6 shows the desired free-body diagram of B.

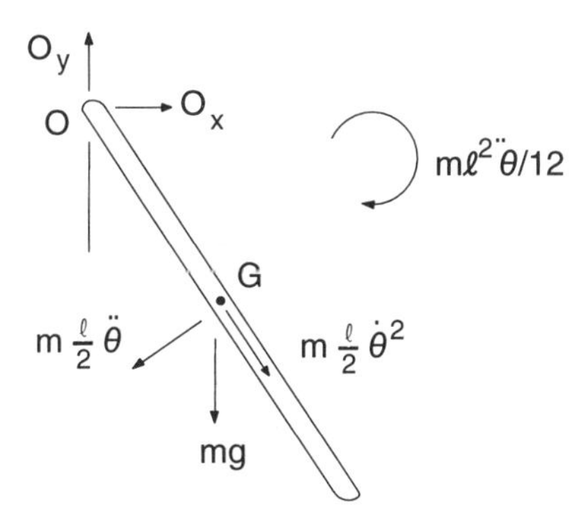

Fig. 10.7.6 A Free-Body Diagram of B

Finally, setting moments about O equal to zero and by adding to zero the horizontal and vertical forces we obtain

$$\ddot{\theta} + (3g/2\ell)\sin\theta = 0 \tag{10.7.10}$$

$$O_x = m(\ell/2)\ddot{\theta}\cos\theta - m(\ell/2)\dot{\theta}^2\sin\theta \tag{10.7.11}$$

$$O_y = mg + m(\ell/2)\dot{\theta}^2\cos\theta + m(\ell/2)\ddot{\theta}\sin\theta \tag{10.7.12}$$

Observe that in comparing Equation (10.7.10) with the equation for the simple pendulum [Equation (5.3.8)] the equations are the same except that the rod pendulum has a higher natural frequency by the factor $\sqrt{3/2}$. Also observe that unlike the simple pendulum the reaction force at O is not in general parallel to the rod.

10.8 Kane's Equations

Consider a rigid body B moving in an inertial reference frame R. Let B be a part of a mechanical system S having n degrees of freedom represented by generalized coordinates q_r $(r = 1,...,n)$. As before, let B be modeled as a set of particles P_i $(i = 1,...,N)$ fixed relative to one another. See Figure 10.8.1.

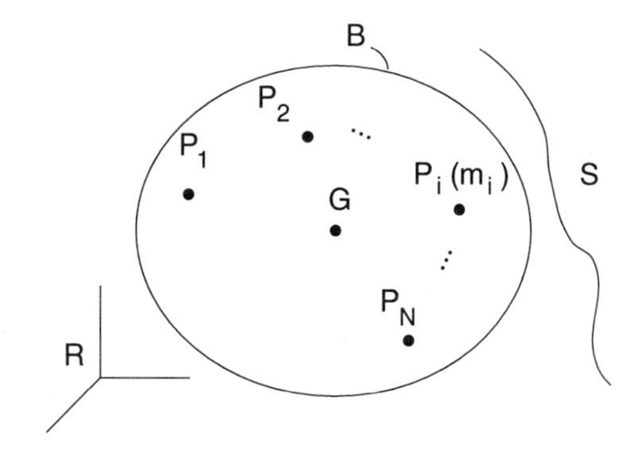

Fig. 10.8.1 A Body B in a Mechanical System S Moving in an Inertial Reference Frame R

Consider a typical particle P_i of B. Newton's law (or d'Alembert's principle) provides a balance between the applied and inertia forces on P_i. That is

$$\mathbf{F}_{P_i} = m_i \mathbf{a}^{P_i} \quad \text{(no sum)} \tag{10.8.1}$$

or

$$\mathbf{F}_{P_i} + \mathbf{F}^*_{P_i} = 0 \tag{10.8.2}$$

where $\mathbf{F}_{P_i}$ is the resultant applied force on P_i, m_i is the mass of P_i, $\mathbf{a}^{P_i}$ is the acceleration of P_i in R, and $\mathbf{F}^*_{P_i}$ is the inertia force on P_i defined as: $-m_i\mathbf{a}^{P_i}$ (no sum).

Kane's equations are developed by projecting the forces of Equation (10.8.2) along the direction of the partial velocity of P_i relative to the generalized coordinates q_r. That is,

$$\mathbf{F}_{P_i} \bullet \mathbf{v}^{P_i}_{\dot{q}_r} + \mathbf{F}^*_{P_i} \bullet \mathbf{v}^{P_i}_{\dot{q}_r} = 0 \quad (r = 1,\ldots,n) \tag{10.8.3}$$

These projections are called "generalized forces" and thus Kane's equations state that the sum of the generalized applied (active) and inertia (passive) forces is zero for each generalized coordinate q_r, or

$$F^{P_i}_{q_r} + F^{*P_i}_{q_r} = 0 \quad (r = 1,\ldots,n) \tag{10.8.4}$$

Kane's equations are readily extended from individual particles to the entire body: Specifically, by adding to forces for the individual particles, we have:

$$\sum_{i-1}^{N} \mathbf{F}_{P_i} \bullet \mathbf{v}^{P_i}_{\dot{q}_r} + \sum_{i=1}^{N} \mathbf{F}_{P_i^*} \bullet \mathbf{v}^{P_i}_{\dot{q}_r} = 0 \quad (r = 1,\ldots,n) \tag{10.8.5}$$

and

$$\sum_{i=1}^{N} F_{q_r}^{P_i} + \sum_{i=1}^{N} F_{q_r}^{*P_i} = 0 \quad (r = 1,\dots,n) \tag{10.8.6}$$

In Chapter 9 we observed that self-equilibrating forces and several other applied forces do not contribute to sums of generalized forces and thus they can be ignored in the analysis. Specifically, if the applied forces on the particles of a body B are equivalent to a set of forces $\mathbf{F}_j$ $(j = 1,\dots,\hat{N})$ applied at particles $\hat{P}_j$ of B (see Figure 10.8.2), then the first sum of Equation (10.8.6) is:

$$\sum_{i=1}^{N} F_{q_r}^{P_i} = \mathbf{F}_1 \bullet \mathbf{v}_{\dot{q}_r}^{\hat{P}_1} + \mathbf{F}_2 \bullet \mathbf{v}_{\dot{q}_r}^{\hat{P}_2} + \dots + \mathbf{F}_N \bullet \mathbf{v}_{\dot{q}_r}^{\hat{P}_N}$$

$$= \sum_{j=1}^{\hat{N}} \mathbf{F}_j \bullet \mathbf{v}_{\dot{q}_r}^{\hat{P}_j} \stackrel{D}{=} F_{q_r} \quad (r = 1,\dots,n) \tag{10.8.7}$$

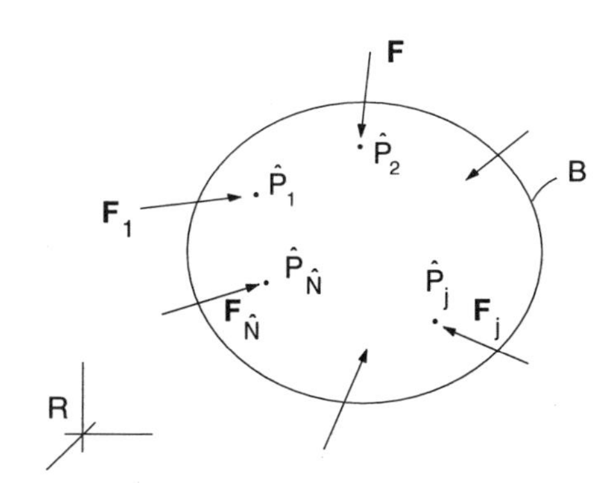

Fig. 10.8.2 Equivalent Applied Force System on Body B

Alternatively if the applied forces on B are equivalent to a single force **F**, passing through a point of B (say the mass center G) together with a couple with torque **T** as in Figure 10.8.3, then from Equation (9.9.4) the first sum of Equation (10.8.6) is

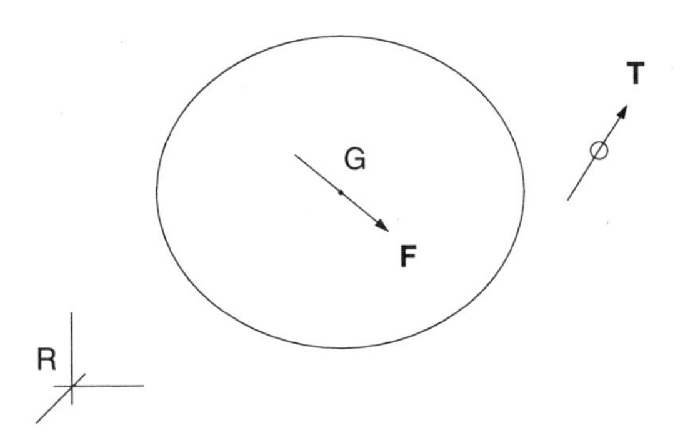

Fig. 10.8.3 Second Equivalent Applied Force System on Body B

$$\sum_{i=1}^{N} F_{q_r}^{P_i} = \mathbf{F} \bullet \mathbf{v}_{\dot{q}_r}^{G} + \mathrm{T} \bullet \boldsymbol{\omega}_{\dot{q}_r} = F_{q_r} \quad (r = 1, \ldots, n) \tag{10.8.8}$$

where $\boldsymbol{\omega}_{\dot{q}_r}$ is the partial angular velocity of B relative to q_r.

Similarly, in Chapter 9 we observed that if the inertia forces on the particles of B are equivalent to a single force $\mathbf{F}^*$ passing through the mass center G, together with a torque $\mathbf{T}^*$ as in Figure 10.8.4, then the second sum of Equation (10.8.6) is [see Equation (9.13.1)]:

$$\sum_{i=1}^{N} F_{q_r}^{*P_i} = \mathbf{F}^* \bullet \mathrm{v}_{\dot{q}_r}^{G} + \mathrm{T}^* \bullet \boldsymbol{\omega}_{\dot{q}_r} = F_{q_r}^* \quad (r = 1, \ldots, n) \tag{10.8.9}$$

From Equation (10.8.6) Kane's equations then become:

$$F_{q_r} + F_{q_r}^* = 0 \quad (r = 1, \ldots, n) \tag{10.8.9}$$

Kane's equations simply state that the sum of the generalized applied (active) and inertia (passive) forces is zero.

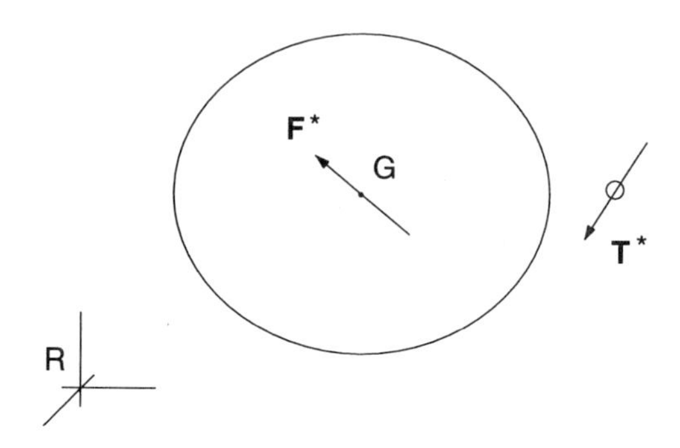

Fig. 10.8.4 Equivalent Inertia Force System on Body B

Example 10.8.1 Rod Pendulum Using Kane's Equations

Consider again the rod pendulum of Example 10.7.1 where a thin rod, or bar B, of length ℓ is supported by a frictionless pin at one end and is allowed to oscillate in a vertical plane as depicted in Figure 10.8.5. As before let θ define the orientation of B. The objective is to determine the governing differential equation for θ.

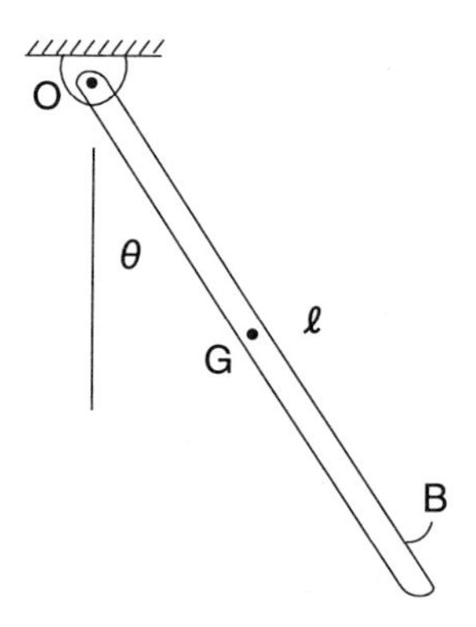

Fig. 10.8.5 A Rod Pendulum

Solution: Let G be the mass center of B. Then the velocity and acceleration of G in a fixed, inertial frame R, may be expressed as:

$$\mathbf{v} = (\ell/2)\dot{\theta}\,\mathbf{n}_\theta \quad \text{and} \quad \mathbf{a} = (\ell/2)\ddot{\theta}\,\mathbf{n}_\theta - (\ell/2)\dot{\theta}^2\,\mathbf{n}_r \qquad (10.8.11)$$

where $\mathbf{n}_r$ and $\mathbf{n}_\theta$ are radial and transversely directed unit vectors as in Figure 10.8.6. The angular velocity and angular acceleration of B are

$$\boldsymbol{\omega} = \dot{\theta}\,\mathbf{n}_z \quad \text{and} \quad \boldsymbol{\alpha} = \ddot{\theta}\,\mathbf{n}_z \tag{10.8.12}$$

where $\mathbf{n}_z$ is the horizontally directed unit vector of Figure 10.8.6.

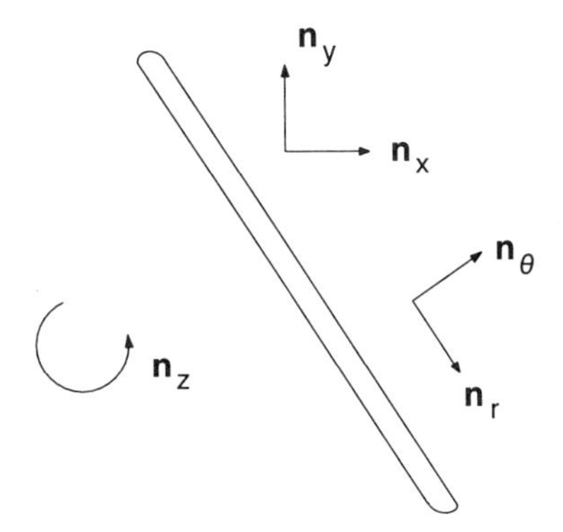

Fig. 10.8.6 Unit Vectors for the Rod Pendulum

With B being pin-supported at O and constrained to move in the vertical plane, it has but one degree of freedom, described by the angle θ. From Equations (10.8.11) and (10.8.12) the partial velocity of G and the partial angular velocity of B are seen to be:

$$\mathbf{v}_{\dot{\theta}} = (\ell/2)\,\mathbf{n}_\theta \quad \text{and} \quad \boldsymbol{\omega}_{\dot{\theta}} = \mathbf{n}_z \tag{10.8.13}$$

The applied forces on B are equivalent to a pin-reaction force **O** at O and a vertical weight force **W** at G given by

$$\mathbf{O} = O_x\,\mathbf{n}_x + O_y\,\mathbf{n}_y \quad \text{and} \quad \mathbf{W} = -mg\,\mathbf{n}_y \tag{10.8.14}$$

where m is the mass of B and $\mathbf{n}_x$ and $\mathbf{n}_y$ are horizontal and vertical unit vectors as in Figure 10.8.6.

The inertia forces on B are equivalent to a single force $\mathbf{F}^*$ passing through G together with a couple with torque $\mathbf{T}^*$ given by [see Equations (10.7.4) and (10.7.5)]:

$$\mathbf{F}^* = -m\mathbf{a} \quad \text{and} \quad \mathbf{T}^* = -I\boldsymbol{\alpha} \tag{10.8.15}$$

where I is the central moment of inertia of B for the $\mathbf{n}_z$ direction $(m\ell^2/12)$.

The generalized applied and inertia forces are then [see Equation (10.8.11)]:

$$F_\theta = \mathbf{v}_{\dot\theta} \bullet \mathbf{W} = -mg(\ell/2)\sin\theta \tag{10.8.16}$$

and

$$F_\theta^* = \mathbf{v}_{\dot\theta} \bullet \mathbf{F}^* + \boldsymbol{\omega}_{\dot\theta} \bullet \mathbf{T}^* = -m(\ell/2)^2\ddot\theta - m(\ell^2/12)\ddot\theta \tag{10.8.17}$$

Kane's equations [Equations (10.8.10)] state that the sum of these generalized forces is zero. That is

$$-mg(\ell/2)\sin\theta - m(\ell/2)^2\ddot\theta - m(\ell^2/12)\ddot\theta = 0 \tag{10.8.18}$$

or

$$\ddot\theta + (3g/2\ell)\sin\theta = 0 \tag{10.8.19}$$

Equation (10.8.19) is seen to be identical to Equation (10.7.10). Observe, however, that expressions for the pin-reaction force components O_x and O_y are not obtained since the pin-reaction force passes through a fixed point and thus does not contribute to the generalized applied force. This is advantageous if the pin-reaction force is not of interest — as is often the case.

10.9 Lagrange's Equations

In most dynamics books, Lagrange's equations are developed from Hamilton's principle using variational calculus to minimize a functional [see Equation (5.2.10) and Section 10.13.5]. However, Lagrange's equations can be directly developed from Kane's equations using Equation (9.4.6): Consider again a rigid body B modeled as a set of particles affixed relative to one another as in Figure 10.9.1. Let B be a part

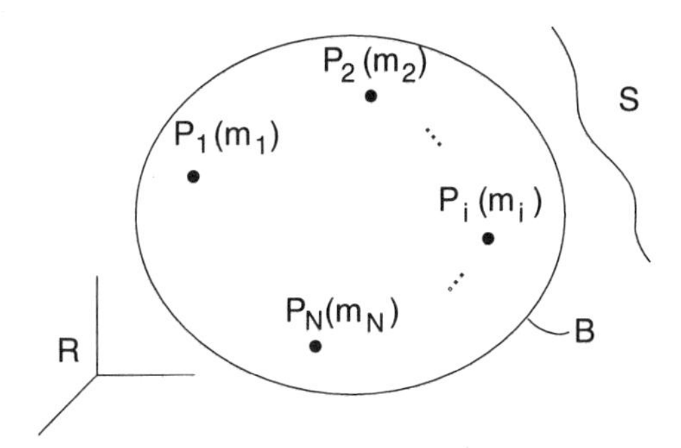

Fig. 10.9.1 A Rigid Body B Modeled as a Set of Particles Moving in an Inertial Reference Frame R.

of a holonomic* mechanical system S with n degrees of freedom, in an inertial reference frame R, represented by generalized coordinates q_r $(r = 1, \dots, n)$. Then for a typical particle P_i Equation (9.4.6) states:

$$\mathbf{a}^{P_i} \bullet \mathbf{v}^{P_i}_{\dot{q}_r} = \frac{1}{2}\frac{d}{dt}\left[\frac{\partial (\mathbf{v}^{P_i})^2}{\partial \dot{q}_r}\right] - \frac{1}{2}\frac{\partial (\mathbf{v}^{P_i})^2}{\partial q_r} \qquad (r = 1, \dots, n) \qquad (10.9.1)$$

where $\mathbf{v}^{P_i}$ and $\mathbf{a}^{P_i}$ are the velocity and acceleration of P_i in R. By multiplying by the mass m_i of P_i (and changing signs) we have:

$$\mathbf{v}^{P_i}_{\dot{q}_r} \bullet (-m_i \mathbf{a}^{P_i}) = -\frac{d}{dt}\left[\frac{\partial\left(\frac{1}{2} m_i (\mathbf{v}^{P_i})^2\right)}{\partial \dot{q}_r}\right] + \frac{\partial\left(\frac{1}{2} m_i \mathbf{v}^{P_i}\right)^2}{\partial q_r} \qquad (r = 1, \dots, n) \qquad (10.9.2)$$

---

*See Chapter 4, Sections 4.5.2, 4.5.3, and 4.5.4.

The term in parenthesis on the left side of Equation (10.9.2) is recognized as the d'Alembert inertia force F^{*P_i} on P_i and the terms in parentheses on the right side as the kinetic energy K^{P_i} of P_i. That is, Equation (10.9.2) may be written as:

$$\mathbf{v}_{\dot{q}_r}^{P_i} \bullet \mathbf{F}^{*P_i} = -\frac{d}{dt}\left(\frac{\partial K^{P_i}}{\partial \dot{q}_r}\right) + \frac{\partial K^{P_i}}{\partial q_r} \quad (r = 1, \dots, n) \tag{10.9.3}$$

The left side of Equation (10.9.3) may in turn be recognized as the generalized inertia force $F_{q_r}^{*P_i}$ on P_i for the generalized coordinate P_i. Hence, we have

$$F_{q_r}^{*P_i} = -\frac{d}{dt}\left(\frac{\partial K^{P_i}}{\partial \dot{q}_r}\right) + \frac{\partial K^{P_i}}{\partial q} \quad (r = 1, \dots, n) \tag{10.9.4}$$

By adding equations of the form of Equations (10.9.4) for all the particles of B we obtain

$$F_{q_r}^{*} = -\frac{d}{dt}\left(\frac{\partial K}{\partial \dot{q}_r}\right) + \frac{\partial K}{\partial q_r} \quad (r = 1, \dots, n) \tag{10.9.5}$$

where $F_{q_r}^{*}$ is the generalized inertia force on B for the generalized coordinate q_r, and where K is the kinetic energy of B.

Finally, by using Kane's equation [Equation (10.8.10)], Equations (10.9.5) may be written as:

$$\frac{d}{dt}\left(\frac{\partial K}{\partial \dot{q}_r}\right) - \frac{\partial K}{\partial q_r} = F_{q_r} \quad (r = 1, \dots, n) \tag{10.9.6}$$

where F_{q_r} is the generalized applied (active) force on B for the generalized coordinate q_r. If, further, S is a "conservative" system such that a potential energy function P exists so that F_{q_r} may be expressed as $-\partial P/\partial q_r$, then Equation (10.9.6) may be expressed as

$$\frac{d}{dt}\left(\frac{\partial L}{\partial \dot{q}_r}\right) - \frac{\partial L}{\partial q_r} = 0 \qquad (r = 1,\ldots,n) \tag{10.9.7}$$

where L (called the "Lagrangian") is defined as $\kappa - P$.

Equations (10.9.6) and (10.9.7) are known as Lagrange's equations.

Example 10.9.1 Rod Pendulum Using Lagrange's Equations

Consider again the rod pendulum of Examples 10.7.1 and 10.8.1 where a thin rod, or bar B, of length ℓ is supported by a frictionless pin at one end and oscillating in a vertical plane as in Figure 10.9.2.

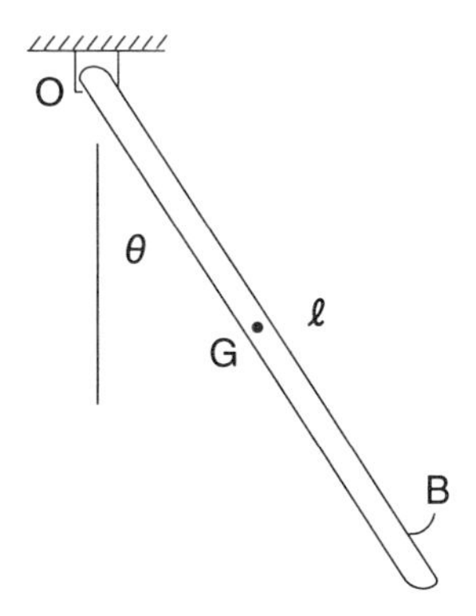

Fig. 10.9.2 A Rod Pendulum

As before, let θ define the orientation of B. The objective is to determine the governing differential equations for θ.

Solution: As before, let G be the mass center of B. Then the velocity of G and the angular velocity of B in a fixed inertia frame R may be expressed as:

$$\mathbf{v} = (\ell/2)\dot{\theta}\,\mathbf{n}_\theta \quad \text{and} \quad \boldsymbol{\omega} = \dot{\theta}\,\mathbf{n}_z \tag{10.9.8}$$

where $\mathbf{n}_\theta$ is a transversely directed unit vector and $\mathbf{n}_z$ is parallel to the axis of the pin support of B as in Figure 10.9.3.

With B being pin-supported at O and constrained to move in the vertical plane, it has but one degree of freedom, described by the angle θ. From Equation (10.9.8) the partial velocity of G is

$$\mathbf{v}_{\dot{\theta}} = (\ell/2)\mathbf{n}_\theta \qquad (10.9.9)$$

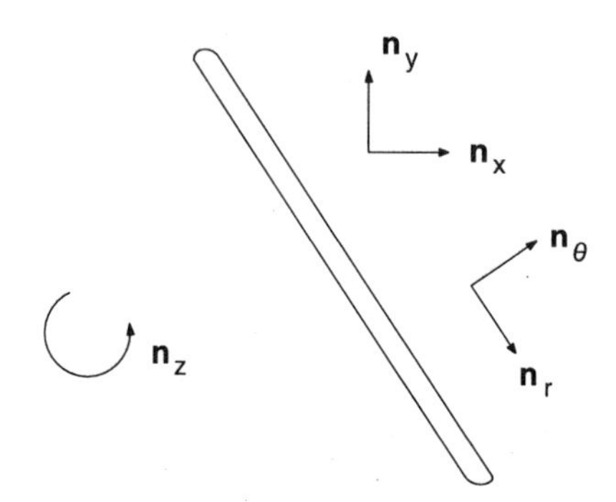

Fig. 10.9.3 Unit Vectors of the Rod Pendulum.

The applied forces on B are equivalent to a pin-reaction force **O** at O and a vertical weight force **W** at G given by

$$\mathbf{O} = O_x\mathbf{n}_x + O_y\mathbf{n}_y \quad \text{and} \quad \mathbf{W} = -mg\mathbf{n}_y \qquad (10.9.10)$$

where m is the mass of B and $\mathbf{n}_x$ and $\mathbf{n}_y$ are horizontal and vertical unit vectors as in Figure 10.9.3.

The generalized applied force is then

$$F_\theta = \mathbf{v}_{\dot{\theta}} \bullet \mathbf{w} = -mg(\ell/2)\sin\theta \qquad (10.9.11)$$

From Equations (10.3.4) and (10.9.8) the kinetic energy of B is:

$$\begin{aligned} K &= (1/2)m\mathbf{v}^2 + (1/2)I\,\boldsymbol{\omega}^2 \\ &= (1/2)m(\ell/2)^2\dot{\theta}^2 + (1/2)(1/12)m\ell^2\dot{\theta}^2 \\ &= (1/6)m\,\ell^2\dot{\theta}^2 \end{aligned} \qquad (10.9.12)$$

where I is the central moment of inertia of the rod $(1/12)m\ell^2$.

By substituting from Equations (10.9.6) and (10.9.12) into (10.9.6) we obtain:

$$\ddot{\theta} + (3g/\partial\ell)\sin\theta = 0 \tag{10.9.13}$$

Equation (10.9.13) is seen to be identical to Equations (10.7.10) and (10.8.19) obtained using d'Alembert's principle and Kane's equations. Observe that here, with Lagrange's equations, it is not necessary to calculate accelerations, thus reducing the computational effort.

10.10 Lagrange's Equations with Simple Non-holonomic Systems

Lagrange's equations as developed in the previous section are restricted to systems that are either unrestrained or are subjected to geometric (position) constraints (that is, "holonomic" systems). Lagrange's equations, however, are not directly applicable with systems with kinematic (velocity, acceleration) constraints, that is "non-holonomic" systems. The reason they are not valid for non-holonomic systems is that their development is based upon the independence which is lost with kinematic constraints. In this section we present an extension of Lagrange's equations for application with simple non-holonomic systems. The analysis closely follows that presented in Reference 10.1.

Consider a mechanical system S containing particles and rigid bodies having n degrees of freedom represented by generalized coordinates q_r $(r = 1, \ldots, n)$. Let S be subjected to m simple kinematic (velocity), or non-holonomic, constraints of the form:

$$\sum_{r=1}^{n} a_{sr}\dot{q}_r + b_s = 0 \qquad s = p+1, \ldots, n \tag{10.10.1}$$

where

$$p = n - m \tag{10.10.2}$$

where the a_{sr} and b_s are functions of the q_r and time t, and where the numbering is used for later simplicity and for consistency with Reference 10.1. Equations (10.10.1) are linear in the $\dot{q}_r$ and are perhaps the simplest of the possible kinematic constraints. Such constraints occur, for example, with rolling bodies. If a body B, of a mechanical system S, rolls on a fixed surface as in Figure 10.10.1, the contact point C of B has zero velocity:

$$\mathbf{v}_C = 0 \tag{10.10.3}$$

Equations (10.10.1) are a generalized form of Equation (10.10.3).

Equations (10.10.1) are also independent in that they may be solved for m, say the last m, of the $\dot{q}_r$ in terms of the remaining $n - m$ $\dot{q}_r$ as

$$\dot{q}_r = \sum_{i=1}^{p} c_{ri}\dot{q}_i + d_r \quad (r = p+1,\ldots,n) \tag{10.10.4}$$

where the c_{ri} and d_r are determined in the solution of the simultaneous equations and are in general functions of the q_r and t. Observe also that with m constraints p (n - m) is now the number of degrees of freedom of the system.

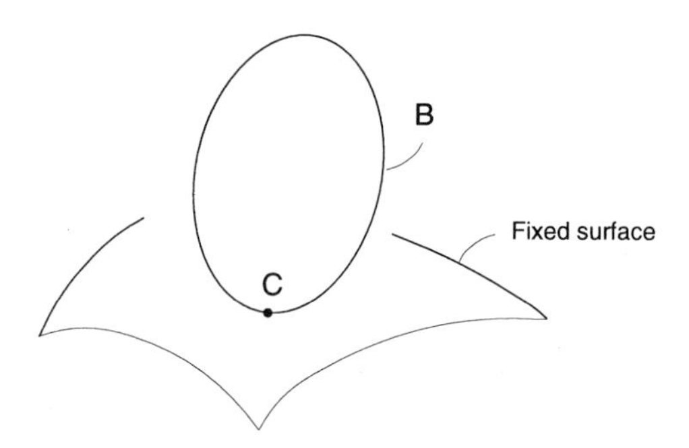

Fig. 10.10.1 A Body B Rolling on a Surface S

To develop the analysis let generalized speeds w_s and parameters a_{sr} and b_s $(s,r - 1,\ldots,n)$ be introduced as

$$w_s = \sum_{r=1}^{n} a_{sr}\dot{q}_r + b_s \quad (s = 1,\ldots,n) \tag{10.10.5}$$

where the last m of these are to be identical to Equations (10.10.1) and where the remaining a_{sr} and b_s are arbitrary provided only that Equations (10.10.5) may be solved for the $\dot{q}_r$ in terms of the w_s (that is, det $a_{rs} \neq 0$). Then since the last m w_s are zero, the $\dot{q}_r$ may be written as:

$$\dot{q}_r = \sum_{s=1}^{n} a_{rs}^{-1}(w_s - b_s) = \sum_{s=1}^{p} a_{rs}^{-1} w_s - \sum_{s=1}^{n} a_{rs}^{-1} b_s \quad (r = 1,\ldots,n) \tag{10.10.6}$$

where the a_{rs}^{-1} are elements of the inverse of the matrix of a_{rs}.

For the unconstrained system s, the velocity $\mathbf{v}$ of a typical point P of S may be written in the form [see Equations (4.5.8) and (4.5.9)]:

$$\mathbf{v} = \sum_{r=1}^{n} \mathbf{v}_{\dot{q}_r}\dot{q}_r + \mathbf{v}_t \tag{10.10.7}$$

By substituting from Equations (10.10.6) into (10.10.7), $\mathbf{v}$ may be expressed as:

$$\mathbf{v} = \sum_{s=1}^{p} \mathbf{v}_{w_s} w_s + \mathbf{v}_b \tag{10.10.8}$$

where $\mathbf{v}_{w_s}$ and $\mathbf{v}_b$ are defined as:

$$\mathbf{v}_{w_s} = \sum_{r=1}^{n} \mathbf{v}_{\dot{q}_r} a_{rs}^{-1} \quad (s = 1,\ldots,p) \tag{10.10.9}$$

and

$$\mathbf{v}_b = \mathbf{v}_t - \sum_{s=1}^{n}\sum_{r=1}^{n} \mathbf{v}_{\dot{q}_r} a_{rs}^{-1} b_s \tag{10.10.10}$$

Then from Equations (4.5.13) and (10.10.8) we have the basic equation:

$$\mathbf{a} \bullet \mathbf{v}_{w_s} = \sum_{r=1}^{n} \frac{1}{2}\left[\frac{d}{dt}\left(\frac{\partial \mathbf{v}^2}{\partial \dot{q}_r}\right) - \frac{\partial \mathbf{v}^2}{\partial q_r}\right] a_{rs}^{-1} \quad (s = 1,\ldots,p) \tag{10.10.11}$$

To complete the analysis left S be subjected to a force field consisting of forces $\mathbf{F}_i$ $(i = 1,\ldots,\hat{N})$ applied at particles P_i of S. Then generalized applied (active) forces K_s relative to the w_s may be defined as:

$$K_s = \sum_{i=1}^{\hat{N}} \mathbf{F}_i \bullet \mathbf{v}_{w_s}^{P_i} = \sum_{i=1}^{\hat{N}}\sum_{r=1}^{n} \mathbf{F}_i \bullet \mathbf{v}_{\dot{q}_r}^{P_i} a_{rs}^{-1} = \sum_{r=1}^{n} F_{q_r} a_{rs}^{-1} \quad (s = 1,\ldots,p) \tag{10.10.12}$$

where the F_{q_r} are the generalized applied forces relative to the q_r [see Section 4.5].

Similarly, generalized inertia (passive) forces K_s^* relative to the w_s may be defined as:

$$K_s^* = \sum_{i=1}^{N} \mathbf{F}_i^* \bullet \mathbf{v}_{w_s}^{P_i} = \sum_{i=1}^{N}\sum_{r=1}^{n} \mathbf{F}_i^* \bullet \mathbf{v}_{\dot{q}_r}^{P_i} a_{rs}^{-1} = \sum_{r=1}^{n} F_{qr}^* a_{rs}^{-1} \quad (s = 1,\ldots,p) \tag{10.10.13}$$

where as before the system is modeled as a system of N particles P_i and where the $\mathbf{F}_i^*$ defined as $-m_i\mathbf{a}^{P_i}$ are inertia forces and the $F_{q_r}^*$ are generalized inertia forces relative to the q_r [see Section (4.7)]. By substituting $-m_i\mathbf{a}^{P_i}$ for $\mathbf{F}_i^*$ into Equation (10.11.13), and by using Equations (10.10.11), K_s^* may be written as:

$$K_s^* = -\sum_{r=1}^{n}\left[\frac{d}{dt}\left(\frac{\partial K}{\partial \dot{q}_r}\right) - \frac{\partial K}{\partial q_r}\right] a_{rs}^{-1} \quad (s = 1,\ldots,p) \tag{10.10.14}$$

Finally, from Kane's equations [Equations (5.2.8)] we have

$$F_{q_r} + F^*_{q_r} = 0 \quad (r = 1,\ldots,n)$$

Then by using Equations (10.10.12), (10.10.13), and (10.10.14) we obtain the expressions:

$$\sum_{r=1}^{n} \left[\frac{d}{dt}\left(\frac{\partial K}{\partial \dot{q}_r}\right) - \frac{\partial K}{\partial q_r} \right] a_{rs}^{-1} = K_s \quad (s = 1,\ldots,p) \tag{10.10.15}$$

Equations (10.10.15) represent the desired extension of Lagrange's equations for simple non-holonomic systems. The equations may be simplified by taking advantage of the arbitrary formation of the w_s in Equations (10.10.5). Specifically, the equations take a more traditional form if the w_s are identified with the $\dot{q}_s$ $(s = 1,\ldots,p)$. To develop this, consider again Equations (10.10.1) but now written in the matrix form as:

$$[A_1 | A_2]\{\dot{q}\} = \{b\} \tag{10.10.16}$$

where by inspection A_1, A_2, $\dot{q}$, and b are

$$A_1 = m[\overset{p}{A_1}] = [a_{sr}] \quad (s = p+1,\ldots,n \ ; \ r = 1,\ldots,p) \tag{10.10.17}$$

$$A_2 = m[\overset{m}{A_2}] = [a_{sr}] \quad (s = p+1,\ldots,n \ ; \ r = p+1,\ldots,n) \tag{10.10.18}$$

$$\dot{q} = n\{\overset{1}{\dot{q}}\} = \begin{bmatrix} \dot{q}_{(1)} \\ \dot{q}_{(2)} \end{bmatrix} = \begin{bmatrix} \dot{q}_r \\ \dot{q}_s \end{bmatrix} \quad (r = 1,\ldots,p \ ; \ s = p+1,\ldots,n) \tag{10.10.19}$$

and

$$b = m\{\overset{1}{b}\} = \{b_s\} \quad (s = p+1,\ldots,n) \tag{10.10.20}$$

Then by substitution into Equation (10.10.16) and by expansion we have

$$[A_1 | A_2]\left[\frac{\dot{q}_{(1)}}{\dot{q}_{(2)}}\right] = A_1\dot{q}_{(1)} + A_2\dot{q}_{(2)} = \{b\} \tag{10.10.21}$$

Solving for $\dot{q}_{(2)}$ we obtain

$$\dot{q}_{(2)} = -A_2^{-1}A_1\dot{q}_{(1)} + A_2^{-1}b \tag{10.10.22}$$

or

$$\dot{q}_{(2)} = C\dot{q}_{(1)} + d \tag{10.10.23}$$

where C and d are $-A_2^{-1}A_1$ and $A_2^{-1}b$ respectively with the elements c_{ri} $(r = p+1,\ldots,n\ ;\ i = 1,\ldots,p)$ and d_r $(r = p+1,\ldots,n)$, as in Equation (10.10.4).

Next, let the matrix A be defined as:

$$A = \begin{matrix} & \begin{matrix} p & m \end{matrix} \\ \begin{matrix} p \\ m \end{matrix} & \left[\begin{array}{c|c} I & O \\ \hline A_1 & A_2 \end{array}\right] \end{matrix} \tag{10.10.24}$$

Then A^{-1} is seen to be

$$A^{-1} = \begin{array}{c} \\ p \\ m \end{array}\!\!\begin{array}{c} \begin{array}{cc} p & m \end{array} \\ \left[\begin{array}{c|c} I & O \\ \hline C & A_2^{-1} \end{array}\right] \end{array} \tag{10.10.25}$$

where as before C is $-A_2^{-1}A_1$.

The elements of A^{-1} are the a_{rs}^{-1} of Equation (10.10.15). Therefore, by using Equation (10.10.25) with (10.10.15) we obtain the equations (see Reference 10.1):

$$\frac{d}{dt}\left(\frac{\partial K}{\partial \dot{q}_s}\right) - \frac{\partial K}{\partial q_s} + \sum_{r=p+1}^{n}\left[\frac{d}{dt}\left(\frac{\partial K}{\partial \dot{q}_r}\right) - \frac{\partial K}{\partial q_r}\right]c_{rs} = K_2 \quad s = 1,..,p \tag{10.10.26}$$

We present an illustrative application of these equations for a rolling disk in the following chapter.

10.11 Momentum Principles

The impulse-momentum principles of Table 10.2.1 are readily developed from Newton's laws and d'Alembert's principle. The linear and angular moments represent the inertia forces and the impulses represent the applied forces.

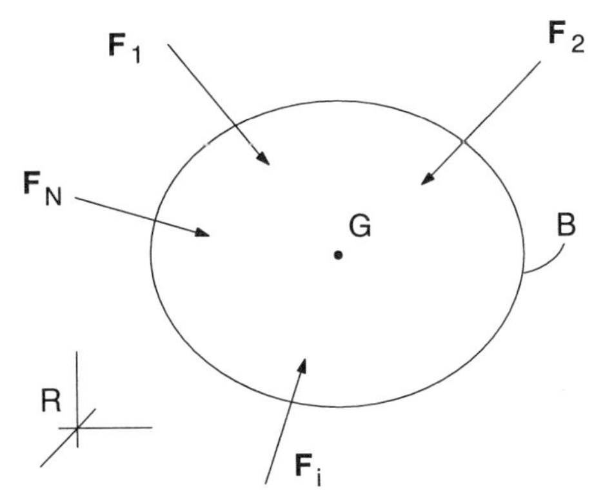

Fig. 10.11.1 A Rigid Body B Moving in a Reference Frame R

To see this consider a rigid body B moving in an inertial reference frame R, as represented in Figure 10.11.1, where G is the mass center of B. Let B be subjected to an applied force field as shown. From Equation (10.5.4) the linear momentum $\mathbf{L}^B$ of B is:

$$\mathbf{L}^B = M\mathbf{v}^G \tag{10.11.1}$$

where M is the mass of B and $\mathbf{v}^G$ is the velocity of G in R. Then by differentiating in R we have

$$d\mathbf{L}^B/dt = Md\mathbf{v}_G/dt = M\mathbf{a}^G \tag{10.11.2}$$

Then from Equation (9.12.6) this momentum derivative is seen to be directly related to the equivalent inertia force $\mathbf{F}^*$ on B as

$$d\mathbf{L}^B/dt = -\mathbf{F}^* \tag{10.11.3}$$

Similarly, from Equation (10.6.7) the angular momentum of B relative to mass center G is:

$$\mathbf{A}^{B/G} = \mathbf{I}^{B/G} \bullet \boldsymbol{\omega} \tag{10.11.4}$$

where $\mathbf{I}^{B/G}$ is the central inertia dyadic of B and $\boldsymbol{\omega}$ is the angular velocity of B in R. The derivative of $\mathbf{A}^{B/G}$ in R is therefore

$$\frac{{}^Rd\mathbf{A}^{B/G}}{dt} = \frac{{}^Rd\mathbf{I}^{B/G}}{dt} \bullet \boldsymbol{\omega} + \mathbf{I}^{B/G} \bullet \frac{{}^Rd\boldsymbol{\omega}}{dt} \tag{10.11.5}$$

Observe that the dyadic $\mathbf{I}^{B/G}$ is dependent upon the orientation of B in R and is therefore not a constant in R. To see this, suppose that $\mathbf{I}^{B/G}$ is expressed in terms of principal unit vectors (see Section 8.10) fixed in B as:

$$\mathbf{I}^{B/G} = I_{11}\mathbf{n}_1\mathbf{n}_1 + I_{22}\mathbf{n}_2\mathbf{n}_2 + I_{33}\mathbf{n}_3\mathbf{n}_3 \tag{10.11.6}$$

where I_{11}, I_{22}, and I_{33} are the central principal moments of inertia of B. Then although these principal moments of inertia are constant, the unit vectors $\mathbf{n}_i$ $(i = 1,2,3)$ are not constant in R since they change orientation in R as B changes orientation in R. Therefore, ${}^R d\mathbf{I}^{B/G}/dt$ is not zero. However, using Equation (6.6.8) we can evaluate ${}^R d\mathbf{I}^{B/G}/dt$ as:

$$\frac{{}^R d\mathbf{I}^{B/G}}{dt} = \frac{{}^B d\mathbf{I}^{B/G}}{dt} + \boldsymbol{\omega} \times \mathbf{I}^{B/G} \tag{10.11.7}$$

From Equation (10.11.6) $\mathbf{I}^{B/G}$ is seen to be constant in B. Therefore ${}^B d\mathbf{I}^{B/G}/dt$ is zero and thus ${}^R d\mathbf{I}^{B/G}/dt$ is simply

$$\frac{{}^R d\mathbf{I}^{B/G}}{dt} = \boldsymbol{\omega} \times \mathbf{I}^{B/G} \tag{10.11.8}$$

Then by substituting into Equation (10.11.5) ${}^R d\mathbf{A}^{B/G}/dt$ becomes:

$$\frac{{}^R d\mathbf{A}^{B/G}}{dt} = \mathbf{I}^{B/G} \bullet \boldsymbol{\alpha} + \boldsymbol{\omega} \times \mathbf{I}^{B/G} \bullet \boldsymbol{\omega} \tag{10.11.9}$$

where as before $\boldsymbol{\alpha}$ is the angular acceleration of B in R $({}^R d\boldsymbol{\omega}/dt)$.

By comparing Equation (10.11.9) with Equation (9.12.11) we see that ${}^R d\mathbf{A}^{B/G}/dt$ is the negative of the torque of the equivalent inertia couple on B. That is,

$$\frac{{}^R d\mathbf{A}^{B/G}}{dt} = -\mathbf{T}^* \tag{10.11.10}$$

Let the applied force field on B be equivalent to a single force **F** passing through G together with a couple with torque $\mathbf{M}_G$. Then from Newton's law/d'Alembert's principle [see Equations (10.7.6) and (10.7.7)] we have

$$\mathbf{F} = M\mathbf{a}^G \tag{10.11.11}$$

and

$$\mathbf{M}_G + \mathbf{T}^* = 0 \tag{10.11.12}$$

By comparing Equations (10.11.2), (10.11.10), (10.11.11), and (10.11.12) we have

$$\mathbf{F} = {}^R d\mathbf{L}^B/dt \tag{10.11.13}$$

and

$$\mathbf{M}_G = {}^R d\mathbf{A}^{B/G}/dt \tag{10.11.14}$$

A principal application of Equations (10.11.13) and (10.11.14) is in studying rigid body response to impulsive forces. "Impulsive" forces, which occur during impact and collision between bodies, are forces with large magnitudes but with short time of application. An "impulse" is defined as the time integral of an impulsive force.

Suppose a body B is subjected to impulsive forces which are equivalent to a single force **F** passing through mass center G together with a couple with torque $\mathbf{M}_G$. Then the impulse **I** and angular impulse $\mathbf{J}_G$ are:

$$\mathbf{I} = \int_{t_1}^{t_2} \mathbf{F}\,dt \quad \text{and} \quad \mathbf{J}_G = \int_{t_1}^{t_2} \mathbf{M}_G\,dt \tag{10.11.15}$$

where (t_1, t_2) is the time interval during which the impulsive forces are applied.

In view of Equations (10.11.15), Equations (10.11.13) and (10.11.14) may be integrated as:

$$\mathbf{I} = \int_{t_1}^{t_2} \mathbf{F}\,dt = \mathbf{L}^B \Big|_{t_1}^{t_2} \tag{10.11.16}$$

and

$$\mathbf{J}_G = \int_{t_1}^{t_2} \mathbf{M}_G\,dt = \mathbf{A}^{B/G} \Big|_{t_1}^{t_2} \tag{10.11.17}$$

or

$$\mathbf{I} = \Delta\mathbf{L}^B \quad \text{and} \quad \mathbf{J}_G = \Delta\mathbf{A}^{B/G} \tag{10.11.18}$$

For a mechanical system containing bodies colliding with one another or otherwise exerting impulsive forces on one another, the net impulse on the system is zero. For such systems $\mathbf{I}$ and I_G are zero and then from Equations (10.10.18) $\Delta\mathbf{L}^B$ and $\Delta\mathbf{A}^{B/G}$ are zero. The linear and angular momentum are thus constant or "conserved" during the short time interval of the impulse. This is sometimes referred to as the "conservation of momentum" principle.

Example 10.11.1 A Rod Pendulum Being Struck by a Moving Object

Consider a rod B with length ℓ and mass M pinned at one end hanging vertically in a gravity field as in Figure 10.10.2. Let a small object, or particle, P with mass m moving horizontally with speed v strike the end of B as indicated. Let the collision be "plastic" so that P adheres to B after the collision. Determine the angular velocity ω of B just after collision.

Solution: Consider free body diagrams showing the momenta of B and P just before and just after the collision, as in Figure 10.11.3, where I_G is the moment of inertia of B for G $[(1/12)M\ell^2]$ and where O_x and O_y are impulsive forces experienced at

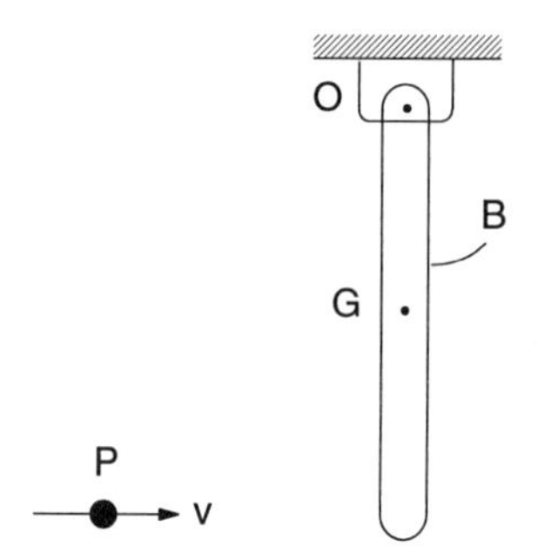

Fig. 10.11.2 A Rod Struck at Its End by a Particle

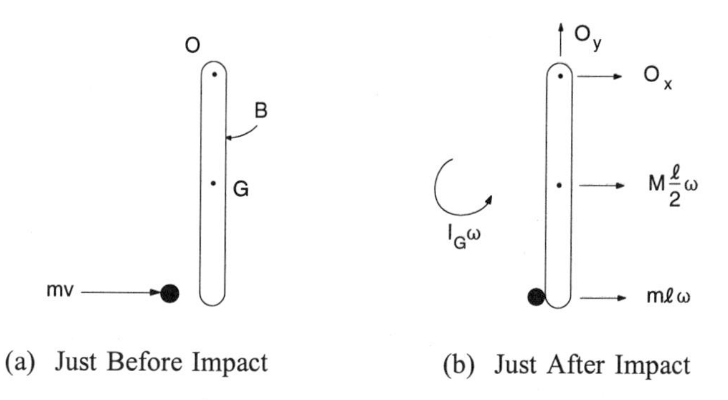

(a) Just Before Impact (b) Just After Impact

Fig. 10.11.3 Free-Body Momentum Diagram of a Rod Struck by a Particle

the hinge O during impact. Assuming the pin to be frictionless it will not provide any resistance to rotation. Therefore, during collision the angular momentum about O will be conserved. By equating the moments of momenta about O just before and just after collision we have

$$m v \ell = m \ell^2 \omega + M\left(\frac{\ell}{2}\right)^2 \omega + \frac{1}{12} M \ell^2 \omega \tag{10.11.19}$$

Solving for ω we have the result

$$\omega = \frac{m(v/\ell)}{[m + (M/3)]} \tag{10.11.20}$$

10.12 Work-Energy

The work-energy principle as symbolized by Equation (5.2.6) is one of the most widely used formulas of elementary mechanics. Its popularity is due to its simplicity, its ease of use, and its quick answers for a broad class of problems. Its ease of use is due to its velocity based formulation (through the kinetic energy) and the availability of simple expressions for work for gravity and spring forces. That is, accelerations need not be calculated and the work is often evaluated by inspection. A disadvantage, however, is that the principle produces only a single equation. Hence, only one unknown can be determined.

The work-energy principle for a particle, as stated by Equation (5.2.6) is readily extended to rigid bodies: As before, let a body B be represented by a set of particles P_i with masses m_i $(i = 1,\ldots,N)$ as in Figure 10.12.1. Let the particles be fixed relative to one another. Consider a typical particle P_i: From Equation (5.2.6) we have

$${}_1W_2^{P_i} = K_2^{P_i} - K_1^{P_i} = \Delta K^{P_i} \quad (i = 1,\ldots,N) \tag{10.12.1}$$

where ${}_1W_2^{P_i}$ is the word done by applied forces on P_i as P_i moves from a position 1 to a position 2, and where K^{P_i} is the kinetic energy of P_i.

Recall from Equation (4.8.11) that the work done on P_i is simply the line integral of the resultant applied force $\mathbf{F}^{P_i}$ on P_i projected onto the curve on which P_i moves in going from position 1 to position 2. That is,

$${}_1W_2 = \int_1^2 \mathbf{F}^{P_i} \bullet d\mathbf{s} \tag{10.12.2}$$

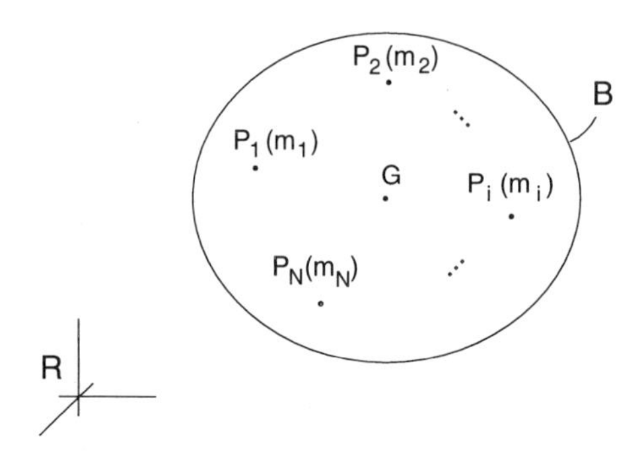

Fig. 10.12.1 A Body B, Represented by Particles, Moving in a Reference Frame R

If Equations (10.12.1) are added for all the particles of B, the total work ${}_1W_2$ on B between configurations 1 and 2 is then simply:

$$ {}_1W_2 = \sum_{i=1}^{N} {}_1W_2^{P_i} \tag{10.12.3} $$

The internal interactive constraint forces between the particles of B cancel one another by Newton's law of action-reaction and therefore they do not contribute to the total work (${}_1W_2$) on B (thus the terminology "non-working" forces). Hence, if the applied forces on B are equivalent to a set of forces $\mathbf{F}_j$ ($j = 1,\ldots,\hat{N}$), or, alternatively, equivalent to a single force **F** passing through the mass center G of B together with a couple with torque **T**, then ${}_1W_2$ may be evaluated simply by computing the work done by the $\mathbf{F}_j$, or equivalently, by computing the work done by **F** and **T**.

Similarly, the sum of the kinetic energies of the particles P_i of B is simply the kinetic energy K of B. This means that if B has angular motion, we can use Equations (10.3.3) to obtain the kinetic energy. That is,

$$K = \frac{1}{2}M(\mathbf{v}^G)^2 + \frac{1}{2}\boldsymbol{\omega}^B \bullet \mathbf{I}^{B/G} \bullet \boldsymbol{\omega}^B \qquad (10.12.4)$$

where M is the mass of B, $\mathbf{I}^{B/G}$ is the central inertia dyadic of B, $\mathbf{v}^G$ is the velocity of G in R, and $\boldsymbol{\omega}^B$ is the angular velocity of G in R.

Therefore, by adding Equations (10.12.1) for all the particles of B the work-energy principle for body B is simply stated as:

$${}_1W_2 = K_2 - K_1 = \Delta K \qquad (10.12.5)$$

Example 10.12.1 A Falling Rod Pendulum

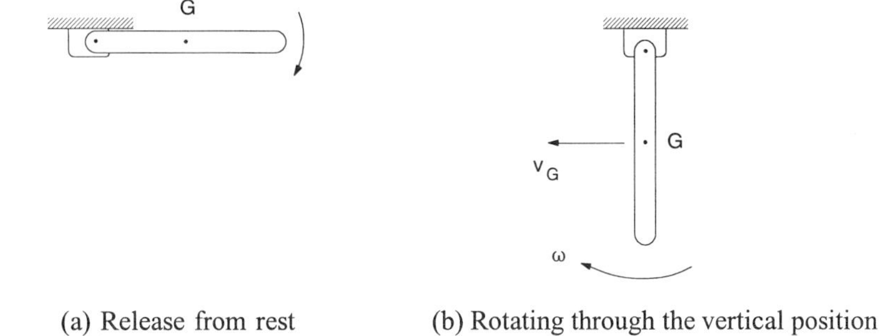

(a) Release from rest (b) Rotating through the vertical position

Fig. 10.12.2 A Falling/Rotating Rod Pendulum

Let a rod B of length ℓ be supported at one end by a frictionless pin O. Let B be held in a horizontal position and released from rest as represented in Figure 10.12.2a. Find the angular speed ω and the speed v_G of the mass center of B as it passed through the vertical position.

Solution: The work W done on B between the horizontal and vertical position is simply the weight mg multiplied by the change in elevation of the mass center G.

That is,

$$W = mg\ell/2 \tag{10.12.6}$$

where m is the mass of B.

The kinetic energy K of B in the horizontal rest position is zero. In the vertical position the kinetic energy is

$$K = \frac{1}{2}m(v_G)^2 + I_G\omega^2 \tag{10.12.7}$$

where I_G is the central moment of inertia of B $[(1/12)m\ell^2]$.

Since B rotates about O, v_G is simply $(\ell/2)\omega$. Therefore, the kinetic energy in the vertical position may be written in the form:

$$K = \frac{2}{3}mv_G^2 = \frac{1}{6}m\ell^2\omega^2 \tag{10.12.8}$$

By using the work-energy principle of Equations (10.12.5), and equating Equations (10.12.6) and (10.11.8), ω and v_G are seen to be:

$$\omega = \sqrt{3g/\ell} \quad \text{and} \quad v_G = \sqrt{3g\ell}/2 \tag{10.12.9}$$

10.13 Other Dynamics Principles and Formulas

There are a number of other dynamics principles, formulas, and methodologies which have been used in dynamic analyses of rigid bodies. In this concluding section we briefly list and discuss a few of these.

10.13.1 Virtual Work

The virtual work principle has been widely used. It is a popular methodology of many analysts. Simply stated, the principle of virtual work asserts that the sum of the virtual work of the applied and inertia forces on a system is zero.

"Virtual work" may be defined and developed as follows: Let S be a mechanical system consisting of particles and rigid bodies and having n degrees of freedom represented by generalized coordinates q_r $(r = 1, \ldots, n)$. Let these coordinates be subjected to incremental values δq_r which are arbitrary provided only that they are consistent with the constraints on the movement of S. Then the virtual works δW and δW^* of the applied and inertia forces on S may be expressed as [10.30]:

$$\delta W = \sum_{r=1}^{n} F_{q_r} \delta q_r \quad \text{and} \quad \delta W^* = \sum_{r=1}^{n} F_{qr}^* \delta q_r \tag{10.13.1}$$

where as before F_{q_r} and $F_{q_r}^*$ are generalized applied and inertia forces on S for q_r $(r = 1, \ldots, n)$.

The principle of virtual work is then embodied in the expression:

$$\delta W + \delta W^* = 0 \tag{10.13.2}$$

Since the δq_r are independent, Equation (10.13.2) together with Equations (10.13.1) are equivalent to

$$F_{q_r} + F_{q_r}^* = 0 \qquad r = 1, \ldots, n \tag{10.13.3}$$

which are Kane's equations [see Equations (10.2.5)].

10.13.2 Virtual Power, Jourdain's Principle

The principle of virtual power is similar to the principle of virtual work. The virtual power principle, also known as Jourdain's principle [10.31], asserts that the virtual power of the applied and inertia forces on a system is zero.

"Virtual power" may be defined and developed as follows: Let S be a mechanical system consisting of particles and rigid bodies and having n degrees of freedom represented by generalized coordinates q_r $(r = 1, \ldots, n)$. Let the coordinate derivatives $\dot{q}_r$ be subjected to incremental values $\delta\dot{q}_r$ which are arbitrary provided only that they are consistent with the constraints on the movement of S. Then the virtual powers δP and δP^* of the applied and inertia forces on S may be written as:

$$\delta P = \sum_{r=1}^{n} F_{q_r} \delta\dot{q}_r \quad \text{and} \quad \delta P^* = \sum_{r=1}^{n} F_{q_r}^* \delta\dot{q}_r \tag{10.13.4}$$

where as before F_{q_r} and $F_{q_r}^*$ are the generalized applied and inertia forces on S for q_r $(r = 1, \ldots, n)$.

The principle of virtual power is then embodied in the expression:

$$\delta P + \delta P^* = 0 \tag{10.13.5}$$

Since the $\delta\dot{q}_r$ are independent, Equations (10.13.4) and (10.13.5) are equivalent to:

$$F_{q_r} + F_{q_r}^* = 0 \quad (r = 1, \ldots, n) \tag{10.13.6}$$

which are Kane's equations [see Equations (10.2.5)].

10.13.3 Comment on the Principles of Virtual Work and Virtual Power

In elementary mechanics work is sometimes defined as "force times distance" and power as "force times velocity." Although Equations (10.13.1) and (10.13.4) do not have those elementary forms, the equations are readily seen to be consistent with the elementary forms. To see this, consider that the work W done by a force **F** applied to a particle Q may be defined as the product of the magnitude of **F** and the accumulated distance moved by Q in the direction of **F**. Specifically W may be defined as

$$W \stackrel{D}{=} \int_C \mathbf{F} \cdot \boldsymbol{\tau} \, ds \tag{10.13.7}$$

where C is the curve on which Q moves, $\boldsymbol{\tau}$ is a unit vector tangent to C (local to where **F** is applied to Q) and where ds is a differential arc length of C.

Similarly, the power P produced by **F** may be defined as

$$P \stackrel{D}{=} \mathbf{F} \cdot \mathbf{V} \tag{10.13.8}$$

where **V** is the velocity of Q in an inertial frame R.

Observe that work is a sum whereas power is an instantaneous value.

The work and power of Equations (10.13.7) and (10.13.8) may be cast into the forms of Equations (10.13.1) and (10.13.4) as follows: Let Q be part of a mechanical system S having n degrees of freedom represented by coordinates q_r $(r = 1,\ldots,n)$. Let O be a fixed point and let **p** locate Q relative to O, as in Figure 10.13.1. Let C be a curve on which Q moves. Then as Q moves to a nearly point $\hat{Q}$, the displacement of Q is $\Delta\mathbf{p}$. We can express $\Delta\mathbf{p}$ as

$$\Delta\mathbf{p} = \boldsymbol{\tau}\Delta s \tag{10.13.9}$$

where $\boldsymbol{\tau}$ is a unit vector tangent to C at Q and Δs is an increment of the arc length of C. For differential movement of Q $\Delta\mathbf{p}$ may be written as $d\mathbf{p}$ and Δs as ds so that

$$d\mathbf{p} = \boldsymbol{\tau} ds \tag{10.13.10}$$

Observe that since the position vector **p** depends upon the coordinates q_r, the differential $d\mathbf{p}$ may be expressed as:

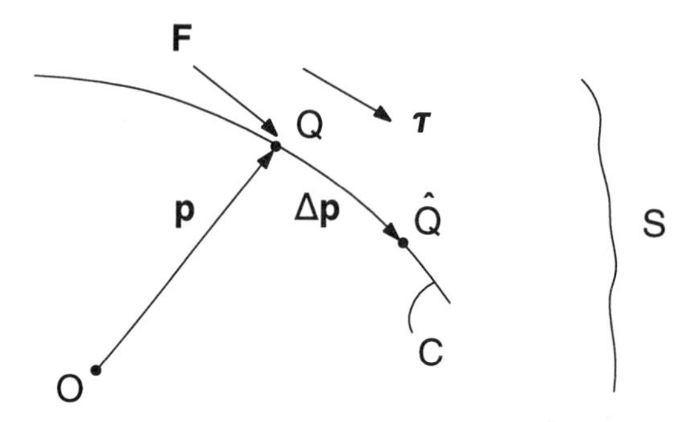

Fig. 10.13.1 A Force **F** Applied to a Particle Q of a Mechanical System S

$$\mathbf{dp} = \sum_{r=1}^{n} \frac{\partial \mathbf{p}}{\partial q_r} dq_r = \sum_{r=1}^{n} \mathbf{v}_{\dot{q}_r} dq_r \tag{10.13.11}$$

[See Equation (4.5.9).] Then by comparing Equations (10.13.7), (10.13.10), and (10.13.11) we see that the differential work of **F** may be written as

$$dW = \mathbf{F} \cdot \boldsymbol{\tau} ds = \mathbf{F} \cdot \mathbf{dp} = \sum_{r=1}^{n} \mathbf{F} \cdot \mathbf{V}_{\dot{q}_r} dq_r = \sum_{r=1}^{n} F_{q_r} dq_r \tag{10.13.13}$$

If the movement of Q is not infinitesimal but instead is incremental due to incremental variation of q_r designated as δq_r then Equation (10.13.12) may be written as:

$$\delta W = \mathbf{F} \cdot \boldsymbol{\tau} \delta s = \mathbf{F} \cdot \delta \mathbf{p} = \sum_{r=1}^{n} \mathbf{F} \cdot \mathbf{V}_{\dot{q}_r} \delta q_r = \sum_{r=1}^{n} F_{q_r} \delta q_r \tag{10.13.13}$$

which is consistent with Equation (10.13.1).

Similarly, since the velocity of Q may be written in the form [see Equation (4.5.10)]:

$$\mathbf{V} = \mathbf{V}_t + \sum_{r=1}^{n} \mathbf{V}_{\dot{q}_r} \dot{q}_r \tag{10.13.14}$$

The power as expressed in Equation (10.13.9) has the form

$$P = \mathbf{F} \bullet \mathbf{V} = \mathbf{F} \bullet \mathbf{V}_t + \sum_{r=1}^{n} \mathbf{F} \bullet \mathbf{V}_{\dot{q}_r} \dot{q}_r = \mathbf{F} \bullet \mathbf{V}_t + \sum_{r=1}^{n} F_{\dot{q}_r} \dot{q}_r \tag{10.13.15}$$

Now, if there are incremental changes in the $\dot{q}_r$ there will be an incremental change in the power given by

$$\delta P = \sum_{r=1}^{n} F_{q_r} \delta \dot{q}_r \tag{10.13.16}$$

which is consistent with Equation (10.13.5).

10.13.4 Gibbs Equations

Although less widely used than other dynamics principles, Gibbs equations present a computationally efficient formulation. Consider again a mechanical system S having n degrees of freedom represented by coordinates q_r $(r = 1, \dots, n)$. Then if B is a rigid body of S, Gibbs equations state that:

$$\frac{\partial G}{\partial \ddot{q}_r} = F_{q_r} \quad (r = 1, \dots, n) \tag{10.13.17}$$

where G is a "kinetic energy of acceleration" defined as:

$$G = \frac{1}{2} \sum_{i=1}^{N} m_i \mathbf{a}_i^2 \tag{10.13.18}$$

where as before B is modeled as a set of particles P_i with masses m_i $(i = 1, \dots, N)$ fixed relative to one another, and also as before, F_{q_r} is the generalized applied force on B for q_r.

G is called the "Gibbs function." As with kinetic energy, the Gibbs function can be extended to include other bodies and particles of S simply by addition. For a single body B, G may be expressed in the convenient form:

$$G = \frac{1}{2}m\mathbf{a}^2 + \frac{1}{2}\boldsymbol{\alpha} \cdot \mathbf{I} \cdot \boldsymbol{\alpha} + \boldsymbol{\alpha} \cdot \boldsymbol{\omega} \times (\mathbf{I} \cdot \omega) \tag{10.13.19}$$

where M is the mass of B, $\mathbf{a}$ is the mass center acceleration in an inertial frame R, $\boldsymbol{\alpha}$ is the angular velocity of B, $\mathbf{I}$ is the central inertia dyadic of B, and $\boldsymbol{\omega}$ is the angular velocity of B.

Equation (10.13.19) is readily developed by recalling that since P_i and the mass center of B are both fixed in B, their accelerations are related by [see Equation (3.15.2)]:

$$\mathbf{a}_i = \mathbf{a} + \boldsymbol{\alpha} \times \mathbf{r}_i + \boldsymbol{\omega} \times (\boldsymbol{\omega} \times \mathbf{r}_i) \tag{10.13.20}$$

where $\mathbf{r}_i$ locates P_i relative to the mass center. Then by substituting into Equation (10.13.18) G takes the form:

$$G = \sum_{i=1}^{N} \frac{1}{2}m_i\{\mathbf{a}^2 + (\boldsymbol{\alpha} \times \mathbf{r}_i)^2 + [\boldsymbol{\omega} \times (\boldsymbol{\omega} \times \mathbf{r}_i)]^2 + 2\mathbf{a} \cdot (\boldsymbol{\alpha} \times \mathbf{r}_i)$$
$$+ 2(\boldsymbol{\alpha} \times \mathbf{r}_i) \cdot [\boldsymbol{\omega} \times (\boldsymbol{\omega} \times \mathbf{r}_i)] + 2\mathbf{a} \cdot [\boldsymbol{\omega} \times (\boldsymbol{\omega} \times \mathbf{r}_i)]\} \tag{10.13.21}$$

where the first term is simply $(1/2)\mathrm{M}\mathbf{a}^2$. The second term may be expressed as:

$$\sum_{i=1}^{N} \frac{1}{2}m_i(\boldsymbol{\alpha} \times \mathbf{r}_i)^2 = \frac{1}{2}\sum_{i=1}^{N} m_i(\boldsymbol{\alpha} \times \mathbf{r}_i) \cdot (\boldsymbol{\alpha} \times \mathbf{r}_i)$$
$$\equiv \frac{1}{2}\boldsymbol{\alpha} \cdot \sum_{i=1}^{N} m_i\mathbf{r}_i \times (\boldsymbol{\alpha} \times \mathbf{r}_i)$$
$$= \frac{1}{2}\boldsymbol{\alpha} \cdot \mathbf{I} \cdot \boldsymbol{\alpha} \tag{10.13.22}$$

where we have used Equation (9.12.9).

The third term of Equation (10.13.21) does not include any terms involving $\ddot{q}_r$ and thus it may be neglected in view of the differentiation of Equation (10.13.17). The fourth term may also be neglected since $\sum_{i=1}^{N} m_i \mathbf{r}_i$ is zero ($\mathbf{r}_i$ is the position vector from the mass center to P_i). That is,

$$\sum_{i=1}^{N} \left(\frac{1}{2} m_i\right) 2\mathbf{a} \bullet [\boldsymbol{\omega} \times (\boldsymbol{\omega} \times \mathbf{r}_i)] = \mathbf{a} \bullet [\boldsymbol{\omega} \times (\boldsymbol{\omega} \times \sum_{i=1}^{N} m_i \mathbf{r}_i)] = 0 \qquad (10.13.23)$$

[See Equation (8.3.2).]

The fourth term of Equation (10.13.21) may be expressed as:

$$\begin{aligned} \sum_{i=1}^{N} \left(\frac{1}{2} m_i\right) 2(\boldsymbol{\alpha} \times \mathbf{r}_i) \bullet [\boldsymbol{\omega} \times (\boldsymbol{\omega} \times \mathbf{r}_i)] &= \boldsymbol{\alpha} \bullet \sum_{i=1}^{N} m_i \mathbf{r}_i \times [\boldsymbol{\omega} \times (\boldsymbol{\omega} \times \mathbf{r}_i)] \\ &\equiv \boldsymbol{\alpha} \bullet \boldsymbol{\omega} \times \left[\sum_{i=1}^{N} m_i \mathbf{r}_i \times (\boldsymbol{\omega} \times \mathbf{r}_i)\right] \\ &= \boldsymbol{\alpha} \bullet \boldsymbol{\omega} \times (\mathbf{I} \bullet \boldsymbol{\omega}) \end{aligned} \qquad (10.13.24)$$

where we have used Equations (9.12.10).

Finally, the last term of Equation (10.13.21) is zero since

$$\sum_{i=1}^{N} \left(\frac{1}{2} m_i\right) 2\mathbf{a} \bullet [\boldsymbol{\omega} \times (\boldsymbol{\omega} \times \mathbf{r}_i)] = \mathbf{a} \bullet [\boldsymbol{\omega} \times \sum_{i=1}^{N} m_i \mathbf{r}_i] = 0 \qquad (10.13.25)$$

[See Equation (8.3.2).]

The results of Equations (10.13.22), (10.13.23), (10.13.24), and (10.13.25), when substituted in Equation (10.13.21), then produces Equation (10.13.19).

Gibbs equations have been shown to be equivalent to Kane's equations [10.33].

10.13.5 Hamilton's Principle

Hamilton's Principle is often used by physicists and theoreticians as a basis for developing dynamics principles and formulations (for example, Lagrange's equations and the principle of least action [10.32]). Alternatively, Hamilton's Principle is often used by structural engineers for approximate and numerical analysis of flexible body dynamics.

Hamilton's Principle states that the movement of a mechanical system S is such that its trajectory or path of motion always minimizes the time integral of the difference of the kinetic and potential energies, consistent with the constraints on S. Specifically,

$$\delta \int_{t_1}^{t_2} (\kappa - P)\, dt = \delta \int_{t_1}^{t_2} L dt = 0 \qquad (10.13.26)$$

where δ is the variational operator [10.34], K and P are kinetic and potential energies and their difference L is called the "Lagrangian."

Let S be a mechanical system with n degrees of freedom represented by coordinates q_r $(r = 1, \ldots, n)$. Then in general, L is a function of the q_r, the $\dot{q}_r$, and time t. By using variational calculus Equation (10.13.26) immediately leads to Lagrange's equations:

$$\frac{d}{dt}\left(\frac{\partial L}{\partial \dot{q}_r}\right) - \frac{\partial L}{\partial q_r} = 0 \qquad (r = 1, \ldots, n) \qquad (10.13.27)$$

10.13.6 Hamilton's Cononical Equations

As before, let S be a mechanical system with n degrees of freedom represented by coordinates q_r $(r = 1, \ldots, n)$. Let K be the kinetic energy of S and let generalized momenta p_r $(r = 1, \ldots, n)$ be defined as

$$p_r \overset{D}{=} \partial K/\partial \dot{q}_r \quad (r = 1,\dots,n) \tag{10.13.28}$$

Let a Hamiltonian function H be defined as:

$$H \overset{D}{=} \sum_{r=1}^{n} p_r \dot{q}_r - L \tag{10.13.29}$$

where, as before, L is the Lagrangian defined as the difference in the kinetic and potential energies $(K - P)$.

Observe in Equation (10.13.28) that with K being a quadratic function of the $\dot{q}_r$, P_r is a linear function of the $\dot{q}_r$. Hence, Equation (10.13.28) could be solved for the $\dot{q}_r$ as functions of the p_r, q_r and time t. That is $\dot{q}_r = \dot{q}_r(p_r, q_r, t)$. As a consequence, the Hamiltonian H of Equation (10.13.29) may be regarded as being a function of the p_r, q_r, and t. That is, $H = H(p_r, q_r, t)$.

Hamilton's canonical equations are then [10.32]

$$\dot{q}_r = \partial H/\partial p_r \quad \text{and} \quad \dot{p}_r = -\partial H/\partial q_r \tag{10.13.30}$$

Finally, observe that these equations form a system of first-order, ordinary differential equations. As such, they are in a convenient form for numerical integrators (or "solvers") of the equations.

References

10.1 C. E. Passerello and R. L. Huston, "Another Look at Nonholonomic Systems," *Journal of Applied Mechanics*, Vol. 40, 1973, pp. 101-104.

10.2 T. R. Kane and D. A. Levinson, "Formulation of Equations of Motion for Complex Spacecraft," *Journal of Guidance and Control*, Vol. 3, No. 2, 1980, pp. 99-112.

10.3 J. G. Papastavridis, "A Panoramic Overview of the Principles and Equations of Motion of Advanced Engineering Dynamics," *Applied Mechanics Reviews*, Vol. 51, No. 4, 1998, pp. 239-265.

10.4 P. Appell, "Sur une Forme Generale des Equations de la Dynamique," *Journal fur die Reine und Angewandte Mathematik*, Vol. 121, 1900, pp. 310-319.

10.5 I. Newton, "Philosophiae Naturalis Principia Mathematica," Societatis Regiae ac Typio Josephi Streator, London, 1687, p. 12.

10.6 J. L. d'Alembert, *Traite de Dynamique*, Chez David l'aine, Paris, 1743.

10.7 J. L. Lagrange, *Mechanique Analitique*, Chez la Veuve Desaint, Paris, 1788.

10.8 W. R. Hamilton, "Second Essay on a General Method in Dynamics," *Philosophical Transactions of the Royal Society of London*, 1835, pp. 95-144.

10.9 J. W. Gibbs, "Fundamental Formulae of Dynamics," *American Journal of Mathematics*, Vol. 2, 1879, pp. 49-64.

10.10 L. Meirovitch, *Methods of Analytical Dynamics*, McGraw Hill, 1970.

10.11 L. A. Pars, *Treatise on Analytical Dynamics*, Ox Bow Press, Woodridge, CT, 1979.

10.12 E. T. Whittaker, *Treatise on the Analytical Dynamics of Particles and Rigid Bodies*, Cambridge University Press, Cambridge, UK, 1937.

10.13 E. J. Haug, *Intermediate Dynamics*, Prentice Hall, Englewood Cliffs, NJ, 1992.

10.14 G. Hamel, *Theoretische Mechanics*, Springer-Verlag, Berlin, 1949.

10.15 H. Goldstein, *Classical Mechanics*, Addison-Wesley, Reading, MA, 1980.

10.16 T. R. Kane and D. A. Levinson, *Dynamics: Theory and Applications*, McGraw Hill, NY, 1985.

10.17 R. P. Beer and E. R. Johnston, Jr., *Vector Mechanics for Engineers*, Sixth Ed., McGraw Hill, NY, 1996.

10.18 J. L. Meriam and L. G. Kraige, *Engineering Mechanics*, Vol. 2, *Dynamics*, Third Ed., John Wiley & Sons, NY, 1992.

10.19 R. C. Hibbler, *Engineering Mechanics: Statics and Dynamics*, Macmillan, NY, 1974.

10.20 A. Higdon and W. B. Stiles, *Engineering Mechanics*, Vol. II, *Dynamics*, Third Ed., Prentice-Hall, Englewood Cliffs, NJ, 1968.

10.21 J. H. Ginsberg, *Advanced Engineering Dynamics*, Harper & Row, 1988.

10.22 E. J. Routh, *Dynamics of a System of Rigid Bodies*, Parts I and II, Sixth Ed., Macmillan, NY, 1905.

10.23 J. L. Synge and B. A. Griffith, *Principles of Mechanics*, Third Ed., McGraw Hill, 1959.

10.24 D. L. Greenwood, *Principals of Dynamics*, Prentice Hall, Englewood Cliffs, NJ, 1965.

10.25 L. D. Landau and E. M Lifshitz, *Mechanics*, Pergamon, Oxford, UK, 1960.

10.26 J. L. Synge, "Classical Dynamics," *Encyclopedia of Physics*, Vol. 3, No. 1, (S. Flügge, Ed.), Springer, New York, NY, 1961.

10.27 R. M. Rosenberg, *Analytical Dynamics of Discrete Systems*, Plenum Press, New York, NY, 1977.

10.28 L. Brand, *Vectorial Mechanics*, John Wiley & Sons, New York, NY, 1930.

10.29 J. G. Papastarridis, "A Panoramic Overview of the Principles and Equations of Motion of Advanced Engineering Dynamics," *Applied Mechanics Reviews*, Vol. 51, No. 4, 1988, pp. 239-265.

10.30 T. R. Kane, *Dynamics*, Holt, Rinehardt and Winston, New York, 1968, pp. 145-146.

10.31 J. G. de Jalon and E. Bayo, *Kinematic and Dynamic Simulation of Multibody Systems*, Springer-Verlag, New York 1993, pp. 126-128.

10.32 J. B. Marion and S. T. Thornton, *Classical Dynamics of Particles and Systems*, Harcourt Brace Jovanovich, San Diego, 1970, pp. 190-193, 222-225.

10.33 R. L. Huston, "On the Equivalence of Kane's Equations and Gibbs' Equations for Multibody Dynamics Formulations," *Mechanics Research Communications*, Vol. 14, 1987, pp. 123-131.

10.34 E. Butkov, *Mathematical Physics*, Addison-Wesley, Reading, MA, 1968, pp. 553-588.

Chapter 11

EXAMPLE PROBLEMS/SYSTEMS

11.1 Introduction

In this chapter we present and discuss a few classical example problems. The objective is to illustrate the principles of the foregoing chapters.

11.2 Double-Rod Pendulum

Consider two identical pin-connected and pin-supported rods moving in a vertical plane as in Figure 11.2.1. Let the orientation of the rods be defined by the angles θ_1 and θ_2, as shown*. Let the pin support be O and the pin connection be Q. Let the rods be called B_1 and B_2 and let their mass centers be G_1 and G_2. Let each rod have length ℓ and mass m. Let the pin connections be frictionless.

The relative simplicity of this system makes it amenable for study by Newton's laws or d'Alembert's principle. To this end, consider free-body diagrams of the rods as in Figures 11.2.2 and 11.2.3, where the inertia forces are represented by forces $\mathbf{F}_1^*$ and $\mathbf{F}_2^*$ passing through the mass centers together with inertia

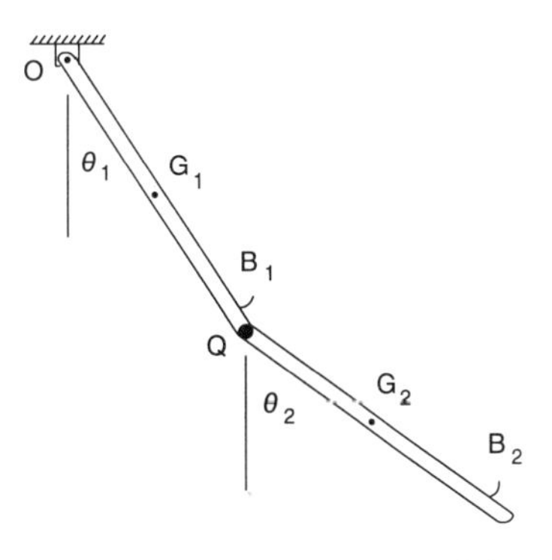

Fig. 11.2.1 Double-Rod Pendulum

*An alternative to using θ_1 and θ_2 would be to use "relative" angles β_1 and β_2 where β_1 is θ_1 and β_2 is $\theta_2 - \theta_1$, the angle between the rods. This approach, while having some intuitive advantages, has the disadvantage of more complicated analysis [see Section 11.3, Equations (11.3.30), (11.3.31), and (11.3.32)].

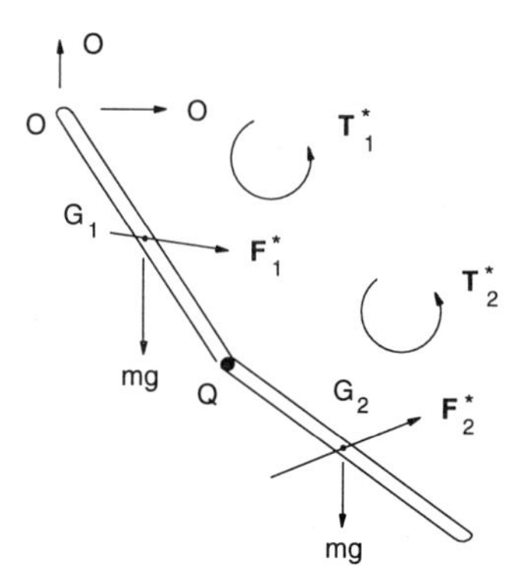

Fig. 11.2.2 Free-Body Diagram of the System of Two Rods

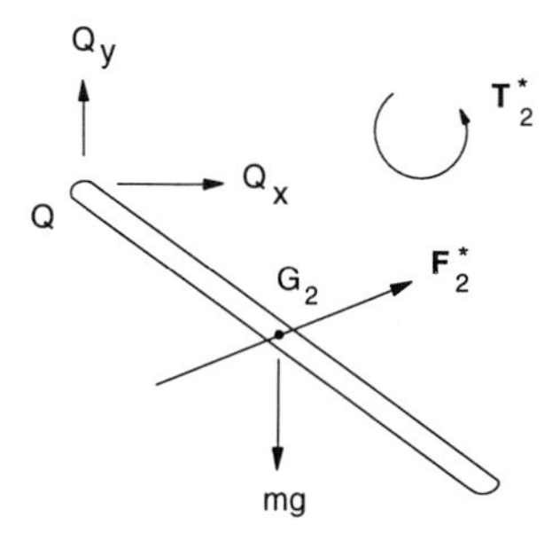

Fig. 11.2.3 Free-Body Diagram of the Lower Rod

torques $\mathbf{T}_1^*$ and $\mathbf{T}_2^*$. These forces and torques may be expressed as [see Equations (5.2.2) and (9.12.11)]:

$$\mathbf{F}_1^* = -m\mathbf{a}^{G_1} \quad \text{and} \quad \mathbf{F}_2^* = -m\mathbf{a}^{G_2} \tag{11.2.1}$$

$$\mathbf{T}_1^* = -\mathbf{I}\boldsymbol{\alpha}^{B_1} \quad \text{and} \quad \mathbf{T}_2^* = -\mathbf{I}\boldsymbol{\alpha}^{B_2} \tag{11.2.2}$$

where I is the central moment of inertia perpendicular to the rods: $(1/12)m\ell^2$, and where the term $\boldsymbol{\omega} \times (\mathbf{I}_G \cdot \boldsymbol{\omega})$ of the inertia torque is zero since the rods have planar motion.

To develop Equations (11.2.1) and (11.2.2) we need expressions for the mass center accelerations and the angular accelerations of the rods. These are:

$$\mathbf{a}^{G_1} = (\ell/2)\ddot{\theta}_1\mathbf{n}_{11} + (\ell/2)\dot{\theta}_1^2\mathbf{n}_{12} \tag{11.2.3}$$

$$\mathbf{a}^{G_2} = \ell\ddot{\theta}_1\mathbf{n}_{11} + \ell\dot{\theta}_1^2\mathbf{n}_{12} + (\ell/2)\ddot{\theta}_2\mathbf{n}_{21} + (\ell/2)\dot{\theta}_2^2\mathbf{n}_{22} \tag{11.2.4}$$

$$\boldsymbol{\alpha}^{B_1} = \ddot{\theta}_1\mathbf{n}_3 \tag{11.2.5}$$

$$\boldsymbol{\alpha}^{B_2} = \ddot{\theta}_2\mathbf{n}_3 \tag{11.2.6}$$

where the unit vectors are parallel and perpendicular to the respective rods and where $\mathbf{n}_3$ is normal to the plane of motion as shown in Figure 11.2.4.

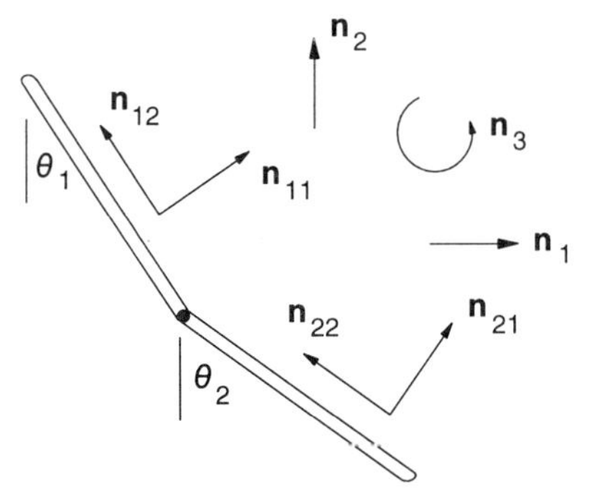

Fig. 11.2.4 Unit Vector Geometry for the Double-Rod System

Referring to the free-body diagrams of Figures 11.2.2 and 11.2.3 and by setting moments about O and Q equal to zero, we obtain (after some analysis and simplifications):

$$\begin{aligned}(4/3)\ddot{\theta}_1 &+ (1/3)\ddot{\theta}_2 + (1/2)\cos(\theta_2 - \theta_1) + (1/2)\ddot{\theta}_2\cos(\theta_2 - \theta_1) \\ &+ (1/2)\dot{\theta}_1^2\sin(\theta_2 - \theta_1) - (1/2)\dot{\theta}_2^2\sin(\theta_2 - \theta_1) + (3/2)(g/\ell)\sin\theta_1 \\ &+ (1/2)(g/\ell)\sin\theta_2 = 0\end{aligned} \tag{11.2.7}$$

and

$$\begin{aligned}(1/3)\ddot{\theta}_2 &+ (1/2)\ddot{\theta}_1\cos(\theta_2 - \theta_1) + (1/2)\dot{\theta}_1^2\sin(\theta_2 - \theta_1) \\ &+ (1/2)(g/\ell)\sin\theta_2 = 0\end{aligned} \tag{11.2.8}$$

Upon examination of Equations (11.2.7) and (11.2.8) we see that the terms of Equation (11.2.8) are all contained within Equation (11.2.7) and may thus be deleted from Equation (11.2.7). The governing equations then become

$$(4/3)\ddot{\theta}_1 + (1/2)\ddot{\theta}_2\cos(\theta_2 - \theta_1) - (1/2)\dot{\theta}_2^2\sin(\theta_2 - \theta_1) + (3/2)(g/\ell)\sin\theta_1 = 0 \tag{11.2.9}$$

and

$$(1/2)\ddot{\theta}_1\cos(\theta_2 - \theta_1) + (1/3)\ddot{\theta}_2 - (1/2)\dot{\theta}_1^2\sin(\theta_1 - \theta_2) + (1/2)(g/\ell)\sin\theta_2 = 0 \tag{11.2.10}$$

We can readily obtain Equations (11.2.9) and (11.2.10) by using Lagrange's equations [Equations (10.2.6)]: Whereas Newton's laws require use of accelerations, Lagrange's equations need only velocities. From Figure 11.2.3 we see that the mass center velocities and the angular velocities of the rods are:

$$\mathbf{V}^{G_1} = (\ell/2)\dot{\theta}_1\mathbf{n}_{11} \tag{11.2.11}$$

$$\mathbf{V}^{G_2} = \ell\dot{\theta}_1\mathbf{n}_{11} + (\ell/2)\dot{\theta}_2\mathbf{n}_{21} \tag{11.2.12}$$

$$\boldsymbol{\omega}^{B_1} = \dot{\theta}_1\mathbf{n}_3 \tag{11.2.13}$$

$$\boldsymbol{\omega}^{B_2} = \dot{\theta}_2\mathbf{n}_3 \tag{11.2.14}$$

[Observe the simplicity of Equations (11.2.11) to (11.2.14) compared with Equations (11.2.3) to (11.2.6).]

The system has two degrees of freedom represented by the angles θ_1 and θ_2. From Equations (11.2.11) and (11.2.12) the partial velocities of G_1 and G_2 for $\dot{\theta}_1$ and $\dot{\theta}_2$ are seen to be:

$$\mathbf{V}_{\dot{\theta}_1}^{G_1} = (\ell/2)\,\mathbf{n}_{11} \qquad \mathbf{V}_{\dot{\theta}_2}^{G_1} = 0 \tag{11.2.15}$$

$$\mathbf{V}_{\dot{\theta}_1}^{G_2} = \ell\,\mathbf{n}_{11} \qquad \mathbf{V}_{\dot{\theta}_2}^{G_2} = (\ell/2)\,\mathbf{n}_{21} \tag{11.2.16}$$

The only "working" active forces are due to gravity which are represented by weight forces $-mg\mathbf{n}_2$ through G_1 and G_2. The generalized applied forces are then:

$$F_{\theta_1} = (-mg\mathbf{n}_2)\bullet\mathbf{V}_{\dot{\theta}_1}^{G_1} + (-mg\mathbf{n}_2)\bullet\mathbf{V}_{\dot{\theta}_1}^{G_2} = -(3/2)mg\ell\sin\theta_1 \tag{11.2.17}$$

$$F_{\theta_2} = (-mg\mathbf{n}_2)\bullet\mathbf{V}_{\dot{\theta}_2}^{G_1} + (-mg\mathbf{n}_2)\bullet\mathbf{V}_{\dot{\theta}_2}^{G_2} = -mg(\ell/2)\sin\theta_2 \tag{11.2.18}$$

From Equations (11.2.11) to (11.2.14) the kinetic energy K of the system is seen to be:

$$\begin{aligned} K &= (1/2)m(\mathbf{V}^{G_1})^2 + (1/2)I(\boldsymbol{\omega}^{B_1})^2 + (1/2)m(\mathbf{V}^{G_2})^2 + (1/2)I(\boldsymbol{\omega}^{B_2})^2 \\ &= (2/3)m\ell^2\dot{\theta}_1^2 + (1/2)m\ell^2\dot{\theta}_1\dot{\theta}_2\cos(\theta_1-\theta_2) + (1/6)m\ell^2\dot{\theta}_2^2 \end{aligned} \tag{11.2.19}$$

Lagrange's equations for the system are [Equations (10.2.6)]:

$$\frac{d}{dt}\left(\frac{\partial K}{\partial\dot{\theta}_1}\right) - \frac{\partial K}{\partial\theta_1} = F_{\theta_1} \quad \text{and} \quad \frac{d}{dt}\left(\frac{\partial K}{\partial\dot{\theta}_2}\right) - \frac{\partial K}{\partial\theta_2} = F_{\theta_2} \tag{11.2.20}$$

Substituting from Equations (11.2.17), (11.2.18), and (11.2.19) into (11.2.20) we obtain

$$(4/3)\ddot{\theta}_1 + (1/2)\ddot{\theta}_2\cos(\theta_1-\theta_2) - (1/2)\dot{\theta}_2^2\sin(\theta_2-\theta_1) + (3/2)(g/\ell)\sin\theta_1 = 0 \tag{11.2.21}$$

and

$$(1/2)\ddot{\theta}_1\cos(\theta_2-\theta_1)+(1/3)\ddot{\theta}_2-(1/2)\dot{\theta}_1^2\sin(\theta_1-\theta_2)+(1/2)(g/\ell)\sin\theta_2 = 0 \quad (11.2.22)$$

Equations (11.2.21) and (11.2.22) are seen to be identical to Equations (11.2.9) and (11.2.10). The computational effort in obtaining Equations (11.2.21) and (11.2.22), however, is about one third that of obtaining Equations (11.2.9) and (11.2.10).

11.3 Triple-Rod Pendulum [11.1, 11.2]

As a natural extension of the double-rod pendulum, consider the triple rod pendulum of Figure 11.3.1. It consists of three, identical pin-connected rods moving in a vertical plane. Let the orientation of the rods be defined by the angles θ_1, θ_2, and θ_3 of Figure 11.3.1. Let the rods be called B_1, B_2, and B_3 and let their mass centers be G_1, G_2, and G_3. Let each rod have length ℓ and mass m. Let the pin connections be frictionless.

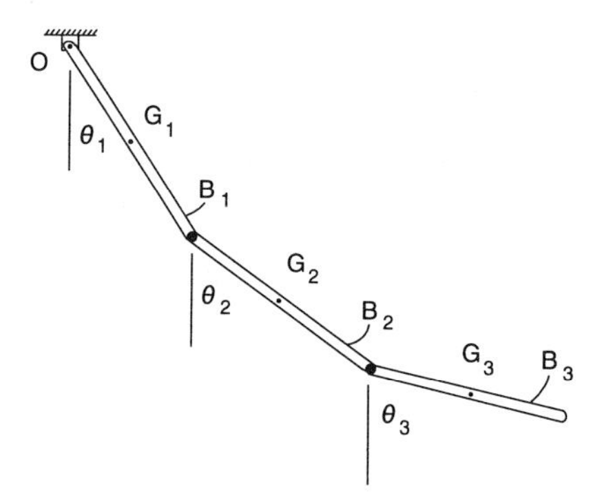

Fig. 11.3.1 Triple-Rod Pendulum

Although this is a relatively simple system, an analysis using Newton's laws or d'Alembert's principle is rather long and tedious. In view of the analysis of the double-rod pendulum, an attractive alternative is to use Lagrange's equations.

Observe that the system has three degrees of freedom represented by the angles θ_1, θ_2, and θ_3. The kinetic energy K of the system is:

$$K = (1/2)m(\mathbf{V}^{G_1})^2 + (1/2)I(\boldsymbol{\omega}^{B_1})^2 + (1/2)m(\mathbf{V}^{G_2})^2 + (1/2)I(\boldsymbol{\omega}^{B_2})^2 + (1/2)m(\mathbf{V}^{G_3})^2 + (1/2)I(\boldsymbol{\omega}^{B_3})^2 \quad (11.3.1)$$

where $\mathbf{V}^{G_i}$ and $\boldsymbol{\omega}^{B_i}$ are the velocities and angular velocities of G_i and B_i $(i = 1,2,3)$, and where I is the central moment of inertia of the rods in a direction perpendicular to the rods: $(1/12)m\ell^2$. As with the double-rod pendulum, it is convenient to introduce and use unit vectors parallel and perpendicular to the rods as well as horizontal and vertical unit vectors as in Figure 11.3.2, where, as before, $\mathbf{n}_3$ is normal to the plane of motion. In terms of these unit vectors $\mathbf{V}^{G_i}$ and $\boldsymbol{\omega}^{B_i}$ are seen to be:

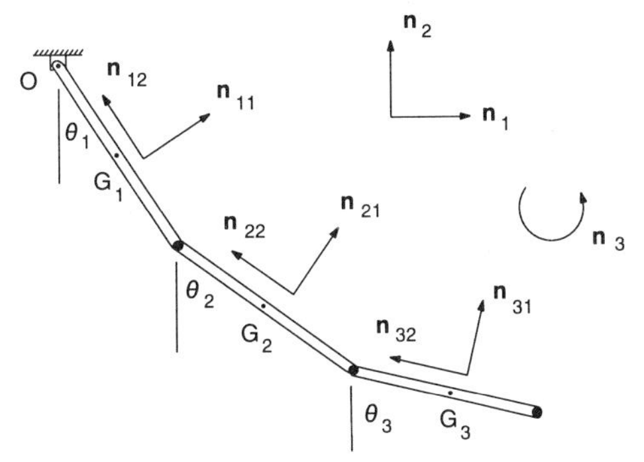

Fig. 11.3.2 Unit Vector Geometry for the Triple-Rod System

$$\mathbf{V}^{G_1} = (\ell/2)\dot{\theta}_1\mathbf{n}_{11} \qquad \mathbf{V}^{G_2} = \ell\dot{\theta}_1\mathbf{n}_{11} + (\ell/2)\dot{\theta}_2\mathbf{n}_{21}$$

$$\mathbf{V}^{G_3} = \ell\dot{\theta}_1\mathbf{n}_{11} + \ell\dot{\theta}_2\mathbf{n}_{21} + (\ell/2)\dot{\theta}_3\mathbf{n}_{31} \quad (11.3.2)$$

and

$$\boldsymbol{\omega}^{B_1} = \dot{\theta}_1\mathbf{n}_3 \qquad \boldsymbol{\omega}^{B_2} = \dot{\theta}_2\mathbf{n}_3 \qquad \boldsymbol{\omega}^{B_3} = \dot{\theta}_3\mathbf{n}_3 \quad (11.3.3)$$

$$K = (1/2)m\ell^2[(7/3)\dot{\theta}_1^2 + (4/3)\dot{\theta}_2^2 + (1/3)\dot{\theta}_3^2 + 3\dot{\theta}_1\dot{\theta}_2\cos(\theta_1 - \theta_2) + \dot{\theta}_2\dot{\theta}_3\cos(\theta_2 - \theta_3) + \dot{\theta}_1\dot{\theta}_3\cos(\theta_1 - \theta_2)] \quad (11.3.4)$$

With the pin connections being frictionless, the only forces contributing to the generalized applied forces are the gravity (or weight) forces. The weight forces may be represented as single vertical forces ($-mg\mathbf{n}_2$) passing through the mass centers G_i $(i = 1,2,3)$ of the rods.

From Equations (11.3.2) the partial velocities of the G_i relative to θ_1, θ_2, and θ_3 are immediately seen to be:

$$\begin{array}{lll} \mathbf{V}_{\dot{\theta}_1}^{G_1} = (\ell/2)\mathbf{n}_{11} & \mathbf{V}_{\dot{\theta}_2}^{G_1} = 0 & \mathbf{V}_{\dot{\theta}_3}^{G_1} = 0 \\ \mathbf{V}_{\dot{\theta}_1}^{G_2} = \ell\mathbf{n}_{11} & \mathbf{V}_{\dot{\theta}_2}^{G_2} = (\ell/2)\mathbf{n}_{21} & \mathbf{V}_{\dot{\theta}_3}^{G_2} = 0 \\ \mathbf{V}_{\dot{\theta}_1}^{G_3} = \ell\mathbf{n}_{11} & \mathbf{V}_{\dot{\theta}_2}^{G_3} = \ell\mathbf{n}_{21} & \mathbf{V}_{\dot{\theta}_3}^{G_3} = (\ell/2)\mathbf{n}_{31} \end{array} \quad (11.3.5)$$

The generalized applied ("active") forces are thus:

$$F_{\theta_1} = (-mg\mathbf{n}_2) \bullet \mathbf{V}_{\dot{\theta}_1}^{G_1} + (-mg\mathbf{n}_2) \bullet \mathbf{V}_{\dot{\theta}_1}^{G_2} + (-mg\mathbf{n}_2) \bullet \mathbf{V}_{\dot{\theta}_1}^{G_3} = -(5/2)mg\ell\sin\theta_1 \quad (11.3.6)$$

$$F_{\theta_2} = (-mg\mathbf{n}_2) \bullet \mathbf{V}_{\dot{\theta}_2}^{G_1} + (-mg\mathbf{n}_2) \bullet \mathbf{V}_{\dot{\theta}_2}^{G_2} + (-mg\mathbf{n}_2) \bullet \mathbf{V}_{\dot{\theta}_2}^{G_3} = -(3/2)mg\ell\sin\theta_2 \quad (11.3.7)$$

$$F_{\theta_3} = (-mg\mathbf{n}_2) \bullet \mathbf{V}_{\dot{\theta}_3}^{G_1} + (-mg\mathbf{n}_2) \bullet \mathbf{V}_{\dot{\theta}_3}^{G_2} + (-mg\mathbf{n}_2) \bullet \mathbf{V}_{\dot{\theta}_3}^{G_3} = -(1/2)mg\ell\sin\theta_3 \quad (11.3.8)$$

Lagrange's equations for the system are [Equations (10.2.6)]:

$$\frac{d}{dt}\left(\frac{\partial K}{\partial \dot{\theta}_i}\right) - \frac{\partial K}{\partial \theta_i} = F_{\theta_i} \quad (i = 1,2,3) \quad (11.3.9)$$

By substituting from Equations (11.3.4), (11.3.6), (11.3.7) and (11.3.8) into (11.3.9), the governing dynamical equations are seen to be:

$$(7/3)\ddot{\theta}_1 + (3/2)\ddot{\theta}_2 \cos(\theta_1 - \theta_2) + (1/2)\ddot{\theta}_3 \cos(\theta_1 - \theta_3)$$
$$+ (3/2)\dot{\theta}_2^2 \sin(\theta_1 - \theta_2) + (1/2)\dot{\theta}_3^2 \sin(\theta_1 - \theta_3) + (5/2)(g/\ell)\sin\theta_1 = 0 \tag{11.3.10}$$

$$(3/2)\ddot{\theta}_1 \cos(\theta_2 - \theta_1) + (4/3)\ddot{\theta}_2 + (1/2)\ddot{\theta}_3 \cos(\theta_2 - \theta_3)$$
$$+ (3/2)\dot{\theta}_1^2 \sin(\theta_2 - \theta_1) + (1/2)\dot{\theta}_3^2 \sin(\theta_2 - \theta_3) + (3/2)(g/\ell)\sin\theta_2 = 0 \tag{11.3.11}$$

$$(1/2)\ddot{\theta}_1 \cos(\theta_3 - \theta_1) + (1/2)\ddot{\theta}_2 \cos(\theta_3 - \theta_2) + (1/3)\ddot{\theta}_3$$
$$+ (1/2)\dot{\theta}_1^2 \sin(\theta_3 - \theta_1) + (1/2)\dot{\theta}_2^2 \sin(\theta_3 - \theta_2) + (1/2)(g/\ell)\sin\theta_3 = 0 \tag{11.3.12}$$

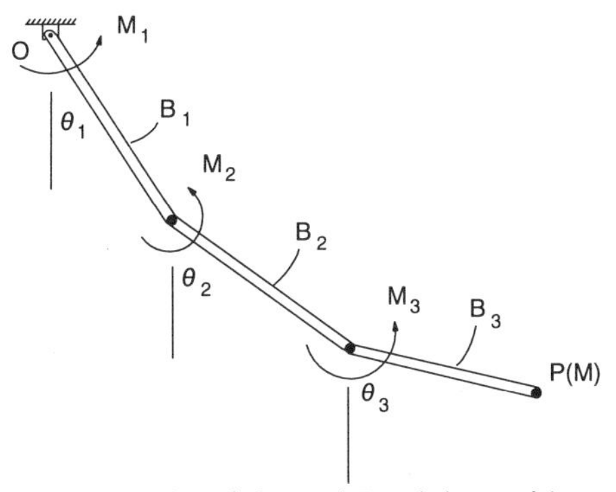

Fig. 11.3.3 A Triple-Rod Pendulum with Concentrated End Mass and Joint Moments

As an extension of this example, consider a triple-rod pendulum consisting of identical pin-connected rods, moving in a vertical plane, and with a concentrated mass P at the end of the third rod as in Figure 11.3.3. As a further extension let there be moments exerted at the joints between adjacent rods as also represented in Figure 11.3.3. The kinetic energy of the system is then the same as in Equation

(11.3.4) with the addition of the term: $(1/2)M(\mathbf{V}^P)^2$ where M is the mass of P. From 11.3.2 the velocity of P is

$$\mathbf{V}^P = \ell\dot{\theta}_1\mathbf{n}_{11} + \ell\dot{\theta}_2\mathbf{n}_{21} + \ell\dot{\theta}_3\mathbf{n}_{31} \tag{11.3.13}$$

The kinetic energy of the system thus becomes:

$$\begin{aligned} K = {} & (1/2)m\ell^2[(7/3)\dot{\theta}_1^2 + (4/3)\dot{\theta}_2^2 + (1/3)\dot{\theta}_3^2 + 3\dot{\theta}_1\dot{\theta}_2\cos(\theta_1 - \theta_2) \\ & + \dot{\theta}_2\dot{\theta}_3\cos(\theta_2 - \theta_3) + \dot{\theta}_1\dot{\theta}_3\cos(\theta_1 - \theta_3)] + (1/2)M\ell^2[\dot{\theta}_1^2 \\ & + \dot{\theta}_2^2 + \dot{\theta}_3^2 + 2\dot{\theta}_1\dot{\theta}_2\cos(\theta_2 - \theta_1) + 2\dot{\theta}_2\dot{\theta}_3\cos(\theta_3 - \theta_2) \\ & + 2\dot{\theta}_3\dot{\theta}_1\cos(\theta_3 - \theta_1)] \end{aligned} \tag{11.3.14}$$

The generalized active forces on the system are the same as in Equations (11.3.6), (11.3.7), and (11.3.8) with the addition of the contributions of 1) the weight force on P and 2) the moments at the joints. Consider first the weight force on P $(-Mg\mathbf{n}_2)$. From Equation (11.3.13) the partial velocities of P relative to θ_1, θ_2, and θ_3 are

$$\mathbf{V}^P_{\dot{\theta}_1} = \ell\mathbf{n}_{11} \qquad \mathbf{V}^P_{\dot{\theta}_2} = \ell\mathbf{n}_{21} \qquad \mathbf{V}^P_{\dot{\theta}_3} = \ell\mathbf{n}_{31} \tag{11.3.15}$$

The contributions of the weight force to F_{θ_1}, F_{θ_2}, and F_{θ_3} are then

$$F_{\theta_1} : \ -Mg\mathbf{n}_2 \bullet \mathbf{V}^P_{\dot{\theta}_1} = -Mg\ell\sin\theta_1 \tag{11.3.16}$$

$$F_{\theta_2} : \ -Mg\mathbf{n}_2 \bullet \mathbf{V}^P_{\dot{\theta}_2} = -Mg\ell\sin\theta_2 \tag{11.3.17}$$

$$F_{\theta_3} : \ -Mg\mathbf{n}_3 \bullet \mathbf{V}^P_{\dot{\theta}_3} = -Mg\ell\sin\theta_3 \tag{11.3.18}$$

Consider next the joint moments: Observe that by the law of action-reaction, the moments on adjacent bodies at a joint will be equal in magnitude but oppositely directed. [For example, at the second joint there is a moment $-M_2\mathbf{n}_3$ on B_1 and a moment $+M_2\mathbf{n}_3$ on B_2 (see Figure 11.3.3).] From Equations (11.3.3) we see that the partial angular velocities of the bodies are:

$$\begin{array}{lll} \boldsymbol{\omega}_{\dot{\theta}_1}^{B_1} = \mathbf{n}_3 & \boldsymbol{\omega}_{\dot{\theta}_2}^{B_1} = 0 & \boldsymbol{\omega}_{\dot{\theta}_3}^{B_1} = 0 \\ \boldsymbol{\omega}_{\dot{\theta}_1}^{B_2} = 0 & \boldsymbol{\omega}_{\dot{\theta}_2}^{B_2} = \mathbf{n}_3 & \boldsymbol{\omega}_{\dot{\theta}_3}^{B_2} = 0 \\ \boldsymbol{\omega}_{\dot{\theta}_1}^{B_3} = 0 & \boldsymbol{\omega}_{\dot{\theta}_2}^{B_3} = 0 & \boldsymbol{\omega}_{\dot{\theta}_3}^{B_3} = \mathbf{n}_3 \end{array} \tag{11.3.19}$$

Then the contributions of the joint moments to the generalized forces are:

$$\begin{aligned} F_{\theta_1} :\ & M_1\mathbf{n}_3 \bullet \boldsymbol{\omega}_{\dot{\theta}_1}^{B_1} - M_2\mathbf{n}_3 \bullet \boldsymbol{\omega}_{\dot{\theta}_1}^{B_1} + M_2\mathbf{n}_3 \bullet \boldsymbol{\omega}_{\dot{\theta}_1}^{B_2} - M_3\mathbf{n}_3 \bullet \boldsymbol{\omega}_{\dot{\theta}_1}^{B_2} + M_3\mathbf{n}_3 \bullet \boldsymbol{\omega}_{\dot{\theta}_1}^{B_3} \\ & = M_1 - M_2 \end{aligned} \tag{11.3.20}$$

$$\begin{aligned} F_{\theta_2} :\ & M_1\mathbf{n}_3 \bullet \boldsymbol{\omega}_{\dot{\theta}_2}^{B_1} - M_2\mathbf{n}_3 \bullet \boldsymbol{\omega}_{\dot{\theta}_2}^{B_1} + M_2\mathbf{n}_3 \bullet \boldsymbol{\omega}_{\dot{\theta}_2}^{B_2} - M_3\mathbf{n}_3 \bullet \boldsymbol{\omega}_{\dot{\theta}_2}^{B_2} + M_3\mathbf{n}_3 \bullet \boldsymbol{\omega}_{\dot{\theta}_2}^{B_3} \\ & = M_2 - M_3 \end{aligned} \tag{11.3.21}$$

$$\begin{aligned} F_{\theta_3} :\ & M_1\mathbf{n}_3 \bullet \boldsymbol{\omega}_{\dot{\theta}_3}^{B_1} - M_2\mathbf{n}_3 \bullet \boldsymbol{\omega}_{\dot{\theta}_3}^{B_1} + M_2\mathbf{n}_3 \bullet \boldsymbol{\omega}_{\dot{\theta}_3}^{B_2} - M_3\mathbf{n}_3 \bullet \boldsymbol{\omega}_{\dot{\theta}_3}^{B_2} + M_3\mathbf{n}_3 \bullet \boldsymbol{\omega}_{\dot{\theta}_3}^{B_3} \\ & = M_3 \end{aligned} \tag{11.3.22}$$

From Equations (11.3.6), (11.3.7), and (11.3.8); (11.3.16), (11.3.17), and (11.3.18); and (11.3.20), (11.3.21), and (11.3.22) the generalized forces are seen to be:

$$F_{\theta_1} = -(5/2)mg\ell \sin\theta_1 - Mg\ell \sin\theta_1 + M_1 - M_2 \tag{11.3.23}$$

$$F_{\theta_2} = -(3/2)mg\ell \sin\theta_2 - Mg\ell \sin\theta_2 + M_2 - M_3 \tag{11.3.24}$$

$$F_{\theta_3} = -(1/2)mg\ell \sin\theta_3 - Mg\ell \sin\theta_3 + M_3 \tag{11.3.25}$$

As before [see Equations (11.3.9)] Lagrange's equations for the system are

$$\frac{d}{dt}\left(\frac{\partial K}{\partial \dot{\theta}_i}\right) - \frac{\partial K}{\partial \theta_i} = F_{\theta_i} \qquad (i = 1,2,3) \tag{11.3.26}$$

By substituting from Equations (11.3.14), (11.3.20), (11.3.21) and (11.3.22) into (11.3.26) the governing equations are seen to be:

$$\begin{aligned}
&(7/3)\ddot{\theta}_1 + (3/2)\ddot{\theta}_2 \cos(\theta_1 - \theta_2) + (1/2)\ddot{\theta}_3 \cos(\theta_1 - \theta_3) + (3/2)\dot{\theta}_2^2 \sin(\theta_1 - \theta_2) \\
&\quad + (1/2)\dot{\theta}_3^2 \sin(\theta_1 - \theta_3) + (M/m)\,[\ddot{\theta}_1 + \ddot{\theta}_2 \cos(\theta_2 - \theta_1) + \ddot{\theta}_3 \cos(\theta_2 - \theta_3) \\
&\quad + \dot{\theta}_2^2 \sin(\theta_1 - \theta_3) + \dot{\theta}_3^2 \sin(\theta_1 - \theta_3)] + (5/2)(g/\ell) \sin\theta_1 - M_1 + M_2 = 0
\end{aligned} \tag{11.3.27}$$

$$\begin{aligned}
&(4/3)\ddot{\theta}_2 + (3/2)\ddot{\theta}_1 \cos(\theta_2 - \theta_1) + (1/2)\ddot{\theta}_3 \cos(\theta_2 - \theta_3) + (3/2)\dot{\theta}_1^2 \sin(\theta_2 - \theta_1) \\
&\quad + (1/2)\dot{\theta}_2^3 \sin(\theta_2 - \theta_3) + (M/m)\,[\ddot{\theta}_2 + \ddot{\theta}_1 \cos(\theta_2 - \theta_1) + \ddot{\theta}_3 \cos(\theta_2 - \theta_3) \\
&\quad + \dot{\theta}_1^2 \sin(\theta_2 - \theta_1) + \dot{\theta}_3^2 \sin(\theta_2 - \theta_3) + (3/2)(g/\ell) \sin\theta_2 - M_2 + M_3 = 0
\end{aligned} \tag{11.3.28}$$

$$\begin{aligned}
&(1/3)\ddot{\theta}_3 + (1/2)\ddot{\theta}_1 \cos(\theta_3 - \theta_1) + (1/2)\ddot{\theta}_2 \cos(\theta_3 - \theta_2) + (1/2)\dot{\theta}_1^2 \sin(\theta_3 - \theta_1) \\
&\quad + (1/2)\dot{\theta}_2^2 \sin(\theta_3 - \theta_1) + (M/m)\,[\ddot{\theta}_3 + \ddot{\theta}_1 \cos(\theta_3 - \theta_1) + \ddot{\theta}_2 \cos(\theta_3 - \theta_2) \\
&\quad + \dot{\theta}_1^2 \sin(\theta_3 - \theta_1) + \dot{\theta}_2^2 \sin(\theta_3 - \theta_2) + (1/2)(g/\ell) \sin\theta_3 - M_3 = 0
\end{aligned} \tag{11.3.29}$$

As noted earlier the angles θ_1, θ_2, and θ_3 define the orientation of the rods relative to the vertical (so-called "absolute orientation angles"). In our selection of variables, or coordinates, to represent the degrees of freedom we might also have chosen "relative" orientation angles defining the orientations of the rods relative to

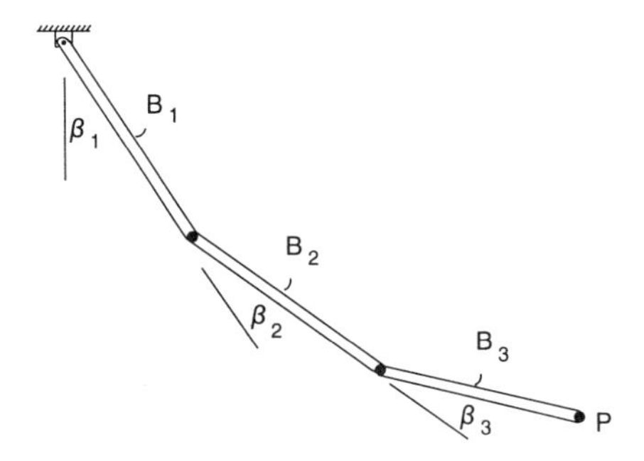

Fig. 11.3.4 A Triple-Rod Pendulum Described by Relative Orientation Angles

each other such as β_1, β_2, and β_3 of Figure 11.3.4. In some contexts, these angles might seem to be more "natural," or more convenient, than absolute orientation angles. The convenience, however, lies more in the intuitive description of the orientation of the rods than in the analysis. By following the same procedures as with Equations (11.3.1) to (11.3.9) we obtain the governing equations, using the β_i, in the forms

$$
\begin{aligned}
&[4 + 3\cos\beta_2 + \cos\beta_3 + \cos(\beta_2 + \beta_3)]\ddot{\beta}_1 + [(5/3) + (3/2)\cos\beta_2 + \cos\beta_3 \\
&(1/2)\cos(\beta_2 + \beta_3)]\ddot{\beta}_2 + [(1/3) + (1/2)\cos\beta_3 + (1/2)\cos(\beta_2 + \beta_3)]\ddot{\beta}_3 \\
&- (3/2)(\dot{\beta}_1 + \dot{\beta}_2)^2 \sin\beta_2 + (3/2)\dot{\beta}_1^2 \sin\beta_2 + (1/2)\dot{\beta}_1^2 \sin(\beta_2 + \beta_3) \\
&+ (1/2)(\dot{\beta}_1 + \dot{\beta}_2)^2 \sin\beta_3 - (1/2)(\dot{\beta}_1 + \dot{\beta}_2 + \dot{\beta}_3)^2 \sin\beta_3 - (1/2)(\dot{\beta}_1 + \dot{\beta}_2 + \dot{\beta}_3)^2 \sin(\beta_2 + \beta_3) \\
&+ (5/2)(g/\ell)\sin\beta_1 + (3/2)(g/\ell)\sin(\beta_1 + \beta_2) + (1/2)(g/\ell)\sin(\beta_1 + \beta_2 + \beta_3) = 0
\end{aligned}
\tag{11.3.30}
$$

$$
\begin{aligned}
&[(5/3) + (3/2)\cos\beta_2 + \cos\beta_3 + (1/2)\cos(\beta_2 + \beta_3)]\ddot{\beta}_1 + [(5/3) + \cos\beta_3]\ddot{\beta}_2 \\
&+ [(1/3) + (1/2)\cos\beta_3]\ddot{\beta}_3 + (3/2)\dot{\beta}_1^2 \sin\beta_2 + (1/2)\dot{\beta}_1^2 \sin(\beta_2 + \beta_3) \\
&+ (1/2)(\dot{\beta}_1 + \dot{\beta}_2)^2 \sin\beta_3 - (1/2)(\dot{\beta}_1 + \dot{\beta}_2 + \dot{\beta}_3)^2 \sin\beta_3 + (3/2)(g/\ell)\sin(\beta_1 + \beta_2) \\
&+ (1/2)(g/\ell)\sin(\beta_1 + \beta_2 + \beta_3) = 0
\end{aligned}
\tag{11.3.31}
$$

$$[(1/3) + (1/2)\cos\beta_3 + (1/2)\cos(\beta_2 + \beta_3)]\ddot{\beta}_1 + [(1/3) + (1/2)\cos\beta_3]\ddot{\beta}_2$$
$$+ (1/3)\ddot{\beta}_3 + (1/2)\dot{\beta}_1^2 \sin(\beta_2 + \beta_3) + (1/2)(\dot{\beta}_1 + \dot{\beta}_2)^2 \sin\beta_3$$
$$+ (1/2)(g/\ell)\sin(\beta_1 + \beta_2 + \beta_3) = 0 \tag{11.3.32}$$

where we have included the effects of joint moments and a concentrated end mass.

Observe the complexity of Equations (11.3.30), (11.3.31), and (11.3.32) when compared with Equations (11.3.27), (11.3.28), and (11.3.29) which use absolute orientation angles. There is, however, one advantage with the relative angles: the joint moment components M_1, M_2, and M_3 occur singly (that is, "uncoupled") in Equations (11.3.30), (11.3.31), and (11.3.32), whereas they are coupled together in Equations (11.3.27), (11.3.28), and (11.3.29).

11.4 The N-Rod Pendulum [11.1]

As a still further extension of the rod pendulum, consider the N-rod pendulum consisting of N identical, pin-connected rods with an end mass, moving in a vertical plane as in Figure 11.4.1. Such a system might model a chain, and in the limit, a cable.

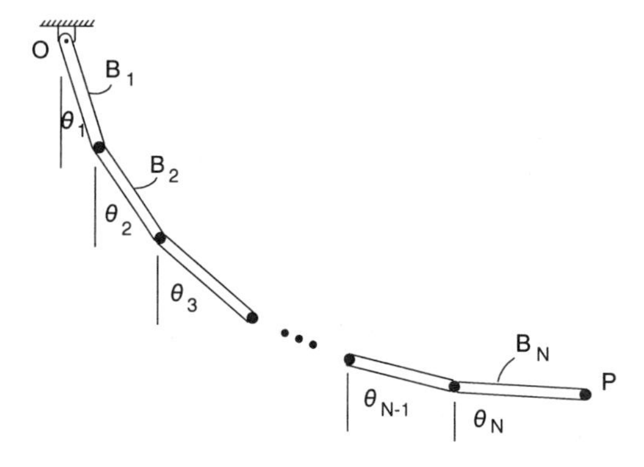

Fig. 11.4.1 An N-Rod Pendulum with an End Mass

Let the orientations of the rods be defined by the angles θ_1, θ_2, ..., θ_N of Figure 11.4.1. Let the rods themselves be called B_1, B_2, ..., B_N and let their mass centers be G_1, G_2, ..., G_N. Let each rod have length ℓ and mass m. Let the pin connections be frictionless. Let the mass at P be M.

We can study this system by taking advantage of the pattern of terms in the analyses of the foregoing sections for the double-rod and triple-rod pendulums. As with those systems, the analysis is conveniently conducted using Lagrange's equations. To this end, observe that the system has N degrees of freedom represented by the angles $\theta_1,\ldots,\theta_N$. The kinetic energy K of the system is:

$$K = (1/2)m(\mathbf{V}^{G_1})^2 + (1/2)I(\boldsymbol{\omega}^{B_1})^2 + (1/2)m(\mathbf{V}^{G_2})^2 + (1/2)M(\mathbf{V}^{P})^2 + (1/2)I(\boldsymbol{\omega}^{B_2})^2 + \cdots + (1/2)m(\mathbf{V}^{G_N})^2 + (1/2)I(\boldsymbol{\omega}^{B_N})^2$$

$$= \sum_{i=1}^{N} [(1/2)m(\mathbf{V}^{G_i})^2 + (1/2)I(\boldsymbol{\omega}^{B_i})^2] \tag{11.4.1}$$

where, as before, $\mathbf{V}^{G_i}$ and $\boldsymbol{\omega}^{B_i}$ are the velocities and angular velocities of G_i and B_i $(i = 1,..,N)$, and where I is the central moment of inertia of the rods in a direction perpendicular to the rods.

As with the double- and triple-rod pendulum it is convenient to introduce and use unit vectors parallel and perpendicular to the rods as well as horizontal and vertical unit vectors as shown for a typical rod as in Figure 11.4.2, where, as before, $\mathbf{n}_3$ is a unit vector perpendicular to the plane of motion.

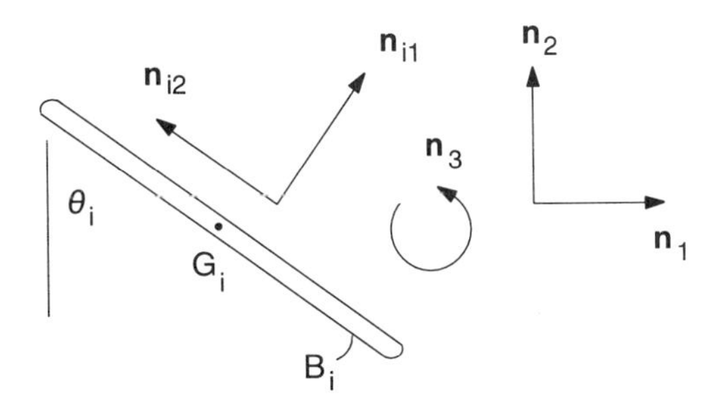

Fig. 11.4.2 A Typical Rod of the N-Rod System

The velocity of typical mass center G_i is readily seen to be:

$$\mathbf{V}^{G_i} = \ell\dot{\theta}_1\mathbf{n}_{11} + \ell\dot{\theta}_2\mathbf{n}_{21} + \cdots + \ell\dot{\theta}_{(i-1)}\mathbf{n}_{(i-1)1} + (\ell/2)\dot{\theta}_i\mathbf{n}_{i1} \tag{11.4.2}$$

The angular velocity of typical rod B_i is

$$\boldsymbol{\omega}^{B_i} = \dot{\theta}_i\mathbf{n}_3 \tag{11.4.3}$$

By substituting from Equations (11.4.2) and (11.4.3) into (11.4.1) the kinetic energy K may be written in the compact form:

$$K = (1/2)m\ell^2 \sum_{i=1}^{N}\sum_{j=1}^{N} c_{ij}\dot{\theta}_i\dot{\theta}_j \tag{11.4.4}$$

where the coefficients c_{ij} are

$$\begin{gathered} c_{ij} = (1/2)\,[1 + 2(N - k) + 2(M/m)]\cos(\theta_j - \theta_i) \\ i \neq j \text{ and } k \text{ is the larger of } i \text{ and } j \end{gathered} \tag{11.4.5}$$

and

$$c_{ii} = N - i + (1/3) + M/m \tag{11.4.6}$$

From Equation (11.4.2) the partial velocity of G_i relative to θ_j is:

$$\mathbf{V}_{\dot{\theta}_j}^{G_i} = \begin{cases} \ell\,\mathbf{n}_{j1} & j < i \\ (\ell/2)\,\mathbf{n}_{j1} & j = i \\ 0 & j > i \end{cases} \tag{11.4.7}$$

With the pin joints being frictionless, the only forces contributing to the generalized applied forces are the gravity (or weight) forces. These weight forces may be represented as single vertical forces $(-mg\mathbf{n}_2)$ passing through the mass centers G_i and a force $-Mg\mathbf{n}_2$ passing through the end mass at P. Using Equation (11.4.7) the generalized applied forces $F_{\theta i}$ are seen to be:

$$F_{\theta i} = -[N - i + (1/2)]\, mg\ell \sin\theta_i - Mg\ell \sin\theta_i \tag{11.4.8}$$

Lagrange's equations for the system are [Equations (10.2.6)]:

$$\frac{d}{dt}\left(\frac{\partial K}{\partial \dot{\theta}_i}\right) - \frac{\partial K}{\partial \theta_i} = F_{\theta i} \quad (i = 1, \ldots, N) \tag{11.4.9}$$

By substituting from Equation (11.4.4) and (11.4.8) into (11.4.9) the governing dynamical equations are seen to be:

$$\sum_{j=1}^{N} [m_{ij}\ddot{\theta}_j + n_{ij}\dot{\theta}_j^2 + (g/\ell)k_{ij}] = 0 \quad i = 1, \ldots, N \tag{11.4.10}$$

where the m_{ij}, n_{ij}, and k_{ij} are:

$$m_{ij} = (1/2)[1 + 2(N - k) + 2(M/m)]\cos(\theta_j - \theta_i) \tag{11.4.11}$$
$$i \neq j \text{ and } k \text{ is the larger of } i \text{ and } j$$

$$m_{ii} = N - i + (1/3) + (M/m) \tag{11.4.12}$$

$$n_{ij} = -(1/2)[1 + 2(N - k) + 2(M/m)]\sin(\theta_j - \theta_i) \tag{11.4.13}$$
$$i \neq j \text{ and } k \text{ is the larger of } i \text{ and } j$$

$$n_{ii} = 0 \tag{11.4.14}$$

$$k_{ij} = 0 \quad i \neq j \tag{11.4.15}$$

$$k_{ii} = [N - i + (1/2) + (M/m)]\sin\theta_i \tag{11.4.16}$$

11.5 Rolling Circular Disk on a Flat Horizontal Surface [11.1, 11.3, 11.4, 11.5]

Consider a sharp-edged circular disk D with mass m and radius r rolling on a flat horizontal surface as in Figure 11.5.1 (see Section 6.12). The objective is to determine the governing equations of motion.

Let the orientation angles of D be θ, ϕ, and ψ as shown where θ is the "lean" angle, ψ is the "rolling" angle, and ϕ is the "turning" or "yaw" angle. Let (x,y,z) be the X,Y,Z coordinates of G, the mass center of D. The rolling requirement restricts D to three degrees of freedom: Recall that an unconstrained rigid body has six degrees of freedom (three in translation and three in rotation). If D were unrestrained, its degrees of freedom could be represented by the parameters

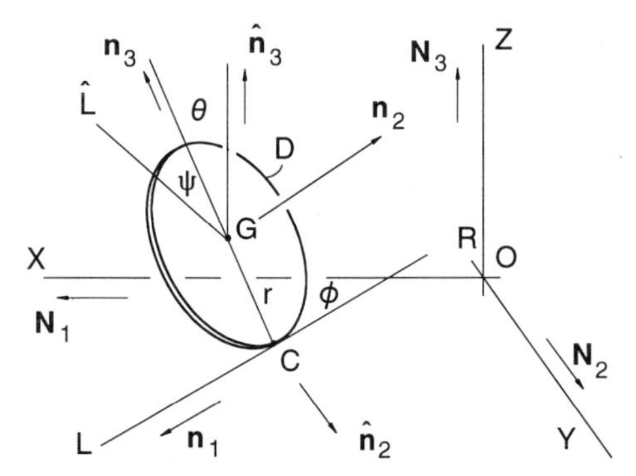

Fig. 11.5.1 A Circular Disk Rolling on a Flat Horizontal Surface

x, y, z (translation) and θ, ϕ, ψ (rotation). The rolling requirement makes these parameters no longer independent. Specifically, the velocity of G relative to the reference frame R (in which X, Y, and Z are fixed) may be expressed in the form:

$$\mathbf{V}^G = \dot{x}\mathbf{N}_1 + \dot{y}\mathbf{N}_2 + \dot{z}\mathbf{N}_3 \tag{11.5.1}$$

and

$$\begin{aligned} \mathbf{V}^G = & \, r(\dot{\theta} c_\theta s_\phi + \dot{\phi} c_\phi s_\theta + \dot{\psi} c_\phi)\mathbf{N}_1 \\ & + r(-\dot{\theta} c_\theta c_\phi + \dot{\phi} s_\theta s_\phi + \dot{\psi} s_\phi)\mathbf{N}_2 \\ & + r(-\dot{\theta} s_\theta)\mathbf{N}_3 \end{aligned} \tag{11.5.2}$$

where, as before, s and c are abbreviations for sine and cosine and where $\mathbf{N}_1$, $\mathbf{N}_2$, and $\mathbf{N}_3$ are unit vectors parallel to X, Y, and Z as in Figure 11.5.1. Equation (11.5.2) is obtained directly from Table 6.12.2. By equating the expressions of Equations (11.5.1) and (11.5.2) we immediately obtain the constraint equations:

$$\dot{x} = r(\dot{\theta} c_\theta s_\phi + \dot{\phi} c_\phi s_\theta + \dot{\psi} c_\phi) \tag{11.5.3}$$

$$\dot{y} = r(-\dot{\theta} c_\theta c_\phi + \dot{\phi} s_\theta s_\phi + \dot{\psi} s_\phi) \tag{11.5.4}$$

$$\dot{z} = r(-\dot{\theta} s_\theta) \tag{11.5.5}$$

These three constraint equations then restrict the degrees of freedom to three.

Observe that Equation (11.5.5) can be integrated to give:

$$z = r c_\theta \tag{11.5.6}$$

This expression shows that D must remain in contact with the rolling surface (the X-Y plane). As such, it is a "geometric" (or "holonomic") constraint. Alternatively, Equations (11.5.3) and (11.5.4) cannot be integrated in closed form. They are "kinematic" (or "non-holonomic") constraints (see Section 10.1).

11.5.1 Use of d'Alembert's Principle

The desired equations of motion are readily obtained using d'Alembert's principle with the aid of a free-body diagram of D as in Figure 11.5.2 where $\mathbf{W}$ is the weight force, $\mathbf{C}$ is the contact force, $\mathbf{F}^*$ is an equivalent inertia force, and $\mathbf{T}^*$ is the corresponding inertia torque, where $\mathbf{W}$, $\mathbf{F}^*$, and $\mathbf{T}^*$ are [see Equations (9.12.6) and (9.12.11)]:

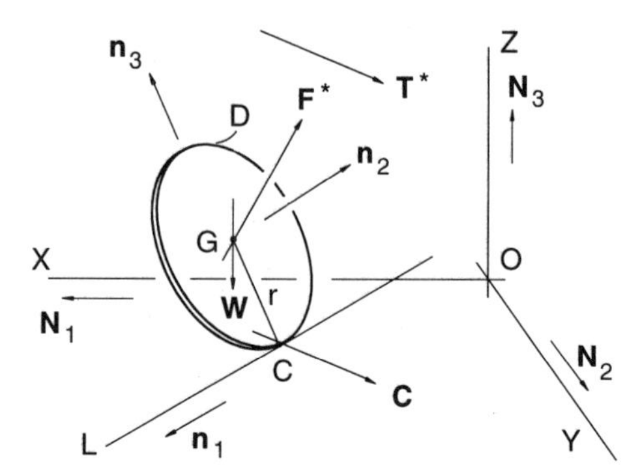

Fig. 11.5.2 Free-Body Diagrams of Rolling Circular Disk

$$\mathbf{W} = -mg\mathbf{N}_3 \tag{11.5.7}$$

$$\mathbf{F}^* = -m\mathbf{a} \tag{11.5.8}$$

and

$$\mathbf{T}^* = -\mathbf{I} \bullet \boldsymbol{\alpha} - \boldsymbol{\omega} \times \mathbf{I} \bullet \boldsymbol{\omega} \tag{11.5.9}$$

where $\mathbf{a}$ is the acceleration of D in R (${}^R\mathbf{a}^D$), $\boldsymbol{\alpha}$ is the angular acceleration of D in R (${}^R\boldsymbol{\alpha}^D$), $\boldsymbol{\omega}$ is the angular velocity of D in R (${}^R\boldsymbol{\omega}^D$), m is the mass of D, and $\mathbf{I}$ is the central inertia dyadic ($\mathbf{I}^{D/G}$). By setting moments about C equal to zero we can eliminate the contact force $\mathbf{C}$ from the analysis resulting in:

$$r\mathbf{n}_3 \times \mathbf{W} + r\mathbf{n}_3 \times \mathbf{F}^* + \mathbf{T}^* = 0 \tag{11.5.10}$$

where $\mathbf{n}_3$ is a unit vector parallel to the diameter at the contact point C.

The relations between the unit vector sets of Figure 11.5.1 may be obtained from the configuration graph of Figure 6.12.3. Of these unit vector sets, the $\mathbf{n}_i$

$(i = 1,2,3)$ remain parallel to principal axes of D during the motion of D. Therefore, the $\mathbf{n}_i$ are convenient for the development of Equations (11.5.10)*. To this end we can express $\mathbf{W}$, $\mathbf{F}^*$, and $\mathbf{T}^*$ as:

$$\mathbf{W} = -mg\,(s_\theta \mathbf{n}_2 + c_\theta \mathbf{n}_3) \tag{11.5.11}$$

$$\mathbf{F}^* = -ma_1\mathbf{n}_1 - ma_2\mathbf{n}_2 - ma_3\mathbf{n}_3 \tag{11.5.12}$$

$$\mathbf{T}^* = T_1\mathbf{n}_1 + T_2\mathbf{n}_2 + T_3\mathbf{n}_3 \tag{11.5.13}$$

where the a_i are the $\mathbf{n}_i$ components of $\mathbf{a}$ and where the T_i $(i = 1,2,3)$ are

$$T_1 = -\alpha_1 I_{11} + \omega_2\omega_3(I_{22} - I_{33}) \tag{11.5.14}$$

$$T_2 = -\alpha_2 I_{22} + \omega_3\omega_1(I_{33} - I_{11}) \tag{11.5.15}$$

$$T_3 = -\alpha_3 I_{33} + \omega_1\omega_2(I_{11} - I_{22}) \tag{11.5.16}$$

where the α_i and ω_i are the $\mathbf{n}_i$ components of $\boldsymbol{\alpha}$ and $\boldsymbol{\omega}$ and where the I_{ii} are central principal moments of inertia given by (see Table 8.13.1):

$$I_{11} = I_{33} = mr^2/4 \tag{11.5.17}$$

$$I_{22} = mr^2/2 \tag{11.5.18}$$

By substituting from Equations (11.5.11), (11.5.12), and (11.5.13) into (11.5.10) and

---

*The d_i (not shown in Figure 11.5.1) might also be considered but a review of the kinematic expressions of Tables 6.12.1 to 6.12.4 quickly shows them to be less convenient than the $\mathbf{n}_i$.

carrying out the indicated operations we obtain the scalar equations:

$$\mathrm{mgrs}_\theta + \mathrm{mra}_2 + \mathrm{T}_1 = 0 \tag{11.5.19}$$

$$-\mathrm{mra}_1 + \mathrm{T}_2 = 0 \tag{11.5.20}$$

$$\mathrm{T}_3 = 0 \tag{11.5.21}$$

Finally, by substituting from Equations (11.5.14), (11.5.15), and (11.5.16) and by using the kinematical expressions of Tables 6.12.1, 6.12.3, and 6.12.4 we obtain the equations:

$$(4g/r)\,s_\theta - 5\ddot{\theta} + 6\dot{\psi}\,\dot{\phi}\,c_\theta + 5\dot{\phi}^2 s_\theta c_\theta = 0 \tag{11.5.22}$$

$$3\ddot{\psi} + 3\ddot{\phi}\,s\theta + 5\dot{\phi}\dot{\theta}\,c_\theta = 0 \tag{11.5.23}$$

$$\ddot{\phi}c_\theta + 2\dot{\psi}\,\dot{\theta} = 0 \tag{11.5.24}$$

11.5.2 Use of Kane's Equations

We can also readily obtain the equations of motion using Kane's equations. Since D rolls on the X-Y plane the forces exerted on D by the plane are non-working and they do not contribute to the generalized applied forces (see Section 9.9). Therefore, these contact forces may be neglected in the analysis with Kane's equations. The only forces contributing to the generalized applied forces are the weight forces on the particles of D which can be represented by a single vertical force: $-\mathrm{mg}\mathbf{N}_3$ passing through the mass center G. In like manner, the inertia forces on D may be represented by a single force $\mathbf{F}^*$ passing through G together with a couple having torque $\mathbf{T}^*$, where $\mathbf{F}^*$ and $\mathbf{T}^*$ are given by Equations (11.5.8) and (11.5.9).

To develop Kane's equations, observe that with the rolling requirements D has three degrees of freedom and that these degrees of freedom may be represented by the angles: θ, ϕ, and ψ. Next, to obtain the generalized forces, observe from Tables 6.12.1 and 6.12.2 that the partial velocities of G and the partial angular velocities of D for θ, ϕ, and ψ are

$$\mathbf{V}_{\dot{\theta}} = -r\mathbf{n}_2 \qquad \mathbf{V}_{\dot{\phi}} = rs_\theta \mathbf{n}_1 \qquad \mathbf{V}_{\dot{\psi}} = r\mathbf{n}_1 \tag{11.5.25}$$

$$\boldsymbol{\omega}_{\dot{i}} = \mathbf{n}_1 \qquad \boldsymbol{\omega}_{\dot{\phi}} = s_\theta \mathbf{n}_2 + c_\theta \mathbf{n}_3 \qquad \boldsymbol{\omega}_{\dot{\psi}} = \mathbf{n}_2 \tag{11.5.26}$$

The generalized applied (active) forces are then:

$$F_\theta = \mathbf{V}_{\dot{\theta}} \bullet (-mg\mathbf{N}_3) = mgr\sin\theta \tag{11.5.27}$$

$$F_\phi = \mathbf{V}_{\dot{\phi}} \bullet (-mg\mathbf{N}_3) = 0 \tag{11.5.28}$$

$$F_\psi = \mathbf{V}_{\dot{\psi}} \bullet (-mg\mathbf{N}_3) = 0 \tag{11.5.29}$$

Similarly, the generalized inertia forces are:

$$F_\theta^* = \mathbf{V}_{\dot{\theta}} \bullet \mathbf{F}^* + \boldsymbol{\omega}_{\dot{\theta}} \bullet \mathbf{T}^* = mra_2 + T_1 \tag{11.5.30}$$

$$F_\phi^* = \mathbf{V}_{\dot{\phi}} \bullet \mathbf{F}^* + \boldsymbol{\omega}_{\dot{\phi}} \bullet \mathbf{T}^* = -mrs_\theta a_1 + s_\theta T_2 + c_\theta T_3 \tag{11.5.31}$$

$$F_\psi^* = \mathbf{V}_{\dot{\psi}} \bullet \mathbf{F}^* + \boldsymbol{\omega}_{\dot{\psi}} \bullet \mathbf{T}^* = -mra_1 + T_2 \tag{11.4.32}$$

Finally, Kane's equations are:

$$F_\theta + F_\theta^* = 0 \quad \text{or} \quad mgrs_\theta + mra_2 + T_1 = 0 \tag{11.5.33}$$

$$F_\phi + F_\phi^* = 0 \quad \text{or} \quad -mrs_\theta a_1 + s_2 T_2 + c_\theta T_3 = 0 \tag{11.5.34}$$

$$F_\psi + F_\psi^* = 0 \quad \text{or} \quad -mra_1 + T_2 = 0 \tag{11.5.35}$$

Equations (11.5.33), (11.5.34), and (11.5.35) are seen to be the same as (or equivalent to) Equations (11.5.19), (11.5.20), and (11.5.21). [Observe that the first two terms of Equation (11.5.34) are zero in view of Equation (11.5.35).] Note further that with Kane's equations, it is not necessary to rely upon a free-body diagram and to select a point to take moments about as we did with d'Alembert's principle.

11.5.3 Use of Lagrange's Equations [11.6]

The equations of motion can also be obtained by using the modified form of Lagrange's equations for application with simple non-holonomic constraints, as developed in Section 10.10 [see Equation (10.10.26)]. To this end, we consider the disk D, as a rigid body, to have six degrees of freedom, but then subjected to the rolling constraints as detailed by Equations (11.5.3), (11.5.4), and (11.5.5). Specifically, let the degrees of freedom be represented by the generalized coordinates q_r $(r = 1,\dots,6)$ defined as:

$$\begin{array}{lll} q_1 = \theta & q_2 = \phi & q_3 = \psi \\ q_4 = x & q_5 = y & q_6 = z \end{array} \tag{11.5.36}$$

Let the constraint equations [Equations (11.5.3), (11.5.4), and (11.5.5)] by written as

$$rc_\theta s_\phi \dot{\theta} + rc_\phi s_\theta \dot{\phi} + rc_\phi \dot{\psi} - \dot{x} = 0 \tag{11.5.37}$$

$$-rc_\theta c_\phi \dot{\theta} + rs_\theta s_\phi \dot{\phi} + rs_\phi \dot{\psi} - \dot{y} = 0 \qquad (11.5.38)$$

$$rs_\theta \dot{\theta} + \dot{z} = 0 \qquad (11.5.39)$$

Then from Equation (10.10.1) if Equations (11.5.37), (11.5.38), and (11.5.39) are written in compact form as:

$$\sum_{r=1}^{G} a_{sr}\dot{q}_r + b_s \quad s = 4,5,6 \qquad (11.5.40)$$

we see from Equations (11.5.36) that the a_{sr} and b_s $(s = 4,5,6;\ r = 1,\ldots,6)$ are:

$$\begin{array}{llllll} a_{41} = rc_\theta s_\phi & a_{42} = rc_\phi s_\theta & a_{43} = rc_\phi & a_{44} = -1\ , & a_{45} = 0 & a_{46} = 0 \\ a_{51} = -rc_\theta c_\phi & a_{52} = rs_\theta s_\phi & a_{53} = rs_\phi & a_{54} = 0 & a_{55} = -1 & a_{56} = 0 \\ a_{61} = rs\theta & a_{62} = 0 & a_{63} = 0 & a_{64} = 0 & a_{65} = 0 & a_{66} = 1 \end{array} \qquad (11.5.41)$$

and

$$b_4 = b_5 = b_6 = 0 \qquad (11.5.42)$$

From Equation (10.10.4) if we solve Equations (11.5.37), (11.5.38) and (11.5.39) [or equivalently Equation (11.5.40)] for $\dot{x}$, $\dot{y}$, and $\dot{z}$ (that is, $\dot{q}_4$, $\dot{q}_5$, and $\dot{q}_6$) we obtain equations of the form:

$$\dot{q}_r = \sum_{i=1}^{3} c_{ri}\dot{q}_i + d_r \qquad r = 4,5,6 \qquad (11.5.43)$$

where the c_{ri} and d_r $(r = 4,5,6$ and $i = 1,2,3)$ are:

$$
\begin{array}{lll}
c_{41} = rc_\theta s_\phi & c_{42} = rc_\phi s_\theta & c_{43} = rc_\phi \\
c_{51} = -rc_\theta c_\phi & c_{52} = rs_\theta s_\phi & c_{53} = rs_\phi \\
c_{61} = -rs_\theta & c_{62} = 0 & c_{63} = 0
\end{array}
\tag{11.5.44}
$$

and

$$
d_4 = d_5 = d_6 = 0 \tag{11.5.45}
$$

Next, from Equation (10.10.5), introduce generalized speeds w_s $(s = 1,...,6)$ of the form:

$$
w_s = \sum_{r=1}^{6} a_{sr}\dot{q}_r + b_s \qquad s = 1,...,6 \tag{11.5.46}
$$

where the individual w_s are:

$$
\begin{aligned}
w_1 &= \dot{\theta} \qquad w_2 = \dot{\phi} \qquad w_3 = \dot{\psi} \\
w_4 &= rc_\theta s_\phi \dot{\theta} + rc_\phi s_\theta \dot{\phi} + rc_\phi \dot{\psi} - \dot{x} \\
w_5 &= -rc_\theta c_\phi \dot{\theta} + rs_\theta s_\phi \dot{\phi} + rs_\phi \dot{\psi} - \dot{y} \\
w_6 &= rs_\theta \dot{\theta} + \dot{z}
\end{aligned}
\tag{11.5.47}
$$

where the last three of these are obtained from Equations (11.5.37), (11.5.38), and (11.5.39).

By comparing Equations (11.5.36), (11.5.40), (11.5.46), and (11.5.47) we see that

$$
b_s = 0 \qquad s = 1,...,6 \tag{11.5.48}
$$

and that the a_{sr} $(s,r = 1,...,6)$ may be regarded as elements of a 6×6 matrix A which may be expressed in partitioned form of 3×3 arrays as:

$$A = \left[\begin{array}{c|c} I & 0 \\ \hline A_1 & A_2 \end{array}\right] \qquad (11.5.49)$$

where the partitioned submatrices are each 3 × 3 arrays, with I, A_1, and A_2 being

$$I = \begin{bmatrix} 1 & 0 & 0 \\ 0 & 1 & 0 \\ 0 & 0 & 1 \end{bmatrix} \qquad (11.5.50)$$

$$A_1 = \begin{bmatrix} rc_\theta s_\phi & rc_\phi s_\theta & rc_\phi \\ -rc_\theta c_\phi & rs_\theta s_\phi & rs_\phi \\ rs_\theta & 0 & 0 \end{bmatrix} \qquad (11.5.51)$$

and

$$A_2 = \begin{bmatrix} -1 & 0 & 0 \\ 0 & -1 & 0 \\ 0 & 0 & 1 \end{bmatrix} \qquad (11.5.52)$$

From Equation (10.10.25) we see that the inverse of A is simply:

$$A^{-1} = \left[\begin{array}{c|c} I & 0 \\ \hline C & A_2^{-1} \end{array}\right] \qquad (11.5.53)$$

where I is the identity array of Equation (11.5.50) and where C is

$$C = -A_2^{-1}A_1 \qquad (11.5.54)$$

From Equation (11.5.52) A_2^{-1} is immediately seen to be:

$$A_2^{-1} = \begin{bmatrix} -1 & 0 & 0 \\ 0 & -1 & 0 \\ 0 & 0 & 1 \end{bmatrix} \tag{11.5.55}$$

Therefore, from Equation (11.5.51) C is

$$C = \begin{bmatrix} rc_\theta s_\phi & rc_\phi s_\theta & rc_\phi \\ -rc_\theta c_\phi & rs_\theta s_\phi & rs_\phi \\ -rs_\theta & 0 & 0 \end{bmatrix} \tag{11.5.56}$$

[Observe that the elements of C are identical to the terms in Equations (11.5.44).]

To obtain the generalized forces, observe that the only forces contributing to the generalized forces are the gravity, or weight forces, on the particles of D, and that these forces are equivalent to the single force: $-mg\mathbf{N}_3$ passing through G. Observe further that from Equations (6.12.8) that the velocity of G may be expressed as:

$$\mathbf{V}^G = r(\dot{\phi} s_\theta + \dot{\psi})\mathbf{n}_1 - r\dot{\theta}\mathbf{n}_2 \tag{11.5.57}$$

But also from Equation (10.10.8) we see that $\mathbf{V}^G$ may be expressed as:

$$\mathbf{V}^G = \sum_{s=1}^{3} \mathbf{V}_{w_s} w_s + \mathbf{V}_b = \mathbf{V}_{\dot{\theta}}\dot{\theta} + \mathbf{V}_{\dot{\phi}}\dot{\phi} + \mathbf{V}_{\dot{\psi}}\dot{\psi} + \mathbf{V}_b \tag{11.5.58}$$

where we have identified w_1, w_2, and w_3 as $\dot{\theta}$, $\dot{\phi}$, and $\dot{\psi}$ respectively as in Equation (11.5.47). Then by comparing Equations (11.5.57) and (11.5.58) we see by inspection that:

$$\mathbf{V}_{w_1} = -r\mathbf{n}_2 \qquad \mathbf{V}_{w_2} = rs_\theta \mathbf{n}_1 \qquad \mathbf{V}_{w_3} = r\mathbf{n}_1 \tag{11.5.59}$$

and

$$\mathbf{V}_b = 0 \tag{11.5.60}$$

Therefore, the generalized forces K_{w_s} $(s = 1,2,3)$ are:

$$K_{w_1} = mgs_\theta \qquad K_{w_2} = 0 \qquad K_{w_3} = 0 \tag{11.5.61}$$

Next, for Lagrange's equations, we need the kinetic energy function K. From Equation (10.3.3) we may express K for the disk D as:

$$K = (1/2)m(\mathbf{V}^G)^2 + \frac{1}{2}\boldsymbol{\omega}^D \bullet \mathbf{I} \bullet \boldsymbol{\omega}^D \tag{11.5.62}$$

where, as before, **I** is the central inertia dyadic of D. In terms of kinematic and inertia components relative to $\mathbf{n}_1$, $\mathbf{n}_2$, and $\mathbf{n}_3$ (principal unit vectors) K then becomes (for the unrestrained disk):

$$K = (1/2)m\dot{x}^2 + (1/2)m\dot{y}^2 + (1/2)m\dot{z}^2 + (1/2)I_{11}\omega_1^2 + (1/2)I_{22}\omega_2^2 + (1/2)I_{33}\omega_3^2 \tag{11.5.63}$$

where the moments of inertia are given by Equations (11.5.17) and (11.5.18) and where the angular velocity components may be obtained from Table 6.12.1. By making the indicated substitutions, K becomes

$$K = (1/2)m(\dot{x}^2 + \dot{y}^2 + \dot{z}^2) + (1/2)(mr^2/4)\dot{\theta}^2 + (1/2)(mr^2/2)(\dot{\phi}s_\theta + \dot{\psi})^2 + (1/2)(mr^2/4)\dot{\phi}^2 c_\theta^2 \tag{11.5.64}$$

Finally, from Equation (10.10.26) Lagrange's equations for the non-holonomic system are

$$\frac{d}{dt}\left(\frac{\partial K}{\partial \dot{q}_s}\right) - \frac{\partial K}{\partial q_s} + \sum_{r=4}^{6}\left[\frac{d}{dt}\left(\frac{\partial K}{\partial \dot{q}_r}\right) - \frac{\partial K}{\partial q_r}\right]c_{rs} = K_s \qquad s = 1,2,3 \tag{11.5.65}$$

where the c_{rs} $(r = 4,5,6;\ s = 1,2,3)$ are given by Equations (11.5.44). Then by substituting from Equations (11.5.61), (11.5.64), and (11.5.44) the governing equations become:

$$\ddot{\theta} - 2(\dot{\phi} s_\theta + \dot{\psi})\dot{\phi} c_\theta + \dot{\phi}^2 s_\theta c_\theta + 4(\ddot{x}/r) c_\theta s_\phi - 4(\ddot{y}/r) c_\theta c_\phi$$
$$- 4(\ddot{z}/r) s_\theta = (4g/r) s_\theta \tag{11.5.66}$$

$$2\ddot{\phi} s_\theta^2 + 4\dot{\phi}\dot{\theta} s_\theta c_\theta + 2\ddot{\psi} s_\theta + 2\dot{\psi}\dot{\theta} c_\theta + \ddot{\phi} c_\theta^2 + 4(\ddot{x}/r) c_\phi s_\theta$$
$$+ 4(\ddot{y}/r) s_\theta s_\phi = 0 \tag{11.5.67}$$

and

$$\ddot{\phi} s_\theta + \dot{\phi}\dot{\theta} c_\theta + \ddot{\psi} + 2(\ddot{x}/r) c_\phi + 2(\ddot{y}/r) s_\phi = 0 \tag{11.5.68}$$

The terms involving $\ddot{x}$, $\ddot{y}$, and $\ddot{z}$ may be eliminated by differentiating in the constraint equations [Equations (11.5.3), (11.5.4), and (11.5.5)] leading to:

$$\ddot{x} = r(\ddot{\theta}\, c_\theta s_\phi - \dot{\theta}^2 s_\theta s_\phi + \dot{\theta}\dot{\phi}\, c_\theta c_\phi + \ddot{\phi}\, c_\phi s_\theta - \dot{\phi}^2 s_\phi s_\theta$$

$$+ \dot{\phi}\dot{\theta}\, c_\phi c_\theta + \ddot{\psi}\, c_\phi - \dot{\psi}\dot{\phi}\, s_\phi) \tag{11.5.69}$$

$$\ddot{y} = r(-\ddot{\theta}\, c_\theta c_\phi + \dot{\theta}^2 s_\theta c_\phi + \dot{\theta}\dot{\phi}\, c_\theta s_\phi + \ddot{\phi}\, s_\theta s_\phi + \dot{\phi}\dot{\theta}\, c_\theta s_\phi$$

$$+ \dot{\phi}^2 s_\theta c_\phi + \ddot{\psi}\, s_\phi + \dot{\psi}\dot{\phi}\, c_\phi) \tag{11.5.70}$$

and

$$\ddot{z} = r(-\ddot{\theta}\, s_\theta - \dot{\theta}^2 c_\theta) \tag{11.5.71}$$

By substituting into Equation (11.5.66), (11.5.67), and (11.5.68) and simplifying, the governing equations are:

$$(4\, g/r)\, s_\theta - 5\ddot{\theta} + 6\,\dot{\psi}\dot{\phi}\, c_\theta + 5\dot{\phi}^2 s_\theta c_\theta = 0 \tag{11.5.72}$$

$$\ddot{\phi}\, c_\theta + 2\,\dot{\psi}\dot{\theta} = 0 \tag{11.5.73}$$

and

$$3\,\ddot{\psi} + 3\ddot{\phi}\, s_\theta + 5\dot{\phi}\dot{\theta}\, c_\theta = 0 \tag{11.5.74}$$

Equations (11.5.72), (11.5.73), and (11.5.74) are seen to be identical to Equations (11.5.22), (11.5.23), and (11.5.24). The analysis effort in obtaining Equations (11.5.72), (11.5.73), and (11.5.74) is considerably greater.

11.5.4 Elementary Solution: Straight Line Rolling [11.3]

Equations (11.5.22), (11.5.23), and (11.5.24) [or equivalently, Equations (11.5.72), (11.5.73), and (11.5.74)] are coupled non-linear differential equations for θ, ϕ, and ψ. Therefore, there are no closed-form solutions in terms of elementary functions except for a few special cases. One of these cases is straight line rolling, as discussed in Example 6.12.2 and as depicted in Figure 11.5.3. In this case the lean angle θ is zero, the turning angle ϕ is a constant, say ϕ_0 and the roll angle ψ represents the single degree of freedom. Then Equations (11.5.72) and (11.5.73) are identically satisfied and Equation (11.5.74) becomes

$$\ddot{\psi} = 0 \tag{11.5.75}$$

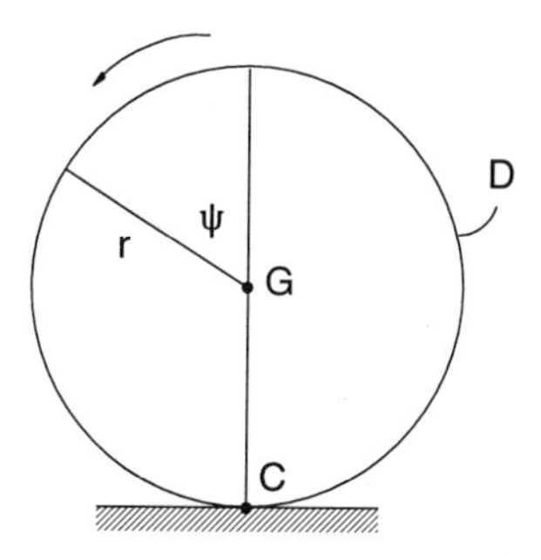

Fig. 11.5.3 A Vertical Disk Rolling on a Straight Line

Hence, the roll rate $\dot{\psi}$ is constant:

$$\dot{\psi} = \dot{\psi}_0 \tag{11.5.76}$$

That is, the disk rolls at a constant rate.

In spite of their complexity, Equations (11.5.72), (11.5.73), and (11.5.74) may nevertheless be used to examine stability of the special solutions cases. To illustrate this, let the solution angles be subject to small disturbances θ^*, ϕ^*, and ψ^* as:

$$\theta = 0 + \theta^* \qquad \phi = \phi_0 + \phi^* \qquad \dot{\psi}_0 + \dot{\psi}^* \tag{11.5.77}$$

Then by substituting into Equations (11.5.72), (11.5.73), and (11.5.74), these disturbances are governed by the expressions:

$$(4g/r)\theta^* - 5\ddot{\theta}^* + 6\dot{\psi}_0\dot{\phi}^* = 0 \tag{11.5.78}$$

$$\ddot{\phi}^* + 2\dot{\psi}_0\dot{\theta}^* = 0 \tag{11.5.79}$$

$$\ddot{\psi}^* = 0 \tag{11.5.80}$$

where we have neglected products of small terms ["starred" (*) quantities].

Equation (11.5.80) shows that a small disturbance $\dot{\psi}^*$ in the rolling speed, or roll angle rate $\dot{\psi}$, remains small — essentially constant. That is,

$$\dot{\psi}^* = \dot{\psi}_0^* \qquad \text{(a constant)} \tag{11.5.81}$$

Next, observe that Equations (11.5.78) and (11.5.79) are independent of ψ^* and that Equation (11.5.79) may be integrated enabling the elimination of $\dot{\phi}^*$. That is, by integrating Equation (11.5.79) we obtain:

$$\dot{\phi}^* + 2\dot{\psi}_0\theta^* = C^* \qquad \text{(a constant)} \tag{11.5.82}$$

Then by substituting into Equation (11.5.78) we have (after rearrangement of terms):

$$\ddot{\theta}^* + \left[(12\,\dot{\psi}_0^2/5) - (4g/5r)\right]\theta^* = 6\dot{\psi}_0 C^*/5 \qquad (11.5.83)$$

The solution for θ^* will be sinusoidal, and hence bounded, if the coefficient of θ^* is positive. Otherwise the solution will grow exponentially in time (or linearly if the θ^* coefficient is zero). Therefore, the straight line rolling disk is stable (that is, it remains vertical and rolling at an essentially constant speed) if

$$(12\dot{\psi}_0^2/5) - (4g/5r) > 0 \qquad (11.5.84)$$

or

$$\dot{\psi}_0 > \sqrt{g/3r} \qquad (11.5.85)$$

11.5.5 Elementary Solution: Pivoting (Spinning) Disk [11.3]

A second special case is a pivoting (or spinning) disk as discussed in Example 6.12.4 and as depicted in Figure 11.5.4. In this case the lean angle θ is zero, the roll angle ψ is a constant, say ψ_0, and the turning angle ϕ represents the single degree of freedom.

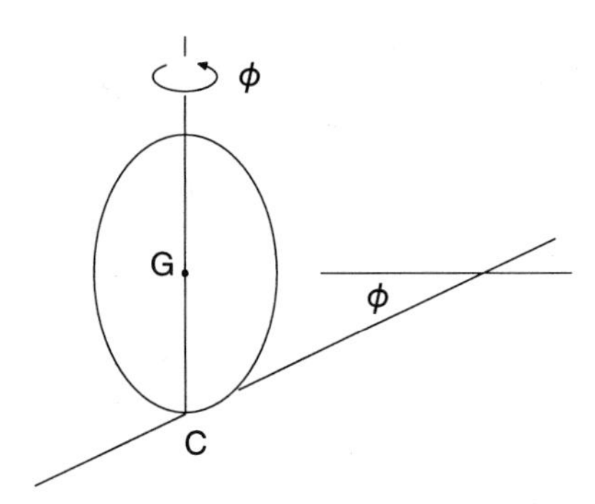

Fig. 11.5.4 A Pivoting (Spinning) Vertical Disk

With these conditions, Equations (11.5.72) and (11.5.74) are identically satisfied and Equation (11.5.73) becomes

$$\ddot{\phi} = 0 \qquad (11.5.86)$$

Hence, the spin rate $\dot{\phi}$ is constant:

$$\dot{\phi} = \dot{\phi}_0 \qquad (11.5.87)$$

That is, the disk spins at a constant rate.

As with the case of straight line rolling, Equations (11.5.72), (11.5.73), and (11.5.74) may be used to study the stability of the spinning disk. Specifically, let the spinning disk be subjected to a small disturbance described as:

$$\theta = 0 + \theta^* \qquad \dot{\phi} = \dot{\phi}_0 + \dot{\phi}^* \qquad \psi = \psi_0 + \psi^* \qquad (11.5.88)$$

Then by substituting these expressions into Equations (11.5.72), (11.5.73), and (11.5.74) and by neglecting products of small terms, we see that these disturbances are governed by the equations:

$$(4g/r)\theta^* - 5\ddot{\theta}^* + 6\dot{\psi}^*\dot{\phi}_0 + 5\dot{\phi}_0^2\theta^* = 0 \qquad (11.5.89)$$

$$\ddot{\phi}^* = 0 \qquad (11.5.90)$$

$$3\ddot{\psi}^* + 5\dot{\phi}_0\dot{\theta}^* = 0 \qquad (11.5.91)$$

Equation (11.5.90) shows that a small disturbance $\dot{\phi}^*$ in the spin rate remains small — essentially constant. That is,

$$\dot{\phi}^* = \dot{\phi}_0^* \quad \text{(a constant)} \qquad (11.5.92)$$

Next, observe that Equations (11.5.89) and (11.5.91) are independent of ϕ^* and that Equation (11.5.91) may be integrated enabling the elimination of $\dot{\psi}^*$. That is, by integrating Equation (11.5.91) we obtain:

$$3\dot{\psi}^* + 5\dot{\phi}_0\theta^* = C^* \qquad \text{(a constant)} \tag{11.5.93}$$

Then by substituting into Equation (11.5.89) and rearranging terms we have:

$$\ddot{\theta}^* + \left[\dot{\phi}_0^2 - (4g/5r)\right]\theta^* = 2\dot{\phi}_0 c^*/5 \tag{11.5.94}$$

As with the straight line rolling case [see Equation (11.5.83)], the solution for θ^* is sinusoidal (and hence, bounded) if the coefficient of θ^* is positive. Otherwise thedisturbance will grow exponentially (or linearly, if the coefficient of θ^* is zero). Therefore, the spinning disk is stable if

$$\dot{\phi}_0^2 > 4g/5r \quad \text{or} \quad \dot{\phi}_0 = \sqrt{4g/5r} \tag{11.5.95}$$

11.5.6 Elementary Solution: Disk Rolling in a Circle

A third special case is that of a disk D rolling at constant speed on a circle as in Example 6.12.3 and as depicted in Figure 11.5.5. In this case the lean angle θ is a constant, say θ_0; the roll rate $\dot{\psi}$ is a constant, say $\dot{\psi}_0$; and the turning rate $\dot{\phi}$ is a constant, say $\dot{\phi}_0$. Then Equations (11.5.73) and (11.5.74) are identically satisfied and Equation (11.5.72) becomes:

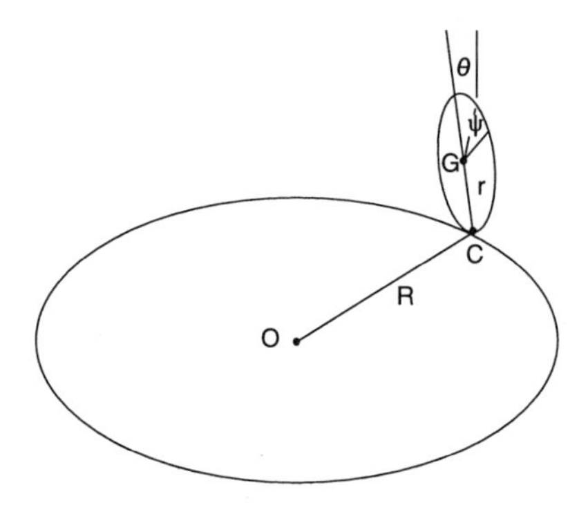

Fig. 11.5.5 A Disk Rolling in a Circle

$$(4g/r)\sin\theta_0 + 6\dot{\psi}_0\dot{\phi}_0\cos\theta_0 + 5\dot{\phi}_0^2\sin\theta_0\cos\theta_0 = 0 \qquad (11.5.96)$$

Equation (11.5.96) is the governing equation of motion for the circular moving disk. The equation provides a relationship between the three angular parameters: θ_0, ϕ_0, and ψ_0.

We can obtain additional relations between these parameters by a further examination of the geometry of the disk and the circular track. Specifically, from Tables 6.12.1, 6.12.2, 6.12.3, and 6.12.4 we have the kinematic relations:

$$\mathbf{V}^G = r(\dot{\psi}_0 + \dot{\phi}_0\sin\theta_0)\mathbf{n}_1 \qquad (11.5.97)$$

$$\mathbf{a}^G = r(\dot{\psi}_0\dot{\phi}_0\cos\theta_0 + \dot{\phi}_0^2\sin\theta_0\cos\theta_0)\mathbf{n}_2 + r(-\dot{\psi}_0\dot{\phi}_0\sin\theta_0 - \dot{\phi}_0^2\sin^2\theta_0)\mathbf{n}_3 \qquad (11.5.98)$$

$$\boldsymbol{\omega}^D = (\dot{\psi}_0 + \dot{\phi}_0\sin\theta_0)\mathbf{n}_2 + \dot{\phi}_0\cos\theta_0\mathbf{n}_3 \qquad (11.5.99)$$

$$\boldsymbol{\alpha}^D = -\dot{\psi}_0\dot{\phi}_0\cos\theta_0\mathbf{n}_1 \qquad (11.5.100)$$

Next, consider an edge view of the disk as in Figure 11.5.6 where R is the radius of the circular track and ρ is the radius of the circle on which the disk center G moves. Then R and ρ are related by the expression

$$R = \rho + r\sin\theta_0 \qquad (11.5.101)$$

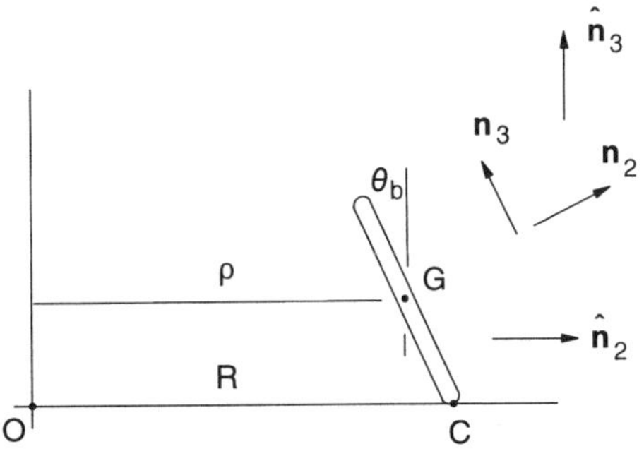

Fig. 11.5.6 Edge View of Rolling Disk

Let D roll clockwise on the circle (when viewed from above). Then by inspection of Figures 11.5.1 and 11.5.5 we have:

$$\dot{\psi}_0 > 0 \qquad \theta_0 > 0 \qquad \dot{\phi}_0 < 0 \tag{11.5.102}$$

But since G moves on a circle with radius ρ we see by inspection of Figure 11.5.4 that $\mathbf{v}^G$ and $\mathbf{a}^G$ are:

$$\mathbf{v}^G = -\rho\dot{\phi}_0\hat{\mathbf{n}}_1 \quad \text{and} \quad \mathbf{a}^G = -v^2/\rho\,\hat{\mathbf{n}}_2 \tag{11.5.103}$$

where v is $|\mathbf{v}^G|$. From Equation (11.5.97) v^2 is

$$v^2 = r^2(\dot{\psi}_0 + \dot{\phi}_0\sin\theta_0)^2 \tag{11.5.104}$$

Then by comparison of Equations (11.5.97), (11.5.101), and (11.5.104) we see that ρ and R are

$$\rho = -r(\dot{\psi}_0 + \dot{\phi}_0\sin\theta_0)/\dot{\phi}_0 \tag{11.5.105}$$

and

$$R = -r\dot{\psi}_0/\dot{\phi}_0 \tag{11.5.106}$$

Finally, by using Table 6.12.1 to express $\mathbf{n}_2$ and $\mathbf{n}_3$ in terms of $\hat{\mathbf{n}}_2$ and $\hat{\mathbf{n}}_3$ we can express $\mathbf{a}^G$ of Equation (11.5.98) as:

$$\mathbf{a}^G = r(\dot{\psi}_0\dot{\phi}_0 + \dot{\phi}_0^2\sin\theta_0)\hat{\mathbf{n}}_2 \tag{11.5.107}$$

Then an inspection of Equations (11.5.104) and (11.5.105) shows that Equations (11.5.104) and (11.5.107) are consistent.

11.6 Disk Striking and Rolling Over a Ledge

Consider a circular disk D with radius r, rolling vertically and in a straight line and encountering a ledge, or step, of height h as depicted in Figure 11.6.1. The objective is to determine the angular speed ω of D, or alternatively the speed v of the mass center G of D, necessary for D to roll over the step.

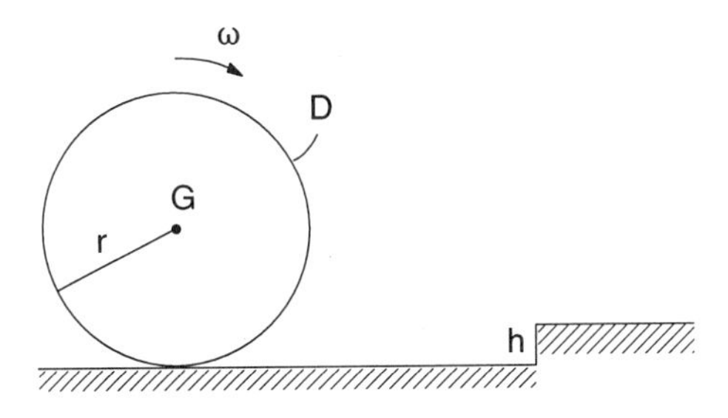

Fig. 11.6.1 A Disk Rolling onto a Step or Ledge

This problem is readily solved using the principles of impulse-momentum (Section 10.11) and work-energy (Section 10.12). Specifically, let the disk D have an angular speed ω at the instant when it encounters the corner O of the step as represented in Figure 11.6.2. Let D have mass m and axial radius of gyration k.

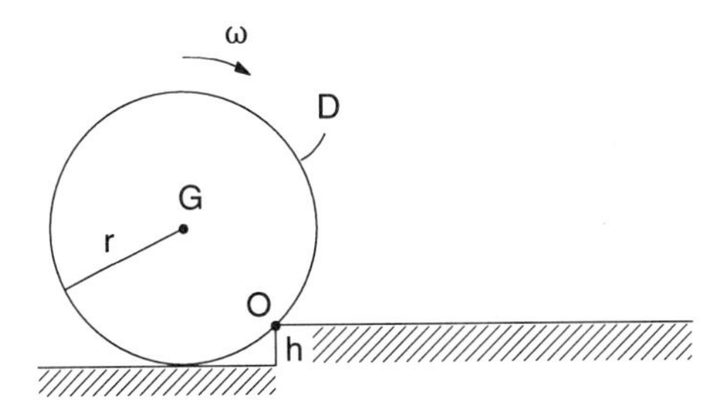

Fig. 11.6.2 Rolling Disk D Encountering Step Corner O

When D encounters O it will begin to rotate about O and if D has sufficient kinetic energy it will rotate up to the top of the ledge as represented in Figure 11.6.3.

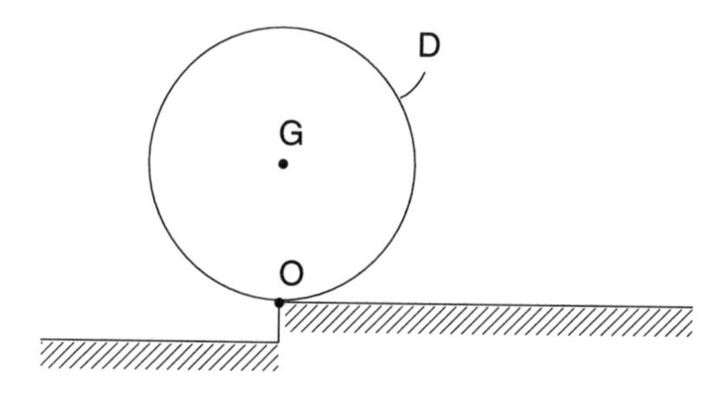

Fig. 11.6.3 Disk D Rolling over a Ledge

When D strikes the ledge at O the angular momentum of D about O, A_O is conserved. Just before the impact A_O is

$$A_{Ob} = mk^2\omega + mv(r - h) \tag{11.6.1}$$

where v is the speed of the mass center G of D just before impact. Since D is rolling v is:

$$v = r\omega \tag{11.6.2}$$

Just after impact D is rotating about O and A_O is then

$$A_{Oa} = mk^2\omega_a + mr^2\omega_a \tag{11.6.3}$$

where ω_a is the angular speed of D just after impact. By equating A_{Oa} and A_{Ob} we have

$$mk^2\omega + mr\omega(r - h) = mk^2\omega_a + mr^2\omega_a \tag{11.6.4}$$

where we have used Equation (11.6.2) to eliminate v. Then by solving for ω_a we obtain

$$\omega_a = \omega(k^2 + r^2 - rh)/(k^2 + r^2) \tag{11.6.5}$$

Next, from the work-energy principle, the work W required to elevate D to the top of the ledge is

$$W = mgh \tag{11.6.6}$$

The kinetic energy K_a of D just after impact is:

$$K_a = (1/2)mv_a^2 + (1/2)I_G\omega_a^2 \tag{11.6.7}$$

where v_a is the speed of mass center G just after impact and I_G is the axial moment of inertia of D (mk^2). Since D rotates about O v_a is simply $r\omega_a$. Hence, K_a is

$$K_a = (1/2)mr^2\omega_a^2 + (1/2)mk^2\omega_a^2 = (1/2)m(r^2 + b^2)\omega_a^2 \tag{11.6.8}$$

If D comes to rest as it reaches the top of the ledge, its kinetic energy of the top of the ledge is zero. Then from Equation (10.12.5), W is K_a, leading to:

$$mgh = (1/2)m(r^2 + k^2)\omega_a^2 \tag{11.6.9}$$

or

$$\omega_a = [2gh/(r^2 + k^2)]^{1/2} \tag{11.6.10}$$

Finally, by eliminating ω_a between Equations (11.6.5) and (11.6.10) the angular speed ω necessary for D to roll over the ledge is

$$\omega = \sqrt{2gh(r^2 + k^2)}/(r^2 - rh + k^2) \tag{11.6.11}$$

The necessary mass center velocity v to roll over the ledge is then

$$v = r\omega = r\sqrt{2gh(r^2 + k^2)}/(r^2 - rh + k^2) \tag{11.6.12}$$

If D is a thin uniform circular disk the radius of gyration k is $r/\sqrt{2}$. Then ω and v become:

$$\omega = \sqrt{3gh}/[(3r/2) - h] \tag{11.6.13}$$

and

$$v = \sqrt{3gh}/[(3/2) - (h/r)] \tag{11.6.14}$$

11.7 Summary of Results for a Thin Rolling Circular Disk

For ready reference we present here a summary of the principal results of the two previous sections. The notation is the same as that of Figure 11.5.1 and as shown in Figure 11.7.1.

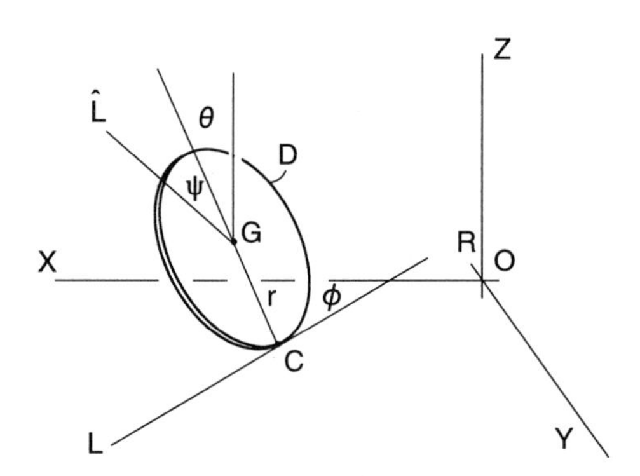

Fig. 11.7.1 A Rolling Circular Disk

11.7.1 Governing Equations

The governing differential equations are [see Equations (11.5.22), (11.5.23), and (11.5.24)]:

$$(4g/r)s_\theta - 5\ddot{\theta} + 6\dot{\psi}\dot{\phi}c_\theta + 5\dot{\phi}^2 s_\theta c_\theta = 0 \tag{11.7.1}$$

$$3\ddot{\psi} + 3\ddot{\phi}s_\theta + 5\dot{\phi}\dot{\theta}c_\theta = 0 \tag{11.7.2}$$

$$\ddot{\phi}c_\theta + 2\dot{\psi}\dot{\theta} = 0 \tag{11.7.3}$$

11.7.2 Stability of Straight Line Rolling

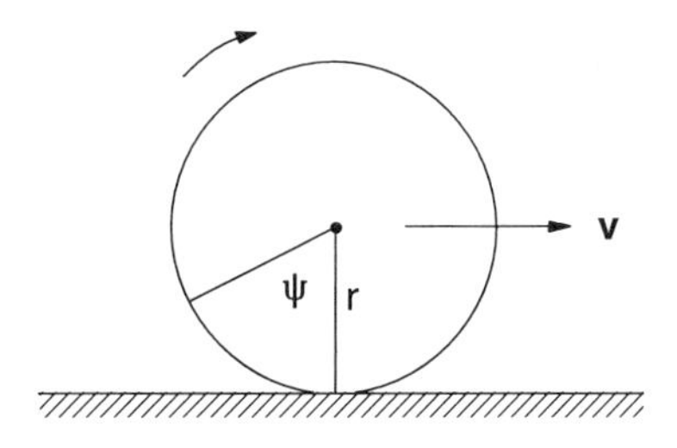

Fig. 11.7.2 A Straight Line Rolling Circular Disk

See Figure 11.7.2. From Equation (11.5.85) the motion is stable providing the center speed and angular speed are or exceed:

$$v \geq \sqrt{rg/3} \quad \text{or} \quad \dot{\psi} \geq \sqrt{g/3r} \tag{11.7.4}$$

11.7.3 Stability of Pivoting or Spinning

See Figure 11.7.3. From Equation (11.5.95) the disk spinning is stable for spinning speeds greater than or equal to:

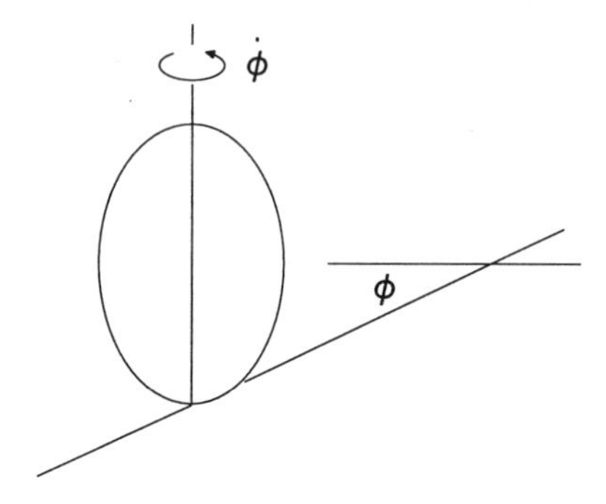

Fig. 11.7.3 A Pivoting (Spinning) Vertical Disk

$$\dot{\phi} \geq \sqrt{4g/5r} \tag{11.7.5}$$

11.7.4 Disk Rolling in a Circle

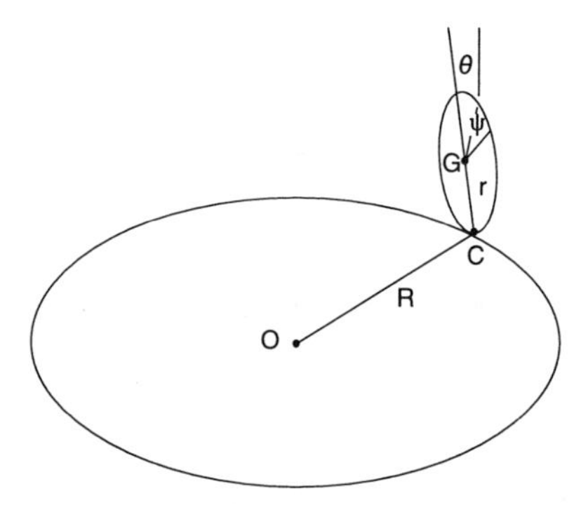

Fig. 11.7.4 A Disk Rolling in a Circle

See Figure 11.7.4. From Equation (11.5.96) the governing differential equation is:

$$(4g/r)\sin\theta + 6\dot{\psi}\dot{\phi}\cos\theta + 5\dot{\phi}^2\sin\theta\cos\theta = 0 \tag{11.7.6}$$

11.7.5 Disk Rolling Over a Ledge or Step

See Figure 11.7.5. From Equations (11.6.13) and (11.6.14), the disk will roll over the ledge if its center speed v and angular speed $\dot{\psi}$ equal or exceed:

$$v = \sqrt{3gh}/[(3/2) - (h/r)] \quad \text{and} \quad \dot{\psi} = \sqrt{3gh}/[(3r/2) - h] \tag{11.7.7}$$

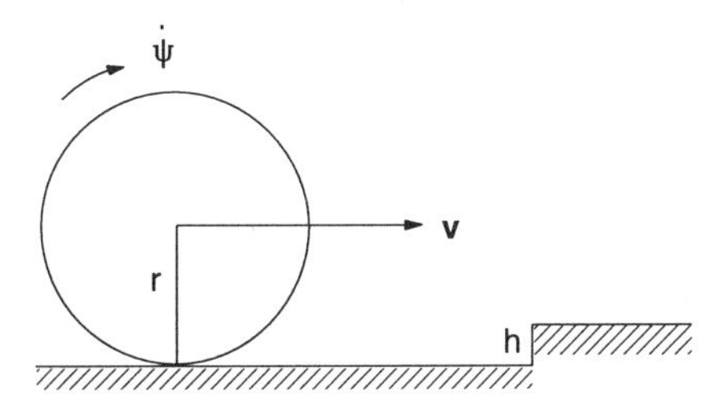

Fig. 11.7.5 Disk Rolling Over a Step

11.8 A Cone Rolling on an Inclined Plane [11.7]

Consider a right circular cone C rolling on an inclined plane Π as depicted in Figure 11.8.1. Let C have altitude h, element length ℓ, base radius r, and half central angle α as shown. Let Π be inclined at an angle β. Finally, let the angle that the contacting element of C makes with a fixed edge or fixed line of Π be ϕ as shown in Figure 11.8.1.

If C is released from rest in a non-equilibrium configuration on Π (that is $\phi \neq 0$), then C will oscillate, rolling back and forth on Π about its apex O. The angle ϕ will satisfy the pendulum equation:

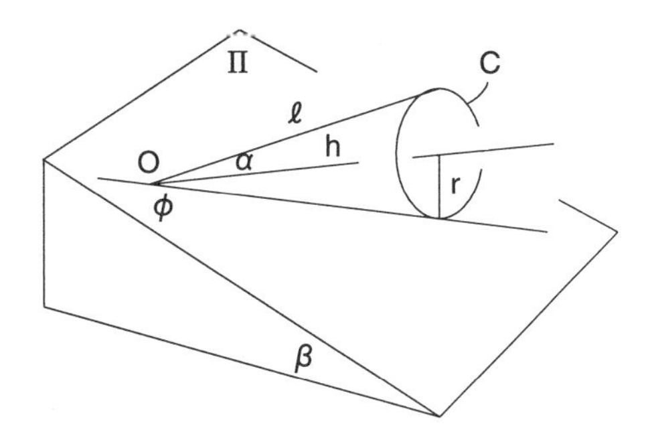

Fig. 11.8.1 A Cone Rolling in an Inclined Plane

$$\ddot{\phi} + k^2 \sin\phi = 0 \tag{11.8.1}$$

The objective is to determine k^2 in terms of the physical and geometrical parameters.

Analysis and Solution. To assist in describing the geometry it is helpful to introduce several perpendicular unit vector sets as in Figure 11.8.2. Let the $\mathbf{N}_i$ and $\hat{\mathbf{N}}_i$ $(i = 1,2,3)$ be fixed relative to the plane Π with $\mathbf{N}_3$ being vertical, $\hat{\mathbf{N}}_3$ being normal to Π, and $\mathbf{N}_1 = \hat{\mathbf{N}}_1$. Let the $\hat{\mathbf{n}}_i$ be able to rotate relative to Π with $\hat{\mathbf{n}}_3$ being normal to Π and $\hat{\mathbf{n}}_2$ parallel to the contact element of C with Π. Finally, let the $\mathbf{n}_i$ also be able to rotate relative to Π with $\mathbf{n}_2$ parallel to the axis of C and $\mathbf{n}_1 = \hat{\mathbf{n}}_1$. [Observe that the $\mathbf{n}_i$ $(i = 1,2,3)$ are principal unit vectors for C (see Section 8.10).]

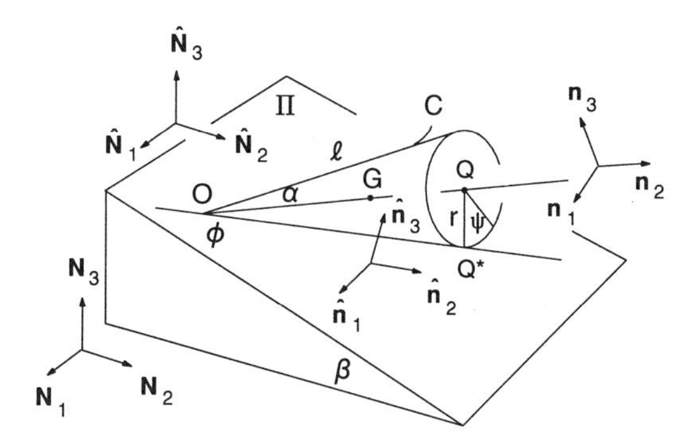

Fig. 11.8.2 Cone/Plane Unit Vector Sets and Geometrical Parameters

The unit vector sets are conveniently related to each other by the configuration graph of Figure 11.8.3 where R is the inertial reference frame in which Π is fixed, L is a reference frame containing the $\hat{\mathbf{n}}_i$, and C^* is a reference frame containing the $\mathbf{n}_i$.

In addition to the unit vector sets it is also convenient to introduce the roll angle ψ at the base of C (See Figure 11.8.2), and the points G, Q and Q^* representing the mass center of C, the base center of C, and the contact point between C and Π at the base of C.

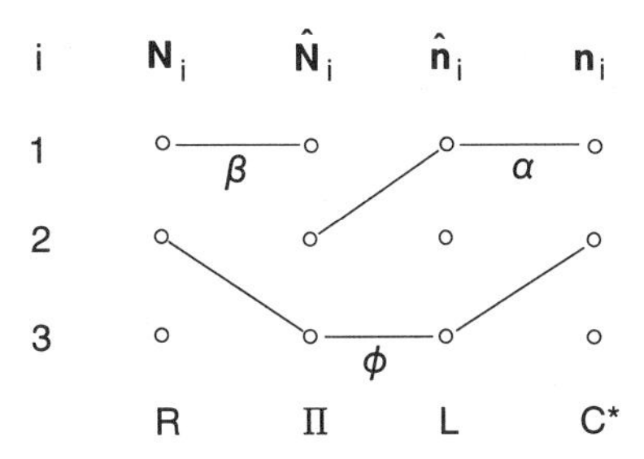

Fig. 11.8.3 Configuration Graph for the Unit Vector Sets of Figure 11.8.2

The angular velocity of C in R may be readily obtained from the configuration graph of Figure 11.8.3 and the addition theorem for angular velocities [Equation (6.7.5)]. That is,

$$\begin{aligned} {}^R\boldsymbol{\omega}^C &= {}^{C^*}\boldsymbol{\omega}^C + {}^L\boldsymbol{\omega}^{C^*} + {}^\Pi\boldsymbol{\omega}^L + {}^R\boldsymbol{\omega}^\Pi \\ &= \dot{\psi}\,\mathbf{n}_2 + \dot{\alpha}\,\mathbf{n}_1 + \dot{\phi}\,\hat{\mathbf{n}}_3 + \dot{\beta}\,\hat{\mathbf{N}}_1 \end{aligned} \tag{11.8.2}$$

Observe that α and β are constants, so that $\dot{\alpha}$ and $\dot{\beta}$ vanish, and then ${}^R\boldsymbol{\omega}^C$ becomes

$${}^R\boldsymbol{\omega}^C = \dot{\psi}\,\mathbf{n}_2 + \dot{\phi}\,\hat{\mathbf{n}}_3 = (\dot{\psi} + \dot{\phi}\,s_\alpha)\mathbf{n}_2 + \dot{\phi}\,c_\alpha\,\mathbf{n}_3 \tag{11.8.3}$$

where as before s and c are abbreviations for sine and cosine.

The system has but one degree of freedom, which may be characterized by the angle ϕ. The rolling criterion may be used to express $\dot{\psi}$ in terms of $\dot{\phi}$: Since C rolls on the contact element both O and Q^* are contact points. Then the velocity of Q can be obtained from two expressions which may be equated. Specifically from Equation (6.12.2) we have

$${}^R\mathbf{v}^Q = {}^R\boldsymbol{\omega}^C \times \mathbf{OQ} \quad \text{and} \quad {}^R\mathbf{v}^Q = {}^Q\boldsymbol{\omega}^C \times \mathbf{Q^*Q} \tag{11.8.4}$$

Observe that **OQ** may be expressed as $h\mathbf{n}_2$ and that $\mathbf{Q^*Q}$ is $r\mathbf{n}_3$. Then by substituting into Equation (11.8.4) and by using Equation (11.8.3) we have

$$\dot{\psi} = -\dot{\phi}/s\alpha \quad (11.8.5)$$

Next, observe from Figure 11.8.2 that the distance from the apex O to the mass center G is (3/4)h or $(3/4)\ell c_\alpha$. Then the velocity of G in R is

$$^R\mathbf{v}^G = {}^R\boldsymbol{\omega}^C \times 3/4h\mathbf{n}_2 = -(3/4)\ell\dot{\phi}c_\alpha^2\mathbf{n}_1 \quad (11.8.6)$$

By routine differentiation in Equations (11.8.3) and (11.8.6) the angular acceleration of C in R and the mass center acceleration of G in R are:

$$^R\boldsymbol{\alpha}^C = -\dot{\psi}\dot{\phi}c_\alpha\mathbf{n}_1 + (\ddot{\psi} + \ddot{\phi}s_\alpha)\mathbf{n}_2 + \ddot{\phi}c_\alpha\mathbf{n}_3 \quad (11.8.7)$$

and

$$^R\mathbf{a}^G = -(3/4)\ell c_\alpha^2\ddot{\phi}\mathbf{n}_1 - (3/4)\ell c_\alpha^3\mathbf{n}_2 + (3/4)\ell\dot{\phi}^2 s_\alpha c_\alpha^2\mathbf{n}_3 \quad (11.8.8)$$

[Recall that the derivatives of the unit vectors $\mathbf{n}_i$ are simply $^R\boldsymbol{\omega}^{C^*} \times \mathbf{n}_i = (\dot{\phi}s_\alpha\mathbf{n}_2 + \dot{\phi}c_\alpha\mathbf{n}_3) \times \mathbf{n}_i \quad (i = 1,2,3)$.]

Consider a free-body diagram of C, as in Figure 11.8.4. The applied forces on C consist of the gravity (or weight) forces and the contact forces exerted across the rolling, contact element of C. Let the gravity forces be represented by the force **W** passing through mass center G. Then **W** may be expressed as:

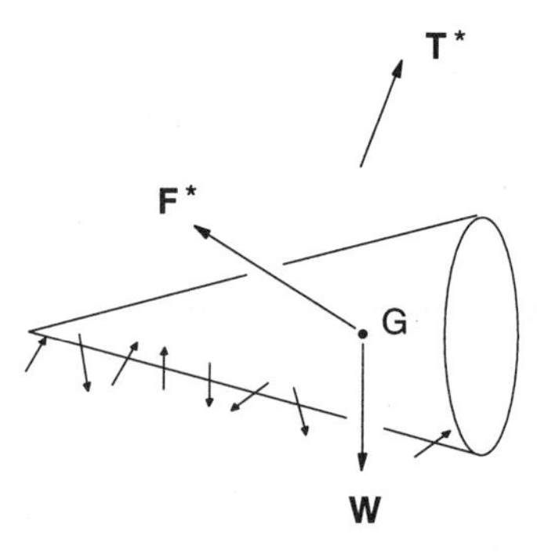

Fig. 11.8.4 Free-Body Diagram of Rolling Cone

$$\mathbf{W} = -mg\mathbf{N}_3 \tag{11.8.9}$$

The contact forces are represented by the arrows along the contact element shown in Figure 11.8.4.

Let the inertia forces on C be represented by a force $\mathbf{F}^*$ passing through G together with a couple with torque $\mathbf{T}^*$ where $\mathbf{F}^*$ and $\mathbf{T}^*$ are [Equations (9.12.6) and (9.12.11)].

$$\mathbf{F}^* = -m\,{}^R\mathbf{a}^G \quad \text{and} \quad \mathbf{T}^* = -\mathbf{I}\cdot{}^R\boldsymbol{\alpha}^C - {}^R\boldsymbol{\omega}^C \times (\mathbf{I}\cdot{}^R\boldsymbol{\omega}^C) \tag{11.8.10}$$

where $\mathbf{I}$ is the central inertia dyadic of C. $\mathbf{I}$ may be expressed in terms of the principal unit vectors $\mathbf{n}_i$ as:

$$\mathbf{I} = I_{11}\mathbf{n}_1\mathbf{n}_1 + I_{22}\mathbf{n}_2\mathbf{n}_2 + I_{33}\mathbf{n}_3\mathbf{n}_3 \tag{11.8.11}$$

where I_{11}, I_{22}, and I_{33} are the central principal moments of inertia of C given by (see Table 8.13.1):

$$I_{11} = I_{33} = 3(4r^2 + h^2)/80 \qquad I_{22} = 3r^2/10 \tag{11.8.12}$$

Let $\mathbf{T}^*$ be written as

$$\mathbf{T}^* = T_1\mathbf{n}_1 + T_2\mathbf{n}_2 + T_3\mathbf{n}_3 \tag{11.8.13}$$

Then T_1, T_2, and T_3 are

$$\begin{aligned} T_1 &= -\alpha_1 I_{11} + \omega_2\omega_3(I_{22} - I_{33}) \\ T_2 &= -\alpha_2 I_{22} + \omega_3\omega_1(I_{33} - I_{11}) \\ T_3 &= -\alpha_3 I_{33} + \omega_1\omega_2(I_{11} - I_{22}) \end{aligned} \tag{11.8.14}$$

where the α_i and ω_i $(i = 1,2,3)$ are the $\mathbf{n}_i$ components of ${}^R\boldsymbol{\alpha}^C$ and ${}^R\boldsymbol{\omega}^C$.

The contact forces on C may be eliminated from the analysis by setting moments about the contacting element equal to zero. Specifically,

$$[\mathbf{OG} \times \mathbf{W} + \mathbf{OG} \times \mathbf{F}^* + \mathbf{T}^*] \bullet \hat{\mathbf{n}}_2 = 0 \tag{11.8.15}$$

Finally, by substituting from Equations (11.8.3), (11.8.5), and (11.8.7) through (11.8.14) into (11.8.15) and simplifying, we obtain the desired governing equation of motion as

$$\ddot{\phi} + k^2 s_\phi = 0 \tag{11.8.16}$$

where k^2 is:

$$k^2 = (g/\ell) s_\beta / [(1/5) + c_\alpha^2] \tag{11.8.17}$$

11.9 A Spinning Rigid Projectile

Consider a rigid body B projected into the air and tumbling as it moves along its trajectory. Neglecting air resistance, the only forces exerted on B are the weight forces and the inertia forces. A free-body diagram of B is shown in Figure 11.9.1 where **W** represents the weight forces and the inertia forces are represented by the force $\mathbf{F}^*$ passing through the mass center G together with a couple with torque $\mathbf{T}^*$. Then **W** and **F** may be expressed as:

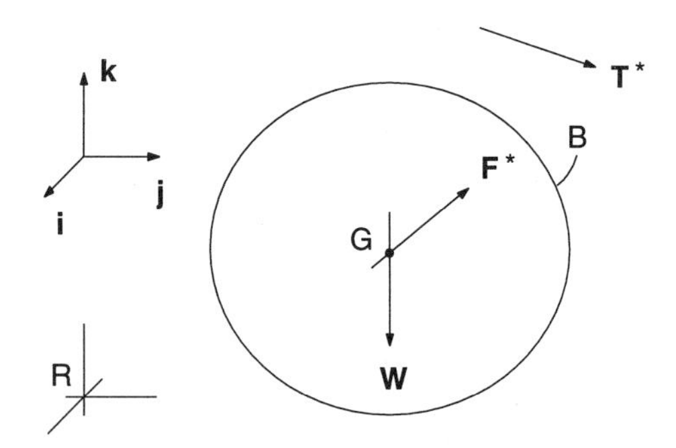

Fig. 11.9.1 Free-Body Diagram of a Projected, Tumbling Body

$$\mathbf{W} = -mg\mathbf{k} \tag{11.9.1}$$

and

$$\mathbf{F}^* = -m\mathbf{a}^G \tag{11.9.2}$$

where $\mathbf{k}$ is a vertical unit vector, m is the mass of B, g is the gravity acceleration and $\mathbf{a}^G$ is the acceleration of G in an inertial frame R [Equation (9.12.6)]. The inertia torque $\mathbf{T}^*$ may be expressed as [Equation (9.12.11)]:

$$\mathbf{T}^* = -\mathbf{I} \cdot \boldsymbol{\alpha} - \boldsymbol{\omega} \times (\mathbf{I} \cdot \boldsymbol{\omega}) \tag{11.9.3}$$

where $\mathbf{I}$ is the central inertia dyadic of B [Section 8.8] and $\boldsymbol{\alpha}$ and $\boldsymbol{\omega}$ are the angular acceleration and angular velocity of B in R.

Since the forces of Figure 11.9.1 form a zero system we have

$$\mathbf{F}^* + \mathbf{W} = 0 \tag{11.9.4}$$

And, by setting moments about G equal to zero, we have:

$$\mathbf{T}^* = 0 \tag{11.9.5}$$

In view of Equations (11.9.1) and (11.9.2), Equation (11.9.4) leads to:

$$\mathbf{a}^G = -g\mathbf{k} \tag{11.9.6}$$

Let $\mathbf{a}^G$ be expressed as:

$$\mathbf{a}^G = \ddot{x}\mathbf{i} + \ddot{y}\mathbf{j} + \ddot{z}\mathbf{k} \tag{11.9.7}$$

where (x,y,z) are the coordinates of G relative to a Cartesian frame in R and where **i** and **j** are horizontal unit vectors. Equation (11.9.6) then in turn leads to

$$\ddot{x} = 0$$
$$\ddot{y} = 0 \qquad (11.9.8)$$
$$\ddot{z} = -g$$

Equations (11.9.8) are the classical projectile equations (see Section 5.6). When integrated they show that G moves in a plane on a parabolic curve. Specifically, and without any significant loss in generality, if B is projected from the origin in the Y-Z plane, the equation of the parabolic curve may be written as:

$$z = (V_{0z}/V_{0y})y - (g/2V_{0y}^2)y^2 \qquad (11.9.9)$$

where V_{0y} and V_{0z} are the components of the initial velocity of G in the Y and Z directions.

Since a parabola is a "smooth" curve, the movement of G will be smooth. That is, as B tumbles, it will tumble, or spin, about G. Then if P is a particle of B, the velocity of P may be expressed as [Equation (6.10.3)]:

$$\mathbf{V}^P = \mathbf{V}^G + \boldsymbol{\omega} \times \mathbf{GP} \qquad (11.9.10)$$

P may be regarded as moving on a circle whose center is on an axis parallel to $\boldsymbol{\omega}$ and passing through G.

If $\mathbf{n}_1$, $\mathbf{n}_2$, and $\mathbf{n}_3$ are principal unit vectors (see Section 8.10) fixed in B, Equation (11.9.5) may be expressed as:

$$T_1\mathbf{n}_1 + T_2\mathbf{n}_2 + T_3\mathbf{n}_3 = 0 \qquad (11.9.11)$$

where the T_i $(i = 1,2,3)$ are the $\mathbf{n}_i$ components of T^*. In view of Equation (11.9.3), the T_i may be expressed as:

$$T_1 = -\dot{\omega}_1 I_{11} + \omega_2 \omega_3 (I_{22} - I_{33}) \tag{11.9.12}$$

$$T_2 = -\dot{\omega}_2 I_{22} + \omega_3 \omega_1 (I_{33} - I_{11}) \tag{11.9.13}$$

$$T_3 = -\dot{\omega}_3 I_{33} + \omega_1 \omega_2 (I_{11} - I_{22}) \tag{11.9.14}$$

where the ω_i $(i = 1,2,3)$ are the $\mathbf{n}_i$ components of $\boldsymbol{\omega}$ and where I_{11}, I_{22}, and I_{33} are the central principal moments of inertia of B. [Observe that since the $\mathbf{M}_i$ are fixed in B the angular acceleration components are simply the derivatives of the angular velocity components. (See Section 6.9.)]

From Equation (11.9.5) we have

$$T_1 = T_2 = T_3 = 0 \tag{11.9.15}$$

or

$$\dot{\omega}_1 I_{11} = \omega_2 \omega_3 (I_{22} - I_{33}) \tag{11.9.16}$$

$$\dot{\omega}_2 I_{22} = \omega_3 \omega_1 (I_{33} - I_{11}) \tag{11.9.17}$$

$$\dot{\omega}_3 I_{33} = \omega_1 \omega_2 (I_{11} - I_{22}) \tag{11.9.18}$$

Equations (11.9.16), (11.9.17), and (11.9.18) form a system three non-linear first-order ordinary differential equations for the ω_i $(i = 1,2,3)$. Although there are no general solutions, we can readily consider two special cases:

CASE 1. NO ROTATION

By inspection of Equations (11.9.16), (11.9.17), and (11.9.18) we see that

$$\omega_1 = \omega_2 = \omega_3 = 0 \tag{11.9.19}$$

in a solution. This means that a body may be projected into the air without rotation.

CASE 2. ROTATION ABOUT A PRINCIPAL AXIS

Also, by inspection of Equations (11.9.16), (11.9.17), and (11.9.18) we see that a solution is

$$\omega_1 = \Omega \quad \text{and} \quad \omega_2 = \omega_3 = 0 \tag{11.9.20}$$

where Ω is a constant. This means that a body may be projected into the air, rotating about one of its principal axes of inertia.

A question arising from Equation (11.9.20) is: What is the stability of this solution? To answer this question, let us introduce a small disturbance to the motion. Specifically, let

$$\omega_1 = \Omega + \omega_1^* \qquad \omega_2 = 0 + \omega_2^* \qquad \omega_3 = 0 + \omega_3^* \tag{11.9.21}$$

where the "starred" $[(\ )^*]$ quantities are small. By substituting into Equations (11.9.16), (11.9.17), and (11.9.18) we have

$$\dot{\omega}_1^* = 0 \tag{11.9.22}$$

$$\dot{\omega}_2^* I_{22} = \omega_3^* \Omega (I_{33} - I_{11}) \tag{11.9.23}$$

$$\dot{\omega}_3^* I_{33} = \Omega \omega_2^* (I_{11} - I_{22}) \tag{11.9.24}$$

where we have neglected products of starred, or small, quantities. Equations (11.9.22), (11.9.23), and (11.9.24) are linear, first-order, ordinary differential equations for ω_1^*, ω_2^*, and ω_3^*. The first of these is readily solved for ω_1^*:

$$\omega_1^* = \omega_{10}^* \qquad \text{a constant} \tag{11.9.25}$$

That is, a small disturbance, increasing or reducing the rotation speed, remains small.

Equations (11.9.23) and (11.9.24) may be solved for ω_2^* and ω_3^* by decoupling the expressions. Specifically, if Equation (11.9.23) is solved for ω_3^* and then is substituted into Equation (11.9.23) we have

$$\ddot{\omega}_2^* + \Omega^2 \left[\frac{(I_{11} - I_{22})(I_{11} - I_{33})}{I_{22} I_{33}} \right] \omega_2^* = 0 \tag{11.9.26}$$

The solution for ω_2^* will be sinusoidal (that is, bounded) if the coefficient of ω_2^* is positive. Since $\Omega^2 > 0$ and each of the moments of inertia is positive, we have a bounded (or "stable") solution if

$$I_{11} > I_{22} \quad \text{and} \quad I_{11} > I_{33}$$

or $$\tag{11.9.27}$$

$$I_{11} < I_{22} \quad \text{and} \quad I_{11} < I_{33}$$

Alternatively, the solution is unstable if

$$I_{22} > I_{11} > I_{33} \quad \text{or} \quad I_{33} > I_{11} > I_{22} \tag{11.9.28}$$

Equations (11.9.27) and (11.9.28) show that a body projected into the air and rotating about a principal inertia axis has stable motion if the axis is associated with either a maximum or minimum moment of inertia. The motion is unstable for rotation about an intermediate inertia axis.

11.10 Law of Gyroscopes

The expression for inertia torques such as in Equation (9.12.11) can be used to obtain an intuitive understanding of the behavior of gyroscopes. To this end, let the gyro of a gyroscope be modeled as a thin spinning disk D as in Figure 11.10.1.

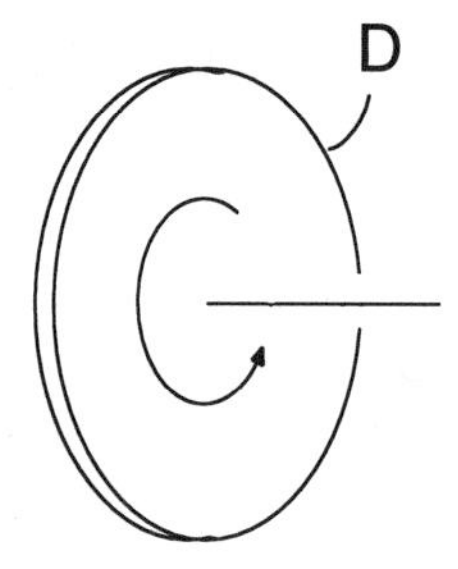

Fig. 11.10.1 A Spinning Disk Representing a Gyro

For the purposes of analysis let D be mounted in a light frame (or gimbal) F, as in Figure 11.10.2. Let the geometric center G be fixed but let F be free to have arbitrary rotation.

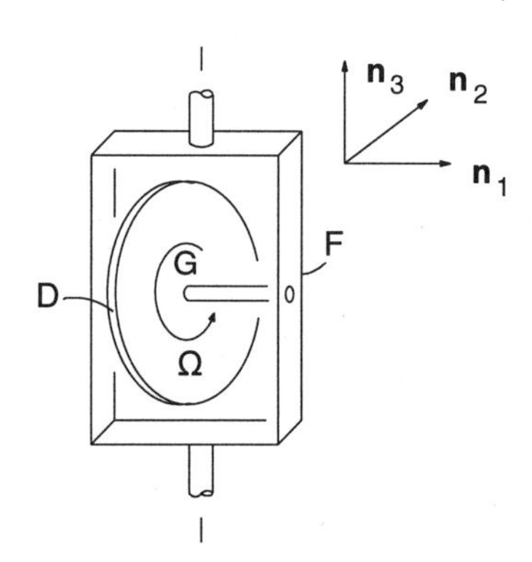

Fig. 11.10.2 A Gyro D Mounted in a Gimbal F

Let $\mathbf{n}_1$, $\mathbf{n}_2$, and $\mathbf{n}_3$ be mutually perpendicular unit vectors fixed in F as in Figure 11.10.2. Then due to the circular symmetry of D, the $\mathbf{n}_i$ are principal unit vectors for D.

Let D rotate with a constant angular speed Ω as indicated in Figure 11.10.2. Then in the configuration shown in Figure 11.10.2 the angular velocity of D in an inertia frame R is simply

$$\boldsymbol{\omega}^D = \Omega \mathbf{n}_1 \tag{11.10.1}$$

Next, let the frame F be given a small constant rotation rate ω^* (small compared with Ω) about $\mathbf{n}_2$. Then the angular velocity of F is

$$\boldsymbol{\omega}^F = \omega^* \mathbf{n}_2 \tag{11.10.2}$$

Consequently, the angular velocity of D then becomes:

$$\boldsymbol{\omega}^D = \Omega \mathbf{n}_1 + \boldsymbol{\omega}^* \mathbf{n}_2 \tag{11.10.3}$$

Observe that under the above conditions, the angular velocity of D will no longer be constant even though Ω and ω^* are constants. Indeed, in view of Equation (11.10.2), $\mathbf{n}_1$ will change orientation. That is, the small rotation of F will create an angular acceleration for D. Specifically, the angular acceleration of D will be

$$\begin{aligned} \boldsymbol{\alpha}^D &= d\boldsymbol{\omega}_D/dt = \Omega d\mathbf{n}_1/dt \\ &= \Omega \boldsymbol{\omega}^F \times \mathbf{n}_1 = -\Omega \omega^* \mathbf{n}_3 \end{aligned} \tag{11.10.4}$$

The inertia forces on D in this relatively simple configuration may be represented by a force $\mathbf{F}^*$ passing through G together with a couple with torque $\mathbf{T}^*$ given by [Equation (5.2.2) and (9.12.11)]:

$$\mathbf{F}^* = -m\mathbf{a}^G \quad \text{and} \quad \mathbf{T}^* = -\mathbf{I}\cdot\boldsymbol{\alpha}^D - \boldsymbol{\omega}^D \times (\mathbf{I}\cdot\boldsymbol{\omega}^D) \tag{11.10.5}$$

where, as before, **I** is the central inertia dyadic.

Since G is fixed, $\mathbf{a}^G$ is zero, and therefore, $\mathbf{F}^*$ is zero.

In terms of the principal unit vectors $\mathbf{n}_i$ $(i = 1,2,3)$, $\mathbf{T}^*$ may be expressed as:

$$\mathbf{T}^* = T_1\mathbf{n}_1 + T_2\mathbf{n}_2 + T_3\mathbf{n}_3 \tag{11.10.6}$$

where the T_i are:

$$T_1 = -\alpha_1 I_{11} + \omega_2\omega_3(I_{22} - I_{33}) \tag{11.10.7}$$

$$T_2 = -\alpha_2 I_{22} + \omega_3\omega_1(I_{33} - I_{11}) \tag{11.10.8}$$

$$T_3 = -\alpha_3 I_{33} + \omega_1\omega_2(I_{11} - I_{22}) \tag{11.10.9}$$

where the α_i and ω_i are the $\mathbf{n}_i$ components of $\boldsymbol{\alpha}^D$ and $\boldsymbol{\omega}^D$ and where I_{11}, I_{22}, and I_{33} are the principal central moments of inertia of D. Recall that for a thin disk (see Table 8.13.1) I_{11}, I_{22}, and I_{33} are:

$$I_{11} = (1/2)mr^2 \quad \text{and} \quad I_{22} = I_{33} = (1/4)mr^2 \tag{11.10.10}$$

By substituting from Equations (11.10.3), (11.10.4), and (11.10.10) into Equations (11.10.7), (11.10.8), and (11.10.9) the inertia torque components are:

$$T_1 = T_2 = 0 \tag{11.10.11}$$

and

$$T_3 = (1/2)mr^2\Omega\omega^* \tag{11.10.12}$$

These results show that with the gyro rotating about the $\mathbf{n}_1$ axis and then subjected to a rotation about the $\mathbf{n}_2$ axis, the gyro responds by rotating about the $\mathbf{n}_3$ axis. This illustrates a principle often referred to as the "law of gyroscopes":

> A spinning gyro will tend to align its axis of rotation with an imposed axis of rotation.

Finally, observe that this law of gyroscopes is valid even for bodies which are not thin disks. For example, if D had been a sphere with $I_{11} = I_{22} = I_{33} = (2/5)mV^2$, then Equations (11.10.11) and (11.10.12) would have been:

$$T_1 = T_2 = 0 \tag{11.10.13}$$

$$T_3 = (2/5)mV^2\Omega\omega^* \tag{11.10.14}$$

11.11 A Translating Rod Striking a Ledge

Consider a rod AB with mass m and length ℓ moving in translation with speed v toward a fixed ledge, in a gravity-free environment, as represented in Figure 11.11.1a. Let the rod strike the ledge with its end A as in Figure 11.11.1b. Upon impact the rod will begin to rotate about A with an angular speed ω. The objective is to determine the relationship between v and ω.

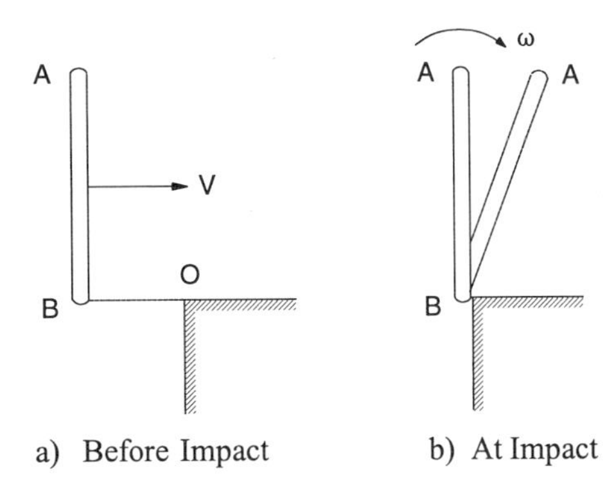

Fig. 11.11.1 A Translating Rod Striking a Ledge at its End

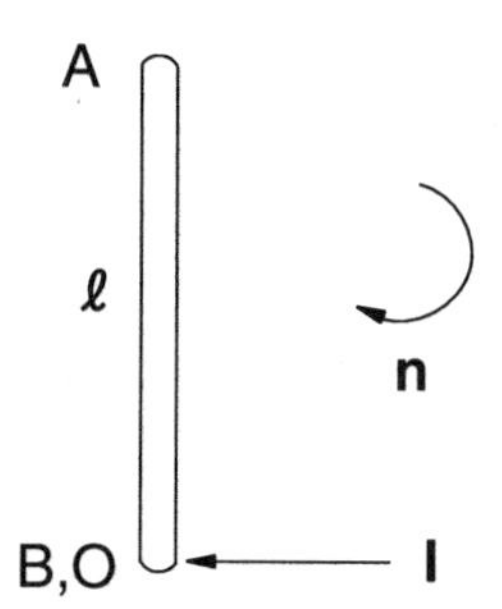

Fig. 11.11.2 Free-Body Diagram of the Rod at Impact with the Ledge

Consider a free-body diagram of the rod as in Figure 11.11.2. In a gravity-free environment, the only forces acting on the rod are the impulsive forces from theledge, which are represented by the impulse **I**. **I** may be eliminated from the analysis by observing that the angular momentum about B (or the ledge corner O) before and after impact, is conserved (see Section 10.2): Specifically, before impact the angular momentum about O is [see Equation (10.6.8)]:

$$\mathbf{A}_O \Big|_{\text{before}} = (mv\ell/2)\,\mathbf{n} \tag{11.11.1}$$

where **n** is a unit vector perpendicular to the plane of motion as in Figure 11.11.1.

After impact, the angular momentum of the rod about O is [see Equation (10.6.10)]:

$$\mathbf{A}_O \Big|_{\text{after}} = I_O \omega\,\mathbf{n} \tag{11.11.2}$$

where I_O is the moment of inertia of the rod about end B $[(1/3)m\ell^2]$.

By equating the expressions in Equations (11.11.1) and (11.11.2) we have

$$mV\ell/2 = m\ell^2\omega/3$$

or

$$\omega = 3v/2\ell \tag{11.11.3}$$

Observe that with the rod temporarily rotating about end B with angular speed ω, the velocity of end A is

$$V^A = \ell\omega = (3/2)v \tag{11.11.4}$$

That is, end A has a 50 percent increase in speed upon impact. Finally, observe that this result is independent of the rod mass m.

11.12 Pinned Double Rods Striking a Ledge in Translation

As a generalization of the foregoing consider two identical rods B_1 and B_2, connected by a frictionless pin, and moving in translation with speed v toward a fixed ledge as in Figure 11.12.1. Let the rods each have mass m and length ℓ, and let the weight forces be neglected. Let the rods be called B_1 and B_2 with ends A,

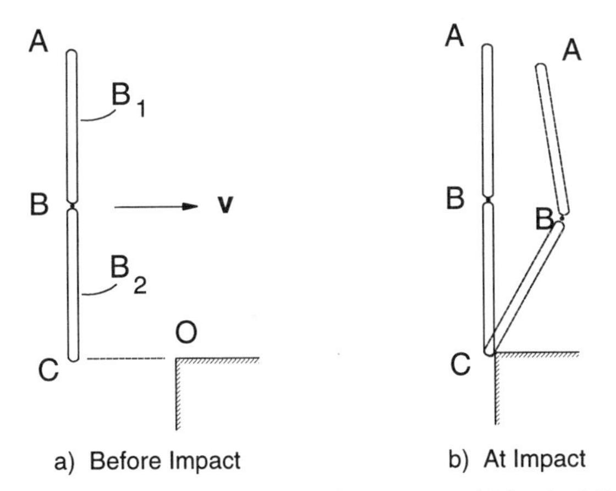

Fig. 11.12.1 Translating, Pin-Connected Identical Rods Striking a Ledge

B, and C as in Figure 11.12.1. The objective is to determine the angular speeds of the rods upon impact as a function of the rod length and initial speed v.

Consider free-body diagrams of the two rods and of the upper rod, B, as in Figures 11.12.2 and 11.12.3. Let G_1 and G_2 be the mass centers of the rods. Let $\mathbf{I}_O$ be the impulse exerted by the ledge at corner O on the lower end C of B_2 and let $\mathbf{I}_B$ be the impulse exerted at the pin on the lower end B of B_1. In a gravity-free environment these are the only forces applied to the rods.

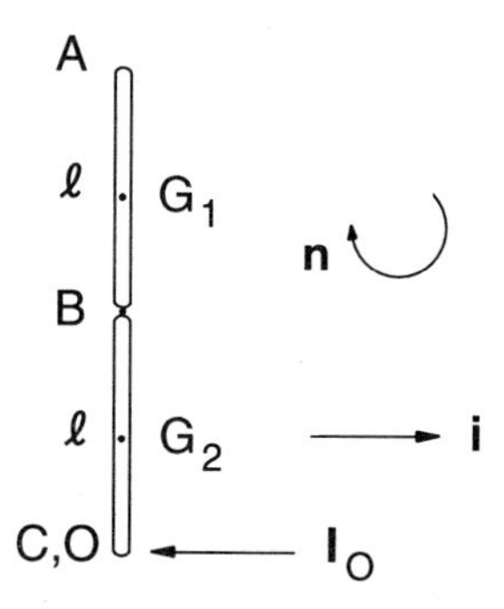

Fig. 11.12.2 Free-Body Diagram of the Rods at Impact with the Ledge

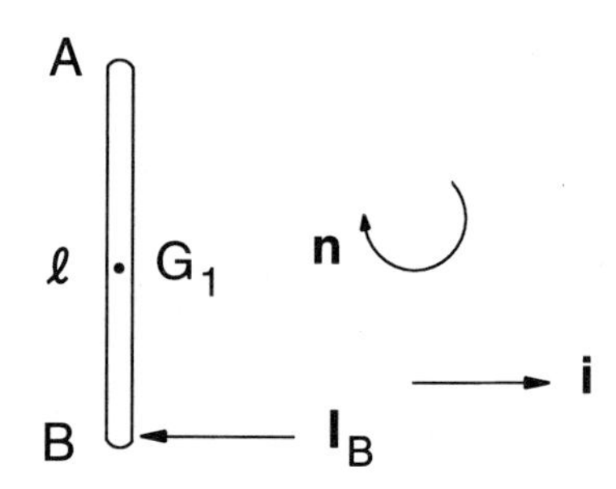

Fig. 11.12.3 Free-Body Diagram of the Upper Rod Upon Impact with the Ledge

Just prior to impact, with the rods moving in translation, their angular velocities are zero and their mass center velocities are simply:

$$\mathbf{v}^{G_1} = \mathbf{v}^{G_2} = v\mathbf{i} \tag{11.12.1}$$

where **i** is a horizontal unit vector. Just after impact the rods will begin to rotate as represented in Figure 11.12.1b. Let the resulting angular velocities and mass center velocities be expressed as:

$$\boldsymbol{\omega}^{B_1} = \omega_1 \mathbf{n} \quad \text{and} \quad \boldsymbol{\omega}^{B_2} = \omega_2 \mathbf{n} \tag{11.12.2}$$

and

$$\mathbf{v}^{G_1} = V^{G_1}\mathbf{i} \quad \text{and} \quad \mathbf{v}^{G_1} = V^{G_2}\mathbf{i} \tag{11.12.3}$$

where **n** is a unit vector normal to the plane of rotation as shown in Figures 11.12.2 and 11.12.3, and where just after impact the mass center velocities are assumed to be nearly horizontal. Then an elementary kinematic analysis shows that

$$V^{G_1} = (\ell/2)\,\omega_1 + \ell\,\omega_2 \quad \text{and} \quad V^{G_2} = (\ell/2)\,\omega_2 \tag{11.12.4}$$

From the free-body diagrams and the linear and angular momentum principles we have:

Horizontal momenta (Figure 11.12.3):

$$-I_B = mV^{G_1} - mv \tag{11.12.5}$$

Moments about G_1 (Figure 11.2.3):

$$I_B(\ell/2) = m(\ell^2/12)\,\omega_1 \tag{11.12.6}$$

and

Moments about O (Figure 11.2.2):

$$mv(3\ell/2) + mv(\ell/2) = mv^{G_1}(3\ell/2) + m(\ell^2/12)\,\omega_1 + mv^{G_2}(\ell/2) + m(\ell^2/12)\,\omega_2 \qquad (11.12.7)$$

where I_B is the magnitude of the impulse $\mathbf{I}_B$ at the pin.

By eliminating I_B between Equations (11.12.5) and (11.12.6) we have

$$mv(\ell/2) = mv^{G_1}(\ell/2) + m(\ell^2/12)\,\omega_1 \qquad (11.12.8)$$

Next, by using the kinematic relations of Equation (11.12.4) to eliminate v^{G_1} and v^{G_2} from Equations (11.12.7) and (11.12.8) we obtain (after simplification):

$$2\omega_1 + 3\omega_2 = 3v/\ell \qquad (11.12.9)$$

and

$$5\omega_1 + 11\omega_2 = 12v/\ell \qquad (11.12.10)$$

Then solving for ω_1 and ω_2 we have:

$$\omega_1 = -(3/7)(v/\ell) \qquad (11.12.11)$$

$$\omega_2 = (9/7)(v/\ell) \qquad (11.12.12)$$

Observe from these results that the angular speed ratio ω_1/ω_2 is:

$$\omega_1/\omega_2 = -1/3 \qquad (11.12.13)$$

Observe further that upon impact the lower rod rotates forward and the upper rod rotates backward, but at a third of the lower rod rotation rate.

Consider a generalization of this problem with bars of unequal lengths and unequal masses as in Figure 11.12.4.

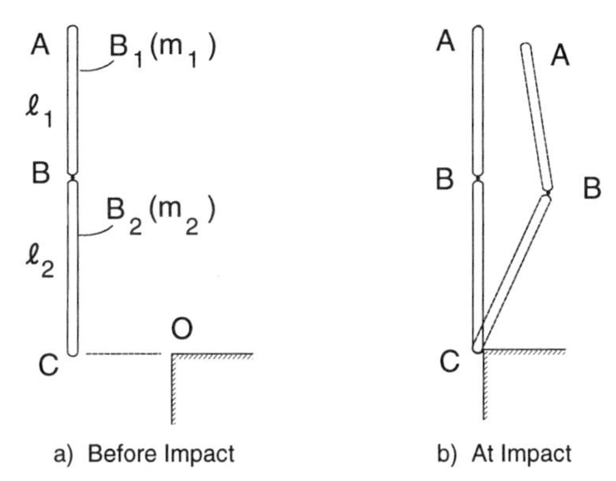

Fig. 11.12.4 Translating, Pin-Connected Rods Striking a Ledge

By following the same procedure as with the identical rods, we find that the equations governing the post impact angular speeds, analogous to Equations (11.12.9) and (11.12.10), are:

$$(2/3)\ell_1\omega_1 + \ell_2\omega_2 = v \tag{11.12.14}$$

and

$$(m_1\ell_1\ell_2/2)\,\omega_1 + (m_1\ell_2^2 + m_2\ell_2^2/3)\,\omega_2 = (m_1\ell_2 + m_2\ell_1/2)\,v \tag{11.12.15}$$

The solutions for ω_1 and ω_2 are:

$$\omega_1 = (v/\Delta)[(m_1\ell_2^2 + m_2\ell_2^2/3) - (m_1\ell_2^2 + m_2\ell_2^2/2)] \tag{11.12.16}$$

$$\omega_2 = (v/\Delta)[(2/3)\ell_1(m_1\ell_2 + m_2\ell_2/2) - m_1\ell_1\ell_2/2] \qquad (11.12.17)$$

where Δ is

$$\Delta = (2\ell_1/3)(m_1\ell_2^2 + m_2\ell_2^2/3) - m_1\ell_1\ell_2^2/2 \qquad (11.12.18)$$

11.13 A Plate Striking a Ledge at a Corner of the Plate [11.8]

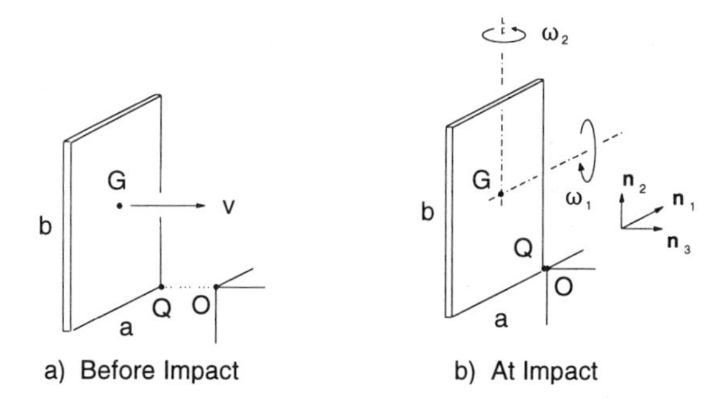

Fig. 11.13.1 A Translating Plate Striking a Fixed Corner at a Corner of the Plate

Consider a thin plate whose sides have lengths a and b moving in translation with speed V toward a fixed corner as depicted in Figure 11.13.1. Let the plate strike the fixed corner O at a corner Q of the plate as shown. Upon impact, the plate will begin to rotate about horizontal and vertical axes in the plane of the plate with angular speeds ω_1 and ω_2 as depicted in Figure 11.13.1b. The objective is to determine ω_1 and ω_2 in terms of a, b, and V.

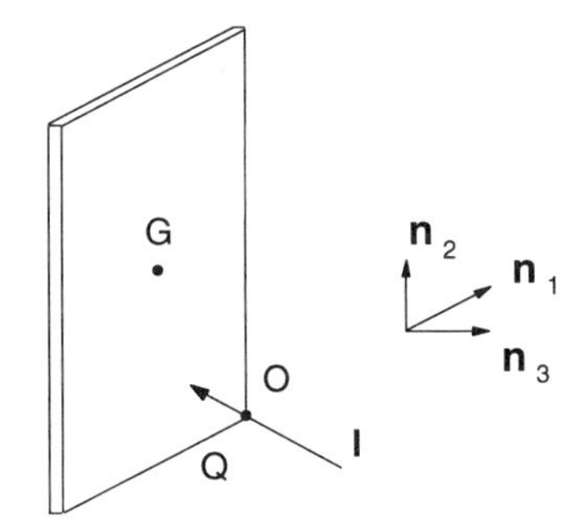

Fig. 11.13.2 Free-Body Diagram of Plate

Consider a free-body diagram of the plate as shown in Figure 11.13.2 where **I** represents the impulse imparted to the plate as it strikes the corner. At the instant of impact, the angular momentum of the plate about O is conserved, that is,

$$\mathbf{A}_O\big|_{\text{before}} = \mathbf{A}_O\big|_{\text{after}} \tag{11.13.1}$$

Just before impact the plate is moving in translation. Therefore, its angular momentum about O before impact is simply:

$$\mathbf{A}_O\big|_{\text{before}} = \mathbf{OG} \times m v \mathbf{n}_3 \tag{11.13.2}$$

where m is the mass of the plate. From Figures 11.13.1 and 11.13.2 we see that **OG** is

$$\mathbf{OG} = -(a/2)\mathbf{n}_1 + (b/2)\mathbf{n}_2 \tag{11.13.3}$$

Then from Equations (11.13.2) $\mathbf{A}_O\big|_{\text{before}}$ becomes:

$$\mathbf{A}_O\big|_{\text{before}} = m v b/2\mathbf{n}_1 + m v a/2\mathbf{n}_2 \tag{11.13.4}$$

Just after impact, the plate is momentarily rotating about corner O. Then from Equation (10.6.10) the angular momentum just after impact is:

$$\mathbf{A}_O\big|_{\text{after}} = \mathbf{I}^Q \bullet \boldsymbol{\omega} \tag{11.13.5}$$

where $\mathbf{I}^Q$ is the inertia dyadic of the plate relative to its corner Q. From the parallel axes theorem [see Equation (8.9.7), $\mathbf{I}^Q$ may be expressed as:

$$\mathbf{I}^Q = \mathbf{I}^G + \mathbf{I}^{G/Q} \tag{11.13.6}$$

where, in matrix form, the components of $\mathbf{I}^G$ and $\mathbf{I}^{G/Q}$ relative to $\mathbf{n}_1$, $\mathbf{n}_2$, and $\mathbf{n}_3$ are:

$$\mathbf{I}^G : \begin{bmatrix} mb^2/12 & 0 & 0 \\ 0 & ma^2/12 & 0 \\ 0 & 0 & m(a^2+b^2)/12 \end{bmatrix} \tag{11.13.7}$$

and

$$\mathbf{I}^{G/Q} : \begin{bmatrix} mb^2/4 & mab/4 & 0 \\ mab/4 & ma^2/4 & 0 \\ 0 & 0 & m(a^2+b^2)/4 \end{bmatrix} \tag{11.13.8}$$

Then from Equation (11.13.6) the matrix form components of $\mathbf{I}^Q$ relative to $\mathbf{n}_1$, $\mathbf{n}_2$, and $\mathbf{n}_3$ are

$$\mathbf{I}^Q : \begin{bmatrix} mb^2/3 & mab/4 & 0 \\ mab/4 & ma^2/3 & 0 \\ 0 & 0 & m(a^2+b^2)/3 \end{bmatrix} \tag{11.13.9}$$

From Equations (11.13.5) and (11.13.9) the post impact angular momentum is

$$\mathbf{A}_O\big|_{\text{after}} = [(mb^2/3)\omega_1 + (mab/4)\omega_2]\mathbf{n}_1 + [(mab/4)\omega_1 + (ma^2/3)\omega_2]\mathbf{n}_2 + [m(a^2+b^2)/3]\omega_3\mathbf{n}_3 \tag{11.13.10}$$

Then by substituting from Equations (11.13.4) and (11.13.10) into Equation (11.13.1) the governing equations for ω_1, ω_2, and ω_3 are found to be:

$$(mb^2/3)\omega_1 + (mab/4)\omega_2 = mvb/2 \tag{11.13.11}$$

$$(mab/4)\omega_1 + (ma^2/3)\omega_2 = mva/2 \tag{11.13.12}$$

$$[m(a^2+b^2)/3]\omega_3 = 0 \tag{11.13.13}$$

Solving for ω_1, ω_2, and ω_3 we have

$$\omega_1 = 6v/7b \tag{11.13.14}$$

$$\omega_2 = 6v/7a \tag{11.13.15}$$

$$\omega_3 = 0 \tag{11.13.16}$$

Comment: Observe that if a is small, as with a tall narrow plate, ω_1 is unchanged whereas ω_2 increases as (1/a). In the limit as a goes to zero, and the plate simulates a rod, ω_1 remains at 6v/7b whereas ω_2 becomes infinite. From the results of Section 11.11, however, for a rod striking a ledge, we see that the post-impact rotational speed is [Equation (11.11.3)]: $3v/2\ell$. This discrepancy can be resolved by further observing that if a is zero Equation (11.13.12) is identically satisfied and then Equation (11.13.11) yields $\omega_1 = 3v/2b$, consistent with Equation (11.11.3).

References

11.1 H. Josephs and R. L. Huston, *Dynamics of Mechanical Systems*, CRC Press, Boca Raton, FL, 2001

11.2 R. L. Huston, *Multibody Dynamics*, Butterworth-Heinemann, Stoneham, MA, 1990.

11.3 T. R. Kane, *Analytical Elements of Mechanics*, Vol. 2, Academic Press, New York, 1961.

11.4 T. R. Kane, *Dynamics*, Holt, Rinehard and Winston, Inc., New York, 1968.

11.5 T. R. Kane and D. A. Levinson, *Dynamics: Theory and Applications*, McGraw Hill, New York, 1985.

11.6 C. E. Passerello and R. L. Huston, "Another Look at Nonholonomic Systems," *Journal of Applied Mechanics*, Vol. 40, 1973, pp. 101-104.

11.7 E. T. Whittaker, *A Treatise on the Analytical Dynamics of Particles and Rigid Bodies*, Fourth Edition, Cambridge, London, 1937, pp. 133, 134.

11.8 F. P. Beer and E. R. Johnston, *Vector Mechanics for Engineers*, Fifth Edition, McGraw Hill, New York, 1988, p. 904.

Chapter 12

MULTIBODY SYSTEMS

12.1 Introduction

Among the most difficult areas in dynamics is the study of multibody systems. A "multibody system" is simply a collection of bodies interacting with one another either through connection joints or by forces exerted between the bodies. In recent years dynamic analyses of multibody systems has been aided by availability of high-speed digital computers and by advances in analytical/ computational procedures. These advances and procedures are analogous to similar advances and procedures in finite element analysis several decades ago.

Prior to the development of multibody analyses were restricted to single body systems or to relatively simple mechanisms containing only a few bodies. With modern multibody analysis there is, at least in principle, no limit to the number of bodies which may be considered nor to the complexity of the system.

In this and in the following two chapters, we summarize the formulas and procedures for multibody dynamics analyses. This chapter provides a description of multibody systems themselves as well as some procedures useful in the sequel. The next chapter focuses upon kinematics of multibody systems. The final chapter discusses kinetics and dynamics.

References 12.1 to 12.11 provide a detailed background for multibody dynamics analysis. References 12.1 and 12.2 provide a summary of pertinent literature up to 1998.

12.2 Types of Multibody Systems

As noted earlier a multibody system is simply a collection of bodies with a given connection configuration. The bodies may have various elastic, viscoelastic, and/or plastic properties. The connecting joints may be of various kinds ranging from simple pins (or "revolute" joints), to sliders ("prismatic" joints), to ball-and-socket (or "spherical" joints), or to general joints allowing for both rotation and translation. Alternatively, the bodies may be related by forces as with spring and damper connectors. Finally, the system may be "open," as a tree, or it may have closed loops. Figure 12.2.1 provides sketches of various multibody systems.

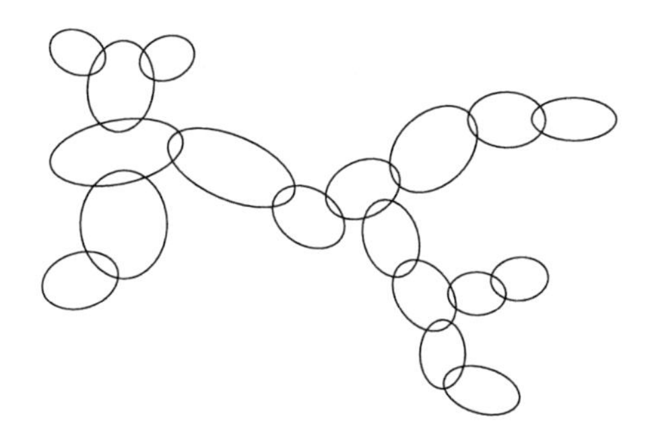

a) An Open-Tree Connected System

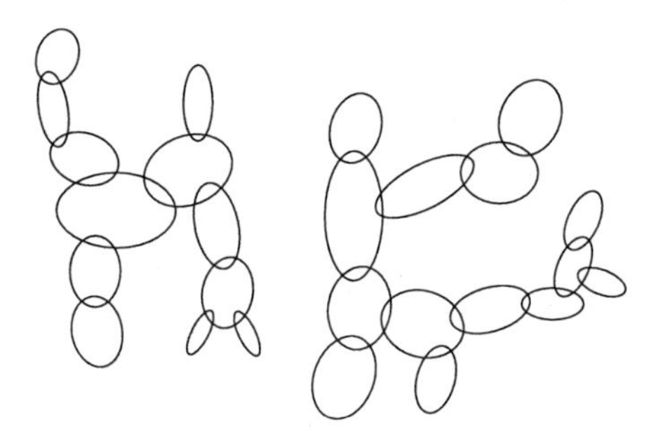

b) An Open, Disconnected System

Fig. 12.2.1 Multibody Systems

Fig. 12.2.1 Continued

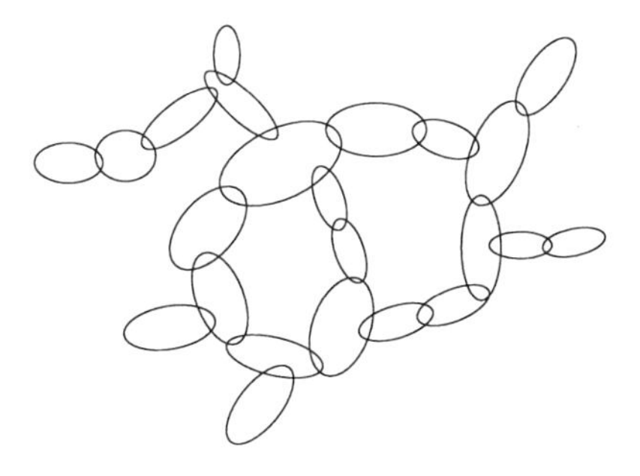

c) A System with Closed Loops

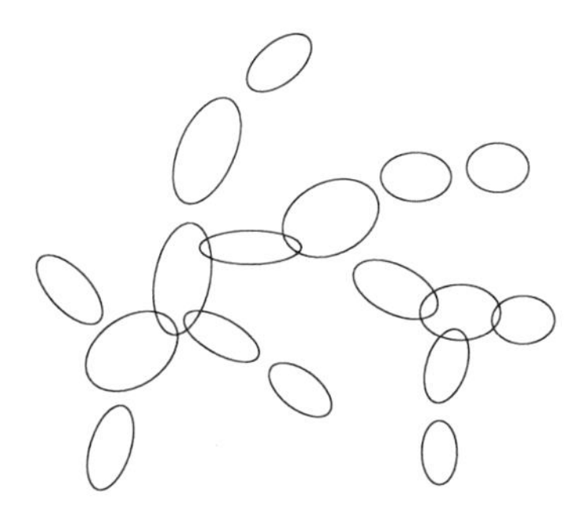

d) A System with Translation at Some of the Joints

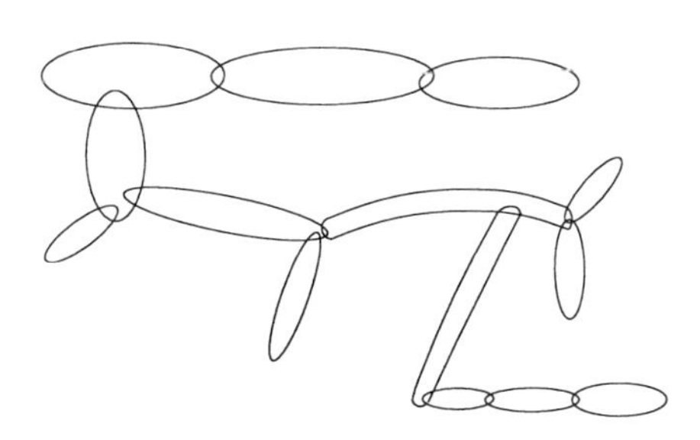

e) A System with Long, Flexible Bodies

Multibody systems are sometimes called "lumped parameter" systems, "chain" systems, or "finite-segment" systems.

Interest in multibody systems arises from the utility of multibody systems in modeling a wide range of physical systems, such as biosystems or human body models, as in Figure 12.2.2, mechanisms, cables or chains, as in Figure 12.2.3, and long, flexible booms, as in Figure 12.2.4.

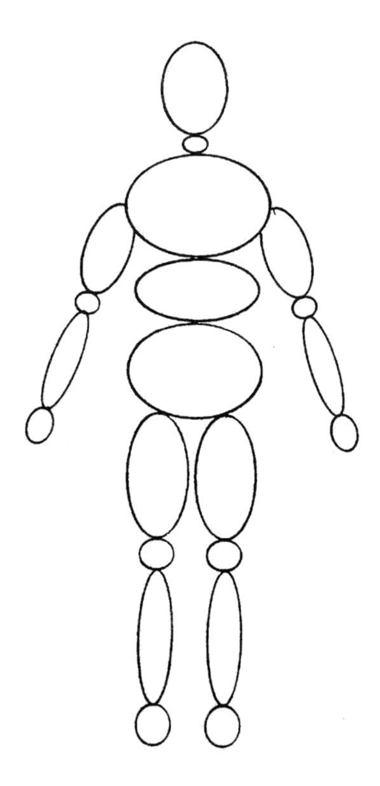

Fig. 12.2.2 A Human Body Model

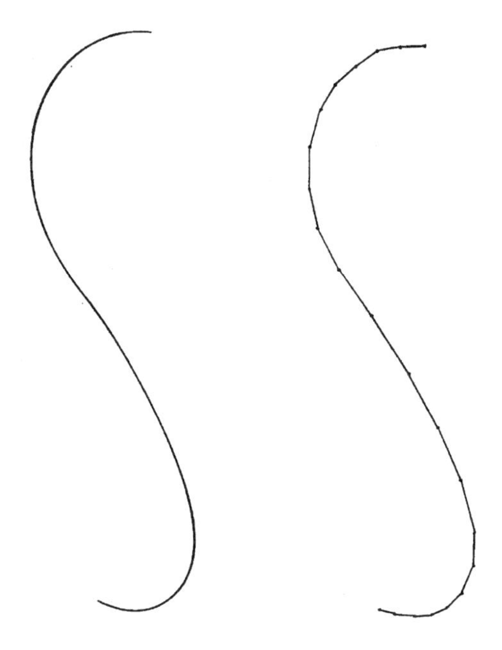

Fig. 12.2.3 A Chain Model of a Cable

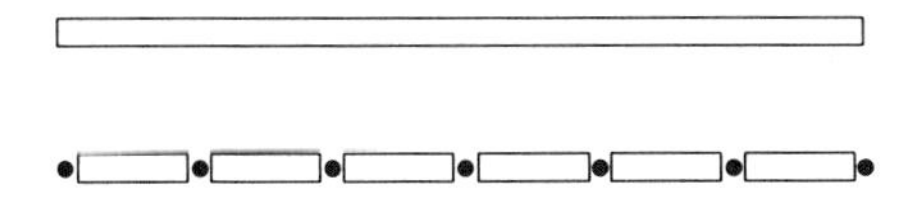

Fig. 12.2.4 A Beam/Boom Model

In the sequel for simplicity in presentation we will restrict our attention to open systems of connected rigid bodies as in Figure 12.2.5. Analysis of more complex systems can be made using the same procedures as presented herein. Reference 12.1 develops and discusses these procedures in detail.

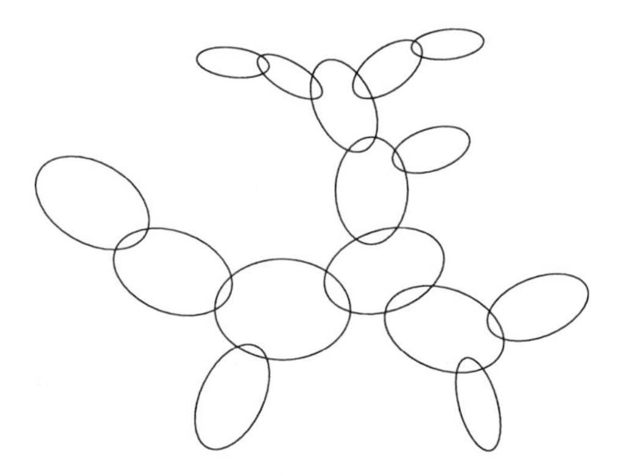

Fig. 12.2.5 An Open-Chain System of Connected Rigid Bodies

12.3 Lower Body Arrays

A principal difficulty in multibody dynamics analyses is describing the complex geometry. For large systems the task could be formidable, even for a system at rest. For a system in motion it is even more difficult. Lower body arrays provide a means for overcoming the difficulty.

A lower body array is simply a row of integers representing the bodies of the system, arranged so as to define the connection configuration. A simple example can illustrate the concept: Consider the multibody system of Figure 12.2.5. Let the bodies of the system be numbered or labeled as follows: Let the profile of the bodies be projected onto a plane. Next, select a body, any body, as a reference, or initial, body and label or number it as B_1 or 1. Then label and number the bodies in ascending progression away from B_1 in either a clockwise or counterclockwise manner through the branches of the tree system. Figure 12.3.1 shows a numbering of the bodies of the system of Figure 12.2.5.

Next, observe that whereas all bodies of the system have unique adjacent lower numbered bodies they do not all have unique adjacent higher numbered bodies. For example, bodies B_1, B_2, B_3, B_6, and B_8 have more than one adjacent higher numbered body and bodies B_4, B_5, B_7, B_{10}, B_{12}, B_{14}, and B_{15} do not have any adjacent higher numbered bodies. Bodies B_1, B_2, B_3, B_6, and B_8 might be called

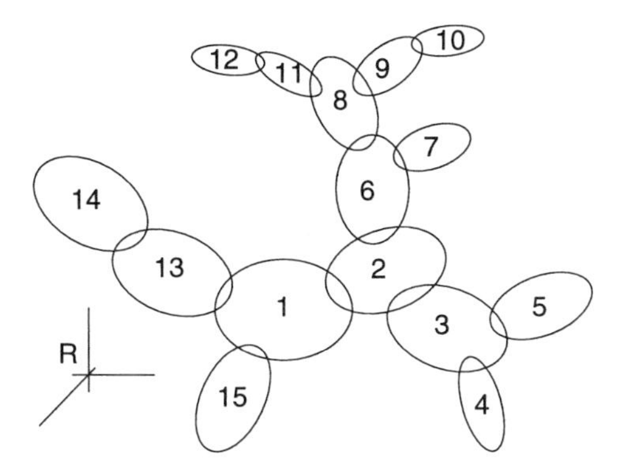

Fig. 12.3.1 A Numbering of the System of Figure 12.2.5

"branching bodies" and bodies B_4, B_5, B_7, B_{10}, B_{12}, B_{14}, and B_{15} might be called "extremity bodies."

Finally, observe in Figure 12.3.1 that some of the bodies, specifically B_9, B_{11}, and B_{13}, have one and only one adjacent higher numbered body. These bodies might be called "intermediate bodies."

Consider the row array (N) of the body numbers of the system of Figure 12.3.1 arranged in the natural sequence or the numbers:

$$(N) = (1,2,3,4,5,6,7,8,9,10,11,12,13,14,15) \tag{12.3.1}$$

Let L(N) be the row array of integers representing (in order) the adjacent lower numbered body for each body represented in the (N) array. For the system of Figure 12.3.1, L(N) is

$$L(N) = (0,1,2,3,3,2,6,6,8,9,8,11,1,13,1) \tag{12.3.2}$$

L(N) is called the "lower body array" or "connection configuration array."

Observe in the L(N) array of Equation (12.3.2) that some body numbers do not appear (specifically, 4, 5, 7, 10, 12, 14, and 15). These numbers correspond to the extremity bodies. Also, in the array, some body numbers appear more than once (specifically, 1, 2, 3, 6, and 8). These correspond to the branching bodies. Finally, some numbers appear once and only once (specifically, 0, 9, 11, and 13). These correspond to the inertial reference frame and the intermediate bodies.

The lower body array of Equation (12.3.2) is equivalent to the sketch of the multibody system in Figure 12.3.1. Equation (12.3.2) may be determined by inspection of Figure 12.3.1 and alternatively Figure 12.3.1 may be sketched by inspection of Equation (12.3.2).

A principal feature of lower body arrays is that they can be directly used in the construction of higher order lower body arrays which in turn can be useful in the development of the kinematics of multibody systems. To demonstrate and develop this consider that L(N) in Equation (12.3.2) may be regarded as an operator L on the array (N) of Equation (12.3.1). That is, for each integer K of (N) there is an integer J of L(N) such that

$$J = L(K) \tag{12.3.3}$$

Using this concept we can apply L to L(N) and form the lower body array of the lower body array — that is, L(L(N)) or $L^2(N)$. Specifically, from Equation (12.2.2) we have

$$L^2(N) = (0,0,1,2,2,1,2,2,6,8,6,8,0,1,0) \tag{12.3.4}$$

where L(0) is assigned the value 0. Observe that the integers of $L^2(N)$ are respectively smaller than (or in the case of 0, equal to) those of L(N). In like manner we can form $L^3(N)$ from $L^2(N)$, $L^4(N)$ from $L^3(N)$, and so on until all zeros occur. That is,

$$
\begin{aligned}
L^3(N) &= (0,0,0,1,1,0,1,1,2,6,2,6,0,0,0) \\
L^4(N) &= (0,0,0,0,0,0,0,0,1,2,1,2,0,0,0) \\
L^5(N) &= (0,0,0,0,0,0,0,0,0,1,0,1,0,0,0) \\
L^6(N) &= (0,0,0,0,0,0,0,0,0,0,0,0,0,0,0)
\end{aligned}
\tag{12.3.5}
$$

Table 12.3.1 summarizes the results of Equations (12.3.1) to (12.3.5), where the array (N) is represented by $L^0(N)$ and L(N) is represented by $L^1(N)$.

Table 12.3.1 Lower Body Arrays for the System of Figure 12.3.1

Body (N)	**1**	**2**	**3**	**4**	**5**	**6**	**7**	**8**	**9**	**10**	**11**	**12**	**13**	**14**	**15**
$L^0(N)$	1	2	3	4	5	6	7	8	9	10	11	12	13	14	15
$L^1(N)$	0	1	2	3	3	2	6	6	8	9	8	11	1	13	1
$L^2(N)$	0	0	1	2	2	1	2	2	6	8	6	8	0	1	0
$L^3(N)$	0	0	0	1	1	0	1	1	2	6	2	6	0	0	0
$L^4(N)$	0	0	0	0	0	0	0	0	1	2	1	2	0	0	0
$L^5(N)$	0	0	0	0	0	0	0	0	0	1	0	1	0	0	0
$L^6(N)$	0	0	0	0	0	0	0	0	0	0	0	0	0	0	0

Consider the columns of Table 12.3.1. Each column is associated with one of the bodies. For example, with body B_{12} we have the numbers: 12, 11, 8, 6, 2, 1, and 0. Observe from Figure 12.3.1 that these numbers correspond exactly with the body numbers of those bodies in the branches leading to B_{12}. That is, the column numbers identify the sequence of bodies connecting B_{12} to the inertial frame R.

The columns of Table 12.3.1 may be used to develop the system kinematics. For example, consider the angular velocity of B_{12} in R. The addition theorem for angular velocity [see Equation (6.7.5)] states that the angular velocity of B_{12} in R may be expressed as:

$$
{}^{R}\boldsymbol{\omega}^{B_{12}} = {}^{B_{11}}\boldsymbol{\omega}^{B_{12}} + {}^{B_8}\boldsymbol{\omega}^{B_{11}} + {}^{B_6}\boldsymbol{\omega}^{B_8} + {}^{B_2}\boldsymbol{\omega}^{B_6} + {}^{B_1}\boldsymbol{\omega}^{B_2} + {}^{R}\boldsymbol{\omega}^{B_1}
\tag{12.3.6}
$$

It is convenient to adopt the notation:

$$ {}^{R}\boldsymbol{\omega}^{B_k} = \boldsymbol{\omega}_k \quad \text{and} \quad {}^{B_j}\boldsymbol{\omega}^{B_k} = \hat{\boldsymbol{\omega}}_k \tag{12.3.7} $$

where B_j is the adjacent lower numbered body of B_k. Then Equation (12.3.6) may be written as:

$$ \boldsymbol{\omega}_{12} = \hat{\boldsymbol{\omega}}_{12} + \hat{\boldsymbol{\omega}}_{11} + \hat{\boldsymbol{\omega}}_{8} + \hat{\boldsymbol{\omega}}_{6} + \hat{\boldsymbol{\omega}}_{2} + \hat{\boldsymbol{\omega}}_{1} \tag{12.3.8} $$

Observe then that the indices on the right side are exactly the same as the column numbers for B_{12} in Table 12.3.1.

Finally, observe that numerical algorithms can be written to generate all of the entries in Table 12.3.1 once the array L(N) is known.

12.4 Orientation Angles and Transformation Matrices

In addition to the connection configuration, lower body arrays are useful in describing and defining the orientations of the bodies of a multibody system. To see this, consider two typical adjoining bodies B_j and B_k as in Figure 12.4.1. To define the relative orientation of the bodies, let $\mathbf{n}_{ji}$ and $\mathbf{n}_{ki}$ $(i = 1,2,3)$ be mutually perpendicular unit vectors fixed in B_j and B_k respectively. Then the orientation of B_k relative to B_j may be defined by a variety of orientation angles between the $\mathbf{n}_{ji}$ and the $\mathbf{n}_{ki}$ as described in Chapter 6 [see Table 6.4.1]. For example, if the unit vector sets are mutually aligned, then B_k may be brought into a general orientation relative to B_j by successive rotations about $\mathbf{n}_{k1}$, $\mathbf{n}_{k2}$, and $\mathbf{n}_{k3}$ through angles θ_{k1}, θ_{k2}, and θ_{k3} (the so-called Bryan orientation angles).

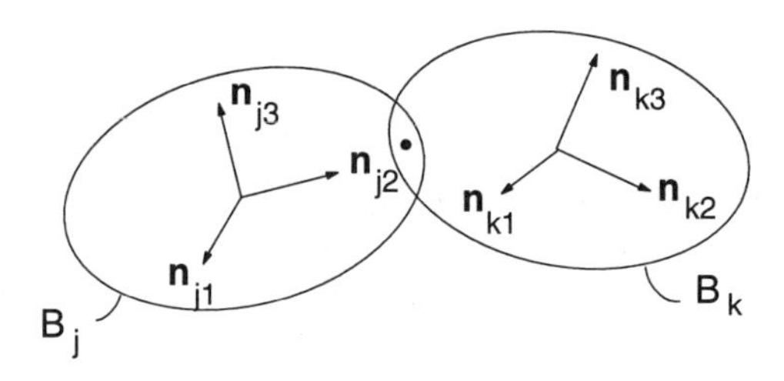

Fig. 12.4.1 Two Typical Adjoining Bodies

The orientation of B_k relative to B_j may be envisioned in terms of the relative inclination of the unit vectors of B_j and B_k. Specifically, we can introduce a transformation matrix SJK between the $\mathbf{n}_{ji}$ and $\mathbf{n}_{ki}$ with elements defined as:

$$SJK_{pq} = \mathbf{n}_{jp} \bullet \mathbf{n}_{kq} \tag{12.4.1}$$

The elements of SJK will involve sines and cosines of the orientation angles as in Table 6.4.1. Alternatively SJK could be expressed in terms of Euler parameters as in Equation (7.6.6).

Recall that the transformation matrices are useful for relating the unit vectors of B_j with those of B_k and conversely, and for relating vector components expressed relative to the $\mathbf{n}_{ji}$ and the $\mathbf{n}_{ki}$ (see Section 6.2). Specifically, we have

$$\mathbf{n}_{jr} = SJK_{rs}\mathbf{n}_{ks} \quad \text{and} \quad \mathbf{n}_{ks} = SJK_{rs}\mathbf{n}_{jr} \tag{12.4.2}$$

and if

$$\mathbf{v} = v_r^{(j)}\mathbf{n}_{jr} = v_s^{(k)}\mathbf{n}_{ks} \tag{12.4.3}$$

then

$$v_r^{(j)} = SJK_{rs}v_s^{(k)} \quad \text{and} \quad v_s^{(k)} = SJK_{rs}v_r^{(j)} \tag{12.4.4}$$

Recall that repeated indices designate a sum. Recall also that the transformation matrices are orthogonal (the inverse in the transpose) so that

$$SJK_{rs}SJK_{rt} = \delta_{st} \quad \text{and} \quad SJK_{rs}SJK_{ts} = \delta_{rt} \tag{12.4.5}$$

or

$$SJK^{-1} = SJK^{T} = SKJ \tag{12.4.6}$$

where as before δ_{st} is Kronecker's delta symbol, forming the elements of the identity matrix.

Next, consider a chain segment of three adjoining bodies B_j, B_k, and B_ℓ as in Figure 12.4.2. As before let $\mathbf{n}_{ji}$, $\mathbf{n}_{ki}$, and $\mathbf{n}_{\ell i}$ be mutually perpendicular unit vectors

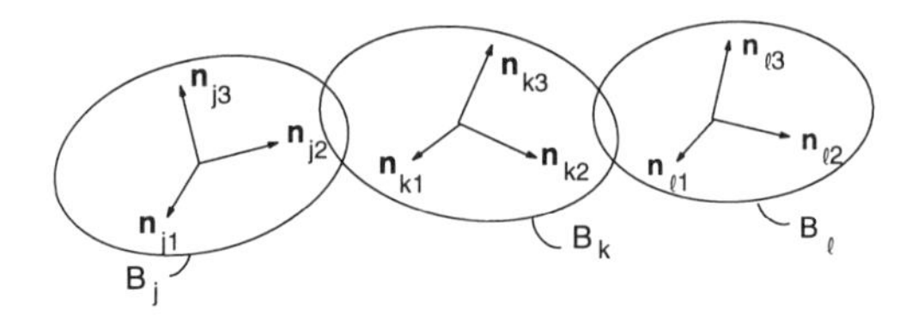

Fig. 12.4.2 Three Typical Adjoining Bodies

fixed in B_j, B_k, and B_ℓ respectively. Then transformation matrices SJK, SKL, and SJL may be introduced with elements given by:

$$SJK_{pq} = \mathbf{n}_{jp} \cdot \mathbf{n}_{kq} \qquad SKL_{pq} = \mathbf{n}_{kp} \cdot \mathbf{n}_{\ell q} \qquad SJL_{pq} = \mathbf{n}_{jp} \cdot \mathbf{n}_{\ell q} \tag{12.4.7}$$

Recall that the identity dyadic **I** may be expressed in terms of the unit vectors of B_j B_k and B_ℓ as

$$\mathbf{I} = \mathbf{n}_{jr}\mathbf{n}_{jr} = \mathbf{n}_{ks}\mathbf{n}_{ks} = \mathbf{n}_{\ell t}\mathbf{n}_{\ell t} \tag{12.4.8}$$

where there is no sum on j, k, or ℓ. (There is a sum from 1 to 3 on r, s, and t.) Then from Equation (12.4.7) the product SJK and SKL is

$$SJK\ SKL = SJL \tag{12.4.9}$$

Equation (12.4.9) is an algorithm for computing transformation matrices between the bodies of a multibody system and particularly between the bodies of the system and the inertia frame R. Specifically, for a typical body B_k the transformation matrix SOK between B_k and R has the elements:

$$SOK_{pq} = \mathbf{n}_{op} \cdot \mathbf{n}_{kq} \tag{12.4.10}$$

where the $\mathbf{n}_{oi}$ $(i = 1,2,3)$ are mutually perpendicular unit vectors fixed in R. then by repeated use of Equation (12.4.9) we have

$$SOK = S01\ S12\ ...\ SIJ\ SJK \tag{12.4.11}$$

where the product is carried out over the transformation matrices between the bodies in the branches containing B_k. For example, for the system of Figure 12.3.1, for B_{12} we have

$$S0,12 = S01\ S12\ S26\ S68\ S8,11\ S11,12 \tag{12.4.12}$$

Observe the sequence of numbers: 0, 1, 2, 6, 7, 11, and 12 in the transformation matrix names of Equations (12.4.12). They are precisely the numbers in column 12 of Table 12.3.1 or the lower body arrays for the example system. That is, the lower body arrays may also be used to write algorithms for the development, computation, and evaluation of the transformation matrices.

12.5 Derivatives of Transformation Matrices

Consider a body B moving in an inertial reference frame R as represented in Figure 12.5.1. Let $\mathbf{n}_i$ and $\mathbf{N}_i$ (i=1,2,3) be mutually perpendicular unit vector sets fixed in B and R. Then a transformation matrix S between the $\mathbf{N}_i$ and the $\mathbf{n}_i$ may be defined with elements S_{ij} given by

$$S_{ij} = \mathbf{N}_i \cdot \mathbf{n}_j \tag{12.5.1}$$

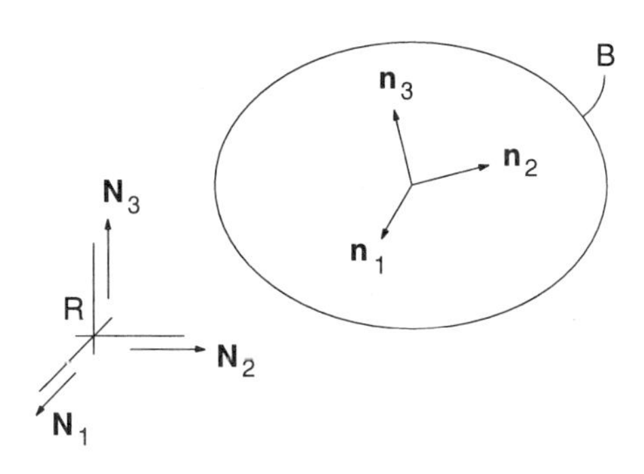

Fig. 12.5.1 A Body B Moving in an Inertial Reference Frame R

Let the angular velocity of B in R referred to the $\mathbf{N}_i$ have the form:

$$^R\boldsymbol{\omega}^B = \Omega_i \mathbf{N}_i \tag{12.5.2}$$

Then from Section 7.7 and Equation (7.7.5) the time derivative of S may be expressed as:

$$dS/dt = \dot{S} = WS \tag{12.5.3}$$

where W is called the "dual matrix" of ${}^R\boldsymbol{\omega}^B$ with elements $W_{i\ell}$ defined as [see Equation (7.7.4)]:

$$W_{i\ell} = -e_{i\ell m}\Omega_m = \begin{bmatrix} 0 & -\Omega_3 & \Omega_2 \\ \Omega_3 & 0 & -\Omega_1 \\ -\Omega_2 & \Omega_1 & 0 \end{bmatrix} \tag{12.5.4}$$

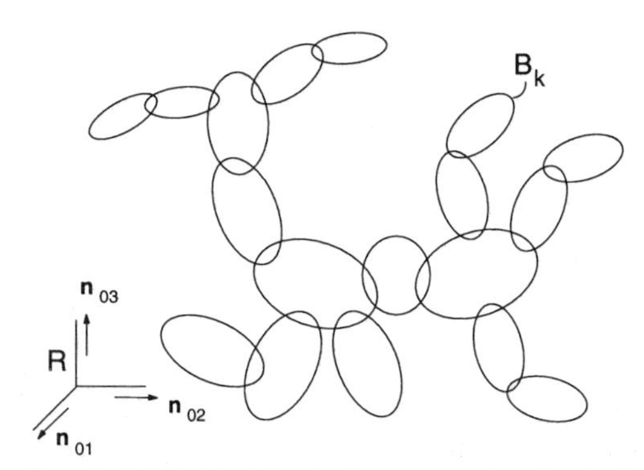

Fig. 12.5.2 A Multibody System with a Typical Body B_k

Consider a typical body B_k of a multibody system as in Figure 12.5.2. Let the system move in an inertial frame R and let $\mathbf{n}_{0i}$ be mutually perpendicular unit vectors fixed in R. Let the angular velocity of B_k in R be expressed in the form:

$$\boldsymbol{\omega}_k = \omega_{km}\mathbf{n}_{0m} \tag{12.5.5}$$

Let S0K be a transformation matrix between unit vectors of B_k and R. Then from Equation (12.5.3) the time derivative of SOK is

$$\dot{SOK} = WK\ SOK \tag{12.5.6}$$

where WK is the dual matrix of $\boldsymbol{\omega}_k$ with elements $WK_{i\ell}$ defined as:

$$WK_{i\ell} = -e_{i\ell m}\omega_{km} = \begin{bmatrix} 0 & -\omega_{k3} & \omega_{k2} \\ \omega_{k3} & 0 & -\omega_{k1} \\ -\omega_{k2} & \omega_{k1} & 0 \end{bmatrix} \tag{12.5.7}$$

Observe that through Equation (12.5.6) the derivatives may be calculated by a multiplication — a useful concept for numerical algorithm development. Observe further that the lower body arrays may be used to develop both the angular velocities and the transformation matrices.

References

12.1 R. L. Huston, "Multibody Dynamics — Modeling and Analysis Methods," *Applied Mechanics Reviews*, Vol. 44, No. 3, 1991, pp. 109-117.

12.2 R. L. Huston, "Multibody Dynamics Since 1990," *Applied Mechanics Reviews*, Vol. 49, No. 10 Part 2, 1996, pp. 535-540.

12.3 R. L. Huston, *Multibody Dynamics*, Butterworth-Heinemann, Stoneham, MA, 1990.

12.4 E. J. Haug, *Computer-Aided Kinematics and Dynamics of Mechanical Systems*, Allyn and Bacon, Boston, MA, 1989.

12.5 R. E. Roberson and R. Schwertassek, *Dynamics of Multibody Systems*, Springer-Verlag, Berlin, Germany, 1988.

12.6 A. A. Shabana, *Dynamics of Multibody Systems*, Wiley, New York, NY, 1988.

12.7 R. M. L. Amirouche, *Computational Methods in Multibody Dynamics*, Prentice Hall, Englewood Cliffs, NJ, 1992.

12.8 J. G. de Jalon and E. Bayo, *Kinematic and Dynamic Simulation of Multibody Systems — The Real-Time Challenge*, Springer-Verlag, New York, NY, 1994.

12.9 H. Rahnejat, *Multi-Body Dynamics — Vehicles, Machines and Mechanisms*, Society of Automotive Engineers, Warrendale, PA, 1998.

12.10 F. C. Moon, *Applied Dynamics — with Applications to Multibody and Mechatronic Systems*, Wiley-Interscience, New York, NY, 1998.

12.11 F. Pfeiffer and C. Glocker, *Multibody Dynamics with Unilateral Contacts*, Wiley-Interscience, New York, NY, 1996.

Chapter 13

MULTIBODY KINEMATICS

13.1 Introduction

In this chapter we summarize the principal equations and formulas for the kinematics of multibody systems. We develop these equations using the modeling and procedures outlined in the foregoing chapter. In the following chapter we similarly summarize the equations and formulas for multibody kinetics and dynamics.

For convenience and brevity we will, for the most part, focus our discussion on open systems of connected rigid bodies as in Figure 13.1.1. We employ Euler parameters and generalized speeds as our fundamental dependent variables.

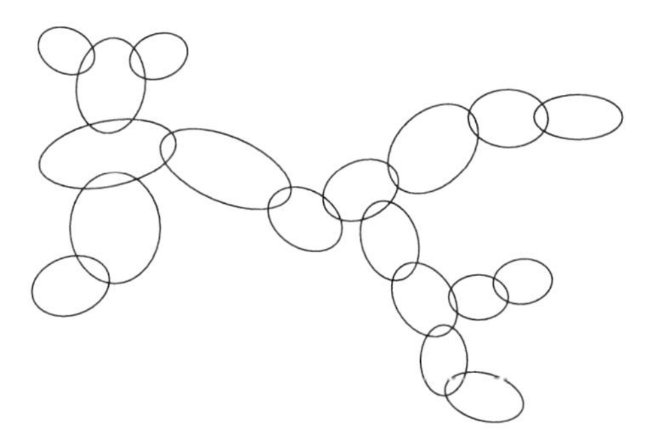

Fig. 13.1.1 An Open-Chain, Open-Tree, Connected Multibody System

The formulas and procedures presented here are readily extended to other kinds of multibody systems (for example, systems with translation between the bodies at the connecting joints; systems with closed loops; or systems with flexible bodies). The procedures are also readily extended or converted to other formulation such as the use of orientation angles (instead of Euler parameters) or absolute orientation variables (instead of relative orientation variables).

13.2 Coordinates, Degrees of Freedom

The "coordinates" of a multibody system are simply the variables or parameters used to define the system configuration. The number of coordinates needed to define the configuration is the number of "degrees of freedom" of the system. That is, there is a one-to-one correspondence between the system coordinates and the degrees of freedom. Recall that an unrestrained rigid body has six degrees of freedom (three for translation and three for rotation). (See Reference 13.1.) It follows that a multibody system with N bodies then has potentially 6N degrees of freedom. If the bodies are connected, however, the number of degrees of freedom is reduced due to the constraining effect of the connecting joints. For example, suppose a system of two bodies is connected by a spherical joint as represented in Figure 13.2.1. Then a requirement that the bodies remain connected

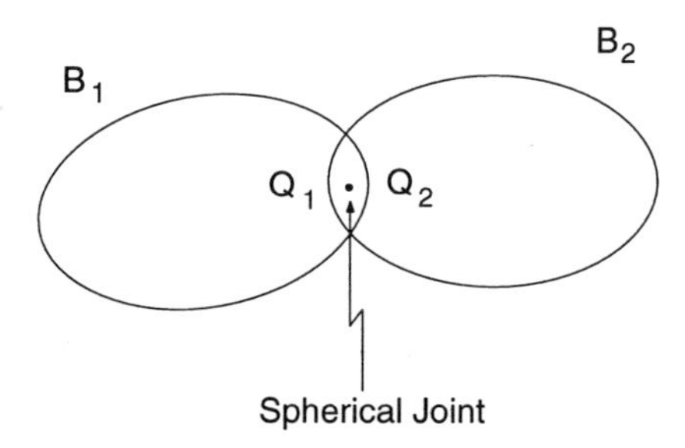

Fig. 13.2.1 Two Bodies Connected by a Spherical Joint

throughout the motion of the system is equivalent to requiring that the bodies share a common point. That is, if the center of the spherical joint is at Q_1 of body B_1 and at Q_2 of body B_2, the points Q_1 and Q_2 must remain coincident throughout the motion. This restriction may be expressed by the position vector constraint equation as:

$$\mathbf{p}_{Q_1} = \mathbf{p}_{Q_2} \tag{13.2.1}$$

This equation is equivalent to three scalar equations. This means that the system of two spherically connected bodies will have twelve minus three, or nine, degrees of freedom. That is, the number of degrees of freedom of a system is the number degrees of freedom of the unrestrained system minus the number of constraint equations.

Consider an open-tree system of N bodies connected by spherical joints as in Figure 13.2.2. With N bodies there are $N-1$ connecting joints, and thus there are $N-1$ position vector constraint equations in the form of Equation (13.2.1). These $N-1$ vector equations are equivalent to $3(N-1)$ scalar equations. Therefore, the N-Body system has $6N-3(N-1)$ or $3N+3$ degrees of freedom. Alternatively, the system may be viewed as having three rotational degrees of freedom for each body and three translational degrees of freedom for one of the bodies, say a "reference" body, B_1.

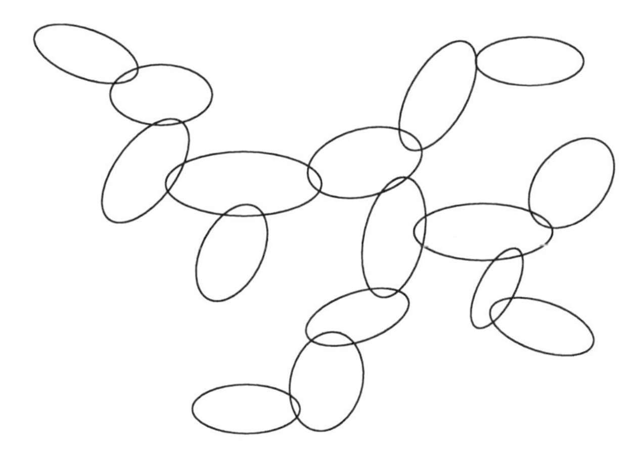

Fig. 13.2.2 An Open-Tree System Connected by Spherical Joints

13.3 Orientation Angles and Euler Parameters

Consider again a typical pair of adjoining bodies such as B_j and B_k as in Figure 13.3.1. Suppose the orientation of B_k relative to B_j is described by orientation angles as discussed in Section 6.2. Observe from Table 7.5.1 that independently of which set of orientation angles is selected, there can occur orientations of B_k such that there is a singularity in expressions for the relative angular velocity components as functions of orientation angle derivatives.

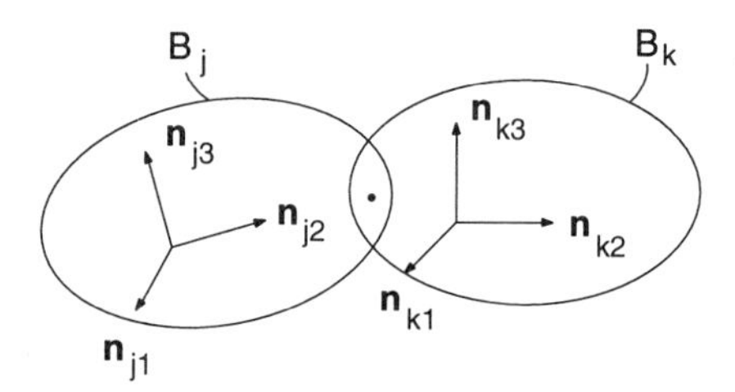

Fig. 13.3.1 Two Typical Adjoining Bodies

In the development of software for the numerical simulation of multibody system dynamics it is important that the software is sufficiently general to incorporate all envisioned simulations. Therefore, it is foreseeable that singularities as in Table 7.5.1 will occur. Fortunately, these singularities can be avoided by employing Euler parameters as in Section 7.6.

Recall that with Euler parameters four variables are used to define the body orientations (as opposed to three orientation angles). This means that there is a redundancy with superfluous parameters. Indeed, the Euler parameters are not independent but instead are related by the expression [see Equation (7.6.5)]

$$\varepsilon_1^2 + \varepsilon_2^2 + \varepsilon_3^2 + \varepsilon_4^2 = 1 \tag{13.3.1}$$

This redundancy, however, allows the equations relating Euler parameters and angular velocity components to be linear, thus avoiding the singularities as in Table

7.5.1. Therefore, in the following paragraphs and sections we will use Euler parameters to define and describe the relative orientation of the bodies.

To develop this, consider again two typical adjoining bodies B_j and B_k as in Figure 13.3.1. Recall that B_k may be brought into a general orientation relative to B_j by a single rotation of B_k through an angle θ_k about an axis fixed in both B_j and B_k, as indicated in Figure 13.3.2 (see Section 7.4). Therefore, let the orientation of

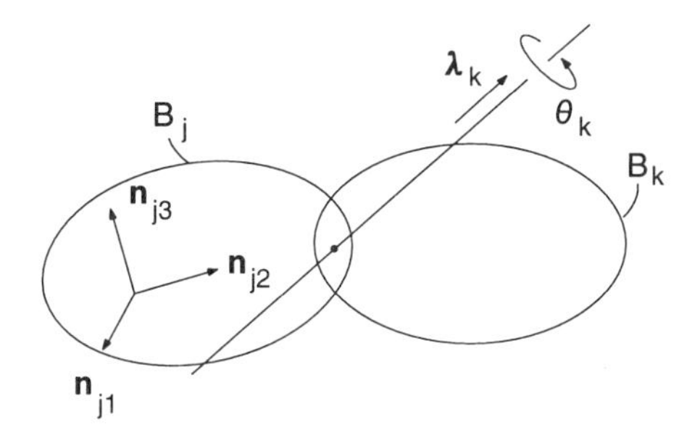

Fig. 13.3.2 Rotation of Typical Body B_k Relative to Adjoining Body B_j About an Axis Fixed in Both B_j and B_k

B_k relative to B_j be represented by the Euler parameters ε_{k1}, ε_{k2}, ε_{k3}, and ε_{k4} defined as:

$$
\begin{aligned}
\varepsilon_{k1} &= \lambda_{k1} \sin(\theta_k/2) \\
\varepsilon_{k2} &= \lambda_{k2} \sin(\theta_k/2) \\
\varepsilon_{k3} &= \lambda_{k3} \sin(\theta_k/2) \\
\varepsilon_{k4} &= \cos(\theta_k/2)
\end{aligned}
\tag{13.3.2}
$$

where the λ_{ki} are the $\mathbf{n}_{ji}$ $(i = 1,2,3)$ components of a unit vector $\boldsymbol{\lambda}_k$ parallel to the rotation axis as shown in Figure 13.3.2, and where as before, the $\mathbf{n}_{ji}$ are unit vectors fixed in B_j. Then from Equation (7.7.6) we can immediately express the transformation matrix SJK in terms of these parameters as:

$$SJK = \begin{bmatrix} (\varepsilon_{k1}^2 - \varepsilon_{k2}^2 - \varepsilon_{k3}^2 + \varepsilon_{k4}^2) & 2(\varepsilon_{k1}\varepsilon_{k2} - \varepsilon_{k3}\varepsilon_{k4}) & 2(\varepsilon_{k1}\varepsilon_{k3} + \varepsilon_{k2}\varepsilon_{k4}) \\ 2(\varepsilon_{k1}\varepsilon_{k2} + \varepsilon_{k3}\varepsilon_{k4}) & (-\varepsilon_{k1}^2 + \varepsilon_{k2}^2 - \varepsilon_{k3}^2 + \varepsilon_{k4}^2) & 2(\varepsilon_{k2}\varepsilon_{k3} - \varepsilon_{k1}\varepsilon_{k4}) \\ 2(\varepsilon_{k1}\varepsilon_{k3} - \varepsilon_{k2}\varepsilon_{k4}) & 2(\varepsilon_{k2}\varepsilon_{k3} + \varepsilon_{k1}\varepsilon_{k4}) & (-\varepsilon_{k1}^2 - \varepsilon_{k2}^2 + \varepsilon_{k3}^2 + \varepsilon_{k4}^2) \end{bmatrix} \quad (13.3.3)$$

The $\mathbf{n}_{ji}$ components, $\hat{\omega}_{ki}$ $(i = 1,2,3)$, of the angular velocity of B_k relative to B_j may then also be readily expressed in terms of Euler parameter derivatives as [see Equation (7.8.1)]:

$$\begin{aligned} \hat{\omega}_{k1} &= 2(\varepsilon_{k4}\dot{\varepsilon}_{k1} - \varepsilon_{k3}\dot{\varepsilon}_{k2} + \varepsilon_{k2}\dot{\varepsilon}_{k3} - \varepsilon_{k1}\dot{\varepsilon}_{k4}) \\ \hat{\omega}_{k2} &= 2(\varepsilon_{k3}\dot{\varepsilon}_{k1} + \varepsilon_{k4}\dot{\varepsilon}_{k2} - \varepsilon_{k1}\dot{\varepsilon}_{k3} - \varepsilon_{k2}\dot{\varepsilon}_{k4}) \\ \hat{\omega}_{k3} &= 2(-\varepsilon_{k2}\dot{\varepsilon}_{k1} + \varepsilon_{k1}\dot{\varepsilon}_{k2} + \varepsilon_{k4}\dot{\varepsilon}_{k3} - \varepsilon_{k3}\dot{\varepsilon}_{k4}) \end{aligned} \quad (13.3.4)$$

Also from Equation (7.8.7) the Euler parameter derivatives may be expressed in terms of the relative angular velocity components as:

$$\begin{aligned} \dot{\varepsilon}_{k1} &= (1/2)(\varepsilon_{k4}\hat{\omega}_{k1} + \varepsilon_{k3}\hat{\omega}_{k2} - \varepsilon_{k2}\hat{\omega}_{k3}) \\ \dot{\varepsilon}_{k2} &= (1/2)(-\varepsilon_{k3}\hat{\omega}_{k1} + \varepsilon_{k4}\hat{\omega}_{k2} + \varepsilon_{k1}\hat{\omega}_{k3}) \\ \dot{\varepsilon}_{k3} &= (1/2)(\varepsilon_{k2}\hat{\omega}_{k1} - \varepsilon_{k1}\hat{\omega}_{k2} + \varepsilon_{k4}\hat{\omega}_{k3}) \\ \dot{\varepsilon}_{k4} &= (1/2)(-\varepsilon_{k1}\hat{\omega}_{k1} - \varepsilon_{k2}\hat{\omega}_{k2} - \varepsilon_{k3}\hat{\omega}_{k3}) \end{aligned} \quad (13.3.5)$$

Observe the linear form of Equations (13.3.4) and (13.3.5). This linearity eliminates zero divisions as can occur in expressions such as those of Table 7.5.1.

Equations (13.3.4) and (13.3.5) may be expressed in the matrix forms:

$$\hat{\omega}_k = 2E_k\dot{\varepsilon}_k \quad \text{and} \quad \dot{\varepsilon}_k = (1/2)E_k^{-1}\hat{\omega}_k \quad (13.3.6)$$

where $\hat{\omega}_k$, ε_k, and E_k are the arrays:

$$\hat{\omega}_k = \begin{bmatrix} \hat{\omega}_{k1} \\ \hat{\omega}_{k2} \\ \hat{\omega}_{k3} \\ 0 \end{bmatrix} \qquad \varepsilon_k = \begin{bmatrix} \varepsilon_{k1} \\ \varepsilon_{k2} \\ \varepsilon_{k3} \\ \varepsilon_{k4} \end{bmatrix} \qquad E_k = \begin{bmatrix} \varepsilon_{k4} & -\varepsilon_{k3} & \varepsilon_{k2} & -\varepsilon_{k1} \\ \varepsilon_{k3} & \varepsilon_{k4} & -\varepsilon_{k1} & -\varepsilon_{k2} \\ -\varepsilon_{k2} & \varepsilon_{k1} & \varepsilon_{k4} & -\varepsilon_{k3} \\ \varepsilon_{k1} & \varepsilon_{k2} & \varepsilon_{k3} & \varepsilon_{k4} \end{bmatrix} \tag{13.3.7}$$

where the fourth rows are obtained by differentiating in Equation (13.3.1):

$$0 = 2(\varepsilon_{k1}\dot{\varepsilon}_{k1} + \varepsilon_{k2}\dot{\varepsilon}_{k2} + \varepsilon_{k3}\dot{\varepsilon}_{k3} + \varepsilon_{k4}\dot{\varepsilon}_{k4}) \tag{13.3.8}$$

Also, it happens that E_k is orthogonal so that E_k^{-1} is E_k^T. That is,

$$E_k^{-1} = \begin{bmatrix} \varepsilon_{k4} & \varepsilon_{k3} & -\varepsilon_{k2} & \varepsilon_{k1} \\ -\varepsilon_{k3} & \varepsilon_{k4} & \varepsilon_{k1} & \varepsilon_{k2} \\ \varepsilon_{k2} & -\varepsilon_{k1} & \varepsilon_{k4} & \varepsilon_{k3} \\ -\varepsilon_{k1} & -\varepsilon_{k2} & -\varepsilon_{k3} & \varepsilon_{k4} \end{bmatrix} \tag{13.3.9}$$

13.4 Generalized Speeds

The geometric complexity of multibody systems creates a need for methods of analysis which are 1) conceptually simple and 2) computationally efficient. Among these methods are the use of lower body arrays to define the connection configuration and the use of Euler parameters to prevent singularities from occurring due to large rotations of the bodies. It is also helpful to use generalized dynamics principles, such as Kane's equations, which by the use of generalized forces, automatically eliminate non-working constraint forces from the analysis. This elimination in turn is accomplished by the use of partial velocity and partial angular velocity vectors [13.2].

Recall that the partial velocity and partial angular velocity vectors are developed through differentiation of velocity and angular velocity vectors with respect to coordinate derivatives. This construction of partial velocity and partial angular velocity vectors can be generalized through the use of generalized speeds leading to further simplification and computational efficiencies.

A "generalized speed" is simply a linear combination of coordinate derivatives. Specifically, suppose a body B moving in an inertial frame R is part of a mechanical system S having n degrees of freedom represented by coordinates q_r $(r = 1,\dots,n)$, as in Figure 13.4.1. Then n generalized speeds y_s $(s = 1,\dots,n)$ may be introduced in the forms:

$$\begin{aligned}
y_1 &= a_{11}\dot{q}_1 + a_{12}\dot{q}_2 + \cdots + a_{1n}\dot{q}_n + b_1 \\
y_2 &= a_{21}\dot{q}_1 + a_{22}\dot{q}_2 + \cdots + a_{2n}\dot{q}_n + b_2 \\
&\vdots \\
y_n &= a_{n1}\dot{q}_1 + a_{n2}\dot{q}_2 + \cdots + a_{nn}\dot{q}_n + b_n
\end{aligned} \tag{13.4.1}$$

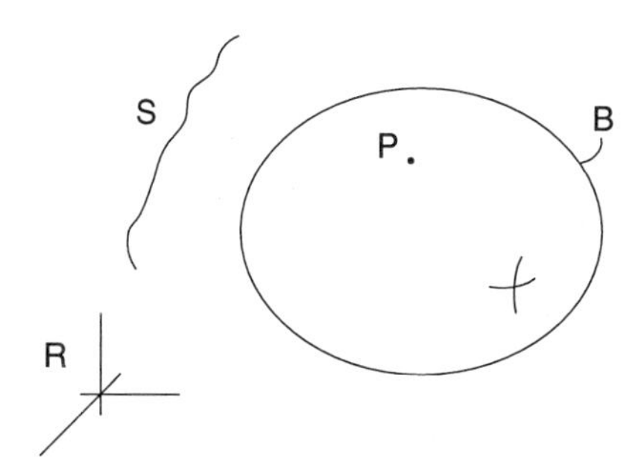

Fig. 13.4.1 A Body B, a Particle P of B, and a Mechanical System S Moving in a Reference Frame R

where the coefficients a_{rs} $(r,s = 1,\dots,n)$ and the b_s $(s = 1,\dots,n)$ may be functions of the q_r. The a_{rs} are arbitrary provided that $\det a_{rs} \neq 0$, so that Equations (13.4.1) may be solved for the $\dot{q}_r$ in terms of the y_s. Alternatively, if Equations (13.4.1) are written as:

$$y_s = \sum_{r=1}^{n} a_{rs}\dot{q}_r + b_s \quad \text{or} \quad y = A^T\dot{q} + b \tag{13.4.2}$$

Then the $\dot{q}_r$ are:

$$\dot{q}_r = \sum_{s=1}^{n} a_{rs}^{-1}(y_s - b_s) \quad \text{or} \quad \dot{q} = A^{-T}(y - b) \tag{13.4.3}$$

where the a_{rs}^{-1} are the elements of A^{-1}, the inverse of the array of a_{rs} elements.

Let P be a typical particle P of body B as in Figure 13.4.1, where as before, S is a mechanical system with n degrees of freedom represented by the coordinates q_r $(r = 1, \ldots, n)$. Then from Equations (4.5.10) and (6.13.5) the velocity of P in r, $\mathbf{V}^P$, and the angular velocity of B in R, $\boldsymbol{\omega}^B$, may be expressed as:

$$\mathbf{V}^P = \mathbf{V}_t + \sum_{r=1}^{n} \mathbf{V}_{\dot{q}_r}\dot{q}_r \tag{13.4.4}$$

and

$$\boldsymbol{\omega}^B = \boldsymbol{\omega}_t + \sum_{r=1}^{n} \boldsymbol{\omega}_{\dot{q}_r}\dot{q}_r \tag{13.4.5}$$

where the $\mathbf{V}_t$, $\mathbf{V}_{\dot{q}_r}$, $\boldsymbol{\omega}_t$, and $\boldsymbol{\omega}_{\dot{q}_r}$ are partial velocity and partial angular velocity vectors for P and B with respect to t and q_r respectively.*

In view of Equations (13.4.3) we immediately see that $\mathbf{V}^P$ and $\boldsymbol{\omega}^B$ may be expressed as linear combinations of the y_s. Specifically,

$$\mathbf{V}^P = \hat{\mathbf{V}}_t + \sum_{s=1}^{n} \hat{\mathbf{V}}_{ys} y_s \quad \text{and} \quad \boldsymbol{\omega}^B = \hat{\boldsymbol{\omega}}_t + \sum_{s=1}^{n} \hat{\boldsymbol{\omega}}_{ys} y_s \tag{13.4.6}$$

---

*The partial velocity and partial angular velocity with respect to time t, $\mathbf{V}_t$, and $\boldsymbol{\omega}_t$ (called "secular terms"), will only occur if S has specified internal motions such as from motors or actuators. For general analyses, it is convenient to temporarily consider such specified motion as degrees of freedom, or variable motion represented by coordinates, and then later specify the coordinates with the given prescribed motion.

where from Equations (13.4.3), (13.4.4), and (13.4.5) $\hat{\mathbf{V}}_t$, $\hat{\mathbf{V}}_{ys}$, $\hat{\boldsymbol{\omega}}_t$, and $\hat{\boldsymbol{\omega}}_{ys}$ are defined as:

$$\hat{\mathbf{V}}_t = \mathbf{V}_t - \sum_{r=1}^{n}\sum_{s=1}^{n} \mathbf{V}_{\dot{q}_r} a_{rs}^{-1} b_s \tag{13.4.7}$$

$$\hat{\mathbf{V}}_{ys} = \sum_{r=1}^{n} \mathbf{V}_{\dot{q}_r} a_{rs}^{-1} \tag{13.4.8}$$

$$\hat{\boldsymbol{\omega}}_t = \boldsymbol{\omega}_t - \sum_{r=1}^{n}\sum_{s=1}^{n} \boldsymbol{\omega}_{\dot{q}_r} a_{rs}^{-1} b_s \tag{13.4.9}$$

$$\hat{\boldsymbol{\omega}}_{ys} = \sum_{r=1}^{n} \boldsymbol{\omega}_{\dot{q}_r} a_{rs}^{-1} \tag{13.4.10}$$

Observe, however, that Equations (13.4.7) need not be obtained via Equations (13.4.7) through (13.4.10). If the y_s are conveniently chosen, Equations (13.4.6) may occur explicitly in the analysis in which case the $\hat{\mathbf{V}}_t$, $\hat{\mathbf{V}}_{ys}$, $\hat{\boldsymbol{\omega}}_t$, and $\hat{\boldsymbol{\omega}}_{ys}$ may be obtained by inspection.

The most common choice of generalized speeds is to define them as components of angular velocity vectors. For example, for a body B of a mechanical system S moving in an inertial reference frame R, as in Figure 13.4.1, let the orientation of B in R be described by "dextral" (or Bryan) angles α , β, and γ. Specifically, let $\mathbf{n}_i$ and $\mathbf{N}_i$ $i = (1,2,3)$ be mutually perpendicular unit vectors fixed in B and R respectively. Let these unit vector sets be mutually aligned and then let B be brought into a general orientation in R by successive dextral rotation of B about $\mathbf{n}_1$, $\mathbf{n}_2$, and $\mathbf{n}_3$ through the angles α, β, and γ. Then in the absence of secular terms (see footnote p. 578) the angular velocity $\boldsymbol{\omega}$ of B in R may be expressed as (see Table 7.5.1):

$$\begin{aligned}\boldsymbol{\omega} = \Omega_i \mathbf{N}_i = (\dot{\alpha} + \dot{\gamma}\sin\beta)\mathbf{N}_1 + (\dot{\beta}\cos\alpha - \dot{\gamma}\cos\beta\sin\alpha)\mathbf{N}_2 \\ + (\dot{\beta}\sin\alpha + \dot{\gamma}\cos\beta\cos\alpha)\mathbf{N}_3\end{aligned} \tag{13.4.11}$$

Generalized speeds y_s $(s = 1,\ldots,6)$ for the motion of B in R are then often defined as:

$$\begin{aligned} y_1 &= \dot{X} \\ y_2 &= \dot{Y} \\ y_3 &= \dot{Z} \\ y_4 &= \dot{\alpha} + \dot{\gamma}\sin\beta \\ y_5 &= \dot{\beta}\cos\alpha - \dot{\gamma}\cos\beta\sin\alpha \\ y_6 &= \dot{\beta}\sin\alpha + \dot{\gamma}\cos\beta\cos\alpha \end{aligned} \tag{13.4.12}$$

where X, Y, and Z are the $\mathbf{N}_1$, $\mathbf{N}_2$, and $\mathbf{N}_3$ components of the position vector of a reference point O_B in R. In terms of these generalized speeds, the angular velocity of B in R then takes the simple form:

$$\boldsymbol{\omega} = y_4\mathbf{N}_1 + y_5\mathbf{N}_2 + y_6\mathbf{N}_3 \tag{13.4.13}$$

The partial angular velocity vectors for the y_s $(s = 1,\ldots,6)$ then also have the simple forms

$$\begin{aligned} \boldsymbol{\omega}_{y_1} &= 0 \\ \boldsymbol{\omega}_{y_2} &= 0 \\ \boldsymbol{\omega}_{y_3} &= 0 \\ \boldsymbol{\omega}_{y_4} &= \mathbf{N}_1 \\ \boldsymbol{\omega}_{y_5} &= \mathbf{N}_2 \\ \boldsymbol{\omega}_{y_6} &= \mathbf{N}_3 \end{aligned} \tag{13.4.14}$$

Observe in Equation (13.4.12) that, as required, the y_s are linear combinations of the $\dot{\alpha}$, $\dot{\beta}$, and $\dot{\gamma}$. Observe further, however, in the last three expressions of Equation (13.4.12) that the y_s (y_4, y_5, and y_6) are not derivative of any elementary functions or coordinates. That is, functions ϕ_s do not exist such that $y_s = \dot{\phi}_s$

($s = 4,5,6$). Therefore, when using generalized speeds as in Equation (13.4.12), the analysis is said to be employing "quasi-coordinates."

The use of generalized speeds is especially convenient when Euler parameters are used to define the orientation configuration. Recall that with Euler parameters four variables are used to define body orientation, thus avoiding geometric singularities. The Euler parameter derivatives and the Euler parameters themselves are linearly related to angular velocity components as shown in Equations (13.3.4) and (13.3.5). Therefore, it is a natural choice to let the Euler parameters play the role of the coordinates if generalized speeds in the form of angular velocity components are used as coordinate derivatives. Then in a numerical analysis, if the generalized speeds (and, hence, the angular velocity components) can be determined, the Euler parameters (and, hence, the body orientation) can be determined through numerical integration of Equations (13.3.5).

13.5 Illustrative Application with a Multibody System

To illustrate the use of generalized speeds with multibody system kinematics, consider again the example system of Figure 12.3.1 and as shown again in Figure 13.5.1. Recall that this system is an open-chain system of 15 bodies connected by spherical pins, and moving in an inertial reference frame R. For convenience, we will label and number this system as in Section 12.3 and as shown in Figure 12.3.1 and as shown again in Figure 13.5.1. The lower body arrays are then given in Table 12.3.1 and as listed again in Table 13.5.1.

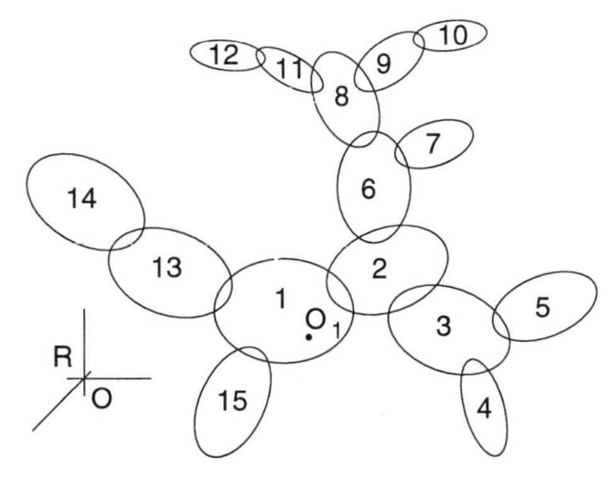

Fig. 13.5.1 A Numbered Multibody System

Table 13.5.1 Lower Body Arrays for the System of Figure 13.5.1

Body (N)	**1**	**2**	**3**	**4**	**5**	**6**	**7**	**8**	**9**	**10**	**11**	**12**	**13**	**14**	**15**
$L^0(N)$	1	2	3	4	5	6	7	8	9	10	11	12	13	14	15
$L^1(N)$	0	1	2	3	3	2	6	6	8	9	8	11	1	13	1
$L^2(N)$	0	0	1	2	2	1	2	2	6	8	6	8	0	1	0
$L^3(N)$	0	0	0	1	1	0	1	1	2	6	2	6	0	0	0
$L^4(N)$	0	0	0	0	0	0	0	0	1	2	1	2	0	0	0
$L^5(N)$	0	0	0	0	0	0	0	0	0	1	0	1	0	0	0
$L^6(N)$	0	0	0	0	0	0	0	0	0	0	0	0	0	0	0

13.5.1 Coordinates and Degrees of Freedom

With N bodies connected by spherical pins the system has $3N + 3$ degrees of freedom (three for the orientation of each body and three for the translation of one of the bodies, say reference body B_1). Let the position of B_1 be defined by coordinates (x, y, z) of a reference point, say O_1 of B_1 relative to the origin O of a Cartesian frame fixed in R. Let the orientation of the bodies be defined by Euler parameters measured relative to the adjacent lower numbered body. For example, for typical body B_k the orientation is measured relative to B_j, the adjacent lower numbered body, by the Euler parameters ε_{k1}, ε_{k2}, ε_{k3}, and ε_{k4}.

As noted in Section 13.4 with the use of Euler parameters, it is useful to then use generalized speeds to describe the motion of the system. For the translation of B_1 the generalized speeds may be defined as simply the derivatives of the coordinates of O_1: $(\dot{x}, \dot{y}, \dot{z})$. For the rotations of the bodies, let the generalized speeds thus be defined as the components of the relative angular velocities of the bodies relative to the adjacent lower numbered bodies. For example, for typical body B_k, the rotation generalized speeds are $\hat{\omega}_{k1}$, $\hat{\omega}_{k2}$, and $\hat{\omega}_{k3}$ where the $\hat{\omega}_{ki}$ are defined by the expression:

$$ {}^{B_j}\boldsymbol{\omega}^{B_k} = \hat{\omega}_k = \hat{\omega}_{k1}\mathbf{n}_{j1} + \hat{\omega}_{k2}\mathbf{n}_{j2} + \hat{\omega}_{k3}\mathbf{n}_{j3} = \hat{\omega}_{km}\mathbf{n}_{jm} \tag{13.5.1} $$

where, as before, the $\mathbf{n}_{jm}$ $(m = 1,2,3)$ are mutually perpendicular unit vectors fixed in B_j.

For the illustrative system of Figure 13.5.1 the generalized speeds may be grouped into sets of triplets as listed in Table 13.5.2.

Table 13.5.2 Generalized Speeds for the System of Figure 13.5.1

$y_1 = \dot{x}$ $y_2 = \dot{y}$ $y_3 = \dot{z}$ (Translation of B_1 in R)	$y_4 = \hat{\omega}_{11}$ $y_5 = \hat{\omega}_{12}$ $y_6 = \hat{\omega}_{13}$ (Rotation of B_1 in R)	$y_7 = \hat{\omega}_{21}$ $y_8 = \hat{\omega}_{22}$ $y_9 = \hat{\omega}_{23}$ (Rotation of B_2 relative to B_1)	$y_{10} = \hat{\omega}_{31}$ $y_{11} = \hat{\omega}_{32}$ $y_{12} = \hat{\omega}_{33}$ (Rotation of B_3 relative to B_2)
$y_{13} = \hat{\omega}_{41}$ $y_{14} = \hat{\omega}_{42}$ $y_{15} = \hat{\omega}_{43}$ (Rotation of B_4 relative to B_3)	$y_{16} = \hat{\omega}_{51}$ $y_{17} = \hat{\omega}_{52}$ $y_{18} = \hat{\omega}_{53}$ (Rotation of B_5 relative to B_3)	$y_{19} = \hat{\omega}_{61}$ $y_{20} = \hat{\omega}_{62}$ $y_{21} = \hat{\omega}_{63}$ (Rotation of B_6 relative to B_2)	$y_{22} = \hat{\omega}_{71}$ $y_{23} = \hat{\omega}_{72}$ $y_{24} = \hat{\omega}_{73}$ (Rotation of B_7 relative to B_6)
$y_{25} = \hat{\omega}_{81}$ $y_{26} = \hat{\omega}_{82}$ $y_{27} = \hat{\omega}_{83}$ (Rotation of B_8 relative to B_6)	$y_{28} = \hat{\omega}_{91}$ $y_{29} = \hat{\omega}_{92}$ $y_{30} = \hat{\omega}_{93}$ (Rotation of B_9 relative to B_8)	$y_{31} = \hat{\omega}_{10,1}$ $y_{32} = \hat{\omega}_{10,2}$ $y_{33} = \hat{\omega}_{10,3}$ (Rotation of B_{10} relative to B_9)	$y_{34} = \hat{\omega}_{11,1}$ $y_{35} = \hat{\omega}_{11,2}$ $y_{36} = \hat{\omega}_{11,3}$ (Rotation B_{11} relative to B_8)
$y_{37} = \hat{\omega}_{12,1}$ $y_{38} = \hat{\omega}_{12,2}$ $y_{39} = \hat{\omega}_{12,3}$ (Rotation of B_{12} relative to B_{11})	$y_{40} = \hat{\omega}_{13,1}$ $y_{41} = \hat{\omega}_{13,2}$ $y_{42} = \hat{\omega}_{13,3}$ (Rotation of B_{13} relative to B_1)	$y_{43} = \hat{\omega}_{14,1}$ $y_{44} = \hat{\omega}_{14,2}$ $y_{45} = \hat{\omega}_{14,3}$ (Rotation of B_{14} relative to B_{13})	$y_{46} = \hat{\omega}_{15,1}$ $y_{47} = \hat{\omega}_{15,2}$ $y_{48} = \hat{\omega}_{15,3}$ (Rotation of B_{15} relative to B_1)

13.5.2 Angular Velocities

The angular velocities of the bodies of a multibody system S are readily obtained using the addition theorem for angular velocity [Equation (6.7.5)]. If B_k is a typical body of S the angular velocity of B_k in R is then simply:

$$^{R}\boldsymbol{\omega}^{B_k} = {}^{R}\boldsymbol{\omega}^{B_1} + \dots + {}^{B_j}\boldsymbol{\omega}^{B_k} \tag{13.5.2}$$

where the sum is taken over the branches of S containing B_k, and where, as before, B_j is the adjacent lower numbered body of B_k. Using the notation of Section 13.3, Equation (13.5.2) may be written as:

$$\boldsymbol{\omega}_k = \hat{\boldsymbol{\omega}}_1 + \cdots + \hat{\boldsymbol{\omega}}_j + \hat{\boldsymbol{\omega}}_k \tag{13.5.3}$$

For the system of Figure 13.5.1, Equation (13.5.3) provides the angular velocities of the various bodies as:

$$\begin{aligned}
\boldsymbol{\omega}_1 &= \hat{\boldsymbol{\omega}}_1 \\
\boldsymbol{\omega}_2 &= \hat{\boldsymbol{\omega}}_2 + \hat{\boldsymbol{\omega}}_1 \\
\boldsymbol{\omega}_3 &= \hat{\boldsymbol{\omega}}_3 + \hat{\boldsymbol{\omega}}_2 + \hat{\boldsymbol{\omega}}_1 \\
\boldsymbol{\omega}_4 &= \hat{\boldsymbol{\omega}}_4 + \hat{\boldsymbol{\omega}}_3 + \hat{\boldsymbol{\omega}}_2 + \hat{\boldsymbol{\omega}}_1 \\
\boldsymbol{\omega}_5 &= \hat{\boldsymbol{\omega}}_5 + \hat{\boldsymbol{\omega}}_3 + \hat{\boldsymbol{\omega}}_2 + \hat{\boldsymbol{\omega}}_1 \\
\boldsymbol{\omega}_6 &= \hat{\boldsymbol{\omega}}_6 + \hat{\boldsymbol{\omega}}_2 + \hat{\boldsymbol{\omega}}_1 \\
\boldsymbol{\omega}_7 &= \hat{\boldsymbol{\omega}}_7 + \hat{\boldsymbol{\omega}}_6 + \hat{\boldsymbol{\omega}}_2 + \hat{\boldsymbol{\omega}}_1 \\
\boldsymbol{\omega}_8 &= \hat{\boldsymbol{\omega}}_8 + \hat{\boldsymbol{\omega}}_6 + \hat{\boldsymbol{\omega}}_2 + \hat{\boldsymbol{\omega}}_1 \\
\boldsymbol{\omega}_9 &= \hat{\boldsymbol{\omega}}_9 + \hat{\boldsymbol{\omega}}_8 + \hat{\boldsymbol{\omega}}_6 + \hat{\boldsymbol{\omega}}_2 + \hat{\boldsymbol{\omega}}_1 \\
\boldsymbol{\omega}_{10} &= \hat{\boldsymbol{\omega}}_{10} + \hat{\boldsymbol{\omega}}_9 + \hat{\boldsymbol{\omega}}_8 + \hat{\boldsymbol{\omega}}_6 + \hat{\boldsymbol{\omega}}_2 + \hat{\boldsymbol{\omega}}_1 \\
\boldsymbol{\omega}_{11} &= \hat{\boldsymbol{\omega}}_{11} + \hat{\boldsymbol{\omega}}_8 + \hat{\boldsymbol{\omega}}_6 + \hat{\boldsymbol{\omega}}_2 + \hat{\boldsymbol{\omega}}_1 \\
\boldsymbol{\omega}_{12} &= \hat{\boldsymbol{\omega}}_{12} + \hat{\boldsymbol{\omega}}_{11} + \hat{\boldsymbol{\omega}}_8 + \hat{\boldsymbol{\omega}}_6 + \hat{\boldsymbol{\omega}}_2 + \hat{\boldsymbol{\omega}}_1 \\
\boldsymbol{\omega}_{13} &= \hat{\boldsymbol{\omega}}_{13} + \hat{\boldsymbol{\omega}}_1 \\
\boldsymbol{\omega}_{14} &= \hat{\boldsymbol{\omega}}_{14} + \hat{\boldsymbol{\omega}}_{13} + \hat{\boldsymbol{\omega}}_1 \\
\boldsymbol{\omega}_{15} &= \hat{\boldsymbol{\omega}}_{13} + \hat{\boldsymbol{\omega}}_1
\end{aligned} \tag{13.5.4}$$

As noted in Section 12.3, the subscripts of the expressions in Equation (13.5.4) follow the pattern of the higher order body arrays of Table 12.3.1. Indeed, the expressions of Equation (13.5.4) can be summarized by the equation:

$$\boldsymbol{\omega}_k = \sum_{p=0}^{r} \hat{\boldsymbol{\omega}}_q \quad \text{where} \quad q = L^p(k) \tag{13.5.5}$$

where r is the index such that

$$L^r(k) = 1 \tag{13.5.6}$$

and where r may be determined by computing $L^p(k)$ and comparing the result to 1 (see References 13.1, 13.3, and 13.4).

13.5.3 Partial Angular Velocity Vectors

In view of Equations (13.4.14), (13.5.4), and (13.5.6), the angular velocities of the bodies may all be expressed as linear combinations of generalized speeds in the form:

$$\boldsymbol{\omega}_k = \omega_{k\ell m} y_\ell \mathbf{n}_{0m} \tag{13.5.7}$$

where the $\omega_{k\ell m}$ $(k = 1,\dots,N;\ \ell = 1,\dots,n;\ m = 1,2,3)$ are components of the partial velocity vectors, N is the number of bodies, n is the number of degrees of freedom, the y_ℓ $(\ell = 1,\dots,n)$ are the generalized speeds, and the $\mathbf{n}_{0m}$ $(m = 1,2,3)$ are vectors fixed in R. Specifically, the partial angular velocity vectors are:

$$\partial \boldsymbol{\omega}_k / \partial y_s = \omega_{k\ell m} \mathbf{n}_{0m} \quad (k = 1,\dots,N;\ s = 1,\dots,n) \tag{13.5.8}$$

It happens that most of the $\omega_{k\ell m}$ are zero since a typical angular velocity will only involve a fraction of the generalized speeds. For example, for the illustrative multibody system of Figure 13.5.1, for a typical body, say Body 7, an inspection and comparison of Equation (13.5.4) and Table 13.5.2 shows that $\boldsymbol{\omega}_7$ involves only the triplets y_4, y_5, y_6; y_7, y_8, y_9; y_{19}, y_{20}, y_{21}; and y_{22}, y_{23}, y_{24}. Indeed, from the seventh expression of Equation (13.5.4), and from Table 13.5.2, $\boldsymbol{\omega}_7$ can be expressed as:

$$\boldsymbol{\omega}_7 = \hat{\boldsymbol{\omega}}_1 + \hat{\boldsymbol{\omega}}_2 + \hat{\boldsymbol{\omega}}_6 + \hat{\boldsymbol{\omega}}_7$$

$$= y_4\mathbf{n}_{01} + y_5\mathbf{n}_{02} + y_6\mathbf{n}_{03} + y_7\mathbf{n}_{11} + y_8\mathbf{n}_{12} + y_9\mathbf{n}_{13} + y_{19}\mathbf{n}_{21} + y_{20}\mathbf{n}_{22} + y_{21}\mathbf{n}_{23} + y_{22}\mathbf{n}_{61} + y_{23}\mathbf{n}_{62} + y_{24}\mathbf{n}_{63} \quad (13.5.9)$$

Then, the partial angular velocities of Body 7 are

$$\partial\boldsymbol{\omega}_7/\partial y_s = 0 \qquad s = 1,\ 2,\ 3,\ 10 \text{ to } 18,\ 25 \text{ to } 48 \quad (13.5.10)$$

and

$$\begin{aligned} \partial\boldsymbol{\omega}_7/\partial y_4 &= \mathbf{n}_{01} = \delta_{m1}\mathbf{n}_{0m} = \omega_{74m}\mathbf{n}_{0m} \\ \partial\boldsymbol{\omega}_7/\partial y_5 &= \mathbf{n}_{02} = \delta_{m2}\mathbf{n}_{0m} = \omega_{75m}\mathbf{n}_{0m} \\ \partial\boldsymbol{\omega}_7/\partial y_6 &= \mathbf{n}_{03} = \delta_{m3}\mathbf{n}_{0m} = \omega_{76m}\mathbf{n}_{0m} \end{aligned} \quad (13.5.11)$$

$$\begin{aligned} \partial\boldsymbol{\omega}_7/\partial y_7 &= \mathbf{n}_{11} = S01_{m1}\mathbf{n}_{0m} = \omega_{77m}\mathbf{n}_{0m} \\ \partial\boldsymbol{\omega}_7/\partial y_8 &= \mathbf{n}_{12} = S01_{m2}\mathbf{n}_{0m} = \omega_{78m}\mathbf{n}_{0m} \\ \partial\boldsymbol{\omega}_7/\partial y_9 &= \mathbf{n}_{13} = S01_{m3}\mathbf{n}_{0m} = \omega_{79m}\mathbf{n}_{0m} \end{aligned} \quad (13.5.12)$$

$$\begin{aligned} \partial\boldsymbol{\omega}_7/\partial y_{19} &= \mathbf{n}_{21} = S02_{m1}\mathbf{n}_{0m} = \omega_{7,19,m}\mathbf{n}_{0m} \\ \partial\boldsymbol{\omega}_7/\partial y_{20} &= \mathbf{n}_{22} = S02_{m2}\mathbf{n}_{0m} = \omega_{7,20,m}\mathbf{n}_{0m} \\ \partial\boldsymbol{\omega}_7/\partial y_{21} &= \mathbf{n}_{23} = S02_{m3}\mathbf{n}_{0m} = \omega_{7,21,m}\mathbf{n}_{0m} \end{aligned} \quad (13.5.13)$$

$$\begin{aligned} \partial\boldsymbol{\omega}_7/\partial y_{22} &= \mathbf{n}_{61} = S06_{m1}\mathbf{n}_{0m} = \omega_{7,22,m}\mathbf{n}_{0m} \\ \partial\boldsymbol{\omega}_7/\partial y_{23} &= \mathbf{n}_{62} = S06_{m2}\mathbf{n}_{0m} = \omega_{7,23,m}\mathbf{n}_{0m} \\ \partial\boldsymbol{\omega}_7/\partial y_{24} &= \mathbf{n}_{63} = S06_{m3}\mathbf{n}_{0m} = \omega_{7,24,m}\mathbf{n}_{0m} \end{aligned} \quad (13.5.14)$$

where for organizational convenience the equations have been grouped together in triplets, with the components expressed in terms of the $\mathbf{n}_{0m}$.

Observe the patterns in Equations (13.5.11) to (13.5.14): By inspection the non-zero partial angular velocity components of Body 7 with respect to the $\mathbf{n}_{0m}$ are

$$\omega_{7,3+n,m} = \delta^T_{mn} \qquad \omega_{7,6+n,m} = S01^T_{mn}$$
$$\omega_{7,18+n,m} = S02^T_{mn} \qquad \omega_{7,21+n,m} = S06^T_{mn} \tag{13.5.15}$$

Observe further from Equations (13.5.11) to (13.5.15) that the partial angular velocity components are elements of transformation matrices. Specifically, for Body 7, they could be listed as in Table 13.5.3.

Table 13.5.3 Partial Angular Velocity Components for Body 7

ℓ	1 2 3	4 5 6	7 8 9	10 11 12	13 14 15	16 17 18	19 20 21	22 23 24	25 26 27	28 29 30	31 32 33	34 35 36	37 38 39	40 41 42	43 44 45	46 47 49
$\omega_{7\ell m}$	0	I	$S01^T$	0	0	0	$S02^T$	$S06^T$	0	0	0	0	0	0	0	0

A similar analysis can be made for the other bodies of the example system. The results for all of the bodies are listed in Table 13.5.4.

Observe the pattern of the entries in Table 13.5.4. A comparison of Table 13.5.1 (lower body arrays), Table 13.5.2 (generalized speed assignments), and Table 13.5.4 show how a table such as Table 13.5.4 could be constructed for any multibody system. Specifically, since most of the entries are zeros, one would start with a table of zeros, and then identify the non-zero entries and replace the zeros accordingly. The non-zero entries are found to be in the lower triangular, region of the table, beginning with the y_4, y_5, y_6 column. The diagonal entries correspond to the L^1 row of Table 13.5.1. The other non-zero entries then follow the pattern of the columns of Table 13.5.1 and are placed in the respective rows.

13.6 Angular Velocity

Once the connection configuration of a multibody system is known the lower body array can be determined by inspection of a sketch of the system. Then Equation (13.5.5) provides an algorithm for determining the angular velocity for each body of the system in an inertial reference frame R. Specifically, for an N-body

Table 13.5.4 Partial Angular Velocity Components ($\omega_{k\ell m}$) for the Example System of Figure 13.5.1.

ℓ B_k	1 2 3	4 5 6	7 8 9	10 11 12	13 14 15	16 17 18	19 20 21	22 23 24	25 26 27	28 29 30	31 32 33	34 35 36	37 38 39	40 41 42	43 44 45	46 47 48
B_1	0	I	0	0	0	0	0	0	0	0	0	0	0	0	0	0
B_2	0	I	$S01^T$	0	0	0	0	0	0	0	0	0	0	0	0	0
B_3	0	I	$S01^T$	$S02^T$	0	0	0	0	0	0	0	0	0	0	0	0
B_4	0	I	$S01^T$	$S02^T$	$S03^T$	0	0	0	0	0	0	0	0	0	0	0
B_5	0	I	$S01^T$	$S02^T$	0	$S03^T$	0	0	0	0	0	0	0	0	0	0
B_6	0	I	$S01^T$	0	0	0	$S02^T$	0	0	0	0	0	0	0	0	0
B_7	0	I	$S01^T$	0	0	0	$S02^T$	$S06^T$	0	0	0	0	0	0	0	0
B_8	0	I	$S01^T$	0	0		$S02^T$	0	$S06^T$	0	0	0	0	0	0	0
B_9	0	I	$S01^T$	0	0	0	$S02^T$	0	$S06^T$	$S08^T$	0	0	0	0	0	0
B_{10}	0	I	$S01^T$	0	0	0	$S02^T$	0	$S06^T$	$S08^T$	$S09^T$	0	0	0	0	0
B_{11}	0	I	$S01^T$	0	0	0	$S02^T$	0	$S06^T$	0	0	$S08^T$	0	0	0	0
B_{12}	0	I	$S01^T$	0	0	0	$S02^T$	0	$S06^T$	0	0	$S08^T$	$S011^T$	0	0	0
B_{13}	0	I	0	0	0	0	0	0	0	0	0	0	0	$S01^T$	0	0
B_{14}	0	I	0	0	0	0	0	0	0	0	0	0	0	$S01^T$	$S013^T$	0
B_{15}	0	I	0	0	0	0	0	0	0	0	0	0	0	0	0	$S01^T$

system, knowing the lower body array L(K), the angular velocities of the bodies in R, are determined by the expression:

$$ {}^{D}\boldsymbol{\omega}^{B_k} \stackrel{D}{=} \boldsymbol{\omega}_k = \sum_{p=1}^{r} \hat{\boldsymbol{\omega}}_q \quad \text{where } q = L^{p}(k) \quad (k = 1,\dots,N) \tag{13.6.1} $$

where $\hat{\omega}_1$ is the angular velocity of reference body B_1 in R.

After such computations have been made the angular velocities will be seen to have the form: [See Equation (13.5.7)]:

$$ \boldsymbol{\omega}_k = \omega_{k\ell m} y_\ell \mathbf{n}_{0m} \quad (k = 1,\dots,N) \tag{13.6.2} $$

where, as before, the y_ℓ ($\ell = 1,\dots,3N+3$) are generalized speeds (see Section 13.4) in the form of relative angular velocity components, except for the first three (y_1, y_2, y_3) which are displacement derivatives of a reference point in reference body B_1, and the $\mathbf{n}_{0m}$ ($m = 1,2,3$) are mutually perpendicular unit vectors fixed in R. For the example system of Figure 13.5.1 the coefficients $\omega_{k\ell m}$ in Equation (13.6.2) are given in Table 13.5.4.

13.7 Angular Acceleration

Once the angular velocities are known in the form of Equations (13.6.2), the angular accelerations may readily be obtained by simple differentiation. Specifically, for an N-body system let the angular velocities of the bodies be expressed in the form:

$$ \boldsymbol{\omega}_k = \omega_{k\ell m} y_\ell \mathbf{n}_{0m} \quad (k = 1,\dots,N) \tag{13.7.1} $$

where, as before, the y_ℓ are generalized speeds, the $\mathbf{n}_{0m}$ are unit vectors fixed in an inertial frame R, and the $\omega_{k\ell m}$ are components of the partial angular velocity vectors of the bodies for the set of the generalized speeds. Then by differentiation, the angular accelerations $\boldsymbol{\alpha}_k$ ($k = 1,\dots,N$) are:

$$\boldsymbol{\alpha}_k = (\omega_{k\ell m}\ddot{y}_\ell + \dot{\omega}_{k\ell m}\dot{y}_\ell)\mathbf{n}_{0m} \tag{13.7.2}$$

With generalized speeds defined as components of relative angular velocity components (see Section 13.3), the $\omega_{k\ell m}$ may be expressed as elements of transformation matrices, as illustrated in Table 13.5.4. Hence the $\dot{\omega}_{k\ell m}$ may be expressed as derivatives of elements of the transformation matrices. Recall from Section 12.5, Equation (12.5.3) that transformation matrix derivatives may be expressed in terms of matrix products. Specifically, for a typical body B_k of a multibody system with transformation S0K between unit vectors fixed in B_k and R, the derivative of S0K may be expressed as

$$d(S0K)/dt = \dot{S0K} = WK\ S0K \tag{13.7.3}$$

where from Equation (12.5.4) the elements of WK are

$$WK_{ij} = -e_{ijm}\Omega_{km} = \begin{bmatrix} 0 & -\Omega_{k3} & \Omega_{k2} \\ \Omega_{k3} & 0 & -\Omega_{k1} \\ -\Omega_{k2} & \Omega_{k1} & 0 \end{bmatrix} \tag{13.7.4}$$

where the Ω_{km} are the $\mathbf{n}_{0m}$ components of $\boldsymbol{\omega}_k$.

For the illustrative system of Figure 13.5.1, the $\dot{\omega}_{k\ell m}$ are immediately obtained from Table 13.5.4. The results are listed in Table 13.7.1.

Table 13.7.1 Derivatives of the Partial Angular Velocity Components ($\dot{\omega}_{k\ell m}$) for the Example System of Figure 13.5.1

ℓ B_k	1 2 3	4 5 6	7 8 9	10 11 12	13 14 15	16 17 18	19 20 21	22 23 24	25 26 27	28 29 30	31 32 33	34 35 36	37 38 39	40 41 42	43 44 45	46 47 48
B_1	0	I	0	0	0	0	0	0	0	0	0	0	0	0	0	0
B_2	0	I	$S\dot{0}1^T$	0	0	0	0	0	0	0	0	0	0	0	0	0
B_3	0	I	$S\dot{0}1^T$	$S\dot{0}2^T$	0	0	0	0	0	0	0	0	0	0	0	0
B_4	0	I	$S\dot{0}1^T$	$S\dot{0}2^T$	$S\dot{0}3^T$	0	0	0	0	0	0	0	0	0	0	0
B_5	0	I	$S\dot{0}1^T$	$S\dot{0}2^T$	0	$S\dot{0}3^T$	0	0	0	0	0	0	0	0	0	0
B_6	0	I	$S\dot{0}1^T$	0	0	0	$S\dot{0}2^T$	0	0	0	0	0	0	0	0	0
B_7	0	I	$S\dot{0}1^T$	0	0	0	$S\dot{0}2^T$	$S\dot{0}6^T$	0	0	0	0	0	0	0	0
B_8	0	I	$S\dot{0}1^T$	0	0		$S\dot{0}2^T$	0	$S\dot{0}6^T$	0	0	0	0	0	0	0
B_9	0	I	$S\dot{0}1^T$	0	0	0	$S\dot{0}2^T$	0	$S\dot{0}6^T$	$S\dot{0}8^T$	0	0	0	0	0	0
B_{10}	0	I	$S\dot{0}1^T$	0	0	0	$S\dot{0}2^T$	0	$S\dot{0}6^T$	$S\dot{0}8^T$	$S\dot{0}9^T$	0	0	0	0	0
B_{11}	0	I	$S\dot{0}1^T$	0	0	0	$S\dot{0}2^T$	0	$S\dot{0}6^T$	0	0	$S\dot{0}8^T$	0	0	0	0
B_{12}	0	I	$S\dot{0}1^T$	0	0	0	$S\dot{0}2^T$	0	$S\dot{0}6^T$	0	0	$S\dot{0}8^T$	$S\dot{0}11^T$	0	0	0
B_{13}	0	I	0	0	0	0	0	0	0	0	0	0	0	$S\dot{0}1^T$	0	0
B_{14}	0	I	0	0	0	0	0	0	0	0	0	0	0	$S\dot{0}1^T$	$S\dot{0}13^T$	0
B_{15}	0	I	0	0	0	0	0	0	0	0	0	0	0	0	0	$S\dot{0}1^T$

13.8 Joint and Mass Center Position Vectors

Consider a series of bodies in a typical branch of a multibody system as in Figure 13.8.1. Let the bodied be labeled as B_h, B_i, B_j, and B_k as shown and let the order of their numbers follow the alphabetic sequence. That is, $h < i < j < k$. Let the

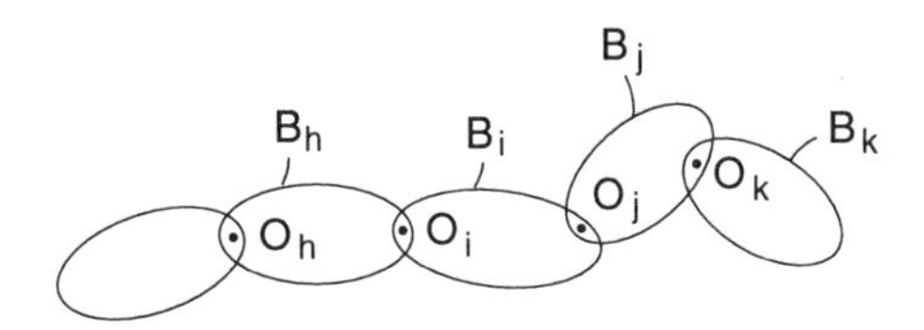

Fig. 13.8.1 A Sequence of Bodies in a Typical Branch of a Multibody System

connecting joints of the bodies with their adjacent lower numbered bodies be labeled similarly. That is, for a typical body, say B_j, the connecting joint with the adjacent lower numbered body B_i, in O_j. Let the connecting joints serve as origins, or reference points, of the bodies respectively. For example, O_j is the reference point of B_j.

Next, focus upon a typical body, say B_j, of the branch as in Figure 13.8.2. Let G_j be the mass center of B_j and let $\mathbf{r}_j$ locate G_j relative to O_j. Let O_k be the connection joint of B_j with an adjacent higher numbered body, say B_k, and let ξ_k locate O_k relative to O_j. Then both $\mathbf{r}_j$ and ξ_k are fixed in B_j.

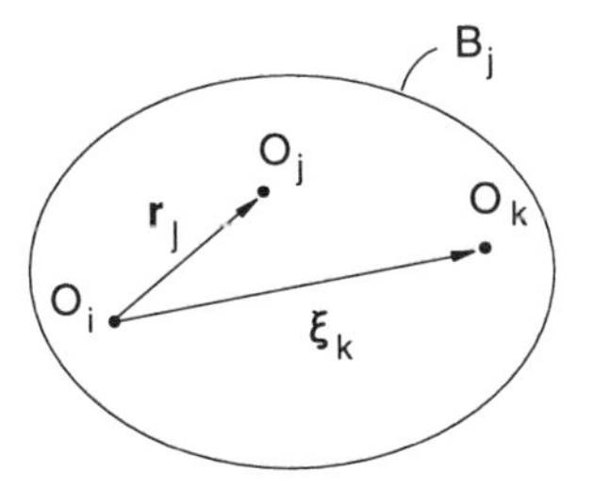

Fig. 13.8.2 A Typical Body of the System

Finally, suppose B_j is a branching body connected to two or more adjacent higher numbered bodies such as B_k and B_ℓ as in Figure 13.8.3. Then the position

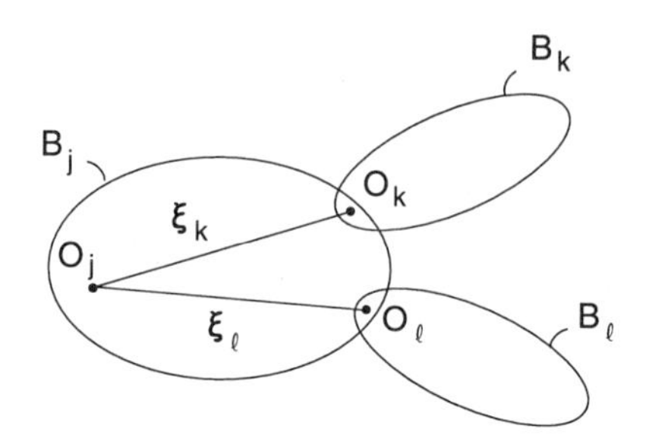

Fig. 13.8.3 A typical Branching Body with Connecting Joints and Position Vectors

vectors locating the connecting joints O_k and O_ℓ relative to O_j are ξ_k and ξ_ℓ as shown. Then both ξ_k and ξ_ℓ are fixed in B_j.

13.8.1 Position Vectors

The foregoing notation may be used to obtain the position vectors locating the body mass centers and the connecting joints relative to a point in a fixed (inertial) frame. To see this consider a multibody system as represented in Figure 13.8.4 where B_1 is the reference body and B_k is a typical body of the system. Let O_1 be the reference point (or origin) of B_1 and let ξ_1 locate O_1 relative to the origin of an inertial frame R. Then unlike the joint position vectors of the foregoing section (that is, ξ_k, $k = 2,\ldots,N$) ξ_1 is not fixed in any body. However, ξ_1 may be expressed in the simple form:

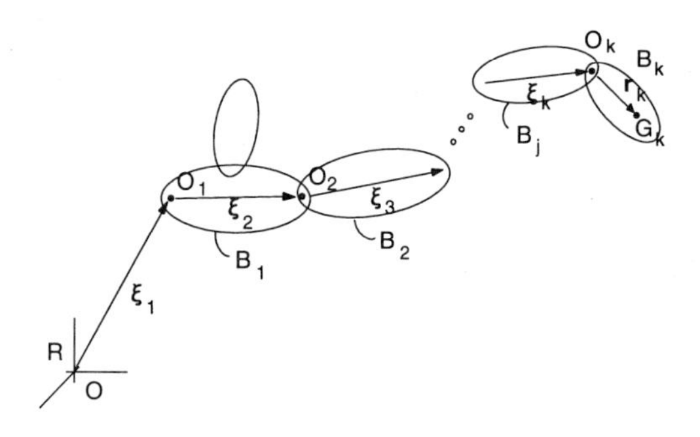

Fig. 13.8.4 Position Vectors in an Open-Chain Multibody System

$$\boldsymbol{\xi}_1 = x\mathbf{n}_{01} + y\mathbf{n}_{02} + z\mathbf{n}_{0z} \tag{13.8.1}$$

where, as before, x, y, and z are the coordinates of O_1 in a Cartesian frame fixed in R, and the $\mathbf{n}_{0i}$ $(i = 1,2,3)$ are mutually perpendicular unit vectors fixed in R.

Let O_k be the center of the connecting joint of B_k with its adjacent lower numbered body B_j. Then O_k may be located relative to O by the vector sum

$$\mathbf{q}_k = \boldsymbol{\xi}_1 + \boldsymbol{\xi}_2 + \cdots + \boldsymbol{\xi}_j + \boldsymbol{\xi}_k \tag{13.8.2}$$

where the sum is taken over the bodies in the branches of the system containing B_k. Then the position vector $\mathbf{p}_k$ locating the mass center G_k of B_k relative to O is simply

$$\mathbf{p}_k = \mathbf{q}_k + \mathbf{r}_k = \boldsymbol{\xi}_1 + \boldsymbol{\xi}_2 + \cdots + \boldsymbol{\xi}_j + \boldsymbol{\xi}_k \tag{13.8.3}$$

where as before $\mathbf{r}_k$ locates G_k relative to O_k.

13.8.2 Mass Center Position Vectors for the Example Multibody System

Consider again the illustrative multibody system of Figure 13.5.1. By following the pattern of Equation (13.8.3) the mass center position vectors are:

$$\begin{aligned}
B_1 &: \mathbf{p}_1 = \boldsymbol{\xi}_1 + \mathbf{r}_1 \\
B_2 &: \mathbf{p}_2 = \boldsymbol{\xi}_1 + \boldsymbol{\xi}_2 + \mathbf{r}_2 \\
B_3 &: \mathbf{p}_3 = \boldsymbol{\xi}_1 + \boldsymbol{\xi}_2 + \boldsymbol{\xi}_3 + \mathbf{r}_3 \\
B_4 &: \mathbf{p}_4 = \boldsymbol{\xi}_1 + \boldsymbol{\xi}_2 + \boldsymbol{\xi}_3 + \boldsymbol{\xi}_4 + \mathbf{r}_4 \\
B_5 &: \mathbf{p}_5 = \boldsymbol{\xi}_1 + \boldsymbol{\xi}_2 + \boldsymbol{\xi}_3 + \boldsymbol{\xi}_5 + \mathbf{r}_5 \\
B_6 &: \mathbf{p}_6 = \boldsymbol{\xi}_1 + \boldsymbol{\xi}_2 + \boldsymbol{\xi}_6 + \mathbf{r}_6 \\
B_7 &: \mathbf{p}_7 = \boldsymbol{\xi}_1 + \boldsymbol{\xi}_2 + \boldsymbol{\xi}_6 + \boldsymbol{\xi}_7 + \mathbf{r}_7 \\
B_8 &: \mathbf{p}_8 = \boldsymbol{\xi}_1 + \boldsymbol{\xi}_2 + \boldsymbol{\xi}_6 + \boldsymbol{\xi}_8 + \mathbf{r}_8 \\
B_9 &: \mathbf{p}_9 = \boldsymbol{\xi}_1 + \boldsymbol{\xi}_2 + \boldsymbol{\xi}_6 + \boldsymbol{\xi}_8 + \boldsymbol{\xi}_9 + \mathbf{r}_9 \\
B_{10} &: \mathbf{p}_{10} = \boldsymbol{\xi}_1 + \boldsymbol{\xi}_2 + \boldsymbol{\xi}_6 + \boldsymbol{\xi}_8 + \boldsymbol{\xi}_9 + \boldsymbol{\xi}_{10} + \mathbf{r}_{10} \\
B_{11} &: \mathbf{p}_{11} = \boldsymbol{\xi}_1 + \boldsymbol{\xi}_2 + \boldsymbol{\xi}_6 + \boldsymbol{\xi}_8 + \boldsymbol{\xi}_{11} + \mathbf{r}_{11} \\
B_{12} &: \mathbf{p}_{12} = \boldsymbol{\xi}_1 + \boldsymbol{\xi}_2 + \boldsymbol{\xi}_6 + \boldsymbol{\xi}_8 + \boldsymbol{\xi}_{11} + \boldsymbol{\xi}_{12} + \mathbf{r}_{12} \\
B_{13} &: \mathbf{p}_{13} = \boldsymbol{\xi}_1 + \boldsymbol{\xi}_{13} + \mathbf{r}_{13} \\
B_{14} &: \mathbf{p}_{14} = \boldsymbol{\xi}_1 + \boldsymbol{\xi}_{13} + \boldsymbol{\xi}_{14} + \mathbf{r}_4 \\
B_{15} &: \mathbf{p}_{15} = \boldsymbol{\xi}_1 + \boldsymbol{\xi}_{15} + \mathbf{r}_{15}
\end{aligned} \tag{13.8.4}$$

13.8.3 Generalization

Observe in Equations (13.8.4) that the subscripts on the $\boldsymbol{\xi}$ position vectors follow the sequences of the columns of the table of lower body arrays. For example, for Body B_{12}, the subscripts are: 1, 2, 6, 8, 11, 12, which are identical to the numbers in column 12 of Table 13.5.1. Then by the pattern established by the subscript sequences and by the lower body arrays, we can express the mass center position vectors in the compact form as:

$$\mathbf{p}_k = \mathbf{r}_k + \sum_{s=0}^{r} \boldsymbol{\xi}_q \qquad k = 1,\ldots,N \tag{13.8.5}$$

where the indices q and r are determined from the expressions:

$$q = L^s(k) \quad \text{and} \quad L^r(k) = 1 \tag{13.8.6}$$

Observe further that $\boldsymbol{\xi}_q$ may be expressed in terms of the $\mathbf{n}_{0m}$ $(m = 1,2,3)$ as:

$$\boldsymbol{\xi}_q = SOP_{mn} \xi_{qn} \mathbf{n}_{0m} \tag{13.8.7}$$

where the ξ_{qn} $(n = 1,2,3)$ are the $\mathbf{n}_{pn}$ components of $\boldsymbol{\xi}_q$. [The $\mathbf{n}_{pn}$ $(n = 1,2,3)$ are unit vectors fixed in B_p.] Then by substituting from Equation (13.8.7) into Equation (13.8.5) the p_k may be expressed as:

$$\mathbf{p}_k = (SOK_{mn} r_{kn} + \sum_{s=0}^{r} SOP_{mn} \xi_{qn}) \mathbf{n}_{0m} \qquad k = 1,\ldots,N \tag{13.8.8}$$

where the r_{kn} $(n = 1,2,3)$ are the $\mathbf{n}_{kn}$ components of $\mathbf{r}_k$ and where as before P, q, and r are given by

$$P = p = L(q) \qquad q = L^s(k) \qquad L^r(k) = 1 \tag{13.8.9}$$

Finally, observe in Equation (13.8.8) that S00 is the identity matrix. Hence, the last term in the sum is simply ξ_{1m}. Therefore, Equation (13.8.8) may be further expanded to the form:

$$\mathbf{p}_k = (S0K_{mn} r_{kn} + \xi_{1m} + \sum_{s=0}^{r-1} S0P_{mn} \xi_{qn})\mathbf{n}_{0m} \tag{13.8.10}$$

where P, q, and r are given by Equation (13.8.9).

13.9 Mass Center Velocities

Expressions for the mass center velocities of a multibody system may be obtained by simply differentiating in the expressions for the mass center positions. To illustrate and develop this, let the mass center position vectors $\mathbf{p}_k$ of an N-body spherically connected open system be given by Equation (13.8.10) as

$$\mathbf{p}_k = [\xi_{1n} + (\sum_{s=0}^{r-1} S0P_{mn} \xi_{qn}) + S0K_{mn} r_{kn}]\mathbf{n}_{0m} \tag{13.9.1}$$

where as before the $\mathbf{n}_{0m}$ are unit vectors fixed in an inertial reference frame R and where from Equations (13.8.9) P, q, and r are given by

$$P - p = L(q) \quad , \quad q = L^s(k) \quad , \quad L^r(k) = 1 \tag{13.9.2}$$

(The $\mathbf{p}_k$ locate the mass centers G_k relative to a fixed point O in R.) Then the velocities $\mathbf{v}_k$ of G_k in R are immediately obtained by differentiating in Equation (13.9.1) as

$$\mathbf{v}_k = \left[\dot{\xi}_{1m} + \left(\sum_{s=0}^{r-1} S\dot{0}P_{mn} \xi_{qn}\right) + S\dot{0}K_{mn} r_{kn}\right]\mathbf{n}_{0m} \tag{13.9.3}$$

where from Equations (12.5.3) and (12.5.4) the transformation matrix derivatives are

$$S\dot{O}K = WK\ SOK \tag{13.9.4}$$

where the elements of the WK arrays are

$$WK_{ij} = -e_{ijm}\Omega_{km} = \begin{bmatrix} 0 & -\Omega_{k3} & \Omega_{k2} \\ \Omega_{k3} & 0 & -\Omega_{k1} \\ -\Omega_{k2} & \Omega_{k1} & 0 \end{bmatrix} \tag{13.9.5}$$

where the Ω_{km} are the $\mathbf{n}_{0m}$ components of the angular velocity $\boldsymbol{\omega}_k$ of the Body B_k in R.

From Equation (13.6.2) the $\boldsymbol{\omega}_k$ are all seen to have the form

$$\boldsymbol{\omega}_k = \omega_{k\ell m} y_\ell \mathbf{n}_{0m} \tag{13.9.6}$$

where the y_ℓ $\ell = 1,\ldots,3N+3$ are the generalized speeds (see Section 13.4) of the system and the $\omega_{k\ell m}$ form arrays of components of the partial angular velocity vectors. Therefore, in Equation (13.9.5) the Ω_{km} are

$$\Omega_{km} = \omega_{k\ell m} y_\ell \tag{13.9.7}$$

Then from Equations (13.9.5) the elements of the WK arrays may be expressed as:

$$WK_{ij} = -e_{ijm}\,\omega'_{k\ell m} y_\ell \tag{13.9.8}$$

Hence, the elements of the arrays of transformation matrix derivatives may be expressed as:

$$S\dot{O}K_{ij} = -e_{inm}\omega_{k\ell m}y_\ell SOK_{nj} \tag{13.9.9}$$

By substituting from Equations (13.9.9) into (13.9.3) we see that the $\mathbf{v}_k$ may be expressed as:

$$\begin{aligned} \mathbf{v}_k = [\dot{\xi}_{1m} - (\sum_{s=0}^{r-1} e_{mij}\,\omega_{p\ell j}y_\ell\,SOP_{in}\,\xi_{qn}) \\ + e_{mij}\,\omega_{k\ell j}y_\ell\,SOK_{in}r_{kn}]\,\mathbf{n}_{0m} \\ (k = 1,\dots,N) \end{aligned} \tag{13.9.10}$$

where P, q, and r are given by Equation (13.9.2).

The detail of Equation (13.9.10) can be reduced by defining the following block arrays:

$$UOK_{m\ell n} \stackrel{D}{=} -e_{mij}\,\omega_{k\ell j}\,SOK_{in} \tag{13.9.11}$$

Then the $\mathbf{v}_k$ may be written as:

$$\begin{aligned} \mathbf{v}_k = [\dot{\xi}_{1m} + (\sum_{s=0}^{r-1} UOP_{m\ell n}y_\ell\,\xi_{qn}) \\ + UOK_{m\ell n}y_\ell r_{kn}]\,\mathbf{n}_{0m} \\ (k = 1,\dots,N) \end{aligned} \tag{13.9.12}$$

where as before P, q, and r are given by Equation (13.9.2).

Finally, by recalling that the $\dot{\xi}_{1m}$ $(m = 1,2,3)$ may be identified with the generalized speeds y_ℓ $(\ell = 1,2,3)$, the $\mathbf{v}_k$ may be written (analogous to the angular velocities) as:

$$\mathbf{v}_k = v_{k\ell m}y_\ell\mathbf{n}_{0m} \qquad (k = 1,\dots,N) \tag{13.9.13}$$

where by comparison with Equations (13.9.12) the $v_{k\ell m}$ are seen to be:

$$v_{k\ell m} = \delta_{km} \qquad k = 1,\ldots,N \ ; \quad \ell = 1,2,3 \ ; \quad m = 1,2,3 \tag{13.9.14}$$

and

$$v_{k\ell m} = \left(\sum_{s=0}^{r-1} UOP_{m\ell n} \xi_{qn} \right) + UOK_{m\ell n} r_{kn} \tag{13.9.15}$$
$$k = 1,\ldots,N \ ; \quad \ell = 4,\ldots,3N+3 \ ; \quad m = 1,2,3$$

where again P, p, and r are given by Equation (13.9.2). (Note that the $v_{k\ell m}$ are the $\mathbf{n}_{0m}$ components of the partial velocity vectors of G_k for y_ℓ.)

13.10 Mass Center Accelerations

Expressions for the mass center acceleration may be obtained by differentiating in the expressions for the mass center velocities. Specifically, from Equations (13.9.13) the mass center velocities are

$$\mathbf{v}_k = v_{k\ell m} y_\ell \mathbf{n}_{0m} \tag{13.10.1}$$

where the $v_{k\ell m}$ are given by Equations (13.9.14) and (13.9.15). Then the mass center accelerations $\mathbf{a}_k$ are:

$$\mathbf{a}_k = (v_{k\ell m} \dot{y}_\ell + \dot{v}_{k\ell m} y_\ell)\mathbf{n}_{0m} \qquad (k = 1,\ldots,N) \tag{13.10.2}$$

where from Equations (13.9.14) and (13.9.15) the $v_{k\ell m}$ and $\dot{v}_{k\ell m}$ are:

$$v_{k\ell m} = \delta_{km} \qquad k = 1,\ldots,N \ ; \quad \ell = 1,2,3 \ ; \quad m = 1,2,3 \tag{13.10.3}$$

$$v_{k\ell m} = \left(\sum_{s=0}^{r-1} UOP_{m\ell n}\xi_{qn}\right) + UOK_{m\ell n}r_{kn} \tag{13.10.4}$$

$$k = 1,\ldots,N;\ \ell = 4,\ldots,3N+3;\ m = 1,2,3$$

$$\dot{v}_{k\ell m} = 0 \quad k = 1,\ldots,N;\ \ell = 1,2,3;\ m = 1,2,3 \tag{13.10.5}$$

$$\dot{v}_{k\ell m} = \left(\sum_{s=0}^{r-1} U\dot{O}P_{m\ell n}\xi_{qn}\right) + U\dot{O}K_{m\ell n}r_{kn} \tag{13.10.6}$$

$$k = 1,\ldots,N;\ \ell = 1,2,3;\ m = 1,2,3$$

where from Equations (13.9.11) the $U\dot{O}K_{m\ell n}$ are seen to be

$$U\dot{O}K_{m\ell n} = -e_{mij}(\dot{\omega}_{k\ell j}SOK_{in} + \omega_{k\ell j}S\dot{O}K_{in}) \tag{13.10.7}$$

where the $S\dot{O}K_{in}$ are given by Equation (13.9.9) and in view of the definition of Equation (13.9.11) may be written as:

$$S\dot{O}K_{in} = UOK_{i\ell n}y_{\ell} \tag{13.10.8}$$

13.11 Summary

The principal kinematic relations for an open-tree, spherical joint connected multibody system are presented by Equations (13.6.2), (13.7.2), (13.9.13), and (13.10.2). Specifically, for a typical body B_k of the system these equations provide expressions for the angular velocity and angular acceleration of the body and for the velocity and acceleration of the mass center G_k of the body.

Remarkably, these equations (repeated here) have similar and relatively simple forms:

1. Angular Velocity $\quad \boldsymbol{\omega}_k = \omega_{k\ell m} y_\ell \mathbf{n}_{0m}$ (13.11.1)
Equation (13.6.2)

2. Angular Acceleration $\quad \boldsymbol{\alpha}_k = (\omega_{k\ell m} \dot{y}_\ell + \dot{\omega}_{k\ell m} y_\ell)\mathbf{n}_{0m}$ (13.11.2)
Equation (13.7.2)

3. Mass Center Velocity $\quad \mathbf{v}_k = v_{k\ell m} y_\ell \mathbf{n}_{0m}$ (13.11.3)
Equation (13.9.13)

4. Mass Center Acceleration $\quad \mathbf{a}_k = (v_{k\ell m} \dot{y}_\ell + \dot{v}_{k\ell m} y_\ell)\mathbf{n}_{0m}$ (13.11.4)
Equation (13.10.2)

where, as before, if there are N bodies in the system $(k = 1, \ldots, N)$ the y_ℓ $(\ell = 1, \ldots, 3N+3)$ are generalized speeds describing the motion of the system, and where the $\mathbf{n}_{0m}$ $(m = 1,2,3)$ are mutually perpendicular unit vectors fixed in an inertia reference frame R.

In Equations (13.11.1) to (13.11.4) the $\omega_{k\ell m}$, $\dot{\omega}_{k\ell m}$, $v_{k\ell m}$, and $\dot{v}_{k\ell m}$ form block arrays with dimensions: $N \times (3N + 3) \times 3$ (for k, ℓ, and m respectively). As seen in the example system the $\omega_{k\ell m}$ may be expressed in terms of transformation matrices (see Table 13.5.4), with the aid of lower body arrays. Once the $\omega_{k\ell m}$ are known, the $\dot{\omega}_{k\ell m}$ may be immediately obtained by differentiation.

Interestingly, the $v_{k\ell m}$ and the $\dot{v}_{k\ell m}$ may also be obtained directly from the $\omega_{k\ell m}$. Specifically, from Equations (13.9.4) and (13.9.15) the $v_{k\ell m}$ are:

$$v_{k\ell m} = \delta_{km} \qquad \ell = 1,2,3$$

and (13.11.5)

$$v_{k\ell m} = \left(\sum_{s=0}^{r-1} \mathrm{U0P}_{m\ell n} \xi_{qn} \right) + \mathrm{U0K}_{m\ell n} r_{kn} \qquad \ell = 4, \ldots, 3N+3$$

where from Equation (13.9.11) the $\mathrm{U0K}_{m\ell n}$ are

$$\mathrm{U0K}_{m\ell n} = -e_{mij}\, \omega_{k\ell j}\, \mathrm{S0K}_{in} \tag{13.11.6}$$

where the S0K are transformation matrices (see Section 12.4) and where as before the P, q, and r are [see Equation (13.9.2)]:

$$P = p = L(q) \qquad q = L^s(k) \qquad L^r(k) = 1 \tag{13.11.7}$$

Similarly, from Equations (13.10.5) and (13.10.6) the $\dot{v}_{k\ell m}$ are:

$$\dot{v}_{k\ell m} = 0 \qquad \ell = 1,2,3$$

and

$$\dot{v}_{k\ell m} = \left(\sum_{s=0}^{r-1} U\dot{0}P_{m\ell n}\,\xi_{qn}\right) + U\dot{0}K_{m\ell n} r_{kn} \qquad \ell = 4,\dots,3N+3 \tag{13.11.8}$$

where from Equation (13.10.7) the $U\dot{0}K_{m\ell n}$ are

$$U\dot{0}K_{m\ell n} = -e_{mij}(\dot{\omega}_{k\ell j} S0K_{in} + \omega_{k\ell j} S\dot{0}K_{in}) \tag{13.11.9}$$

and where from Equation (13.10.8) the $S\dot{0}K_{in}$ are

$$S\dot{0}K_{in} = U0K_{i\ell n} y_\ell \tag{13.11.10}$$

[In Equations (13.11.5) through (13.11.10) k and m have the ranges: 1,...,N and 1,2,3 respectively.]

Equations (13.11.5) and (13.11.8) and their subsequent defining equations show the principal role played by the $\omega_{k\ell m}$ and the lower body arrays in providing for an automated (numerical) development of multibody system kinematics (see Reference 13.1). Finally, recall that the $\omega_{k\ell m}$ and the $v_{k\ell m}$ are the $\mathbf{n}_{0m}$ components of the partial angular velocity of body B_k and the partial velocity of its mass center G_k for the generalized speeds y_ℓ.

References

13.1 R. L. Huston, *Multibody Dynamics*, Butterworth Heinemann, Boston, 1990, pp. 292-293.

13.2 T. R. Kane and D. A. Levinson, *Dynamics, Theory and Applications*, McGraw Hill, New York, 1985.

13.3 R. L. Huston, C. E. Passerello, and M. W. Harlow, "Dynamics of Multirigid-Body Systems," *Journal of Applied Mechanics*, Vol. 45, No. 4, 1978, pp. 889-894.

13.4 H. Josephs and R. L. Huston, *Dynamics of Mechanical Systems*, CRC Press, Boca Raton, FL, 2001.

Chapter 14

MULTIBODY KINETICS AND DYNAMICS

14.1 Introduction

In this final chapter we outline the procedures for determining the kinetics (forces) and the subsequent dynamics of multibody systems. We use the same notation and nomenclature as in Chapter 13. Specifically, we focus our attention on open-chain systems with spherical connecting joints.

We use Kane's equations and their associated procedures for determining the kinetics (generalized applied and inertia forces) and the subsequent dynamical equations of motion. That is, we use Euler parameters, generalized speeds, partial angular velocities and partial velocities as in Sections 13.3, 13.4, 13.5, and 13.9.

14.2 Generalized Applied (Active) Forces

Consider a multibody system S subjected to a general applied force field as in Figure 14.2.1. The force field could consist of forces external to the system, such

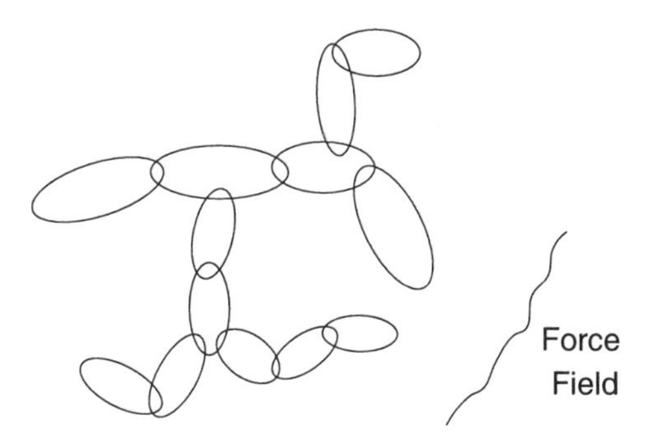

Fig. 14.2.1 A Multibody System in a Generalized Force Field

as gravity and contact forces, and also of forces internal to the system such as contact forces between the bodies arising from joint constraints and/or actuators. Also, internal forces could arise from springs and dampers between the bodies.

In this section we consider the effects of externally applied forces. We will consider internal forces in the next section.

As before let S be an open system consisting of N bodies, connected with spherical joints. If otherwise unconstrained, S will then have $3N+3$ degrees of freedom. Let these degrees of freedom be represented by generalized speeds y_s $(s = 1,\dots,3N+3)$ (see Section 13.4).

Consider a typical body B_k of S as in Figure 14.2.2. Let B_k be subjected to a force system as represented in Figure 14.2.2a and let the force system in turn be represented by an equivalent force system consisting of a single force $\mathbf{F}_k$ passing through the mass center G_k of B_k together with a couple with torque $\mathbf{M}_k$, as represented in Figure 14.2.2b. Then the generalized applied (active) force $F_\ell^{B_k}$ on B_k for y_ℓ may be expressed as [see Equation (9.5.7)]:

$$F_\ell^{B_k} = \mathbf{F}_k \cdot \mathbf{V}_{y_\ell}^{G_k} + \mathbf{M}_k \cdot \boldsymbol{\omega}_{y_\ell}^{B_k} \tag{14.2.1}$$

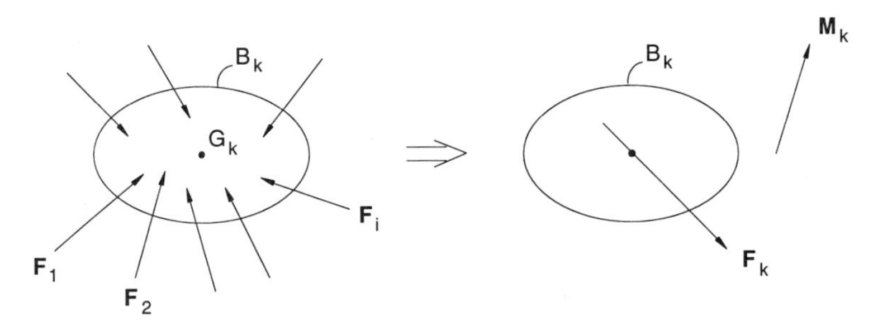

Fig. 14.2.2 Applied Forces on a Typical Body and an Equivalent Force System

where, as before, $\mathbf{V}_{y_\ell}^{G_k}$ and $\boldsymbol{\omega}_{y_\ell}^{B_k}$ are the partial velocity of G_k for y_ℓ and the partial angular velocity of B_k for y_ℓ.

From Equations (13.9.13) and (13.9.4) we see that $\mathbf{V}_{y\ell}^{G_k}$ and $\boldsymbol{\omega}_{y\ell}^{B_k}$ may be written in compact forms:

$$\mathbf{V}_{y\ell}^{G_k} = V_{k\ell m}\mathbf{n}_{0m} \quad \text{and} \quad \boldsymbol{\omega}_{y\ell}^{B_k} = \omega_{k\ell m}\mathbf{n}_{0m} \tag{14.2.2}$$

where as before the $\mathbf{n}_{0m}$ $(m = 1,2,3)$ are mutually perpendicular unit vectors fixed in an inertial frame R.

In Equation (14.2.1) $\mathbf{F}_k$ and $\mathbf{M}_k$ may also be expressed in terms of the $\mathbf{n}_{0m}$ as:

$$\mathbf{F}_k = F_{km}\mathbf{n}_{0m} \quad \text{and} \quad \mathbf{M}_k = M_{km}\mathbf{n}_{0m} \tag{14.2.3}$$

By substituting from Equations (14.2.2) and (14.2.3) into Equation (14.2.1) we see that the generalized applied force $F_\ell^{B_k}$ on B_k may be expressed as

$$F_\ell^{B_k} = F_{km}V_{k\ell m} + M_{km}\omega_{k\ell m} \quad (k = 1,\ldots,N;\ \ell = 1,\ldots,3N+3) \tag{14.2.4}$$

where there is no sum on k.

Observe in Equation (14.2.4) that the generalized force is computed for body B_k (and hence, of course, there is then no sum on k). To obtain the generalized applied force F_ℓ for the entire system S, for any given generalized speed, we need simply add the generalized forces for the individual bodies of S for that generalized speed. That is, for S, the generalized applied forces are given by

$$F_\ell = \sum_{k=1}^{N} F_\ell^{B_k} = F_{km}V_{k\ell m} + M_{km}\omega_{k\ell m} \quad \ell = 1,\ldots,3N+3 \tag{14.2.5}$$

14.3 Applied Forces Between Bodies and at Connecting Joints

Consider forces exerted by adjoining bodies on one another. These forces could arise as joint forces, as spring or damper forces, or as actuator forces. Such forces could also be used to model flexibility of bodies of the system.

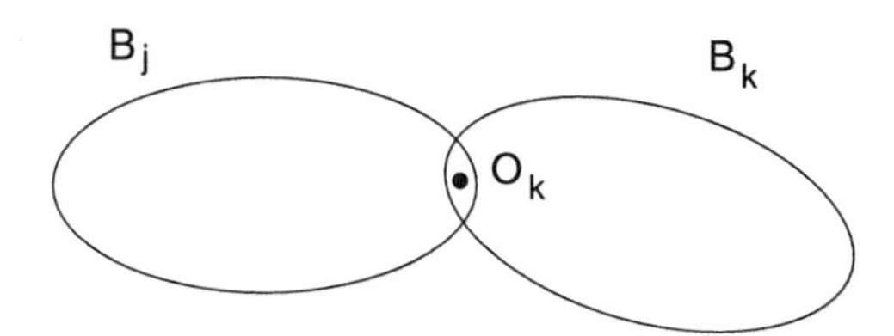

Fig. 14.3.1 Two Typical Adjoining Bodies of a Multibody System

Consider two typical adjoining bodies such as B_j and B_k as in Figure 14.3.1. Let O_k be the center of the spherical connecting joint. Let the force system exerted on B_k by B_j be represented by a single force $\hat{\mathbf{F}}_k$ passing through O_k together with a couple with torque $\hat{\mathbf{M}}_k$. Similarly, let the force system exerted by B_k on B_j be represented by a single force $\hat{\mathbf{F}}_j$ passing through O_k together with a couple with torque $\hat{\mathbf{M}}_j$ as in Figure 14.3.2a and b. Then by the law of action and reaction $\hat{\mathbf{F}}_j$ and $\hat{\mathbf{F}}_k$, and $\hat{\mathbf{M}}_j$ and $\hat{\mathbf{M}}_k$ are related by the expressions:

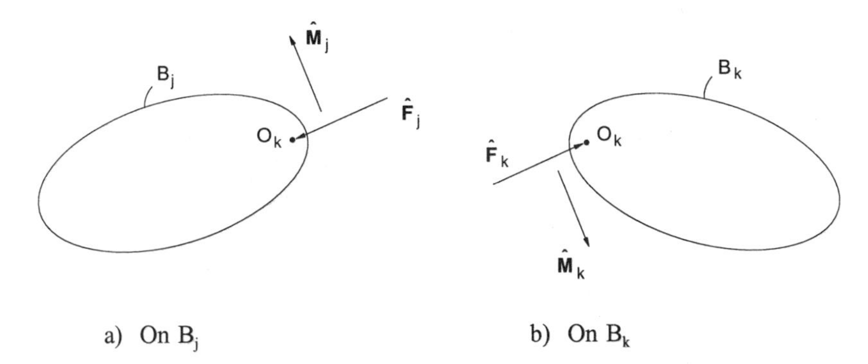

a) On B_j b) On B_k

Fig. 14.3.2 Equivalent Force Systems on Bodies B_j and B_k

$$\hat{\mathbf{F}}_k = -\hat{\mathbf{F}}_j \quad \text{and} \quad \hat{\mathbf{M}}_k = -\hat{\mathbf{M}}_j \tag{14.3.1}$$

From Equation (9.9.4) the contribution $\hat{F}_s$ of these forces to the generalized applied forces may be expressed as:

$$\hat{F}_s = \mathbf{V}_{y_s}^{O_k} \bullet \hat{\mathbf{F}}_j + \boldsymbol{\omega}_{y_s}^{B_j} \bullet \hat{\mathbf{M}}_j + \mathbf{V}_{y_s}^{O_k} \bullet \hat{\mathbf{F}}_k + \boldsymbol{\omega}_{y_s}^{B_k} \bullet \hat{\mathbf{M}}_k \quad s = 1, \ldots, 3N+3 \tag{14.3.2}$$

In view of Equation (14.3.1) $\hat{F}_\ell$ may be expressed as:

$$\hat{F}_s = (\boldsymbol{\omega}_{y_s}^{B_k} - \boldsymbol{\omega}_{y_s}^{B_j}) \bullet \hat{\mathbf{M}}_k = {}^{B_j}\boldsymbol{\omega}_{y_s}^{B_k} \bullet \hat{\mathbf{M}}_k \tag{14.3.2}$$

Suppose that as before [see Equation (13.5.1)] the angular velocity of B_k relative to B_j is expressed as:

$${}^{B_j}\boldsymbol{\omega}^{B_k} = \hat{\boldsymbol{\omega}}_k = \hat{\omega}_{k1}\mathbf{n}_{j1} + \hat{\omega}_{k2}\mathbf{n}_{j2} + \hat{\omega}_{k3}\mathbf{n}_{j3} \tag{14.3.3}$$

where as before the $\mathbf{n}_{jm}$ $(m = 1,2,3)$ are mutually perpendicular unit vectors fixed in B_j. Then as before the generalized speeds y_s (except for the translation variables of B_1) are identified with the relative angular velocity components $\hat{\omega}_{km}$. Specifically,

$$y_s = \hat{\omega}_{km} \quad \text{where} \quad s = 3k + m \tag{14.3.4}$$

Then in Equation (14.3.2) the ${}^{B_i}\boldsymbol{\omega}_{y_s}^{B_k}$ are

$${}^{B_j}\boldsymbol{\omega}_{y_s}^{B_k} = \begin{cases} 0 & s \neq 3k+m \\ \mathbf{n}_{jm} & s = 3k+m \end{cases} \quad (m = 1,2,3) \tag{14.3.5}$$

Let the joint torque components $\hat{\mathbf{M}}_k$ be expressed as:

$$\hat{\mathbf{M}}_k = \hat{\mathbf{M}}_{km}\mathbf{n}_{jm} = \hat{M}_{k1}\mathbf{n}_{j1} + \hat{M}_{k2}\mathbf{n}_{j2} + \hat{M}_{k3}\mathbf{n}_{j3} \tag{14.3.6}$$

Then in view of Equations (14.3.5) the non-zero contributions to the generalized applied forces are simply components of the torque components. Specifically, the $\hat{F}_s$ are

$$\hat{F}_s = \begin{cases} 0 & s \neq 3k+m \\ \hat{M}_{km} & s = 3k+m \end{cases} \quad (m = 1,2,3) \tag{14.3.7}$$

Observe that there is a one-to-one uncoupled relation between the joint torque components and the contribution to the generalized forces.

14.4 Generalized Inertia (Passive) Forces

The generalized inertia forces for each generalized speed can be developed in a similar manner. Consider again a typical body B_k of a multibody system S moving in an inertial reference frame R as in Figure 14.4.1. Let the inertia forces on the particles of B_k be represented by an equivalent force system consisting of a single force $\mathbf{F}_k^*$ passing through the mass center G_k together with a couple with torque $\mathbf{M}_k^*$ as represented in Figure 14.4.1. Then from Equation (9.12.6) and (9.12.11) $\mathbf{F}_k^*$ and $\mathbf{M}_k^*$ may be expressed as:

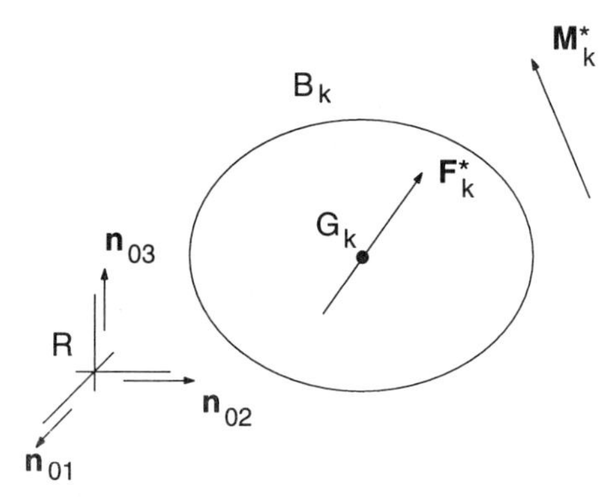

Fig. 14.4.1 Equivalent Inertia Force System on a Typical Body B_k of a Multibody System S

$$\mathbf{F}_k^* = -m_k\mathbf{a}_k \quad \text{(no sum on k)} \tag{14.4.1}$$

and

$$\mathbf{M}_k^* = -\mathbf{I}_k \bullet \boldsymbol{\alpha}_k - \boldsymbol{\omega}_k \times \mathbf{I}_k \bullet \boldsymbol{\omega}_k \quad \text{(no sum on k)} \tag{14.4.2}$$

where m_k is the mass of B_k, $\mathbf{I}_k$ is the central inertia dyadic, $\mathbf{a}_k$ is the acceleration of G_k in R, $\boldsymbol{\omega}_k$ is the angular velocity of B_k in R, and $\boldsymbol{\alpha}_k$ is the angular acceleration of B_k in R.

Let $\mathbf{F}_k^*$ and $\mathbf{M}_k^*$ be expressed in terms of mutually perpendicular unit vectors $\mathbf{n}_{0m}$ $(m = 1,2,3)$ fixed in R as:

$$\mathbf{F}_k^* = F_{km}^* \mathbf{n}_{0m} \quad \text{and} \quad \mathbf{M}_k^* = M_{km}^* \mathbf{n}_{0m} \tag{14.4.3}$$

The generalized inertia force F_ℓ^* on B_k for generalized speed y_ℓ may be expressed as [see Equations (9.5.7) and (14.2.1)]:

$$F_\ell^* = \mathbf{F}_k^* \bullet \mathbf{V}_{y_\ell}^{G_k} + \mathbf{M}_k^* \bullet \boldsymbol{\omega}_{y_\ell}^{B_k} \tag{14.4.4}$$

where, as before, $\mathbf{V}_{y_\ell}^{G_k}$ and $\boldsymbol{\omega}_{y_\ell}^{B_k}$ are the partial velocity of G_k for y_ℓ and the partial angular velocity of B_k for y_ℓ.

From Equations (13.9.13) and (13.9.6) we see that $\mathbf{V}_{y_\ell}^{G_k}$ and $\boldsymbol{\omega}_{y_\ell}^{B_k}$ may be written in the compact forms:

$$\mathbf{V}_{y_\ell}^{B_k} = v_{k\ell m} \mathbf{n}_{0m} \quad \text{and} \quad \boldsymbol{\omega}_{y_\ell}^{B_k} = \omega_{k\ell m} \mathbf{n}_{0m} \tag{14.4.5}$$

By substituting from Equations (14.4.3) and (14.4.5) into Equation (14.4.4) the generalized inertia force F_ℓ^* on B_k for generalized speed y_ℓ may be expressed as:

$$F_\ell^* = F_{km}^* v_{k\ell m} + M_{km}^* \omega_{k\ell m} \quad (k = 1,\dots,N;\ \ell = 1,\dots,3N+3) \tag{14.4.6}$$

where there is no sum on k.

Observe in Equation (14.4.6), as in Equation (14.2.4), that the generalized force is computed for Body B_k and thus there is no sum on k. However, to obtain the generalized force F_ℓ^* for the entire system S we need simply add the corresponding generalized forces for the individual bodies of S. That is, for the entire system S the generalized inertia force for the generalized speed y_ℓ is also given by Equation (14.4.6), but with a sum on k for the N bodies of S.

Using Equations (13.11.1) to (13.11.4) and Equations (14.4.1) and (14.4.2) we can obtain more explicit expressions for the generalized inertia forces: From Equations (13.11.1), (13.11.2), and (13.11.4) the angular velocity, the angular acceleration, and the mass center acceleration of a typical body B_k may be expressed as:

$$\boldsymbol{\omega}_k = \omega_{k\ell m} y_\ell \mathbf{n}_{0m} \tag{14.4.7}$$

$$\boldsymbol{\alpha}_k = (\omega_{k\ell m} \dot{y}_\ell + \dot{\omega}_{k\ell m} y_\ell)\mathbf{n}_{0m} \tag{14.4.8}$$

and

$$\mathbf{a}_k = (v_{k\ell m} \dot{y} + \dot{v}_{k\ell m} y_\ell)\mathbf{n}_{0m} \tag{14.4.9}$$

where the $\mathbf{n}_{0m}$ $(m = 1,2,3)$ are mutually perpendicular unit vectors fixed in the inertial reference frame R.

The central inertia dyadic $\mathbf{I}_k$ of Equation (14.4.2) may be written in the form:

$$\mathbf{I}_k = I_{kmn} \mathbf{n}_{0m} \mathbf{n}_{0n} \tag{14.4.10}$$

By substituting from Equations (14.4.7) to (14.4.10) into Equations (14.4.1) and (14.4.2) the inertia force $\mathbf{F}_k^*$ and inertia torque $\mathbf{T}_k^*$ may be expressed as:

$$\mathbf{F}_k^* = -m_k(v_{k\ell m} \dot{y}_\ell + \dot{v}_{k\ell m} y_\ell)\mathbf{n}_{0m} \quad \text{(no sum on k)} \tag{14.4.11}$$

and

$$\mathbf{T}_k^* = -(I_{kmn}\omega_{k\ell n}\dot{y}_\ell + I_{kmn}\dot{\omega}_{k\ell n}y_\ell + e_{tsm}I_{ksn}\omega_{k\ell t}\omega_{kpn}y_\ell y_p]\mathbf{n}_{0m} \tag{14.4.12}$$

$$\text{(no sum on k)}$$

By substituting from Equations (14.5.11) and (14.4.12) into Equations (14.4.6), the generalized inertia forces may be expressed as:

$$\begin{aligned} F_\ell^* = & -m_k v_{k\ell m} v_{kpm} \dot{y}_p - m_k v_{k\ell m} \dot{v}_{kpm} y_p \\ & - I_{kmn} \omega_{k\ell m} \omega_{kpn} \dot{y}_p - I_{kmn} \omega_{k\ell m} \dot{\omega}_{kpn} y_p \\ & - e_{tsm} I_{ksn} \omega_{k\ell m} \omega_{kpt} \omega_{kqn} y_p y_q \end{aligned} \tag{14.4.13}$$

14.5 Multibody Dynamics and Equations of Motion

Once the generalized forces (both the applied and inertia forces) are known, Kane's equations (see Section 10.8) are ideally suited to obtain the governing dynamical equations of motion.

To see this, consider again a multibody system S consisting of an open-chain of bodies connected with spherical joints moving in an inertial reference frame R as represented in Figure 14.5.1. As before, let S have N bodies with 3N+3 degrees

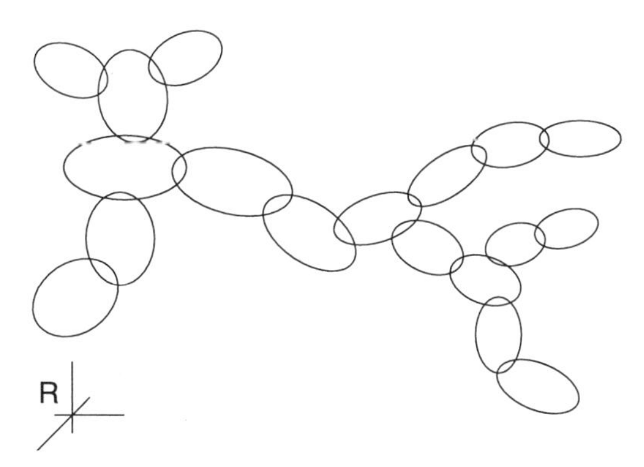

Fig. 14.5.1 An Open-Chain ("Open-Tree") Multibody System with Spherical Joints

of freedom described by generalized speeds y_s $(s = 1,...,3N+3)$. Aside from the first three of these, which describe the translation variables, let the generalized speeds be components of angular velocities of the bodies. Then from Equations (14.2.5) and (14.4.6) the generalized applied and inertia forces, F_ℓ and F_ℓ^*, for generalized speed y_ℓ may be expressed as:

$$F_\ell = F_{km} v_{k\ell m} + M_{km} \omega_{k\ell m} \tag{14.5.1}$$

and

$$F_\ell^* = F_{km}^* v_{k\ell m} + M_{km}^* \omega_{k\ell m} \tag{14.5.2}$$

where ℓ ranges from 1 to $3N+3$ and there is a sum on k from 1 to N and on m from 1 to 3. As in Equation (14.2.5) and (14.4.6) the F_{km}, F_{km}^*, M_{km}, and M_{km}^* are $\mathbf{n}_{0m}$ components of the equivalent applied and inertia force and couple torques where the $\mathbf{n}_{0m}$ are mutually perpendicular unit vectors fixed in R. Also, as before, the $v_{k\ell m}$ and the $\omega_{k\ell m}$ are $\mathbf{n}_{0m}$ components of the partial velocities of the mass centers and the partial angular velocities of the bodies themselves.

Kane's equations state that the sum of the generalized applied and inertia forces are zero for each generalized speed [14.1]. That is,

$$F_\ell + F_\ell^* = 0 \qquad (\ell = 1,...,3N+3) \tag{14.5.3}$$

Recall from Equations (14.4.11), (14.4.12), and (14.5.13) that by substituting expressions for the equivalent inertia force and torque components into Equation (14.5.2) the generalized inertia force F_ℓ^* may be expressed as [see Equation (14.4.13)]:

$$\begin{aligned} F_\ell^* = & -m_k v_{k\ell m} v_{kpm} \dot{y}_p - m_k v_{k\ell m} \dot{v}_{kpm} y_p \\ & - I_{kmn} \omega_{k\ell m} \omega_{kpn} \dot{y}_p - I_{kmn} \omega_{k\ell m} \dot{\omega}_{kpn} y_p \\ & - e_{tsm} I_{ksn} \omega_{k\ell m} \omega_{kpt} \omega_{kqn} y_p y_q \end{aligned} \tag{14.5.4}$$

Then using this expression, Equations (14.5.3) may be written in the compact form

$$a_{\ell p}\dot{y}_p = f_\ell \qquad \ell = 1,\ldots,3N+3 \tag{14.5.5}$$

where by inspection of Equations (14.5.3) and (14.5.4) $a_{\ell p}$ and f_ℓ are seen to be

$$a_{\ell p} = m_k v_{k\ell m} v_{kpm} + I_{kmn}\omega_{k\ell m}\omega_{kpn} \tag{14.5.6}$$

and

$$\begin{aligned} f_\ell = F_\ell - (m_k v_{k\ell m}\dot{v}_{kpm} y_p + I_{kmn}\omega_{k\ell m}\omega_{kpn}\dot{y}_p \\ + I_{kmn}\omega_{k\ell m}\dot{\omega}_{kpn} y_p + e_{tsm} I_{ksn}\omega_{k\ell m}\omega_{kpt}\omega_{kqn} y_p y_q) \end{aligned} \tag{14.5.7}$$

Equations (14.5.6) may also be written in the matrix form:

$$A\dot{y} = f \tag{14.5.8}$$

where A is an $n \times n$ array with elements $a_{\ell p}$ given by Equation (14.5.6); y is an $n \times 1$ column array of generalized speeds; and f is an $n \times 1$ array with elements given by Equation (14.5.7), when n is $3N+3$.

Observe in Equations (14.5.6) and (14.5.7) the central role played by the partial velocity and partial angular velocity components ($v_{k\ell m}$ and $\omega_{k\ell m}$). Also, in Equation (14.5.8) the A array is at times called the "generalized inertia matrix."

14.6 Constrained Multibody Dynamics

In the foregoing sections we have developed and illustrated the dynamics of open-chain, spherical-joint connected multibody systems. While such systems are useful in modeling a wide variety of physical systems they are not good models for many systems such as those with closed loops, constrained systems, and systems with non-spherical joints. In this section we will consider ways to modify and extend the foregoing analysis so that it can be applied with a wider variety of physical systems.

Consider again an open-chain multibody system connected with spherical joints as represented in Figure 14.6.1. This system can be constrained in several ways: First, some or all of the joint degrees of freedom may be reduced. For example, a typical spherical joint may be replaced by a revolute or pin joint. Next, extremity bodies may be joined to one another forming loops in the system. Finally, extremity bodies may have a specified or desired motion. Alternatively, there may be specified motion at a joint.

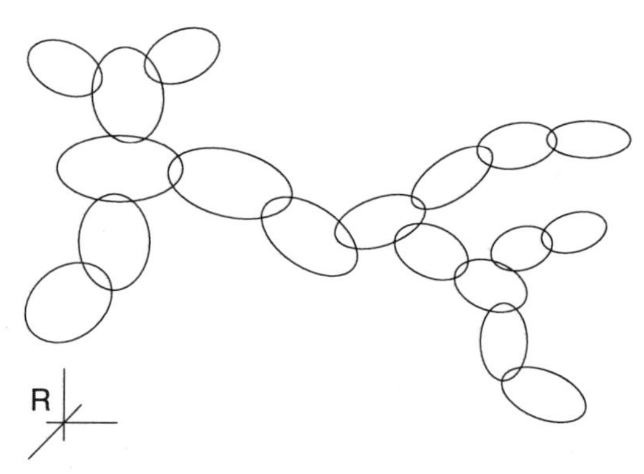

Fig. 14.6.1 An Open-Chair ("Open-Tree") Multibody System with Spherical Joints

The first two of these constraints are geometric (or "holonomic") constraints whereas the third is a kinematic (or "non-holonomic") constraint.

Let the holonomic constraints be represented by equations of the form:

$$\psi_i(q_r,t) = \phi_i(t) \quad (i = 1,\dots,p) \tag{14.6.1}$$

where the q_r $(r = 1,\dots,n)$ are generalized coordinates with n being the number of degrees of freedom of the unrestrained system, and p is the number of holonomic constraints.

Similarly, let the non-holonomic constraints be expressed in the form:

$$\psi_i(q_r, \dot{q}_r, t) = \phi_i(t) \qquad (i = p+1, \dots, m) \tag{14.6.2}$$

where m is the total number of constraint equations.

As before, let y_s $(s = 1, \dots, n)$ be generalized speeds (see Section 14.5). Then there are linear and invertible relations between the generalized speeds and the generalized coordinate derivatives $\dot{q}_r$ $(r = 1, \dots, n)$. [See Equations (13.4.1), (13.4.2), and (13.4.3).]

If Equations (14.6.1) are differentiated with respect to time we obtain expressions of the forms:

$$d\psi_i/dt = \dot{\psi}_i = \frac{\partial \psi_i}{\partial t} + \sum_{r=1}^{n} \frac{\partial \psi_i}{\partial q_r} \dot{q}_r = \dot{\phi}_i(t) \qquad (i = 1, \dots, p) \tag{14.6.3}$$

Then by using expressions as in Equation (13.4.3) we may express Equation (14.6.3) in the form

$$\sum_{s=1}^{n} b_{is} y_s = g_i \qquad (i = 1, \dots, p) \tag{14.6.4}$$

Suppose further that the non holonomic constraints of Equations (14.6.2) are linear in the generalized coordinate derivatives to that with the use of Equations (13.4.3), these non-holonomic constraints may be similarly expressed as:

$$\sum_{s=1}^{n} b_{is} y_s = q_i \qquad (i = p+1, \dots, m) \tag{14.6.5}$$

Finally, by combining Equations (14.6.4) and (14.6.5), the constraint equations may be expressed as

$$\sum_{s=1}^{n} b_{is} y_s = g_i \qquad (i = 1, \ldots, m) \tag{14.6.6}$$

In matrix form, these equations may be expressed in the compact form as:

$$By = g \tag{14.6.7}$$

where B is an m $\times$ n array with elements b_{is}.

Recall from Equation (14.5.8) that for an unconstrained system the dynamical equations are

$$A\dot{y} = f \tag{14.6.8}$$

where A is the n $\times$ n array will elements given by Equation (14.5.6). To constrain variables of the system, constraint forces need to be imposed. These forces in turn will contribute to the generalized forces. Interestingly, these contributions can be expressed by the constraint matrix B as [14.2]

$$F' = B^T \lambda \tag{14.6.9}$$

where F' is an n $\times$ 1 column array of generalized constraint forces F'_ℓ $(\ell = 1, \ldots, n)$ and λ is an m $\times$ 1 column array of parameters (or "multipliers") which may be identified with physical components of the constraint forces (see [14.2, 14.3, 14.4]).

In view of Equation (14.6.8) the governing dynamical equations then become

$$A\dot{y} = f + F' = f + B^T \lambda \tag{14.6.10}$$

Finally, by differentiating, the constraint equations may be expressed in the same form as the dynamical equations leading to:

$$B\dot{y} = \dot{g} - \dot{B}y \tag{14.6.11}$$

Equations (14.6.10) and (14.6.11) form a system of $m + n$ equations for the n y_ℓ and the m constraint force and moment components of the λ array. Solution methods are presented in the following section.

14.7 Solution Procedures for Constrained System Equations

Observe that although Equations (14.6.10) and (14.6.11) represent $m + n$ equations for the n y_ℓ and the m λ_s, they are non-linear differential-algebraic equations. That is, the y_ℓ appear in non-linear and differentiated forms [see Equations (14.5.5) to (14.5.7)], and the λ_s appear linearly. This mixed nature of the equations greatly complicates the solution procedures even with numerical methods.

There are many occasions, however, where the primary interest is in the movement of the system, as opposed to the constraint force and moment components. Where this is the case, the λ_s $(s = 1, \ldots, m)$ may be eliminated from Equations (14.6.10) by use of an orthogonal complement C of the constraint array B. Specifically, let C be an $n \times (n - m)$ array with rank $n - m$ such that:

$$BC = 0 \tag{14.7.1}$$

Then we also have

$$C^T B^T = 0 \tag{14.7.2}$$

where C^T is an $(n - m) \times n$ array. By premultiplying Equation (14.6.9) by C^T, we have

$$C^T A \dot{y} = C^T f + C^T B^T \lambda \tag{14.7.3}$$

where the last term is zero in view of Equation (14.8.2). Then by combining Equations (14.8.3) and (14.6.11) we have the set

$$
\begin{aligned}
C^T A\dot{y} &= C^T f \\
B\dot{y} &= \dot{g} - \dot{B}y
\end{aligned}
\tag{14.7.4}
$$

or

$$
D\dot{y} = h \tag{14.7.5}
$$

where D is the $n \times n$ array defined as

$$
D = n[\overset{n}{D}] = \begin{bmatrix} C^T A \\ ---- \\ B \end{bmatrix} = \begin{matrix} n-m \\ \\ m \end{matrix}\overset{n}{\begin{bmatrix} C^T A \\ ---- \\ B \end{bmatrix}} \tag{14.7.6}
$$

and h is the $n \times 1$ column array defined as

$$
h = n[\overset{1}{h}] = \begin{bmatrix} C^T f \\ ----- \\ \dot{g} - \dot{B}y \end{bmatrix} = \begin{matrix} n-m \\ \\ m \end{matrix}\overset{1}{\begin{bmatrix} C^T f \\ ----- \\ \dot{g} - \dot{B}y \end{bmatrix}} \tag{14.7.7}
$$

With the proper choice of C^T in satisfying Equation (14.7.2) the rank of C^T and hence also of $C^T A$ will be $n - m$. Then the rank of D will in general be n so that D is non-singular and thus may be inverted.

Equation (14.7.5) then forms a system of n linear algebraic equations for the n $\dot{y}_\ell$ which may be solved as:

$$
\dot{y}_\ell = D^{-1}h \tag{14.7.8}
$$

Equation (14.7.8) is in a form ideally suited for numerical integration. To this end it may be combined with expressions such as in Equations (13.3.5) for Euler parameter derivatives. The combined system then forms a set of first-order ordinary differential equations for the generalized speeds, the Euler parameters, and the translation variables of the reference body. Given suitable initial conditions this set of equations may then be integrated to obtain a time history of the movement of the bodies of the multibody system.

Observe that a key step in this procedure is the determination of the orthogonal complement array C, and specifically, its transpose C^T. Although C is not unique, and although there may be several ways of obtaining C [see, for example, Reference 14.5], C may also be obtained from a zero eigenvalues theorem as documented by Walton and Steeves [14.6] and others [14.7, 14.8]. In this latter procedure it is observed that since m is less than n the rank of the array B^TB is at most m. Then in the eigenvalue equation

$$B^TB\xi = \nu\xi \tag{14.7.9}$$

there are $n - m$ zero eigenvalues ν. Then there are in turn $n - m$ eigenvectors $\hat{\xi}$ associated with the zero eigenvalues. That is, for a given zero eigenvalue ($\nu = 0$) if $\hat{\xi}$ is an associated eigenvector, we have

$$B^TB\hat{\xi} = 0 \tag{14.7.10}$$

Then by premultiplying by C^T we have

$$C^TB^TBC = 0 \tag{14.7.11}$$

or

$$(BC)^T(BC) = 0 \quad \text{or} \quad (BC)^2 = 0 \quad \text{or} \quad BC = 0 \tag{14.7.11}$$

Therefore, C is the desired orthogonal complement array.

14.8 Comments and Closure

Observe in the foregoing paragraphs that the constraint force and moment components λ_s ($s = 1,\ldots,m$) are eliminated from the analysis by multiplication by the orthogonal complement array. These eliminated force and moment components, however, may be determined, if desired, by back substitution of the generalized speeds into Equations (14.6.10).

We can use a simple geometric figure to obtain further insight into the nature of the generalized forces and the solution procedure of the constrained governing equations [14.9]. Specifically, if by Kane's equations the sum of the generalized forces is zero, we have

$$F + F^* + F' = 0 \tag{14.8.1}$$

where F, F^*, and F' are the generalized applied, inertia, and constraint force arrays with each array having n elements, and where from Equation (14.6.8) the constraint force array F' may be expressed as:

$$F' = B^T\lambda \tag{14.8.2}$$

where B is the m $\times$ n array of constrained equation coefficients. Then Equation (14.8.1) may be represented geometrically by the force triangle of Figure 14.8.1. Then the multiplication of Equation (14.8.1) by C^T, the transpose of the orthogonal complement of B, may be interpreted as a projecting of F and F^* onto a direction orthogonal to F' as represented in Figure 14.8.2. This projection eliminates λ resulting in the equation

$$C^TF + C^TF^* = 0 \tag{14.8.3}$$

or

$$K + K^* = 0 \tag{14.8.4}$$

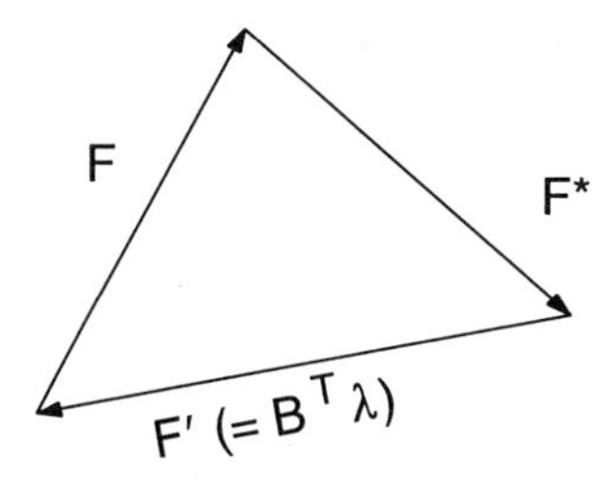

Fig. 14.8.1 A Force Triangle of Generalized Forces

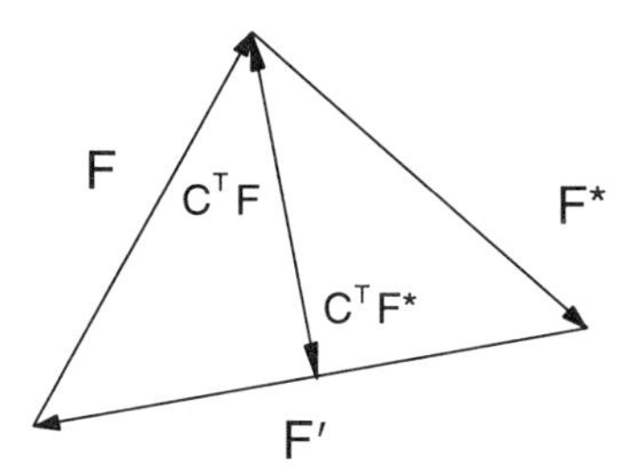

Fig. 14.8.2 Projection of Generalized Forces Along the Orthogonal Direction to F′

where K and K*, defined by inspection, are reduced generalized forces in the "direction of freedom" of the constrained system. Equation (14.8.4) may be described as a reduced (or "projected") form of Kane's equations.

As noted earlier, many physical systems can be modeled as open-chain systems of spherical joint connected bodies. Most physical systems, however, are not that simple. Instead, the joints may be more like hinges (for example, an elbow in a human body model), or the bodies may form a closed loop (for example, clasped hands of a human body model), or there may be small separation between the bodies (for example, with cervical vertebrae of a human body model). Two approaches may be taken to study such systems: The first, best applied with relatively simple systems, is to construct the multibody system with only those degrees of freedom needed to model the physical system. The second approach, best applied with large complex systems, is to model the physical systems by an open-chain system of spherical joint connected bodies and then apply constraints as outlined in the immediate foregoing paragraphs. This latter approach is also better suited for algorithm development for numerical computation.

On some occasions, it is desirable to be able to model translation between adjoining bodies. For these cases an even more general formulation is required, but as noted earlier, the analysis is conceptually similar to that outlined herein.

Finally, the use of multibody systems for modeling physical systems is virtually unlimited — ranging from robots, to mechanisms, to cables, to chains, to rotating blades, to space stations, to biosystems. Additional information is readily obtained in the references (see specifically References 14.7, 14.10, and 14.11).

References

14.1 T. R. Kane and D. A. Levinson, *Dynamics: Theory and Applications*, McGraw Hill, New York, 1985.

14.2 R. L. Huston, "Constraint Forces and Undetermined Multipliers in Constrained Multibody Systems," *Multibody System Dynamics*, Vol. 3, 1999, pp. 381-389.

14.3 J. T. Wang and R. L. Huston, "Kane's Equations with Undetermined Multipliers — Application to Constrained Multibody Systems," *Journal of Applied Mechanics*, Vol. 43, 1987, pp. 424-429.

14.4 M. Xu, C. Q. Liu, and R. L. Huston, "Analysis of Non-Linearly Constrained Non-Holonomic Multibody Systems," *International Journal of Non-Linear Mechanics*, Vol. 25, No. 5, 1990, pp 511-519.

14.5 C. L. Lawson and R. J. Hanson, *Solving Least Squares Problems*, Prentice Hall, Englewood Cliffs, NJ, 1974.

14.6 W. C. Walton, Jr., and E. C. Steeves, "A New Matrix Theory and Its Application for Establishing Independent Coordinates for Complex Dynamical Systems with Constraints," NASA Technical Report TR-326, 1969.

14.7 R. L. Huston, *Multibody Dynamics*, Butterworth-Heinemann, Boston, MA, 1990.

14.8 J. W. Kamman and R. L. Huston, "Dynamics of Constrained Multibody Systems," *Journal of Applied Mechanics*, Vol. 51, 1984, pp. 899-904.

14.9 J. T. Wang and R. L. Huston, "Computational Methods in Constrained Multibody dynamics: Matrix Formalisms," *Computers and Structures*, Vol. 29, No. 2, 1988, pp. 331-338.

14.10 R. L. Huston, "Multibody Dynamics — Modeling and analysis Methods," Feature Article, *Applied Mechanics Reviews*, Vol. 44, NO. 3, 1991, pp. 109-117.

14.11 R. L. Huston, "Multibody Dynamics since 1990," *Applied Mechanics Reviews*, Vol. 49, No. 10, 1996, pp. 535-540.

Index

D

E